Nano Physical Pharmaceutics

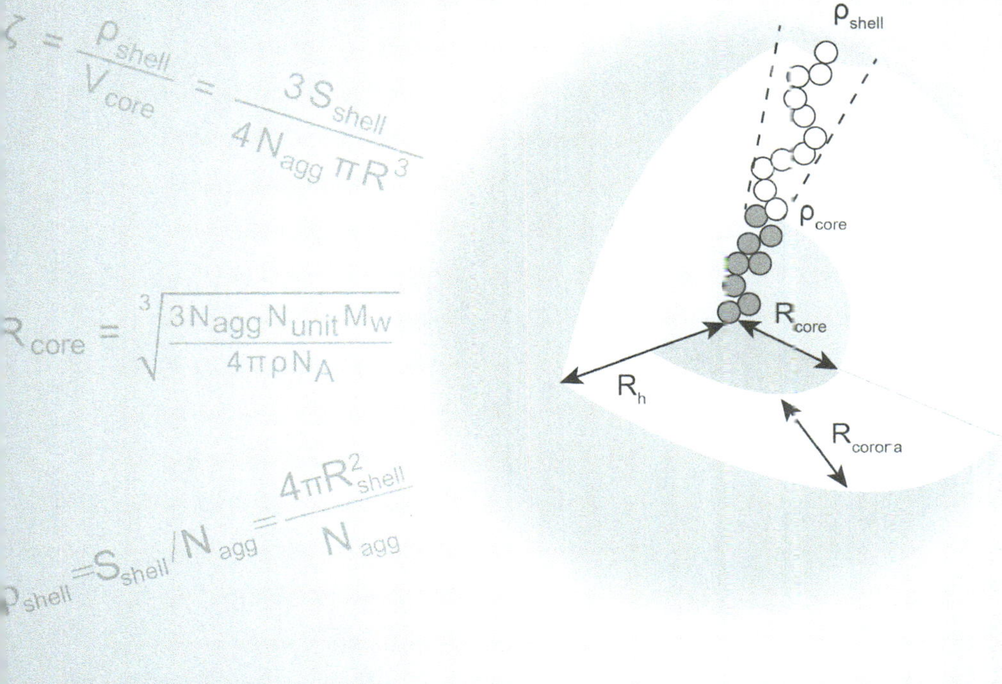

Nano Physical Pharmaceutics

edited by

Wei Li

JENNY STANFORD
PUBLISHING

Published by

Jenny Stanford Publishing Pte. Ltd.
101 Thomson Road
#06-01, United Square
Singapore 307591

Email: editorial@jennystanford.com
Web: www.jennystanford.com

British Library Cataloguing-in-Publication Data
A catalogue record for this book is available from the British Library.

Nano Physical Pharmaceutics

ISBN 978-981-4968-52-2 (Hardcover)
ISBN 978-1-003-51393-3 (eBook)

Contents

Preface

Nanomedicine is an emerging and rapidly evolving field, which could significantly help in early disease diagnosis, therapeutic improvement, development of novel therapeutic modalities, and reduction in medical expenditures. The challenge, however, is that specially designed and carefully synthesized nanomedicines function poorly in animal models despite great performance *in vitro*. We believe a deeper understanding of the physicochemical properties of nanomedicine, that is, nano physical pharmaceutics, is the key.

This book addresses the "bottleneck" of nano-based formulations by focusing on clinical translation and applies physical theories and models to determine the parameters for controlling the physicochemical properties of nanomedicines, including micelles, liposomes, and inorganic nanoparticles. Qualitative and quantitative relationships are established to guide nanomedicine design, characterization, and analysis. The book also compiles cutting-edge research in nanomedicine from the interdisciplinary team of the Department of Nanomedicine at Naval Medical University, which is currently establishing the characteristic discipline called Nano Physical Pharmaceutics. Edited by Wei Li, a prominent nanotechnology researcher, this book will appeal to anyone involved in nanotechnology, medicine, macromolecular science, biology, and chemistry, especially those with an interest in drug delivery or cancer therapy. The book has been accomplished with contributions from the team members, including Dr Xiaoling Pan, Dr Zhiwen Qiu, Dr Xiaodong Zhu, Dr Ziya Xia, Dr Yening Xia, Dr Mengxin Zhao, Dr Man Wang, Dr Jing Liu, Dr Xiangling Gu, Dr Beibei Liang, Dr Di Chen, Dr Fei Duan, Dr Junfang Li, and Dr. Hanwen Sun.

The guidance from Prof. Chi Wu, Prof. Huiming Hou, Prof. Teruo Okano, Prof. Ngai To, Prof. Masamichi Nakayama, Prof. Hao Wang and Prof. Jian Wang and the grant of the National

Natural Science Foundation of China, the Program of Shanghai Academic Research Leader et al. are appreciated.

The book is interdisciplinary and involves the fields of chemistry, polymer, physicochemistry, biology, and pharmaceutics. The context and the logical illustration of the mechanism of the kinetics and dynamics of polymer chains in solutions have been cited from references. The theory and application of laser light scattering have been cited from the editor's PhD thesis. Both citations are appreciated [1–3]. However, as some misspellings or other mistakes in the text cannot be ruled out, the readers' suggestions for any improvements are welcome.

Wei Li
Department of Nanomedicine
Naval Medical University
Shanghai, China
August 2024

References

1. Teraoka I, *Polymer Solutions*, John Wiley & Sons, Inc, New York, 2002.
2. Wu C, *A Concise Textbook of Thermodynamics*, Higher Education Press, 2019.
3. Li W, *Reexamination of Dynamic of Semidilute Polymer Solution*, PhD thesis, The Chinese University of Hong Kong, 2006.
4. Fu XC, Hou WH, *Physical Chemistry*, Higher Education Press, 2022.
5. Hou HM, Wang H, Zhang GJ, *Application Technology of Pharmaceutical Excipients*, 2nd ed., China Medical Science Press, 2002.

Chapter 1

The Fundamentals and Powerful Tool for Nano Physical Pharmaceutics

Ziya Xia, Yening Xia, and Wei Li

Department of Nanomedicine,
Naval Medical University, Shanghai 200433, China

liwei_dds@163.com

1.1 Polymer Chains Behavior in Solutions

For a certain polymer, some solvents (good solvents) dissolve it well, some solvents (bad solvents) cannot, and solvents whose solubility is between the good and bad solvents will dissolve it to a certain extent. With the variation of temperature or concentration of the polymers, phase separation may take place, resulting in aggregation of the polymers. In this section, we will introduce the compatibility theory of polymers-solvent, as well as the self-assembly mechanism of polymers.

1.1.1 Basic Interaction of Polymer Chain in Solution [1–3]

1.1.1.1 Introduction to the χ parameter

From the viewpoint of thermodynamics, if the polymers are soluble in solvents, the dissolution process must reduce the free energy of

Nano Physical Pharmaceutics
Edited by Wei Li

ISBN 978-981-4968-52-2 (Hardcover), 978-1-003-51393-3 (eBook)
www.jennystanford.com

the system. The free energy of mixing of the system is as follows [1]:

$$\Delta G_{\mathrm{m}} = \Delta H_{\mathrm{m}} - T\Delta S_{\mathrm{m}} \tag{1.1}$$

where ΔG_{m} is the free energy of mixing of the system, ΔH_{m} is the enthalpy of mixing of the system, ΔS_{m} is the entropy of mixing of the system and T is the absolute temperature. The change of enthalpy of the system comes from the interaction change upon mixing. Note that only short-ranged interactions are considered here, containing van der Walls interactions (also referred to as dispersions), hydrogen bonding, and dipole-dipole interactions. The change of entropy is because the order of the system changes. Since dissolution is always a process of entropy increasing ($\Delta S_{\mathrm{m}} > 0$), the enthalpy of mixing determines the sign of the free energy of the system, that is, either the enthalpy of the system decreases ($\Delta H_{\mathrm{m}} < 0$) due to dissolution, or the product of temperature and entropy of mixing must be larger than the enthalpy of mixing, then dissolution occurs. Compared with the low-molecular-weight solutes, the degree of freedom of the polymer monomer is lower, so the increase of entropy of mixing is far less than that of the low-molecular-weight solutes-solvent system and the compatibility of the polymers-solvent system will be much lower. The solvents dissolving a certain polymer are only that which surround the polymer chain.

Flory–Huggins parameter, i.e., χ parameter, is commonly used to describe the interaction change upon mixing of polymer–solvent system. A lot of solubility theories are based on the concept of molecular "lattices". The definition of χ parameter using the lattice model which considers the interaction between adjacent ones only is described below [2, 3]. The lattice model only considers interactions resulting from the contacts of adjacent molecules in the sites. As shown in Fig. 1.1, the interactions for a polymer–polymer (P–P) contact (i.e., contact of adjacent polymer monomers), a solvent–solvent (S–S) contact, and a polymer–solvent (P–S) contact are represented by $\varepsilon_{\mathrm{PP}}$, $\varepsilon_{\mathrm{SS}}$ and $\varepsilon_{\mathrm{PS}}$, respectively. Due to the rearrangement of contacts as polymers are mixed with the solvent, the total interaction energy changes.

For example, there are four P–P and four S–S contacts in respective lattice sites of the polymers and solvent before mixing, while after mixing, two P–P and two S–S contacts are replaced by four P–S contacts. Thus, the bond energy changes from $4\varepsilon_{SS}$ + $4\varepsilon_{PP}$ to $4\varepsilon_{PS}$ + $2\varepsilon_{SS}$ + $2\varepsilon_{PP}$ and the variation is $4\varepsilon_{PS} - 2(\varepsilon_{SS} + \varepsilon_{PP})$. For one contact, the variation is $\varepsilon_{PS} - (\varepsilon_{SS} + \varepsilon_{PP})/2$. Then, χ parameter is defined as the product of the lattice coordinate Z and the energy change reduced by $k_B T$:

$$\chi = \frac{Z\left[\varepsilon_{PS} - \frac{1}{2}(\varepsilon_{PP} + \varepsilon_{SS})\right]}{k_B T} \tag{1.2}$$

A negative χ indicates that polymer–solvent contact is more likely to occur when mixing, meaning that a negative χ can promote the polymer to dissolve. In contrast, A positive one shows that the polymer–solvent contacts are more difficult than the polymer–polymer and solvent–solvent contacts.

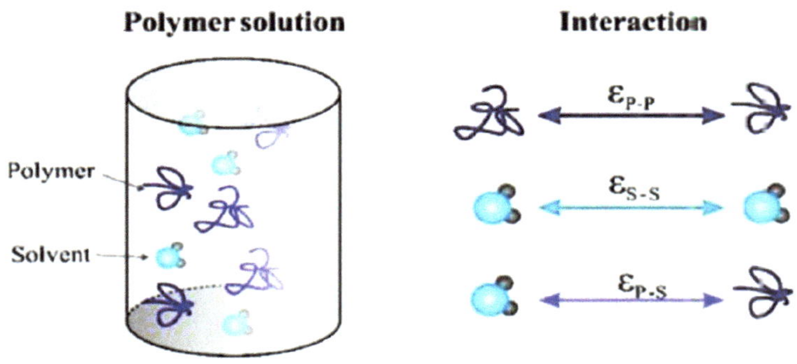

Figure 1.1 Scheme describes the basic interactions of polymers in solution.

1.1.1.2 Relationship between χ parameter and stability of the solution [2, 3]

In order to describe the stability of polymer solutions, the concept of replacement chemical potential ($\Delta\mu_{rep}$) is introduced. Replacement chemical potential is the change of the free energy due to the removal of solvent molecules and the replacement of polymers resulting in an increase in the polymer concentration. When a polymer chain with degree of polymerization N is added to a system, it will replace N solvent molecules to maintain the total volume unchanged, the replacement chemical potential of the system is:

$$\Delta\mu_{\text{rep}} = \Delta\mu_{\text{P}} - N\Delta\mu_{\text{S}} \tag{1.3}$$

where $\Delta\mu_{\text{P}}$ is the difference of chemical potential between polymer chains in solution and melt. $\Delta\mu_{\text{S}}$ is the difference of chemical potential between the solvent molecules in solution and pure solvent. So $\Delta\mu_{\text{rep}}$ can be expressed as

$$\frac{\Delta\mu_{\text{rep}}}{k_{\text{B}}T} = \ln\varphi + 1 - N - N\ln(1-\varphi) + \chi N(1-2\varphi) \tag{1.4}$$

where φ is the volume fraction of the polymers. Let $\frac{\partial\Delta\mu_{\text{rep}}}{\partial\varphi} = 0$, we can find the inflection points of the curve $(\Delta\mu_{\text{rep}}/(k_{\text{B}}T))$, which corresponds to the boundary of instability of the solution. The expression of the boundary is:

$$\frac{1}{\varphi} + \frac{N}{1-\varphi} = 2\chi N \tag{1.5}$$

The instable condition of the solution is:

$$\frac{1}{\varphi} + \frac{N}{1-\varphi} < 2\chi \text{N} \tag{1.6}$$

The dividing line between the stable and unstable regions of the system, described by Equation 1.5, is called the spinodal line. The relationship between the stability of the solution and χ as well as φ is shown in Fig. 1.2.

The shaded part of Fig. 1.2 represents the unstable region, and the minimum value (χ_c) of the spinodal line is the critical point. When $\chi < \chi_c$, Equation 1.5 has no real root and the polymer solutions are stable throughout the concentration interval. When $\chi > \chi_c$, it has two real roots. If φ is between two roots, the polymer solution is unstable and will be separated spontaneously into two phases with different polymer volume fractions.

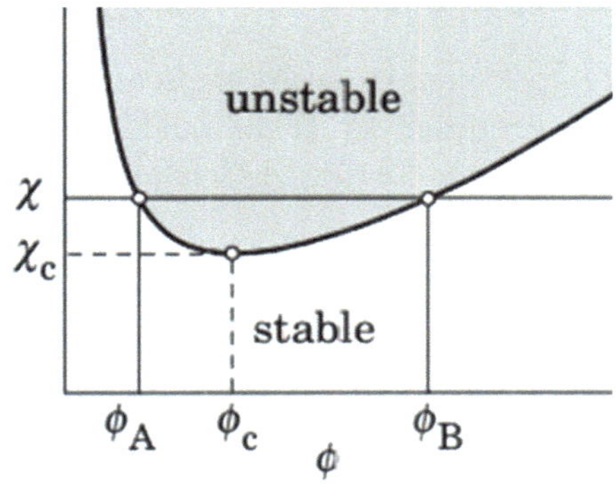

Figure 1.2 Relation between the stability of the solution and χ as well as φ, when $\frac{\partial \Delta\mu_{\text{rep}}}{\partial\varphi} = 0$. Reproduced with permission from [2].

Taking the derivative of both sides of Equation 1.5 with respect to φ, and let $\frac{\partial\chi}{\partial\varphi} = 0$, we get

$$\chi_c = \frac{(1+N^{\frac{1}{2}})^2}{2N} \approx \frac{1}{2} + N^{-\frac{1}{2}} \tag{1.7}$$

$$\varphi_c = \frac{1}{1+N^{\frac{1}{2}}} \approx N^{-\frac{1}{2}} \tag{1.8}$$

With increasing the degree of polymerization N, the critical point χ_c gradually approaches 1/2. It should be noted that when $0 < \chi < 1/2$, although the polymer–solvent contact is unfavorable, the small increase in enthalpy caused by mixing can be offset by the entropy of mixing, and the polymer can still dissolve. For $\chi > 1/2$, the entropy of mixing was not enough to offset the increase of the interaction energy, so the solubility of the polymer became worse. Therefore, a solvent that makes χ of the system sufficiently small compared with 1/2 (including $\chi < 0$) is a good solvent, while one that makes $\chi > 1/2$ is a bad solvent. Moreover, with χ increasing, the solvent can no longer dissolve the polymer, so the solvent is nonsolvent.

1.1.2 Solubility of Polymer

1.1.2.1 Solubility parameter

The solubility parameter (also known as the Hildebrand parameter) was proposed by Hildebrand et al. [4, 5] and it is defined as the square root of the cohesive energy density, i.e.,

$$\delta_i = (\Delta E_i / V_i)^{1/2} \tag{1.9}$$

where ΔE_i is the molar energy of vaporization of substance i and V_i is its molar volume. The unit of solubility parameter δ_i is $(\text{MPa})^{1/2}$ or $(\text{cal/cm}^3)^{1/2}$, and they satisfy $1\ (\text{MPa})^{1/2} = 2.0455\ (\text{cal/cm}^3)^{1/2}$. The solubility parameter is often used to express the interaction between a polymer and a solvent. For binary systems, using simple thermodynamics, the χ parameter can be represented with the solubility parameter:

$$\chi = \frac{V_S}{RT}(\delta_S - \delta_P)^2 + 0.34 \tag{1.10}$$

where δ_S and δ_P are solubility parameters of solvent and polymer, respectively. V_S is the molar volume of the solvent, and R is the gas constant. Together with the relationship between χ and dissolvability of the solvent (described in Section 1.2), it can be seen that polymers and solvents are miscible when their solubility parameters are close to each other, otherwise they are not. The so-called "like dissolves like" means that the solubility parameters are similar.

1.1.2.2 Real polymer solutions

Because the above definition of solubility parameter only takes into account the dispersion interaction, losing sight of the hydrogen bonding and dipole-dipole interaction, it is only applicable to the non-polar system. For polar substances, there will be large errors in the estimation of the solubility properties. In response to this, many researchers try to revise or improve it. In this section, we will introduce the Hansen solubility parameter (also known as the three-dimensional solubility parameter) briefly.

Hansen thinks that the total energy of vaporization of substance *i* is composed of three components: dispersion interaction (ΔE_{id}), polarity interaction (ΔE_{ip}) and hydrogen bonding (ΔE_{ih}) [6, 7], namely

$$\Delta E_i = \Delta E_{id} + \Delta E_{ip} + \Delta E_{ih} \tag{1.11}$$

Accordingly, the solubility parameter is split into dispersion effect, polarity effect and hydrogen bonding interaction parameters. The relation between the solubility parameter and its three components is:

$$\delta_i = \sqrt{\delta_d^2 + \delta_p^2 + \delta_h^2} \tag{1.12}$$

For the Hansen solubility parameter, the relative energy difference (RED) is used to judge the compatibility of solute and solvent. The relative energy difference is defined as the ratio of the "spatial distance" of the solubility parameters of solute and solvent (Ra) to the radius of the interaction (Ro), i.e., RED = Ra/Ro. When RED < 1, it indicates that the solvent is good, while RED > 1, the solvent is bad. For more detailed introductions or applications of the Hansen solubility parameter, please refer to other relevant materials [8–11].

1.1.3 Self-Assembly Behavior of Polymers in Solution

When amphiphilic polymers are dispersed in a solvent that is selectively soluble for only one of the polymer components the solvophobic part of the polymer will tend to aggregate together to minimize contact with the solvent, thereby reducing the interfacial energy. Under these conditions, phase separation will take place in solutions, and the polymers aggregate to form nanoparticles. This process occurs spontaneously and is called self-assembly in solutions. Water is a commonly used as selective solvent in current studies. Entropy, interaction free energy and molecule geometry of the system are combined to provide a framework for self-assembly theory.

1.1.3.1 Dynamics of self-assembly

The free energy of a micelle composed of *n* amphiphilic polymers is nG_n^0. Assuming that the dilute solution theory is applicative, then equilibrium thermodynamics requires that the molar fraction x_n of the amphiphilic polymers of the micelles above should satisfy [12, 13]:

$$G_n^0 + \frac{kT}{n}\ln\left(\frac{x_n}{n}\right) = G_1^0 + kT\ln(x_1) = \text{const} \tag{1.13}$$

where k is Boltzmann's constant, and the subscript 1 represents an isolated amphiphilic polymer. Or Equation 1.13 can be rewritten as

$$x_n = nx_1 e^{[n(G_1^0 - G_n^0)/kT]} \tag{1.14}$$

If $G_n^0 - G_1^0 < 0$, aggregations are energetically favorable. The transition from the dispersed state to the aggregated state is due to the competition between entropy and enthalpy.

1.1.3.2 Free energy of self-assembly systems

The self-assembly bulk is in a thermodynamically stable state, and the system free energy is also the lowest, correspondingly. The morphology of self-assembly is mainly controlled by three kinds of free energy in the system:

1. The extension degree of the hydrophobic chain segment;
2. The interfacial tension between the hydrophobic chain segment and solvent;
3. The repulsive force between the hydrophilic chain segments.

Assembly morphology is essentially determined by the balance of these three kinds of energy. So, factors that may affect this balance, such as the composition and structure of the polymer participating in self-assembly or conditions for self-assembly, will change the final assembly morphology directly or indirectly. In the experiment, the assembly morphology is often adjusted by adjusting these factors on purpose.

1.1.3.3 Basic morphology of self-assembly systems

Packing parameter theory [14–16], proposed based on the micellization of amphiphilic small molecules in solutions, can explain the assembly morphology of polymers. Self-assembly behavior of amphiphilic molecules in solutions is controlled by two opposing forces, i.e., an attractive force between the hydrophobic segments leading to aggregation and a repulsive force between the hydrophilic segments preventing unlimited growth of the self-assembly bulk into a distinct macroscopic phase. Self-assembled structures are stabilized in a solution thanks to the interaction between the hydrophilic segments and the solvent. For an amphiphilic polymer, the packing parameter P can be calculated by the following formula:

$$P = \frac{v}{a_0 l_c} \tag{1.15}$$

where v is the volume of the hydrophobic chain segment, a_0 is the equilibrium area of the hydrophilic part, and l_c is the length of the hydrophobic chain segment. As shown in Fig. 1.3, the value of P determines the final morphology of the self-assembly bulk. Spherical micelles are obtained with $0 < P < 1/3$, while cylindrical micelles with $1/3 < P < 1/2$ and vesicles with $½ < P < 1$. When $P = 1$, one gets a planar membrane and the inverse-phase assembly will be observed with $P > 1$. Note that the aggregation number of polymers n increases as spherical micelles → rod-shaped micelles → spherical vesicles → planar bilayer.

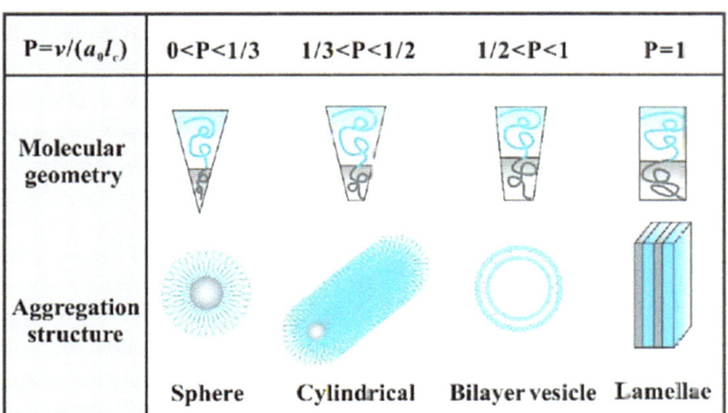

Figure 1.3 Relationship between the packing parameters, the microstructure of amphiphilic block copolymers and their self-assembly morphology in solution. The formula related to the table is shown in 1.10, 1.15.

1.2 Self-Assembly of Amphiphilic Copolymers in Solution

1.2.1 Self-Assembly of Block Copolymers

Self-assembly of molecules is a process during which molecules assemble or organize to form regular structures under no artificial factors. It is one of the most important methods to prepare nanomaterials. Self-assembly of molecules will transform the system from a disordered state to an ordered state, a common phenomenon in nature such as various biomolecules self-assembling into cells, phospholipids self-assembling into cell membranes, and surfactants into soap bubbles. Self-assembly of polymers, an important branch of self-assembly of molecules, utilizes the interaction between polymers, or polymers and solvent (hydrogen bonding, hydrophilic/hydrophobic interaction, electrostatic interaction, etc.) to assemble polymers into highly ordered microphase structure. Compared with the self-assembly of small molecules, aggregation products of polymers have higher stability and better mechanical properties, and have been widely used in the fields of nanomedicine, biochemistry and materials science. The self-assembly bulk in solutions mainly includes micelles, liposomes, microspheres and gels. In this section, the morphologies and phase transitions of amphiphilic assembly units (block copolymers and lipids) during the self-assembly process are introduced.

Due to the obvious advantages of realizing the controllability of self-assembly morphology and size, block copolymers have become the most widely studied polymers. Microphase separation may occur for most block copolymers because of the incompatibility of the constituent components [17]. The self-assembly process of block copolymers is driven by combinatorial enthalpy and combinatorial entropy, and the covalent bonds connecting the blocks prevent macrophase separation. The microphase separation depends mainly on three factors [18]: (1) the volume fraction of each block (f); (2) the total number of the segments (or degree of polymerization) (N); (3) the Flory–Huggins interaction parameter between the blocks (χ). χ parameter, which is temperature-dependent (see Equation 1.2), describes the compatibility of different blocks. The value of χN denotes the segregation power. With temperature increasing (or χN decreasing), the rising of the combinatorial entropy results in a higher compatibility between blocks, and thus the copolymer may undergo a transition from order to disorder. The critical temperature for order-to-disorder transition (ODT) is referred to as the T_{ODT}.

The phase behavior of diblock copolymers in molten state has been investigated theoretically and experimentally [17, 19–26]. According to χN, the so-called weak segregation limit (WSL) ($\chi N < 10$) and the strong segregation limit (SSL) ($\chi N \gg 10$) have been introduced, while the self-consistent mean-field (SCMF) theory was developed as $10 < \chi N < 100$. For SSL and WSL, both T_{ODT} and the microdomain size (D) can be easily obtained. For the former, the difference of free energy between ordered and disordered states (ΔG_{m}) can be expressed as [25]

$$\Delta G_{\mathrm{m}} = \Delta H_{\mathrm{m}} - T\Delta S_{\mathrm{m}} = C_1 \chi^{\frac{1}{2}} \left(\frac{N}{D} \right) + \frac{C_2 D^2}{N} - C_3 \quad (1.16)$$

where C_1, C_2 and C_3 are positive constants independent of χ, N and D. The derivative of Equation 1.16 with respect to D gives the value of D at the equilibrium (denoted as D_{eq}):

$$D_{\mathrm{eq}} \sim N^{\frac{2}{3}} \chi^{\frac{1}{6}} \quad (1.17)$$

The temperature at which ΔG_{m} (D_{eq}) = 0 is the critical temperature T_{ODT}. For WSL, in terms of the theory from Leibler [19], one can get:

$$D \sim N^{\frac{1}{2}} \chi^{0} \quad (1.18)$$

The phase behavior of the diblock copolymer is predicted by SCMF theory as follows. When the value of χN (>10.5) is low, with f_{A} increasing, transition starts from the disordered state to body-centered cubic spheres (S phase), hexagonally packed cylinders (C phase), bicontinuous gyroids (G phase) and lamellae (L phase), successively. For a higher χN, the G phase is unstable compared to C or L phase.

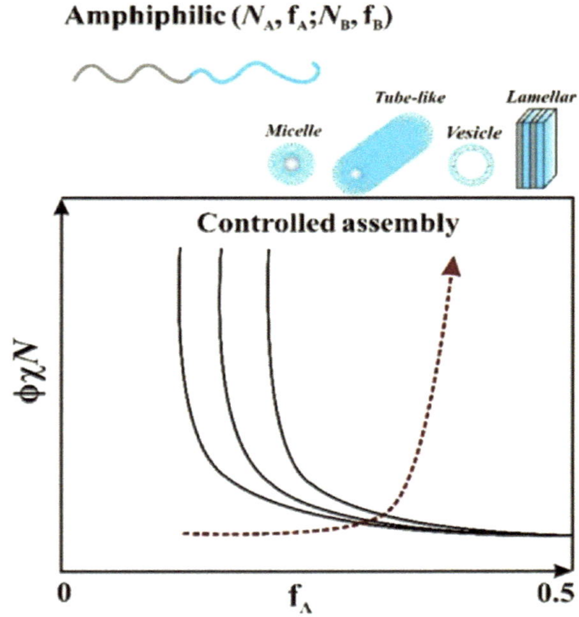

Figure 1.4 Phase diagram of assembly control of amphiphilic block copolymers in solutions.

As the amphiphilic polymers have been dispersed in a selective solvent, only one block of the polymers is soluble in the solvent, and the solvophobic blocks tend to aggregate together to reduce the contact with the solvent, resulting in phase separation between the polymers and the solvent to form nanoparticles (polymer micelles). The process occurs spontaneously. Compared with self-assembly in bulk, self-assembly of block copolymers in solutions is more complex. Besides temperature, the self-assembly of block copolymers into nanoparticles in solutions depends on the concentration of polymers in solutions as well. Therefore, the degree of incompatibility has been described by $\phi\chi N$, where ϕ represents the descriptive concentration of polymer in solution. The controlled factors of assembly of amphiphilic polymers in solutions are shown in Fig. 1.4. The variation of the self-assembled morphology of polymers in solutions with the solution concentration ($\boldsymbol{C}_{\text{polymer}}$), the length ratio of the head to tail ($\boldsymbol{L}_{\text{head/tail}}$) and the χ parameter is displayed in Fig. 1.5. Once the critical micelle concentration (CMC) is reached, self-assembly of the block copolymers takes place and the nanostructures in the solution include spherical micelles, rod-like micelles, vesicles and lamellar.

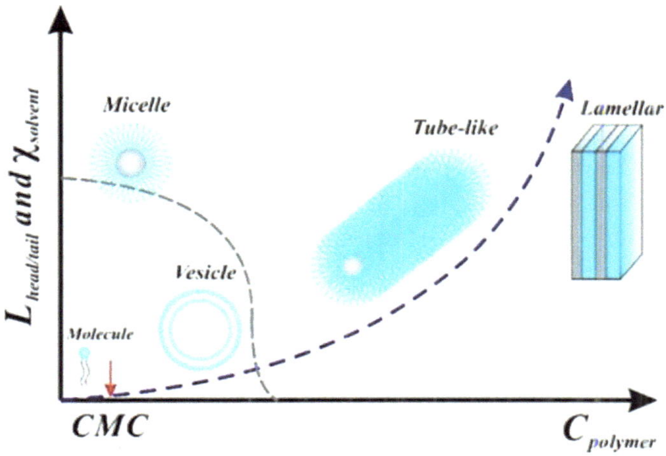

Figure 1.5 The concentration ($C_{polymer}$), $L_{head/tail}$ and χ dependence of the self-assembled morphology of amphiphilic copolymers in solutions.

1.2.2 Self-Assembly of Liposomes

1.2.2.1 Formation of liposomes

Besides amphiphilic block copolymers, lipids could self-assemble to form vesicle structures in aqueous solution as well. Self-assembly of block copolymers results from interactions of solvent-polymer and polymer–polymer, while that of lipids is mainly based on the hydrophobic combination between acyl chains. When lipid molecules are in the water environment, due to the characteristics of amphiphilicity, the hydrophilic heads will be exposed outside and the hydrophobic tails are coated inside, forming micelles, vesicles, regular bilayer or other structures. Lipids with large head groups and flexible double chains are cylindrical, which is the geometric configuration of most membrane lipids. These lipid molecules can assemble into spherical vesicles or planar bilayers (Referring to Fig. 1.3). Liposomes are spherical vesicles composed of one or more layers of phospholipid bilayer. Lipid assemblies are typical representatives of the self-assembly phenomenon in organisms. Lipid molecules are bonded noncovalently, so they can undergo rapid rearrangement, for instance, changing positions through lateral diffusion or flip-flop, which is more complex than other biological constituents whose molecules bond covalently, such as polynucleotides, proteins, and polysaccharides.

1.2.2.2 Phase behaviors of lipid bilayers

For a one-component lipid bilayer, as the temperature is lower than the melting temperature (T_m) of phospholipids, the bilayers are in the solid-gel state (L_β), and their tail acyl chains are arranged in order. With the temperature increasing, the lipids would undergo phase transition from L_β to the fluid state (L_α). At ambient temperatures, the phase of lipid with the high melting point is usually L_β (such as sphingomyelins) or $L_{\beta'}$ (chain-inclined solid phase) (such as DPPC or DSPC). For low-melting lipids, such as DOPC, POPC, or SOPC, their phase is L_α. In order to increase the assembled efficiency of liposomes, the temperature should be higher than the melting temperature of phospholipids.

The phase behaviors of binary lipid bilayers have been depicted as a function of compositions and temperatures. In two-component membranes, containing high and low-melting-temperature lipids, L_β and L_α phases could coexist. The two lipids are miscible in both solid and fluid states when the chains are similar sufficiently; conversely, as the chains are dissimilar adequately, they are immiscible in both the solid and solid+fluid states [27].

Membranes containing three or more lipids are closer to biological membranes than binary mixtures. Since the content of cholesterol in animal cell membranes is about 35–45 mole% of total lipids, the phase behaviors of mixtures containing cholesterol are more attractive [28]. Ternary mixtures of lipids, i.e., high-melting lipid + low-melting lipid + cholesterol, are the combination with least components that can produce rich phase behaviors [29]. At present, the phase diagrams of three-component lipid mixtures: DPPC/DOPC/cholesterol at temperatures from 15° to 45° [30, 31], DPPC/diphytanoyl-PC/cholesterol at temperatures from 10° to 60° [32] and DSPC/DOPC/cholesterol at ambient temperature [33], have been determined. Results show that a high cholesterol concentration can strongly change the fluid structure of bilayers and transforms the L_β phase into a new phase, i.e., the liquid ordered phase (L_o). The coexistence of two macroscopic liquid phases (L_o + liquid disordered phase L_d) emerges over a wide range of compositions and temperatures. Here, the L_o phase is rich in cholesterol, while L_d is poor in cholesterol. In addition, it seems to be that cholesterol could increase the ordering of the L_d phase, but decrease that of the L_o phase. That is, compared to the binary mixtures, cholesterol could reduce the difference in properties of coexisting domains in the ternary mixture.

1.3 Stability of Nanosystems in Solutions

Due to the high surface area and surface energy, the self-assembled nanoparticles are thermodynamically unstable in the water-dispersed system. The particles tend to aggregate together to reduce their surface energy, that is, they are prone to coagulating and become instable. So, stabilizers are needed necessarily. On the other hand, the intense Brown motion of nanoparticles makes coagulation difficult to occur in the gravitational field, so they have kinetic stability. Scholars believe that the former is more important, because once the thermodynamic stability is lost, the particles will aggregate together and grow up, resulting in the system losing the kinetic stability. The stability of nanosystems in solutions has always been an important issue in colloidal science. In this section, the stability theory of colloids, i.e., DLVO theory will be introduced.

1.3.1 DLVO Theory

In the 1940s, Soviet scholars Derjaguin and Landau, as well as Dutch scholars Verwey and Overbeek, proposed the calculation methods of mutual attraction energy and repulsion energy between particles with various shapes under different conditions, respectively. They are used to describe the interaction between colloidal particles quantitatively. This theory has been called the DLVO theory simply.

1.3.1.1 Interaction energy between nanoparticles

The attractive energy-making nanoparticles aggregate together and the repulsive energy preventing them from aggregating exists simultaneously. These two energies, both of which are related to the distance between particles, determine the stability of the system.

The attraction between particles is essentially the same as the van der Waals attraction between molecules, but it is the mutual attraction between particles made up of many molecules and is the sum of each molecule's contributions. The mutual attractive force between particles is a long-range force, inversely proportional to the third power of the distance. The repulsive force between particles comes from the electrical double-layer structures around the particles. The nanoparticles in contact with the liquid will adsorb certain ions from the liquid selectively. So, the solid and liquid phases have different symbols of charge, forming an electrical double-layer structure on the interface as shown in Fig. 1.6. When the distance between particles is large, the electrical double layers have not overlapped, so the repulsive force does not act. However, once the particles are so close that the electrical double layers overlap, the concentration of ions in the overlapping part increases, and the osmotic pressure of the excess ions will hinder the proximity of the particles, resulting in repulsive interaction.

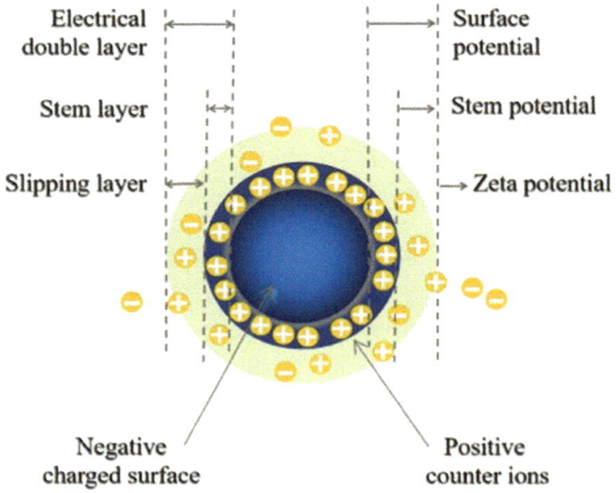

Figure 1.6 Schematic diagram of the electrical double layer of negatively charged nanoparticles in aqueous solution.

The relationship between the total interactive energy and the distance between particles is shown in Fig. 1.7. When the particles are far apart from each other, the electrical double layers do not overlap. Under the circumstances, only the attractive interaction works, and the total potential energy is negative. As the distance between particles is so close that the electrical double layers overlap, the repulsive force plays a major role and the total potential energy increases significantly. At the same time, with the distance decreasing, the attractive force between the particles also increases. For the distance short to a certain extent, the attractive force starts to dominate again and the potential energy decreases. As can be seen from Fig. 1.7(c), the particles must overcome an energy barrier to aggregate together, which is the reason why the nanoparticles do not coagulate in solutions. Therefore, although the particles might collide due to the Brown motion, so long as the distance between the particles gets shorter enough for the overlap of the electrical double layers, they are then separated by repulsive interaction and do not aggregate together.

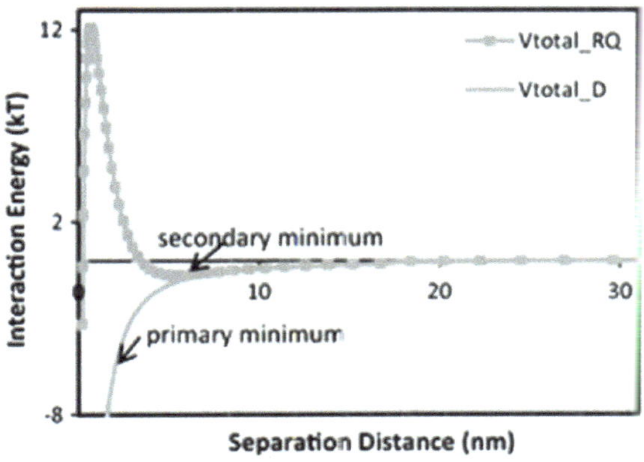

Figure 1.7 The relation curve of the interactive energy and distance between particles in an electrolyte solution as described by DLVO theory. Reproduced from [34] under the terms of Creative Commons Attribution 4.0 International License.

External factors, such as the electrolyte concentration in a disperse system, have little influence on van der Waals interaction, but can strongly affect the repulsive energy between particles. For the highly charged surface in a dilute electrolyte solution, the strong long-range repulsive force results in a high energy barrier to be overcome for particle aggregation. So, the nanosystem is stable. For higher concentrated electrolyte solution, a secondary minimum emerges along the curve like the cure shown in Fig. 1.7, and the energy barrier decreases somewhat. However, it is still difficult for the particles to overcome the energy barrier. In this case, the particles are either at the weak secondary potential energy corresponding to reversible aggregation (thermodynamic equilibrium at the primary minimum corresponding to the irreversible aggregation), or completely dispersed in the solution. The latter case has been known as the kinetic stable state. As the surface charge density or potential decreases, or the electrolyte concentration increases), the energy barrier continues to drop. When the energy barrier drops below the axis of U = 0, the particles will rapidly coagulate and the nanosystem is unstable. When the surface charge or potential approaches zero, the two surfaces attract each other at any distance, and the system rapidly coagulates. In addition to the influence of electrolyte concentration, surface charge density or potential on the stability of the nanosystems, the solvent layer around particles is beneficial to prevent them from coagulating and improves the stability of the system.

1.3.1.2 Effects of DLVO theory

When discussing the stability of nanoparticles, DLVO theory considers the total effect of two aspects, i.e., the attractive energy between particles (U_{vdW}) which promotes the coagulation and the repulsive energy (U_{EL}) which hinders it. The theory provides the calculation methods of the repulsive energy and attractive energy between colloid particles, based on which the stability of colloids is quantitatively treated. The relationship between the critical coagulation concentration (also known as the condensation value) and the valence number of ions with opposite charges to the colloidal particle is obtained. And the Schulze-Hardy rule (i.e., the coagulation value is inversely proportional to the sixth power of the valence number of counterions) is formulated theoretically.

Under the Derjaguin integration approximation, the interaction energy between two spherical particles with radii R_1 and R_2 can be expressed as [35, 36]:

$$U = U_{\text{vdW}} + U_{\text{EL}} \tag{1.19}$$

$$U_{\text{vdW}} = \frac{A_{\text{H}}}{6}\left[\frac{8R_{\text{R}}^2}{D(8R_{\text{R}}+D)} + \frac{8R_{\text{R}}^2}{(4R_{\text{R}}+D)^2} + \ln\frac{D(8R_{\text{R}}+D)}{(4R_{\text{R}}+D)^2}\right] \tag{1.20}$$

$$U_{\text{EL}} = 4\pi\varepsilon\varepsilon_0 R_{\text{R}}\left[\varphi_1\varphi_2\exp(-\kappa D) - \frac{1}{4}(\varphi_1^2+\varphi_2^2)\exp(-2\kappa D)\right] \tag{1.21}$$

$$\kappa^{-1} = \sqrt{\frac{\varepsilon\varepsilon_0 k_{\text{B}}T}{2N_{\text{A}}Ie^2}} \tag{1.22}$$

$$R_{\text{R}} = \frac{R_1R_2}{R_1+R_2} \tag{1.23}$$

where A_{H} is the Hamaker constant. D is the distance between the surfaces of the two spheres. ε is the dielectric constant of water. ε_0 is the dielectric constant of vacuum, and φ_1, φ_2 are the surface potentials of the two interacting particles, approximately equal to their ζ potentials. κ^{-1} is Debye length, representing the relationship between the thickness of the ionic diffusion layer and ionic strength, as well as electrolyte. k_{B} is Boltzmann's constant, T is absolute temperature, N_{A} is Avogadro's number, I is ionic strength and e is unit charge.

The variation of the total interaction energy U with the distance D between particles has been shown in Fig. 1.7. The height of the barrier is a sign of the stability of the system. When the height of the barrier is zero, the system becomes unstable. The curve, for which the barrier drops to zero due to the addition of electrolytes, satisfies

$$\frac{\text{d}U}{\text{d}D} = 0,\ U = 0 \tag{1.23}$$

Correspondingly, the added electrolyte concentration c is the critical coagulation concentration, which is usually called the coagulation value. By several steps of treatment, one can get a simplified expression of DLVO theory with water as a medium:

$$c = K\frac{\gamma^4}{A_{\mathrm{H}}^2 z^6} \tag{1.25}$$

where K is a constant, and γ is a physical quantity related to the surface potential. For a certain sol, K and γ have fixed values. Obviously, the condensation value is inversely proportional to the sixth power of the valence number z of the antisign ions, which theoretically illustrates the Schulze-Hardy rule.

1.3.2 Experimental Verification of DLVO

DLVO force was first confirmed by measuring the thickness of soap film [37]. Later, experiments between mercury and mica surfaces in dilute electrolyte solutions demonstrated that DLVO theory is completely correct. According to the literature [34, 38], reporting the relationship of the measured DLVO force and the distance between mica surfaces in different dilute electrolyte solutions, the maximum force is reached when the surface spacing is 20–50 Å and the two micas could attract together eventually below this interval. These results show that the DLVO theory is basically correct. However, when the distance is very small, the theory seems not to hold.

In addition, numerous studies have been conducted on the electrical double layers or DLVO forces in various monovalent, divalent and multivalent electrolyte solutions by means of SFA, AFM or osmotic pressure. Related works include the force between surfactant and lipid bilayer, the internal and external tension of soap film, the forces between the surfaces of silica, sapphire, metal or metal oxide, and so on. Overall, the results are very consistent with the DLVO theory.

1.3.3 Limitations of Classical DLVO

The success of the DLVO theory in describing the stability of colloids is unquestionable. However, some experimental results have shown substantial deviations from the theory [36]. These differences mainly include: the DLVO theory can only describe short distance (0.1–10 nm) interaction, when the distance is less than 0.1 nm, the experimental results are inconsistent with the theory; the

DLVO theory is best used for monovalent salts with concentrations below 5 × 10^{-2} M, and above 0.1 or 0.2 M (range of biological interest or physiological concentrations), the DLVO theory loses the predictability. These deviations are often attributed to the presence of non-DLVO forces, such as hydration force, hydrophobic, oscillatory, membrane fluctuations, steric forces and so on.

Considering the above limitations of the classical DLVO theory, the DLVO model has been improved. Extended DLVO (or EDLVO) theory is known as a model that considers Lewis acid-base interactions or other non-DLVO forces. In most cases where the EDLVO model is applied, the Lewis acid-base interaction is considered to be an important supplement to the DLVO theory, and the predicted results are in agreement with the coagulation experiment.

1.4 The Powerful Tool for Study of Nano Physical Pharmaceutics (NPP) [39]

When a beam of monochromatic, coherent light hits a dilute macromolecule solution or suspension of colloidal particles and the solvent refractive index is different from that of the solute (macromolecules or colloidal particles), the incident light is scattered by each illuminated macromolecule or colloidal particle in all directions [40, 41]. The scattered light waves from different macromolecules or particles mutually interfere, or combine, at a distant from the fast photomultiplier tube (PMT) or avalanche-photodiode (APD) detector and produce a net scattered intensity $I(t)$ or photon counts $n(t)$ which is not uniform on the detection plane. If all the macromolecules or particles are stationary, the scattered light intensity in each direction would be a constant i.e., independent of time. However, in reality, all the scatters in the solution are undergoing constant Brownian motions, and this fact leads to both fluctuations of the scattered intensity pattern on the detection plane and the fluctuations of $I(t)$ with time if the detection area is sufficiently small. The fluctuation rates can be related to different relaxation processes such as translational and rotational diffusions as well as internal motions of the macromolecules. The faster the relaxation process, the faster the intensity fluctuations will be.

In a broad definition, laser light scattering (LLS) could be classified as inelastic (e.g., Raman, fluorescence, and phosphorescence) and elastic (no absorption) light scattering. However, in polymer and colloid science, light scattering is normally referred to in terms of static (elastic) or dynamic (quasi-elastic) measurements, or both, of the scattered light [40]. Static LLS as a classical and absolute analytical method measures the angular distribution of time-average scattered intensity. On the other hand, dynamic LLS measures the intensity fluctuations instead of the average light intensity (this is where the word dynamic comes from), and its essence may be explained as follows. When the incident light is scattered by one moving macromolecule or particle, the detected frequency of the scattered light will be slightly higher or lower than that of the original incident light due to the Doppler Effect, depending on whether the particle moves towards or away from the detector. Thus, the frequency distribution of the scattered light is slightly broader than that of the incident light. This is why dynamic LLS is also called quasi-elastic light scattering (QELS). The frequency broadening ($\approx 10^5$–10^7 Hz) is so small in comparison with the incident light frequency ($\approx 10^{15}$ Hz) that it is very difficult to detect the broadening directly in the frequency domain. However, it can be effectively recorded in the time domain through a time correlation function. For this reason, dynamic light scattering is sometimes known as intensity fluctuation spectroscopy. If we use digital photons to measure the intensity fluctuations, the term photon correlation spectroscopy (PCS) is then used to refer to the technique described here.

The recorded observation of the light scattering can be traced back to 1802 when Richard, J. B noticed the light path of the gold colloid. However, the first man-made experimental investigation on light scattering was Tyndall. He observed the scattering of the natural light when it passed through a colloid amphiphilic. In 1881, based on Maxwell's theory of electromagnetic field, Rayleigh derived that the intensity of the scattered light by the non-absorption, non-interaction and optically isotropic small particles is reversely proportional to the fourth power of the incident wavelength. In 1944, Debye measured the molecular weight of macromolecules from a dilute solution using light scattering method. Later, Zimm [42] proposed the famous Zimm plot by extrapolating both concentration and angular angle to zero value at a single coordinate. Since then, light scattering, strictly, static light scattering as a classical and absolute analytical method has been widely used to characterize both synthetic and natural macromolecules. However, light scattering at that time was limited in measuring scattered intensity at different angles and concentrations, from which three parameters of macromolecules, namely the weight-average molecular weight (M_w), z-average root-mean-square radius of gyration ($\langle R_z^2 \rangle^{1/2}$, simply $\langle R_g \rangle$) and the second-order virial coefficient (A_2) can be obtained. This situation was changed in the 1960s with the invention of the laser. In 1964, Cummins [43] first reported using lasers as source of incident radiation for the study of macromolecular solutions, poly(styrene), and during the last two decades, thanks to the advance of stable laser, ultrafast electronics and personal computers, LLS, especially dynamic LLS has evolved from a very special instrument for physicists and physical chemists to a routine analytical tool in polymer laboratories or even to a daily quality-control device in production lines [44, 45]. Commercially available research-grade LLS instruments are normally capable of making static and dynamic measurements simultaneously for studies of colloidal particles in suspension or macromolecules in solution as well as in gels and viscous media. Note: Sections 1.1 and 1.2 are quoted from the doctoral thesis of the author Wei Li [39].

1.4.1 Static Laser Light Scattering (SLLS)

1.4.1.1 Scattering by a small particle

The electric field of a light wave acting on a particle induces in it a dipole that oscillates with the same frequency as the incident light (Fig. 1.8). An oscillating dipole produces a secondary oscillating field and it radiates electromagnetic energy. In other words, the particle scatters the incident light. Consider a single, optical isotropic particle (a macromolecule or a colloid particle) with a polarizability α at origin o (Fig. 1.8) in vacuum. When the particle is much smaller than the wavelength of the incident light λ_0 (in practice, smaller than $\lambda_0/20$), the electric field of the incident light

$$\vec{E}=\vec{E_0}\exp[i(2\pi\nu t-\varphi)] \tag{1.26}$$

is homogeneous within the particle; it induces in it a dipole $\vec{\mu}=\alpha\vec{E}$. The electric field oscillates with frequency ν and so does the induced dipole. When the electric field is not too strong, the induced dipole is proportional to the field:

$$\vec{\mu}=\alpha\vec{E}=4\pi\varepsilon_0\alpha'\vec{E} \tag{1.27}$$

here α' is the polarizability volume [46] of the particle and ε is the dielectric coefficient at vacuum.

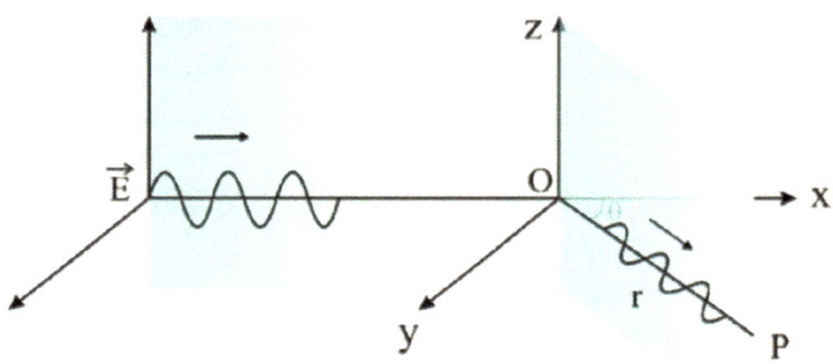

Figure 1.8 Three-dimensional coordinates where origin O represents a scatter and P is the observation point at the xOy plane. The vertically polarized incident beam causes polarization in the scatter, which radiates in different directions.

According to Maxwell's electromagnetic equations, the electric field generated by the oscillating dipole at point p with distance r from origin o and angle of θ from the incident light is

$$\vec{E_s}=\frac{d^2\vec{p}/dt^2}{4\pi\varepsilon_0 rc^2}=\frac{d^2(\vec{\mu_0}+\vec{\mu})/dt^2}{4\pi\varepsilon_0 rc^2}=-\frac{4\pi^2\alpha'}{r\lambda_0^2}\vec{E} \tag{1.28}$$

where c is the velocity of light in vacuum and α' is the polarizability volume. $\vec{p}$ is the total dipole of the particle, i.e., the summation of the permanent $\overrightarrow{\mu_0}$ and the induced dipole $\vec{\mu}$. Since $\overrightarrow{\mu_0} \ll \vec{\mu}$ at room temperature and $\overrightarrow{\mu_0}$ is independent of high frequency electromagnetic field, we have adopted $d^2\vec{p}/dt^2 = d^2\vec{\mu}/dt^2$ in Equation 1.28. Thus, the time-average scattered intensity i (kJ·m^{-2}·s^{-1}) of the particle at point p is

$$i = \varepsilon_o c \left\langle E_s^2 \right\rangle = \frac{16\pi^4 \alpha'^2}{\lambda_0^4 r^2}\left(\varepsilon_o c \left\langle E^2 \right\rangle\right) = \frac{16\pi^4 \alpha'^2}{\lambda_0^4 r^2} I_o \tag{1.29}$$

where I_0 is the intensity of primary light. Equation 1.29 shows that the scattered intensity is proportional to the square of molecular weight since for particles made from a given isotropic material, α' is proportional to their molecular weight. On the other hand, $i/I_0 \propto \lambda_0^{-4}$, means that the scattering is much stronger for light of a shorter wavelength. This explains why the sky is blue because the scattering is stronger toward the short-wavelength end of the visible spectrum.

1.4.1.2 Scattering by many small-particle systems

When there are N-independent same small particles in volume V, the total scattered intensity per unit scattering volume I is the simple summation of the scattered intensity of N/V particles:

$$I = i(N/V) = \frac{16\pi^4 N \alpha'^2}{V \lambda_0^4 r^2} I_o \tag{1.30}$$

Define $R = Ir^2/I_0$ (known as the Rayleigh ratio named after the "father" of the theory of light scattering), we have

$$R = \frac{16\pi^4 N \alpha'^2}{V \lambda_0^4} \tag{1.31}$$

where R (dimension of L) is dependent on the concentration, the size and the nature of the particles.

1.4.1.3 Scattering by real polymer solution

For a real system, two kinds of interference must be considered. One is the intraparticle interference (i.e., the particle is not so small, dimension > $\lambda/20$, that light scattered from two scattering elements within the volume of the same particle has a significant phase difference), and the other is from the interference between different particles. In the first case, a larger particle with volume V' is convenient to assume that it is composed of N' scattering units each with equivalent volume and polarizability volume α_0'. For each scattering unit, its scattered intensity at observation point p can still be expressed by Equation 1.28. Thus, the scattered field of the whole particle at point p is the superposition of the electric field of all N' units:

$$\overrightarrow{E_s} = \sum_{l=1}^{N'} \overrightarrow{E_{s,l}} = -\frac{4\pi^2}{\lambda_0^2 r} \alpha_o' \overrightarrow{E_0} \sum_{l=1}^{N'} \exp[i(2\pi\nu t - \phi_l)] \tag{1.32}$$

Further, the time-average scattered intensity at point p becomes

$$i = \varepsilon_o c \left\langle E_s^2 \right\rangle = \frac{16\pi^4 \alpha_o'^2}{\lambda_0^4 r^2} I_o \left\langle \sum_{l=1}^{N'} \sum_{m=1}^{N'} \exp[i(\vec{q} \cdot \overrightarrow{r_{lm}})] \right\rangle \tag{1.33}$$

where $\Delta\phi_{lm} = \phi_l - \phi_m$ is phase difference between scattering unit l and m at point p.

$$\Delta\phi_{lm} = \vec{q} \cdot \overrightarrow{r_{lm}} \tag{1.34}$$

where $\overrightarrow{r_{lm}} = \overrightarrow{r_m} - \overrightarrow{r_l}$ and $\vec{q} = \overrightarrow{r_i} - \overrightarrow{r_f}$ is called scattering vector which is the difference of the unit vector in the direction of the incident beam $(\overrightarrow{r_f})$ and the one along the scattered beam $(\overrightarrow{r_f})$, and the vector module

$$q = \left|\vec{q}\right| = \frac{4\pi n}{\lambda_o} \sin\left(\frac{\theta}{2}\right) \tag{1.35}$$

where n is the refractive index of the medium. With an increasing θ, q increases and q^{-1} has been used as a spatial resolution ruler with which static LLS is able to probe the size of colloidal particles and macromolecules at a finite angle. By averaging over all possible orientations $(\overrightarrow{r_{lm}})$ of the particle, we finally have

$$i(\theta)=\frac{16\pi^4}{r^2\lambda_0^4}I_o\alpha_o'^2\sum_{l=1}^{N'}\sum_{m=1}^{N'}\frac{\sin(qr_{lm})}{qr_{lm}} \tag{1.36}$$

where θ is the inclined angle from the direction of the primary beam to that of line OP, as is shown in Fig. 1.8. When $\theta \to 0$,

$$i(\theta\to 0)=\frac{16\pi^4}{r^2\lambda_0^4}I_o\alpha_o'^2N'^2 \tag{1.37}$$

Note $N'\alpha_0' = \alpha'$ so that Equation 1.37 is the same as Equation 1.29. We then define an angular scattering function $P(\theta)$ as

$$P(\theta)=\frac{R(\theta)}{R(0)}=\frac{1}{N'^2}\sum_{l=1}^{N'}\sum_{m=1}^{N'}\frac{\sin(qr_{lm})}{qr_{lm}} \tag{1.38}$$

After developing $\sin(qr_{lm})$ into a Taylor series ($\sin x = x - x^3/6 + \ldots$) and retaining the two leading terms, the result reads

$$P(\theta)=1-\frac{q^2}{6N'^2}\sum_{l=1}^{N'}\sum_{m=1}^{N'}r_{lm}^2+\cdots \tag{1.39}$$

where $\frac{1}{2N'^2}\sum_{l=1}^{N'}\sum_{m=1}^{N'}r_{lm}^2$ is defined as the mean square of radius of gyration R_g^2. Thus

$$P(\theta)=1-(1/3)q^2R_g^2+\cdots \tag{1.40}$$

which is related to the conformation of the larger particles and that is why it is also called the structure factor ($q\langle R_g\rangle < 1$) or form factor ($q\langle R_g^2\rangle > 1$). For N-independent larger particles in volume V, Equation 1.36 can be rewritten as

$$R(\theta)=\frac{16\pi^4N}{V\lambda_0^4}\alpha'P(\theta) \tag{1.41}$$

It is interesting to note that when the above discussion is applied to a homogeneous pure gas or liquid, we will find that the scattered intensity is zero in all directions except in the direction of the incident beam. This is because for any selected small scattering element in the scattered volume, we can always find another one whose electric field at point p is the counterweight of the former by interference. Then the scattered waves will exactly cancel each other. All the volume elements can be paired up in this way. However, light scattering does exist, even for pure gas and liquid of small molecules. It is because the molecules are undergoing random movement all the time and the properties of individual volume elements will fluctuate from the average properties. The light waves scattered by individual elements will not have identical amplitudes and thus they will not be canceled totally by interference. As one of the many properties, polarization volume α_l' of a scattering unit l will also more or less deviate its most probable value α_0' as: $\alpha_l' = \alpha_0' + \delta\alpha_0'$, and it is the existence of $\delta\alpha_0'$ that leads to the light scattering of pure gas and liquid. Following the above treatment, we divide scattering volume V into N scattering units and substitute α_0' in Equation 1.32 with $\alpha_0' + \delta\alpha_0'$. We know the term α_0' will contribute nothing to scattered intensity because of the interference of all units so that only the term $\delta\alpha_0'$ needs to be considered. Equation 1.32 then reads

$$\overrightarrow{E_s} = \sum_{l=1}^{N} \overrightarrow{E_{s,l}} = -\frac{4\pi^2}{\lambda_0^2 r} \overrightarrow{E_0} \sum_{l=1}^{N} \delta\alpha_l' \exp[i(2\pi\nu t - \phi_l)] \tag{1.42}$$

Similar to deriving Equation 1.33, taking time-averaged scattered intensity and following the definition of Rayleigh ratio, we can rewrite Equation 1.42 as

$$R(\theta) = \frac{16\pi^4}{V\lambda_0^4} \sum_{l=1}^{N} \sum_{m=1}^{N} \delta\alpha_l' \delta\alpha_m' \exp[i\vec{q}(\overrightarrow{r_m} - \vec{r_l})] \tag{1.43}$$

where $\delta\alpha_l'$ and $\delta\alpha_m'$ are still random functions of time and space. We need to further average Equation 1.43 over time. For this purpose, it is convenient to separate the terms in the double sum into terms for which $l = j$ and $l \neq j$.

$$R(\theta) = \frac{16\pi^4}{V\lambda_0^4} \left[\sum_{l=1}^{N} \sum_{m=1}^{N} \left\langle (\delta\alpha_l')^2 \right\rangle + \sum_{l \neq m}^{N} \sum_{m=1}^{N} \left\langle \delta\alpha_l' \delta\alpha_m' \right\rangle \exp[i\vec{q}\overrightarrow{r_m} - \vec{r_l})] \right] \tag{1.44}$$

The fluctuation of $\delta\alpha'_l$ and $\delta\alpha'_m$ are independent of each other and $\langle\delta\alpha\rangle = 0$. Therefore, $\langle\delta\alpha'_l\,\delta\alpha'_m\rangle = \langle\delta\alpha'_l\rangle\,\langle\delta\alpha'_m\rangle = 0$ and Equation 1.44 turns to be

$$R(\theta) = \frac{16\pi^4}{V\lambda_0^4}\left[\sum_{l=1}^{N}\sum_{m=1}^{N}\left\langle(\delta\alpha'_l)^2\right\rangle\right] \quad (1.45)$$

The time-average fluctuation of polarization volume of the small scattering unit, $\left\langle(\delta\alpha'_i)^2\right\rangle$ should be a constant and is written as $\left\langle(\delta\alpha'_l)^2\right\rangle$. Thus, $\sum_{l=1}^{N}\sum_{m=1}^{N}\left\langle(\delta\alpha'_l)^2\right\rangle = N^2\left\langle(\delta\alpha'_l)^2\right\rangle$ and Equation 1.45 turns to be

$$R = \frac{16\pi^4}{V\lambda_o^4}N^2\cdot\left\langle(\delta\alpha'_o)^2\right\rangle = \frac{16\pi^4}{V\lambda_o^4}\cdot\left\langle(\delta\alpha')^2\right\rangle \quad (1.46)$$

where we have used the relation $\delta\alpha' = N\delta\alpha_0'$. Now the remained question is to find the expression of $\langle(\delta\alpha')^2\rangle$. Consider a macromolecule solution or a colloid dispersion, α' is a function of concentration C and density ρ. $\delta\alpha' = (\partial\alpha'/\partial C)\,\partial C + (\partial\alpha'/\partial\rho)\partial\rho$. Since C and ρ are independent of each other, we have

$$\left\langle(\delta\alpha')^2\right\rangle = \left(\frac{\partial\alpha'}{\partial C}\right)^2\left\langle(\delta C)^2\right\rangle + \left(\frac{\partial\alpha'}{\partial\rho}\right)^2\left\langle(\delta\rho)^2\right\rangle \quad (1.47)$$

Thus, for a dilute solution, Equation 1.46 becomes

$$R_{\text{solution}} = \frac{16\pi^4}{V\lambda_o^4}\left(\frac{\partial\alpha'}{\partial C}\right)^2\left\langle(\partial C)^2\right\rangle + \frac{16\pi^4}{V\lambda_o^4}\left(\frac{\partial\alpha'}{\partial\rho}\right)^2\left\langle(\partial\rho)^2\right\rangle \quad (1.48)$$

where $\frac{16\pi^4}{\lambda_o^4}\left(\frac{\partial\alpha'}{\partial C}\right)^2\left\langle(\partial C)^2\right\rangle$ and $\frac{16\pi^4}{\lambda_o^4}\left(\frac{\partial\alpha'}{\partial\rho}\right)^2\left\langle(\partial\rho)^2\right\rangle$ are the excess Rayleigh ratio of the solution (R_{excess}) and that of the solvent (R_{solvent}) respectively. R_{excess} is the net scattering intensity of the solute by subtracting the intensity of the solvent from that of the solution. According to Clausius–Mossotti equation [46], $\varepsilon_r - 1 = 4\pi\alpha'/V$ and $\varepsilon_r = n^2$, we have

$$\left(\frac{\partial\alpha'}{\partial C}\right)=\left(\frac{\partial\alpha'}{\partial\varepsilon_r}\right)^2\left(\frac{\partial\varepsilon_r}{\partial n}\right)\left(\frac{\partial n}{\partial C}\right)=\frac{n}{2\pi}\left(\frac{dn}{dc}\right) \tag{1.49}$$

On the other hand, we know from thermodynamics that the concentration fluctuation can be expressed as $\left\langle(\delta C)^2\right\rangle=\frac{k_B T}{(\partial^2 A/\partial C^2)_{T,V}}$ and $\left(\frac{\partial^2 A}{\partial C^2}\right)_{T,V}=-\frac{1}{CV_m}\left(\frac{\partial\mu}{\partial C}\right)_{T,V}$, where V_m and μ are respectively the partial volume and chemical potential of the solvent. And the change in concentration can cause a change in osmotic pressure, namely

$$\left(\frac{\partial\mu}{\partial C}\right)_{T,V}=\left(\frac{\partial\mu}{\partial\pi}\right)_{T,V}\left(\frac{\partial\pi}{\partial C}\right)_{T,V}=-\left(\frac{\partial\mu}{\partial P}\right)_{T,V}\left(\frac{\partial\pi}{\partial C}\right)_{T,V}=-V_m\left(\frac{\partial\pi}{\partial C}\right)_{T,V}$$

In dilute macromolecular solution, $\frac{\pi}{C}=\frac{RT}{M}(1+A_2CM+\cdots)$ where A_2 is the second-order virial coefficient. Thus

$$\left(\frac{\partial\pi}{\partial C}\right)_{T,V}=\frac{RT}{M}(1+2A_2CM+\cdots) \text{ and } \left\langle(\delta C)^2\right\rangle=\frac{CM}{N_A V(1+2A_2CM+\cdots)}$$

Now substitute $\langle(\delta C)^2\rangle$ and Equation 1.49 into the expression of R_{excess}, we have

$$R_{\text{excess}}=\frac{4\pi^2n^2}{\lambda_o^4 N_A}\left(\frac{dn}{dC}\right)^2\frac{CM}{(1+2A_2CM+\cdots)} \tag{1.50}$$

Rearrange Equation 1.50 by defining $K = 4\pi^2n^2(dn/dC)^2/(N_A\lambda_o^4)$, we get

$$\frac{KC}{R}=\frac{1}{M}+2A_2C+\cdots \tag{1.51}$$

where we have omitted the footnote "excess" in R_{excess}. For larger macromolecules, a construction factor must be introduced, thus

$$\frac{KC}{R(\theta)} = \frac{1}{MP(\theta)} + 2A_2C \tag{1.52}$$

The last question in deriving the basic equation of static LLS is the polydispersity in real cases. From Equation 1.40 we know $P(\theta) = 1 - (1/3)q^2R_g^2 + \cdots$. Thus, in the limit of vanishing concentration $C \to 0$,

$$R(\theta) = KCMP(\theta) = KCM[1 - (1/3)q^2R_g^2 + \cdots] \tag{1.53}$$

Considering the additive nature of the excess Rayleigh ratio, for a polydispersed polymer solution at $C \to 0$

$$R(\theta) = \sum_i R(\theta)_i = \sum_i KC_iM_i[1 - (1/3)q^2R_{g,i}{}^2 + \cdots] \tag{1.54}$$

If we divide Equation 1.54 by the total polymer concentration $C = \sum_i C_i$, we get, after a slight rearrangement,

$$\frac{R(\theta)}{KC} = \frac{\sum_i C_iM_i}{\sum_i C_i}\left(1 - \frac{q^2\sum_i C_iM_iR_{g,i}{}^2}{3\sum_i C_iM_i} + \cdots\right) \tag{1.55}$$

or

$$\frac{R(\theta)}{KC} = M_w\left[1 - (1/3)q^2\left\langle R_g^2\right\rangle_z + \cdots\right] \tag{1.56}$$

Now, come back to Equation 1.52. When $q^2R_g^2 \ll 1$, omitting the higher order terms in series, we get

$$\frac{KC}{R(\theta)} = \frac{1}{M_w}\left[1 + \frac{1}{3}q^2\left\langle R_g^2\right\rangle_z\right] + 2A_2C \tag{1.57}$$

This is the basic equation of static LLS which is frequently shown in scientific papers. Naturally, the molar mass in the equation, $M_w = \sum_i C_iM_i / \sum_i C_i$, is weight-average; and the mean square radius of gyration, $\left\langle R_g^2\right\rangle_z = \sum_i C_iM_iR_{g,i}^2 / \sum_i C_iM_i$, is z-average. It shows that with $R(\theta)$ measured at a series of C and q, we are able to determine $\langle R_g^2 \rangle_z$ from the slope of $[KC/R(\theta)]_{C\to 0}$ versus q^2; A_2 from the slope of $[KC/R(\theta)]_{\theta\to 0}$ versus C; and M_w from $[KC/R(\theta)]_{C\to 0,\ K\to 0}$. The Zimm plot, i.e., $KC/R(\theta)$ versus (q^2 + kC) with k being an adjustable constant, allows both q and C extrapolations to be made on a single coordinate plane [42, 47]. It should be mentioned that Equation 1.57 is valid under the restriction that the polymer solution exhibits no adsorption, no fluorescence, and no depolarized scattering. For branched structures, the Berry [48] plot ($[KC/R(\theta)]^{1/2}$ vs q^2 + kC) is more adequately used because it often removes much of the curvature observed in the angular dependence of the Zimm plot. If the structure is expected to be large and globular, the Berry plot is not linearized but still shows an upturn. In these cases, it is often appropriate to apply a modified Guinier [49, 50] plot, i.e., $\ln[KC/R(\theta)]$ vs q^2 + kC that removes the upturn even more efficiently.

In practice, the Rayleigh ratio is determined by a relative method; namely, by measuring the scattering intensity of a standard such as benzene or toluene, we can calculate the Rayleigh ratio of a given solution by

$$R_{vv}(\theta) = R_{vv}^{o}(\theta)\frac{I(\theta)_{\text{solution}} - I(\theta)_{\text{solvent}}}{I(\theta)_{\text{standard}}}\left(\frac{n_{\text{solvent}}}{n_{\text{standard}}}\right)^{\gamma} \quad (1.58)$$

where the subscript "vv" means both the incident and the scattered light are vertically (z-axis direction in Fig. 1.8) polarized; I and n are, respectively, the time-averaged scattered light intensity and the refractive index. The term $(n_{\text{solvent}}/n_{\text{standard}})^{\gamma}$ is a refraction correction for the scattering volume and γ is a constant between 1 and 2, depending on the detection geometry of the light scattering instrument, because we should compare the same scattering volume from the solution and the reference standard. If we take the incident light as the x-direction and the scattered light as the y-direction (i.e., θ = 90°), we only need to have a linear correction of the refraction in the x-direction if a slit is used to determine the scattering volume, i.e., γ = 1 because we have already seen all the scattered lights in the z-direction (vertical). On the other hand, if a pinhole with a size much smaller than the diameter of the incident beam at the center of the scattering cell, we have to correct the refraction in both the x- and z-directions, i.e., $\gamma = 2$. However, if the pinhole size is comparable to the beam diameter, $1 < \gamma < 2$. In practice, we should avoid this situation by choosing either a slit or a smaller pinhole [41, 51].

1.4.2 Dynamic Laser Light Scattering (DLLS)

Motions (translational, rotational or internal motion) of macromolecules or colloidal particles in solution can be conveniently studied by using dynamic LLS. Measurement at a single scattering angle (or at a finite q) gives information on the dimension of macromolecules or colloidal particles in the solution with reasonable accuracy. Unlike the static LLS version, dynamic LLS does not rely on the excess scattering intensity between the pure solvent and a dilute solution. The signal from the slowly moving polymer is unambiguously separated from the signal that originates from the rest of the solution. The principle of dynamic LLS has been utilized in some commercial particle-sizing systems for many years. The measurement and data analysis are automated. Users only need to prepare clean solutions by filtration. In recent years, dynamic LLS has been also used as an online detector in size exclusion chromatography (SEC) [51]. Nowadays, the most commonly used method in QELS is the digital technique of photon correlation spectroscopy (or optical mixing) which measures the intensity fluctuation of scattered light in time domain. Practically there are two basic forms of optical mixing: heterodyne and homodyne (self-beating). By heterodyne mixing we refer to the mix of the scattered light with a reference beam (local oscillator) unshifted or shifted in frequency from the incident light beam. In self-beating optical mixing, the scattered wave is not mixed with a reference signal but directly detected. Here we only consider the self-beating intensity-intensity time correlation spectroscopy.

1.4.2.1 Power spectrum of scattered light

Now we consider again an N particle (macromolecule or colloidal particles) scattering system with scattering volume V. We view

N particles as N scatters. Thus, the scattered field as well as the scattered intensity at point p in Fig. 1.8 can still be expressed using Equations 1.32 and 1.33 and they are,

$$\vec{E}(t)=\sum_{i=1}^{N}\vec{E}_i(t)=-\frac{4\pi^2}{\lambda_0^2 r}\alpha_o'\vec{E}_0\sum_{i=1}^{N}\exp i[2\pi vt-\phi_i(t)] \tag{1.59}$$

and

$$I(t)\propto\sum_{i=1}^{N}\sum_{j=1}^{N}\exp i\,\vec{q}[\vec{r}_i(t)-\vec{r}_j(t)] \tag{1.60}$$

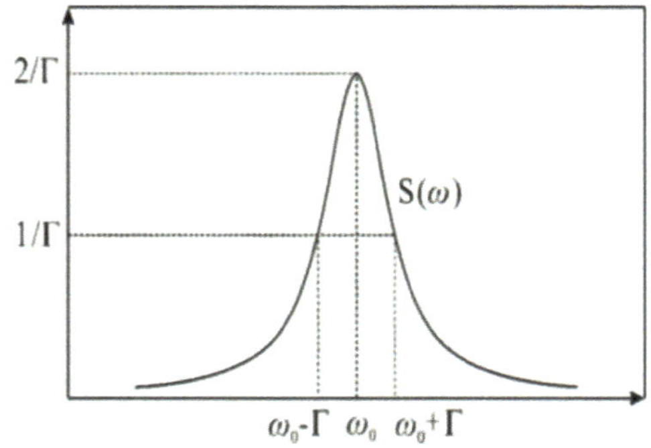

Figure 1.9 The power spectrum $S(\omega)$ of scattered light: a Lorentzian optical frequency distribution centered by the angular frequency of the incident light ω_0 with line width of Γ.

Note $\phi_i(t)$ represents the phase term of the ith particle and is now a function of time due to the motion of the particles. The same situation should be applied to $I(t)$ and $\Delta\phi_{ij}(t)$ because $\Delta\phi_{ij}(t) = \vec{q}\cdot\vec{r}_{ij}(t)$ and $\vec{r}_{ij}(t)$ will have different orientations at different time t.

When the scatters are undergoing Brownian motion, $\vec{E}(t)$ has a randomly modulated phase. The scattered light is broadened in frequency with an optical frequency distribution, or, power spectrum $S(\omega)$ as illustrated in Fig. 1.9. Since the motion of the particles has no preferred direction, the optical spectrum of scattered light contains a continuous distribution of frequencies, i.e., Lorentzian distribution centered by ω_0, the angular frequency of the incident light:

$$S(\omega) = \frac{2\Gamma}{\Gamma^2 + (\omega - \omega_o)^2} \tag{1.61}$$

It can be seen from Fig. 1.9 and Equation 1.61 that when $\omega = \omega_0$, $S(\omega) = 2/\Gamma$; and when $\omega = \omega_0 \pm \Gamma$, $S(\omega) = 1/\Gamma$. That means when the scattered frequency shifts a distance of Γ from ω_0, the density function of power spectrum $S(\omega)$ is half the value that of its peak value. For this reason, Γ is called half-width at half-height, or simply, line width. As stated before, it is difficult to measure Γ (or $S(\omega)$) directly in frequency domain because $\Gamma \ll \omega_0$. It is known from mathematics that $S(\omega)$ and the field-field autocorrelation function $\langle E(0)E^*(t)\rangle$ are a pair of Fourier transform and inverse Fourier transform:

$$\left\langle E(0)E^*(t)\right\rangle = \int_{-\infty}^{\infty} S_x(\omega)\exp(-i\omega t)\mathrm{d}\omega \tag{1.62}$$

$$S(\omega) = \frac{1}{2\pi}\int_{-\infty}^{\infty}\left\langle E(0)E^*(t)\right\rangle \exp(i\omega t)\,\mathrm{d}t \tag{1.63}$$

These two equations are known as Wiener–Khintchine theory. Thus, $S(\omega)$ and $\langle E(0)E^*(t)\rangle$, two functions originally located in the frequency domain and time domain respectively, are now connected with each other.

1.4.2.2 Siegert relation [41]

Another important equation in dynamic LLS is the Siegert relation. Without a local oscillator (i.e., a constant fraction of the incident light reaching the detector from various intentional sources, such as surface scratching or reflection), the self-beating of the scattered electric field leads to normalized intensity-intensity autocorrelation function, $g^{(2)}(q,t)$ based in essence on the Sieget relation:

$$g^{(2)}(q,t) = 1 + |g^{(1)}(q,t)|^2 \tag{1.64}$$

where $g^{(2)}(q,t)$ $(\equiv\langle I(q,0)I(q,t)\rangle/\langle I(q,0)\rangle^2)$ and $g^{(1)}(q,t)$ $(\equiv\langle E(q,0)E^*(q,t)\rangle/\langle E(q,0)E^*(q,0)\rangle)$ is the normalized field-field time correlation functions. Thus, the intensity-intensity time correlation function

$$G^{(2)}(q,t) = \langle I(q,0)I(q,t)\rangle = \langle I(q,0)\rangle^2 g^{(2)}(q,t)$$
$$= \langle I(q,0)\rangle^2[1 + |g^{(1)}(q,t)|^2] \quad (1.65)$$

The significance of introducing $g^{(2)}(q,t)$ and $G^{(2)}(q,t)$ lies in the fact that $G^{(2)}(q,t)$ and $\langle I(q,0)\rangle$ can be measured experimentally. In practice, the detection area cannot be zero no matter small it is. Therefore, the scattered light detected cannot be purely coherent and an instrument parameter, β (<1), is introduced in Equation 1.65:

$$G^{(2)}(q,t) = A(1+\beta\,|\,g^{(1)}(q,t)\,|^2) \quad (1.66)$$

where $A(\equiv\langle I(q,0)\rangle^2)$ is the baseline, t is the delay time, β is a parameter depending on the coherence of the detection optics, and $I(t)$ is the detected scattered intensity or photon counts at time t, including contributions from the solvent and the solute. Therefore, $G^{(2)}(q,t) = \langle[I_{\text{solvent}}(q,0)\rangle + I_{\text{solute}}(q,0)]\ [I_{\text{solvent}}(q,t)\rangle + I_{\text{solute}}(q,t)]\rangle$ and Equation 1.66 becomes

$$G^{(2)}(q,t) = A\left\{1+\beta\left[\frac{I_{\text{solvent}}}{I_{\text{solution}}}\,|\,g^{(1)}_{\text{solvent}}(q,t)\,| + \frac{I_{\text{solute}}}{I_{\text{solution}}}\,|\,g^{(1)}_{\text{solute}}(q,t)\,|\right]^2\right\} \quad (1.67)$$

where all the cross terms have been dropped by assuming that the light scattered by solvent molecules and particles is not correlated. It should be noted that $|\,g^{(1)}_{\text{solute}}(q,t)\,|^2$ decays much faster than $|\,g^{(1)}_{\text{solute}}(q,t)\,|^2$ because small solvent molecules diffuse much faster than larger particles. Thus, after a very short delay time, Equation 1.67 becomes

$$G^{(2)}(q,t) \cong A\left[1+\beta\left(\frac{I_{\text{solute}}}{I_{\text{solution}}}\right)^2 |\,g^{(1)}_{\text{solute}}(q,t)\,|^2\right]$$
$$= A\left[1+\beta_{\text{app}}\left|g^{(1)}_{\text{solute}}(q,t)\right|^2\right] \quad (1.68)$$

where $\beta_{\text{app}} = \beta\,(I_{\text{solute}}/I_{\text{solution}})^2$. For a dilute solution, the scattered intensity from solvent molecules could become appreciable (i.e., $I_{solute} \leq I_{solution}$) and thus the apparent coherence β_{app} would be lower, i.e., $G^{(2)}(q,0)$ appears to have a very low value than expected. We should be aware of this situation, especially for weakly scattered dilute and low-molar-mass polymer solutions. For example, if $I_{solute} = I_{solvent}$, $\beta_{app} = \beta/4$. Noted that β is constant for each particular optical geometry of the scattering instrument. In fact, I_{solute} can be estimated from β_{app} if the values of β at different scattering angles have been pre-calibrated with a narrowly distributed latex standard whose scattering intensity is much stronger than water (solvent), as first demonstrated by Sun et al. [52].

1.4.2.3 Translational diffusions

Now we will see how to get the information about the motion of the particles from the measured intensity-intensity time correlation function $G^{(2)}(q,t)$. Generally, the relaxation of $|g^{(1)}(q,t)|$ includes both diffusion (translation and rotation) and internal motions. Let us first consider the translational diffusion relaxation of the particles. For monodispersed spherical scatters, $|g^{(1)}(q,t)|$ is theoretically represented as an exponential decay function:

$$|g^{(1)}(q,t)| = G_{\exp}(-\Gamma t) \tag{1.69}$$

where G and Γ are the factor of proportionality and the line width, respectively, and $\Gamma = \tau_c^{-1}$, the characteristic decay time representing the rate of dynamic relaxation in self-beating. For a polydispersed polymer sample with a continuous distribution of molar mass M, Equation 1.69 may be generalized as

$$|g^{(1)}(q,t)| = \int_0^\infty G(\Gamma)\exp(-\Gamma t)d\Gamma \tag{1.70}$$

where $G(\Gamma)$ is called the line width distribution and $G(\Gamma)d\Gamma$ is the statistic weight of the particles or macromolecules which possess line width Γ. For a dilute solution, Γ measured at a finite scattering angle is related to C and q by [53]

$$\Gamma = q^2 D(1 + k_d C)(1 + fq^2 \langle R_g^2 \rangle_z) \tag{1.71}$$

where D is the translational diffusion coefficient of the solute molecule at $C \to 0$, k_d is the diffusion second virial coefficient, and f is a dimensionless parameter depending on polymer chain structure and solvent (For polymers with flexible chains in a good solvent, f is between 0.1 and 0.2). Hence, for small C and q, $D \approx \Gamma/q^2$ and it is apparent that $|g^{(1)}(q,t)|$ decays faster at a higher scattering angle. Figure 1.10 demonstrates the linear dependence of the line with Γ on the scattering vector $q(k)$ and shows that this relaxation mode has a diffusive character.

It should be noted that by the definition of $|g^{(1)}(q,t)|$, $G(D) = q^2G(\Gamma)$, the translational diffusion coefficient distribution, is an intensity distribution. And, since $|g^{(1)}(q,t)|$ approaches unity as $t \to 0$, we have

$$|g^{(1)}(q \to 0, t \to 0)| = \frac{\langle E(q,0)E^*(q,t \to 0)\rangle}{\langle E(q,0)E^*(q,0)\rangle}$$
$$= \int_0^\infty G(\Gamma)d\Gamma = \int_0^\infty G(D)dD = 1 \quad (1.72)$$

The average diffusion coefficient $\langle D \rangle$ is defined as

$$\langle D \rangle = \int_0^\infty G(D)DdD \quad (1.73)$$

Further, the translational diffusion coefficient D may be related to the molecular friction factor f through the Stokes–Einstein relation

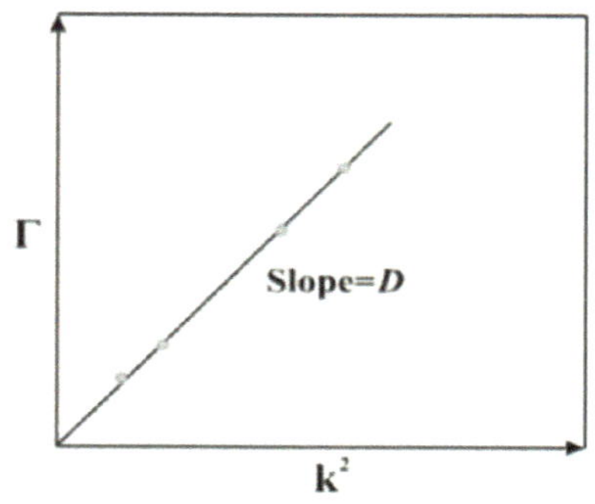

Figure 1.10 Linear dependence of the characteristic line width Γ on scattering vector q.

$$D = k_B T / f \tag{1.74}$$

where k_B and T are the Boltzmann constant and the absolute temperature respectively. For a hard sphere with a radius of R, $f = 6\pi\eta R$, where η is the viscosity of the solvent. For a polymer coil, R is replaced by its hydrodynamic radius R_h, so that

$$R_h = \frac{k_B T}{6\pi\eta D} \tag{1.75}$$

1.4.2.4 Analysis of the correlation function profile

Equation 1.70 indicates that once $|g^{(1)}(q,t)|$ is determined from $G^{(2)}(q,t)$ through Equation 1.66, $G(\Gamma)$ and then $G(D)$ can be computed from the Laplace inversion of $|g^{(1)}(q,t)|$ [45, 54, 55]. In the last three decades, many computation programs have been developed. At the earlier stage, the computation speed was a very important factor in the development of the programs. This constraint has gradually been removed because the personal computer has become faster and faster in the last 10 years. Among many programs, the CONTIN program developed by Provencher [56] is still one of the most widely used and accepted programs for this computation. However, it should be noted that Equation 1.66 is one of the first kind of Fredholm integral equations. Its inversion is an ill-conditioned problem because of the bandwidth limitation of photon correlation instruments, some unavoidable noises in the measured time correlation function, and a limited number of data points. In other words, the data of $g^{(1)}(q,t)$ do not always provide information necessary and sufficient to determine $G(\Gamma)$ uniquely. Thus, in practice, reducing the noise in the measured intensity-intensity time correlation function becomes more important than choosing a program for data analysis. For this reason, it is crucial that the sample solution is cleaned (i.e., made "dust-free") very thoroughly before it is subjected to LLS measurements. A common guideline is to keep the relative difference between the measured and calculated baselines not exceeding 0.1%. The error analysis related to the above problem can be found elsewhere [57, 58].

It is worth noting that there is a temptation among the users of dynamic LLS to extract too much information from the measured intensity-intensity time correlation function. In the literature, three or even four peaks in $G(D)$ were often reported. It is meaningless because they were actually extracting "data" from experimental noises. It has to be warned that even a bimodal distribution of $G(D)$ has to be well justified by other physical evidence or pre-experimental knowledge. This does not mean that many of the Laplace inversion programs developed in the past are useless. On the contrary, they have been quite successful in retrieving the desired information, especially in terms of the average line width $\langle\Gamma\rangle \left(\equiv \int_0^\infty \Gamma G(\Gamma) d\Gamma\right)$ and the relative width $(\mu_2/\langle\Gamma\rangle^2)$ of the line width distribution $(G(\Gamma))$ with $\mu_2 = \int_0^\infty (\Gamma - \langle\Gamma\rangle)^2 G(\Gamma) d\Gamma$. Therefore, the Laplace inversion as a very helpful method in the analysis of the line width distribution $G(\Gamma)$, should be used with a clear understanding of its ill-conditioned nature and its limitations.

In practice, if one is only interested in the determination of $\langle\Gamma\rangle$ and $\mu_2/\langle\Gamma\rangle^2$, a fast but more limited cumulants analysis adopted by Koppel [59] can be used, wherein $[G^{(2)}(q,t) - A]/A$ is expanded as

$$\ln \frac{G^{(2)}(q,t) - A}{A} = (1 + \ln\beta) - \langle\Gamma\rangle t + \frac{\mu_2 t^2}{2!} - \frac{\mu_3 t^3}{3!} + \cdots \tag{1.76}$$

where

$$\mu_m = \int_0^\infty (\Gamma - \langle\Gamma\rangle)^m G(\Gamma) \mathrm{d}\Gamma \tag{1.77}$$

is the mth moment of the line width distribution $G(\Gamma)$. A mth order cumulants fit means that all the terms higher than t^{m} in Equation 1.78 are terminated in the data analysis. The first cumulant Γ, sometimes also called the initial slope, is an important quantity, since it can be calculated for many physical systems and situations [60, 61]. The second cumulant μ_2 is a measure of the width of the distribution. For unimodal distributions of slightly polydispersity polymers in solution, the following relation has been derived [61]:

$$\mu_2/\Gamma^2 \approx (M_z/M_w - 1)/4 \tag{1.78}$$

It is worth noting that, in practice, the cumulants fit can be used for relatively narrow characteristic line width distributions. For $\mu_2/\langle\Gamma\rangle^2 < \sim 0.2$, the second-order cumulants fit is normally sufficient, while when $\mu_2/\langle\Gamma\rangle^2$ is in the range ~0.2–0.3, the third-order cumulants fit is required. For even higher values of $\mu_2/\langle\Gamma\rangle^2$, higher order expansions should be used. However, it is often difficult to find how many terms are sufficient to obtain meaningful results because using too many terms in the cumulants fit might lead to an over-fitting of experimental noises. Therefore, for a broadly distributed sample, the use of cumulants fit is very tedious. On the other hand, the use of CONTIN could yield reliable $\langle\Gamma\rangle$ and $\mu_2/\langle\Gamma\rangle^2$ values under all conditions as long as the measured time correlation function was obtained within a proper bandwidth range and the photon counts have sufficient statistics, e.g., the baseline (A) has a total count over 10^6. However, it warns that the line width distribution obtained from the Laplace inversion is only an estimate. One should be aware of the limitations, yet must realize that the Laplace inversion methods can provide useful information and distinguish between unimodal and multimodal line width distributions, especially, when the peak positions are separated by a factor of more than 2. It is pointed out that function 1.76 was always used for the analysis of the intensity-intensity time correlation function in our study of semidilute solutions. Because in most previous literature, analyzing the autocorrelation function is so arbitrary. However, selecting the fitting function is very important. In this study, by changing exponents of β_f and β_f in function as follows, and comparing the Chi-squared of different fitting methods, we find that one diffusion plus one-stretched mode is the most suitable that is, $\beta_f = 1$ and $0 < \beta_S (= \beta_{coupling}) < 1$, as:

$$g^{(2)}(q,t) = \beta_{apparent}\left\{A_f \exp\left[-\left(t/\tau_f\right)^{\beta_f}\right] + A_s \exp\left[-\left(t/\tau_s\right)^{\beta_s}\right]\right\}^2 \tag{1.79}$$

1.4.3 Typical Application of Laser Light Scattering in NPP Accomplished by Our Group

The light scattering technique is one of the important characterization methods in the polymer field. The combination of static light scattering and dynamic light scattering can obtain a wealth of information about the polymer, such as the weight-average molecular weight, z-average root-mean-square radius of gyration, the second-order virial coefficient, radius of hydrodynamics, size distribution, molecular chain conformation and so on. In addition to polymers, light scattering technology is also suitable for studying the behavior of biomacromolecules, colloids, vesicles, nanoparticles and so on in solutions. Several typical applications of light scattering are described below.

1.4.3.1 The special *in-house* designed powerful tool for NPP study

Figure 1.11 shows the *in vivo* blood conditions mimic testing system by a special *in-house* designed equipment with *in vivo* barriers panel (left), the viscous flow conditions tuning panel (middle) and the detecting model with flowing cell (right), which generates the *in vitro* viscous flow conditions for NPP study in solution. This equipment is special for the study of nanomedicines or formulations in viscous fluids, which can help us to indicate and evaluate the real state, structure and properties of nanoformulations post administration.

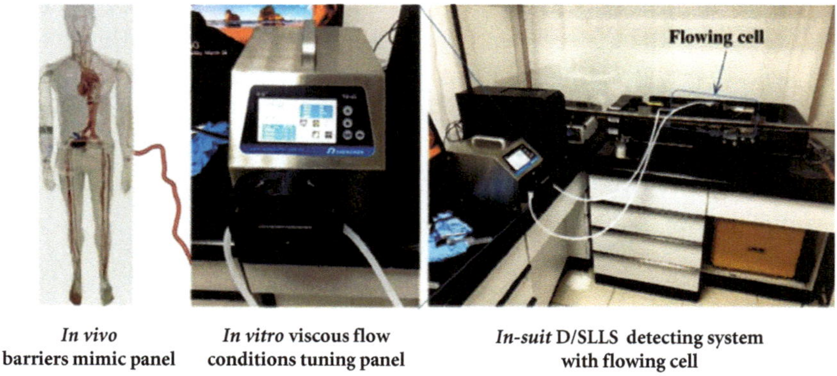

Figure 1.11 The special *in-house* built equipment designed for the NPP study.

1.4.3.2 Characterization of polymer chains dynamics in solution

Figure 1.12 shows the variation of the polymer concentration with evaporation time as measured by light scattering. As time goes on, the solvent volatilizes and polymer chain entanglement increases, resulting in the increase of polymer concentration. The polymer solution gradually changes from dilute to sub-concentrated solution. This *in-suit* time-dependence of polymer concentration tuning helps us to avoid dust and other artificial errors, which is a big trouble for the LLS study.

Based on the smart LLS experimental design, the semidiluted solutions of polystyrene in toluene are successfully obtained. It is easy to know that the polymer chains will entangle each other as the C increases. The fast diffusion of chain segments in between the entangled points, that is the chain blob, is thus recorded and plotted. And a so-called scalding law between the average scattering intensity (⟨*I*⟩) and the polymer concentration (C) is obtained in the experiments as shown in the inserted figure. The experiments tell us that the LLS is a powerful tool for studying the polymers in solutions.

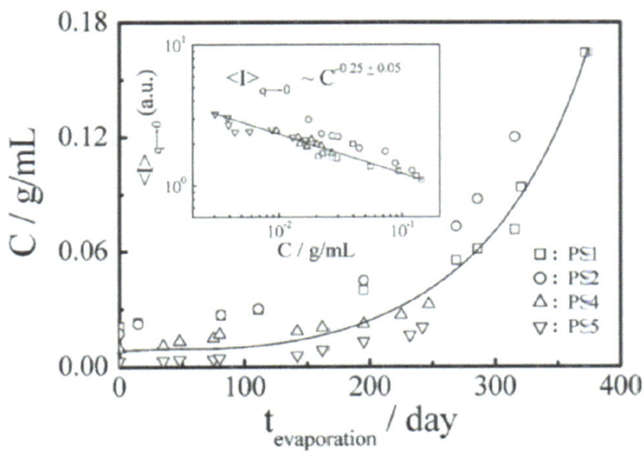

Figure 1.12 The variation of the polymer concentration with evaporation time. Reproduced with permission from [62].

1.4.3.3 Characterization of self-assembly of polymer aggregates in solutions

Figure 1.13 describes the self-assembly process during which the polymers aggregate from small particles into large particles with the increase in temperature, and the number of chains in the particles increases gradually. This figure tells us that the LLS can be used to indicate the *in-suit* inter-particles/chains aggregations process or self-assembly process in solutions [63].

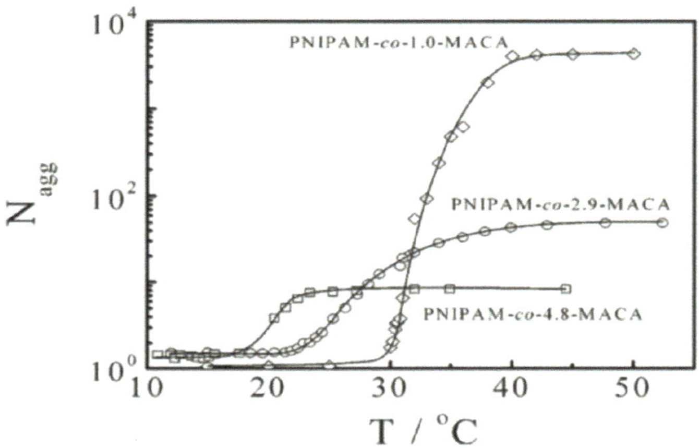

Figure 1.13 Polymer self-assembly with temperature increasing. Reproduced with permission from [63].

1.4.3.4 Characterization of the morphology of nanoparticles in physiological media

Figure 1.14 shows the change in particle size and morphology of nanoparticles characterized by light scattering in our previous work. With the increase of the surface chain density, the stability of particles increases, forming a single particle, which cannot be adsorbed by serum proteins. The well-separated peaks of BSA and delivery systems indicate the stability of the drug delivery in serum solution with high concentration. This is a useful evaluation method for evaluating the *in vivo* serum stability of nanoformulations.

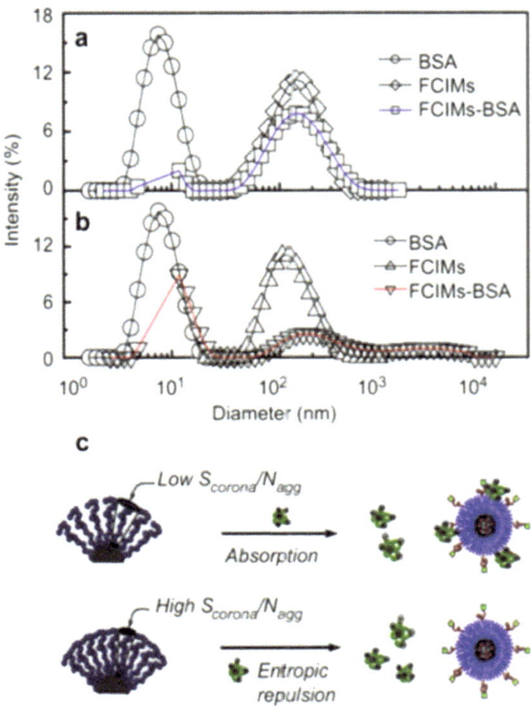

Figure 1.14 Characterization of the morphology of nanoparticles in physiological media by light scattering. Reproduced with permission from [64].

1.4.3.5 Well-defined nanogel *in-suit* preparation by radical co-polymerization and indicated by LLS

Figure 1.15 shows the nanogel with core–shell structure, narrow particle size and well-defined morphology which is regulated by light scattering in our group. The core and shell of nanogel in the TEM image is quantitatively analyzed by red circle (core) and blue circle (shell) with marked radius R, R_{core}, R_{shell}, the density of core and shell (ρ_{core}, ρ_{shell}) on left panel. Then in the SLLS testing, the dot lines of structure factor (p(q)) are well fitted by the solid line as introducing the microstructural parameters into the function of p(q), which indicates that the experimental data is well observed with the theoretic curve. On the other hand, the R_{core} and R_{shell} can be designed by tuning the feeding ratio of PEI and PNIPAM monomers (A = W_{PEI}/W_{NIPAM}) by Equation 1.80 in the reaction [65]. Thus, according to Equation 1.80, the core–shell structure can be easily tailored in the experiment.

$$\frac{R_g}{R_h} = \sqrt[0.5]{\frac{3(Ax^2 - (1+A)x^5 + 1)}{5(1+A)(1-x^2)}} \tag{1.80}$$

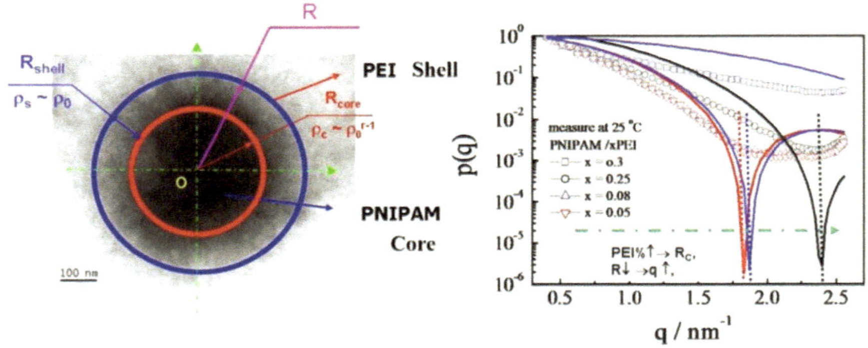

Figure 1.15 Nanogel with core–shell structure tailoring by LLS in our group.

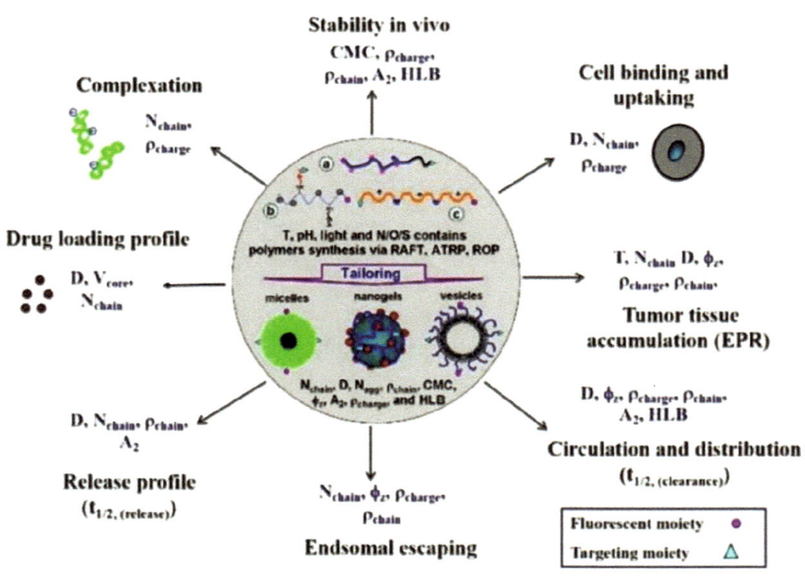

Figure 1.16 Correlations between the NPP and the *in vitro/vivo* performance. Where: ϕ_Z: Surface potential; $\rho_{\text{surface chain}}$: Surface chain density; ρ_{charge}: Surface charge density; A_2: Second virial efficiency; CMC: Critical micellization concentration; *D*: Diameter; EPR: Enhanced permeability and retention; HLB: Hydrophilic lipophilic balance; N_{agg}: Aggregation number; N_{Chain}: Chain number in particle; $t_{1/2}$: Half time of circulation; *T*: Temperature; V_{core}: Core volume.

1.4.3.6 NPP of different delivery systems obtained by SLLS

Figure 1.16 shows the correlations between the NPP of different delivery systems and the *in vitro*/*vivo* performance, which was conducted and summarized by our group. These key parameters, which are the basis of NPP discussed in this book, are discussed in detail in the following chapters.

References

1. Wu C, *A Concise Textbook of Thermodynamics*, Higher Education Press, 2019.
2. Teraoka I, Polymer Solutions, A John Wiley & Sons, Inc, New York, 2002.
3. Zhang G, Jin F, Ye X, Hong L, *Polymer Solutions*, China Science Publishing & Media Ltd. 2014 (Translated version by Zhang G. et al.).
4. Hildebrand JH, *Chem Rev* **1949**, *44*, 37–45.
5. Hildebrand JH, Scott RL, *The Solubility of Nonelectrolytes*, Reinhold, New York, 1950.
6. Hansen CM, *J Paint Technol* **1967**, *39*, 104–117.
7. Hansen CM, *J Paint Technol* **1967**, *39*, 505–510.
8. Hancock BC, York P, Rowe RC, *Int J Pharm* **1997**, *148*, 1–21.
9. Wieneke JU, Kommoß B, Gaer O, Prykhodko I, Ulbricht M, *Ind Eng Chem Res* **2011**, *51*, 327–334.
10. Louwerse MJ, Maldonado A, Rousseau S, Moreau-Masselon C, Roux B, Rothenberg G, *ChemPhysChem* **2017**, *18*, 2999–3006.
11. Gao M, Zhang Z, Zhang W, Cao Q, Tang Z, Zhao W, *FlatChem* **2022**, *32*, 100346.
12. Israelachvili JN, Mitchell DJ, Ninham BW, *J Chem Soc, Faraday Trans 2* **1976**, *72*, 1525.
13. Israelachvili JN, Mitchell DJ, Ninham BW, *Biochimica et Biophysica Acta* **1977**, *470*, 185–201.
14. Nagarajan R, *Langmuir* **2002**, *18*, 31–38.
15. Blanazs A, Armes SP, Ryan AJ, *Macromol Rapid Commun* **2009**, *30*, 267–277.
16. Israelachvili JN, *Intermolecular and Surface Forces*, Elsevier, Amsterdam, 2011.
17. Bates FS, *Science* **1991**, *251*, 898–905.

18. Bates FS, Fredrickson GH, *Physics Today* **1999**, *52*, 32–38.
19. Leibler L, *Macromolecules* **1980**, *13*, 1602–1617.
20. Matsen MW, Schick M, *Phys Rev Lett* **1994**, *72*, 2660–2663.
21. Matsen MW, Bates FS, *Macromolecules* **1996**, *29*, 1091–1098.
22. Bates FS, Fredrickson GH, *Annu Rev Phys Chem* **1990**, *41*, 525–557.
23. Fredrickson GH, Bates FS, *Annu Rev Mater Sci* **1996**, *26*, 501–550.
24. Abbott S, *Solubility Science: Principles and Practice,* University of Leeds: Leeds, UK, 2017.
25. Kim JK, Lee JI, Lee DH, *Macromol Res* **2008**, *16*, 267–292.
26. Mai Y, Eisenberg A, *Chem Soc Rev* **2012**, *41*, 5969–5985.
27. Binder WH, Barragan V, Menger FM, *Angew Chem* **2003**, *42*, 5802–5827.
28. Silvius JR, del Giudice D, Lafleur M, *Biochemistry* **1996**, *35*, 15198–15208.
29. Feigenson GW, *Biochim Biophys Acta* **2009**, *1788*, 47–52.
30. Veatch SL, Keller SL, *Biochim Biophys Acta* **2005**, *1746*, 172–185.
31. Yoon YZ, Hale JP, Petrov PG, Cicuta P, *J Phys Condens Matter* **2010**, *22*, 062101.
32. Veatch SL, Gawrisch K, Keller SL, *Biophys J* **2006**, *90*, 4428–4436.
33. Zhao J, Wu J, Heberle FA, Mills TT, Klawitter P, Huang G, Costanza G, Feigenson GW, *Biochim Biophys Acta* **2007**, *1768*, 2764–2776.
34. Lakshmanan S, Holmes WM, Sloan WT, Phoenix VR, *Int J Environ Sci Technol* **2015**, *12*, 3373–3384.
35. Zetasizer Nano Series User Manual (Chapter 15 Zeta Potential Theory), Malvern Instruments Ltd, 2003.
36. Adjei IM, Sharma B, Labhasetwar V, *Nanomaterial: Impacts on Cell Biology and Medicine*, Springer, Dordrecht, Netherlands, 2014.
37. Lyklema J, Mysels KJ, *J Am Chem Soc* **1965**, *87*, 2539–2546.
38. Israelachvili JN, *Adv Colloid Interface Sci* **1982**, *16*, 31–47.
39. PhD. thesis of Li W, *Reexamination of Dynamic of Semidilute Polymer Solution*. The Chanese University of Hong Kong, **2006**.
40. Chu B, *Laser Light Scattering: Basic Principles and Practice*, 2nd edn, Academic Press, New York, 1991.
41. Berne BJ, Pecora R, *Dynamic Light Scattering*, John Wiley & Sons Press, New York, 1972.

42. Zimm BH, *J Chem Phys* **1948**, *16*, 1099.

43. Cummins HZ, Knable N, Yeh Y, *Phys Rev Lett* **1964**, *12*, 150.

44. Brown W, ed, *Light Scattering: Principles and Development*, Clarendon Press, Oxford, 1996.

45. Brown W, ed, *Dynamic Light Scattering: The Method and Some Applications*, Clarendon Press, Oxford, 1993.

46. Atkins PW, *Physical Chemistry*, 6th edn, Oxford University Press, Oxford, 1998.

47. Burchard W, Light scattering techniques, in *Physical Techniques for Study of Food Biopolymers* (Ross-Murphy SB, ed), Blackie Academic & Professional, London, 1995, Chapter 4.

48. Berry GC, *J Chem Phys* **1966**, *44*, 4550.

49. Guinier A, *Ann Phys* **1939**, *12*, 161.

50. Guinier A, Fournet G, *Small Angle Scattering of X-Rays*, Wiley, New York, 1955.

51. Wu C, Chu B, Light scattering, in *Experimental Methods in Polymer Science* (Tanaka T, ed), Academic Press, San Diego, 2000. pp. 1–56.

52. Sun ST, Nishio I, Swislow G, Tanaka T, *J Chem Phys* **1980**, *73*, 5971.

53. Stockmayer WH, Schmidt M, *Macromolecules* **1984**, 17, 509.

54. Chu B, Ford JR, Dhadwal HS, *Methods Enzymol* **1983**, *117*, 256.

55. Chu B, Wu C, Ford JR, *J Colloid Interface Sci* **1985**, *105*, 473.

56. Provencher SW, *J Chem Phys* **1976**, *64*, 2772.

57. Raczek J, *Eur Polym J* **1983**, *19*, 607.

58. Nordmeier E, Lechner MD, *Polym J* **1989**, *21*, 623.

59. Koppel DE, *J Chem Phys* **1972**, *57*, 4814.

60. Akcasu Z, *Polymer* **1980**, *21*, 866.

61. Akcasu Z, Han CC, *Polymer* **1981**, *22*, 1019.

62. Li JF, Li W, Huo H, Luo SZ, Wu C, *Macromolecules* **2008**, *41*, 901–911.

63. Wu C, Li W, Zhu XX, *Macromolecules* **2004**, *37*, 4989–4992.

64. Li W, Zhao H, Qian W, Li H, Zhang L, Ye Z, Zhang G, Xia M, Li J, Gao J, Li B, Kou G, Dai J, Wang H, Guo Y, *Biomaterials* **2012**, *33*, 5349–5362.

65. Yang C, Li W, Wu C, *J Phys Chem B* **2004**, *108*, 11866.

Chapter 2

Nano Physical Pharmaceutics of Micelle-Based Systems

Yening Xia, Ruochen Huang, Fei Duan, and Wei Li

Department of Nanomedicine,
Naval Medical University Shanghai 200433, China

liwei_dds@163.com

2.1 Classification of Micelle

Micelle refers to the ordered aggregate obtained by self-assembly when the concentration of surfactant in the solution exceeds critical micelle concentration (CMC). The compounds that form micelles are generally amphiphilic molecules, which contain both hydrophilic and hydrophobic chemical structures [1–2]. A single surfactant molecule is surrounded by water molecules after being dissolved in water. The hydrophilic group in the molecule is attracted by water and the hydrophobic group is repelled by water, so it is first adsorbed on the surface of the solution and extends the hydrophobic group to the air. When the concentration of the surfactant increases to the point where the surface of the solution reaches saturation and cannot continue to be adsorbed, it transfers to the inside of the solution. After reaching a certain concentration, the hydrophobic groups in the molecule attract each other and

Nano Physical Pharmaceutics
Edited by Wei Li

ISBN 978-981-4968-52-2 (Hardcover), 978-1-003-51393-3 (eBook)
www.jennystanford.com

associate into groups, while the hydrophilic groups extend into the water and combine with the water molecules to form aggregates with hydrophobic kernel and hydrophilic shell, namely micelles. Micelles and monomers in the solution are in kinetic equilibrium, constantly breaking and recombining. A micelle is generally formed by dozens or hundreds (50~200) of surfactant molecules. The number of surfactant molecules in micelles is called the aggregation number (m), and the aggregation number is multiplied by the relative molecular mass of the surfactant to obtain the micelle amount, that is, the relative micelle mass [1, 3].

Surfactants are classified into ionic surfactants and non-ionic surfactants according to the dissociation properties of their polar groups. Table 2.1 lists the chemical structures of several common surfactant hydrophilic groups [4]. The ionic surfactants usually contain functional groups, such as thiol, ester, sulfate, sulfonate, phosphate and amide. In non-ionic surfactants, the hydrophilic group does not dissociate in aqueous solution, and these are commonly polyethylene glycol (PEG, $(-CH_2CH_2O-)_n$)-based structures. Micelles can be classified into three categories based on the surfactants.

Table 2.1 Chemical structures of common surfactant hydrophilic groups [4]

Name	Type of Surfactant	Structure
Carboxylate	Anionic	$R—(COO^-)_nM^{n+}$
Sulfonate	Anionic	$R—SO_3^-M^+$
Thiocarboxylate	Anionic	$-O_3S—RCOO—M^{2+}$
Ammonium salt	Cationic	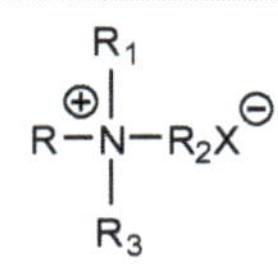
Sulfhydryl group	Cationic	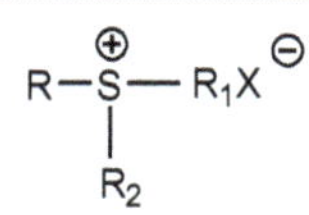
Phosphate	Cationic	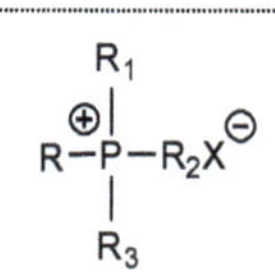

Table 2.1 (*Continued*)

Name	Type of Surfactant	Structure
Pyridine	Cationic	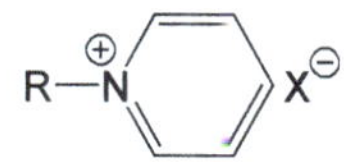
Polyoxyethylene fatty alcohol ether	Non-ionic	$R\text{-}(OCH_2CH_2)_n\text{-}OH$
Polyoxyethylene ester	Non-ionic	$R\text{-}COO\text{-}(CH_2CH_2)_n\text{-}OH$
Polyoxyethylene alkyl thioether	Non-ionic	$R\text{-}COO\text{-}(CH_2CH_2O)_n\text{-}OH$
Triglycine	Amphoteric	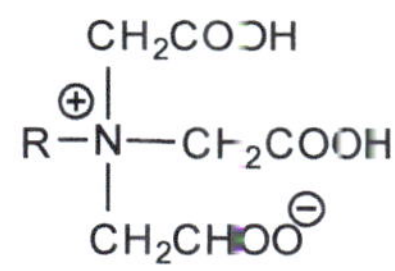

2.1.1 Ionic Surfactant Micelle

The hydrophobic portions of surfactants attract each other and associate with each other to form the inner core, while the hydrophilic groups are arranged on the surface of the core, and attract ions with opposite charges from the solution. Hartley first discovered that micelles were spherical when the concentration of surfactant was slightly greater than CMC without the presence of any other additives [5]. The light scattering method confirms that micelles are spherical and the degree of association remains unchanged when the concentration is greater than CMC within a certain range. The shape also may vary from spherical depending upon the chain length and the area occupied per polar group on the surface. Debye et al. found that micelles are rod-shaped in concentrated solution, which has higher thermodynamic stability. When the concentration increases further, micelles aggregate as bundles or superimpose as layers [6].

2.1.2 Non-Ionic Surfactant Micelle

Non-ionic surfactants can also form micelles of different shapes. For example, the micelle is reticulated when the degree of polymerization of PEG is high at room temperature, because water only exists in the thin layer among micelles, while the high-concentration aqueous solution is extremely viscous. As the temperature increases, the hydrogen bond between PEG and water molecules is destroyed, causing water loss and spherical micelle formation. Becher et al. studied the shape of micelles by measuring the light scattering data and viscosity of polyethylene glycol lauryl ether aqueous solution and found that when the number of aggregation of molecules is different, the shape of micelles is also different [7]. When the aggregation number is greater than 150, the surfactant molecules are longitudinally arranged into round rod-shaped micelles.

2.1.3 Mixed Micelle

Mixed micelles are mixtures of amphiphile systems (including surfactants, polymers, and copolymers) that aggregate into the micellar form, exhibiting characteristic properties different from the individual amphiphile. Alcohol with six or more carbon atoms can be added to the micelle solution, and embedded between the surfactant molecules to form a mixed micelle. Since the total surface charge of the micelle remains unchanged, while the surface area expands due to the embedding of alcohol molecules, the surface charge density of the micelle decreases, and hence the decrease of ionic repulsion and CMC value. The mixed micelle formed by two different ionic surfactants is similar to the micelle mentioned above.

2.2 Preparation of Micelles

Polymeric micelles (PMs) are formed from amphiphilic block copolymers, which have hydrophobic polymer chains linked to hydrophilic polymeric chains. Their good biocompatibility, stability and drug-loading capability have led to the development of nanomedicines, and some are already industrialized or currently being tested in clinical trials. The preparation of PMs, which are the focus of this chapter, can be summarized into two methods.

The first method involves using a selective solvent for one of the segments to dissolve the bulk sample. Notably, this method only works under the premise that the total M_W of the copolymer is small enough with a short-length insoluble block. Other ways including prolongated stirring, thermal, or ultrasound treatments are used to “anneal” the solution for better solubility. However, research shows that these techniques have the disadvantage that they easily lead to non-equilibrium micelles, especially when the core-forming chains are below phase transition temperature (T_g). The annealing treatment can only lead to the dispersion of samples into the selective solvent, rather than forming micelles based on a unimer–aggregate equilibrium. In other words, the characteristic properties of the final micelles mainly hinge on the morphology of the starting bulk sample [8].

The second method is to first dissolve block copolymers in a non-selective solvent to achieve complete dissolution at the molecular level. Then, a selective solvent for one segment is added in order to trigger aggregation of the dissolved chains and hence micelle formation. To remove the initial non-selective solvent, the method of evaporation or dialysis against selective solvent was first introduced by Tuzar and Kratochvil, which became the most common way to prepare micelle solution [9]. Other ways, such as changing pH value or temperature to reach the unimer–micelle equilibrium, can be used to prepare stimulus-responsive micelles that undergo morphology change with applied stimulus as will be discussed in Section 2.5.

2.3 Effects on Micelle Assembly

2.3.1 Critical Micelle Concentration

The minimum concentration of surfactant molecules that associate to form micelles in the solvent is the CMC. When the concentration of the surfactant is below CMC, it has a strong adsorption tendency at the air/water interface, forming a monomolecular film oriented such that the hydrophilic group is dissolved in water and the hydrophobic group is exposed to the air to reduce the surface tension of the air/water interface. As the concentration increases, the adsorbed molecules at the interface increase until saturation. Then the surfactant enters the solution, and when reaching CMC, micelles are formed through intermolecular association. If so, the properties of the solution, such as osmotic pressure, density, interfacial tension, and molar conductivity, all show sudden changes.

Therefore, when the concentration of many aqueous solutions of amphiphilic substances increases, the properties of the solution deviate significantly from the theory, and the deviation is usually much larger than that caused by the increase of strong electrolytes in the aqueous solution. For example, by plotting the surface tension, conductivity or light scattering intensity against concentration, it is easy to see the non-ideal deviation through the change of the slope of the curve, and the root cause is the formation of micelles. As the concentration of surfactant molecules increases, their structures can change from single molecules to spherical, rod-shaped and layered micelles. It is generally believed that the concentration when spherical micelles are formed is the first CMC, and the concentration when spherical micelles are transformed into rod-shaped micelles is the second CMC.

2.3.2 Mechanism of Micellization

The micellization of surfactants can reduce the total free energy of the solution. The aggregation of the hydrophobic moiety makes it lose a certain degree of freedom, which means that the entropy will decrease [4]. Compared with single molecules, the translational entropy and rotational entropy of micelles are both reduced. On the other hand, the interaction between the hydrophilic group in the surfactant molecule and water reduces the free energy of the system, while the free energy of the system increases by reducing the contact between the hydrophobic moiety and water molecules through micellization. In addition, the hydrophilic groups may be ionized and charged, and the resulting electrostatic repulsion will not be conducive to the generation of micelles. It can be seen that the occurrence of micellization is affected by multiple forces. The following figure shows the general process of spherical micelle assembly from block copolymer.

Micellization is mainly driven by the entropy change caused by the hydrophobic effect. In order to minimize the contact between the hydrophobic moiety and water, water molecules have a tendency to be driven from the hydrophobic region to the bulk aqueous solution, for which the disorder increases, so the entropy of the system increases. The micellization process can be expressed by free energy and CMC through the following equation:

$$\Delta G^0 = RT\ln(\text{CMC}) \tag{2.1}$$

where ΔG^0 is the standard free energy change of micellization, R is the gas constant, T is the absolute temperature of the system, and CMC can also be expressed by enthalpy:

$$\ln(\text{CMC}) = \frac{\Delta H^0}{RT} + \text{Constant} \tag{2.2}$$

It is assumed that the standard enthalpy change of micellization is ΔH^0, and is independent of the aggregation number and temperature.

From the perspective of thermodynamics, there is a dynamic equilibrium in the process of forming micelles with an aggregation number of "m" from a single surfactant molecule (Z), where the equilibrium constant K is:

$$\text{mZ} \leftrightarrow \text{Zm} \tag{2.3}$$

$$K_{\text{m}} = \frac{[\text{Zm}]}{[\text{Z}]^{\text{m}}} \tag{2.4}$$

$\Delta G\text{m}^0$ refers to the free energy for transferring a single surfactant molecule from water medium to a micelle.

$$\text{m}\Delta G\text{m}^0 = -RT\ln K \tag{2.5}$$

In the basic equation of micellization, $RT\ln K$ can be replaced equivalently by $\text{m}\Delta G\text{m}^0$, where the concentration is expressed in mole fraction.

$$\ln X_m = -\frac{m\Delta Gm^0}{RT} + m\ln X_1 + \ln m \tag{2.6}$$

X_m refers to the number of moles of surfactant molecules in a micelle with aggregation number of m, X_1 is the mole fraction of the monomer, and X_m is a function of m at a specific value of X_1.

This equation can be regarded as a distribution function of the micelle volume, and the collaboration of micelle formation is reflected in $m\ln X_1$ in the equation. X_m obtained by X_1 in a narrow range close to the CMC is more reasonable.

$$\Delta Gm^0 = \Delta Um^0 + Wm \tag{2.7}$$

ΔUm^0 represents the change of the free energy of hydrophobic hydrocarbyl groups in the surfactant molecule in the transfer from water to the micelle core. Wm represents the free energy resulting from repulsion between polar head groups. For a given surfactant system, the micellization and CMC depend on the relative balance of forces that favor and delay molecular aggregation, and also affect the size, shape and stability of micelles [4].

2.3.3 Influence of the Surfactant Structure on CMC

The micellization of surfactants in aqueous solution is controlled by two opposite forces. The hydrophobic force, which drives hydrophobic moiety out of the aqueous environment and aggregate to form the nucleus, is the main driving force for micellization. The polar head group extends into the water to form a shell after being dissolved, and the electrostatic repulsion or steric hindrance is not conducive to the formation of micelles. There is a delicate balance between the driving force and the counterforce. In addition, factors such as the volume of hydrophobic groups, the structure of polar groups, co-solutes, pH and temperature can affect the micellization process [10].

2.3.3.1 Volume of hydrophobic groups

The volume of the hydrophobic group in surfactant has the greatest effect on micellization. For example, by increasing the size of the hydrophobic moiety, it is possible to enhance the hydrophobic effect and promote the formation of micelles. For example, the CMC of surfactant decreases as the length of the carbon chain in the molecule increases. For compounds with the same polar group, the relationship is described as Equation 2.8:

$$\lg(\text{CMC}) = A - Bm \quad (2.8)$$

where m is the number of carbon atoms in the carbon chain, A and B are constants, which are related to the structure of surfactants, Table 2.2 lists corresponding values to some surfactants.

Table 2.2 Value A & B of various surfactants

Surfactant	Temperature/°C	A	B
C_mCOOK	25	1.92	0.290
C_mSO_2Na	40	1.59	0.294
C_mSO_4Na	45	1.42	0.265
C_mNH_3HCl	25	–	0.275

Note: C_m represents the carbon chain with m carbon atoms.

The effect of substituents on the hydrophilicity and hydrophobicity of surfactants can be estimated by Table 2.3 and can be used to speculate and compare the micellization of similar structures. Table 2.4 lists a variety of amphiphilic diphenylmethane drugs and the properties of their micelles. The comparison shows that as the hydrophobicity of the substituents gradually elevate, the CMC value of the drug drops and the aggregation number of the micelles increases.

Table 2.3 Properties of substituent groups

Substituent Group	Classification
$-CH_3$	Hydrophobic
$-CH_2-$	Hydrophobic
–Cl, –Br, –F	Hydrophobic
$-N(CH_3)_2$	Hydrophobic

Table 2.3 (*Continued*)

Substituent Group	Classification
$-SCH_3$	Hydrophobic
$-OCH_2CH_3$	Hydrophobic
$-OCH_3$	Slightly hydrophilic
$-NO_2$	Slightly hydrophilic
–CHO	Hydrophilic
–COOH	Slightly hydrophilic
–COO	Slightly hydrophilic
$-NH_2$	Hydrophilic
$-NH_3^+$	Slightly hydrophilic
–OH	Slightly hydrophilic

Table 2.4 Effect of substituent on diphenylmethane drugs and properties of micelles

Drug	R	R′	R″	CMC ($mol \cdot kg^{-1}$)	Aggregation Number
Diphenylamine	H	H	H	0.132	3
O-methyl-diphenhydramine	CH_3	H	H	0.096	7
Brominated diphenylamine	Br	H	H	0.053	11
Oxy Mephenhydramine	Cl	H	CH_3	0.045	13

Regardless of ionic or non-ionic surfactants, the CMC decreases with longer carbon chain. Generally speaking, for each additional carbon atom in the carbon chain, the CMC of ionic surfactant is reduced to about 1/2 of the original one, and the CMC of non-ionic surfactant is reduced to about 1/3 of the original one. As for surfactants with the same number of carbon atoms, the CMC of the former is generally much smaller than the latter. If the hydrophilic group is the same and the hydrophobic group contains the same value of carbon atoms, the CMC of surfactants with branched chains in the carbon chain is higher than the CMC of surfactants without branched chains. The benzene ring is slightly different, as it is actually equivalent to only 3.5 methylene units despite being composed of six carbon atoms.

2.3.3.2 Chemical structure and volume of hydrophilic groups

The chemical structure of hydrophilic groups can also significantly affect the aqueous behavior of surfactants. For example, if the hydrophobic groups are the same, non-ionic surfactant generally shows lower CMC value than ionic surfactant, and the micellar aggregation number is higher. As for non-ionic surfactants with PEG as the hydrophilic group, their properties are affected by the length of PEG chain segments. As shown in Table 2.5, with the increase of the PEG chain length, the hydrophilicity and CMC values both rise.

Table 2.5 CMC and micelle volume of hexadecyl PEG

n	CMC × 10^4 (mol·kg^{-1})	Micelle Volume × 10^{-1}	Aggregation Number
6	1.7	12.3	2430
7	1.7	3.27	590
9	2.1	1.4	220
12	2.3	1.17	150
21	3.9	0.82	150

2.3.3.3 Synergy of hydrophilic groups and hydrophobic groups

During micellization, due to the simultaneous existence of driving force and counterforce in the system, the delicate balance between the two also has a great influence on the formation of micelles. Table 2.6 lists the corresponding micellar aggregation number of various surfactants in water.

The following Fig. 2.1 shows the general principal and the process of micelle assembly from the block copolymers. The detail mechanism and structural tailoring of such micelle. Here, for creating a well-defined core–shell structural micelles, the block

Table 2.6 Aggregation number of different surfactants in water

Surfactant	Temperature/°C	Aggregation Number
$C_8H_{17}SO_4Na$	Room temperature	20
$C_{10}H_{21}SO_4Na$	Room temperature	50
$C_{12}H_{25}SO_4Na$	23	71
$C_{12}H_{25}N(CH_3)_3Br$	–	50
$C_{14}H_{29}N(CH_3)_3Br$	–	73
$C_{12}H_{25}O(C_2H_4O)_8H$	25	123
$C_{12}H_{25}O(C_2H_4O)_{12}H$	25	81
$C_{12}H_{25}O(C_2H_4O)_{18}H$	25	51
$C_9H_{19}(C_6H_4)O(C_2H_4O)_{10}H$	25	276
$C_9H_{19}(C_6H_4)O(C_2H_4O)_{15}H$	25	80
$C_9H_{19}(C_6H_4)O(C_2H_4O)_{20}H$	25	62

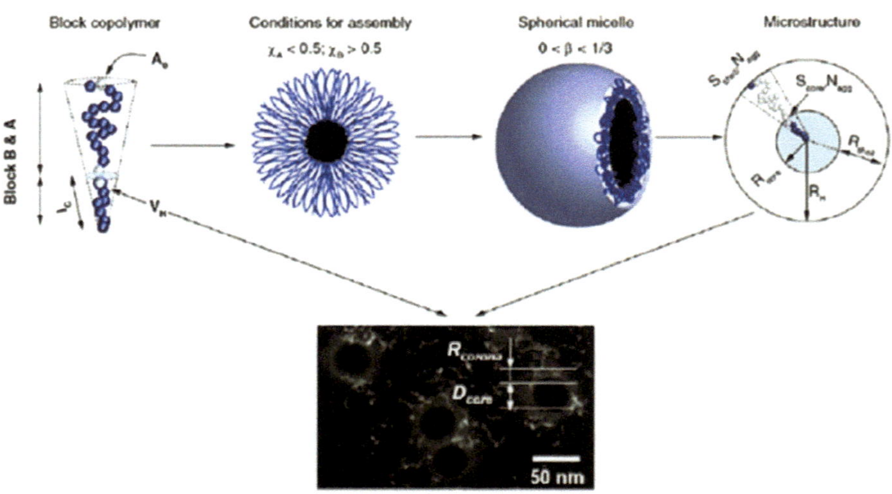

Figure 2.1 The spherical micelle and its microstructure assembly from block copolymer. The formula related to the table is shown in 1.10, 1.15.

length of amphiphilic block copolymer should be tuned in the polymerization for getting the Flory parameter $\chi_{hydrophobic} > 0.5$ while $\chi_{hydrophilic} < 0.5$ by the function 2.18. This is the thermodynamic condition for the self-assembly of amphiphilic block copolymers in solutions. Then, the composition of the copolymer should be further regulated for tuning the packing parameter (β) in the range of 0~1/3 according to function 2.20, which results in a spherical morphology.

2.3.3.4 Effects of counterions

The counterion of ionic surfactants can also affect the micellization process. For example, the micelle volume of cationic surfactants changes with the order of the counterion $I^- > Br^- > Cl^-$, while the micelle size corresponding to anionic surfactants varies with the order of counterion $Cs^+ > K^+ > Na^+$. In general, the weaker hydration ability of the counterion, the larger the micelles formed by the surfactant. This is due to the fact that ions with weak hydration are more easily adsorbed on the surface of the micelles, thereby reducing the electrostatic repulsion between polar groups. The CMC of surfactants under organic counterion (e.g., maleic acid) conditions decreases significantly and the micelle volume increases significantly compared to that of inorganic counterion.

2.3.4 Influence of External Conditions on CMC

2.3.4.1 Temperature

The solubility of ionic surfactant in water is limited, it rises up slowly as the temperature increases, resulting in the slight increase of CMC. In the meantime, the molecular thermal movement is aggravated, which is not conducive to the formation of micelles, leading to higher CMC as well. However, for non-ionic surfactants, the CMC decreases with increasing temperature.

Krafft Point. For ionic surfactant, the solubility in water varies with temperature. As the temperature rises, the solubility increases sharply at a certain point, and the temperature corresponding to the turning point K is called the Krafft point, and the corresponding solubility is called the ionic surfactant CMC [11].

The higher the Krafft point, the smaller the CMC. Krafft point is also the lower limit of the surfactant's application temperature, or only when the temperature is higher than Krafft point can the surfactant play a better role. For example, the Krafft point of sodium dodecyl sulfate is 8 °C, and the Krafft point of sodium dodecyl sulfonate is 70 °C. At room temperature, the former has a better solubilization effect, while the Krafft point of the latter is too high for satisfactory results.

Cloud Point. The solubility of non-ionic surfactants in aqueous solution decreases with increasing temperature, and the solution becomes cloudy when the temperature reaches a specific value, which is called the cloud point [12]. Cloud point is the eigenvalue of non-ionic surfactant, generally ranging from 70 to 100 °C. For example, for Tween surfactants, the cloud point of Tween 20 is 90 °C, 76 °C for Tween 60, and 93 °C for Tween 80. The reason for the occurrence of cloud point is that as the temperature rises, the hydrogen bond between polyethylene glycol and water is destroyed and the solubility decreases. The hydrogen bond can be formed again after cooling, and the solution becomes clear. When the chain length of polyethylene glycol is the same, the longer the carbon chain length, the lower the cloud point of the surfactant. When the carbon chain length is the same, the cloud point increases with the growth of the polyethylene glycol segment.

2.3.4.2 Electrolytes

The addition of a strong electrolyte to ionic surfactant solution can reduce CMC because of the electrostatic interaction between the electrolyte ions and the oppositely charged surfactant ions, which reduces the electrostatic repulsion between the polar groups and reduces the electrical work consumed during micellization. Table 2.7 lists the effect of sodium chloride on the properties of the dodecyl trimethyl ammonium bromide (DTAB) micelle, a cationic surfactant. The properties of non-ionic surfactant micelles are relatively less affected by the electrolyte.

Table 2.7 Effect of the electrolyte on the properties of DTAB micelle

NaCl concentration/mol · L^{-1}	CMC/mol · L^{-1}	Aggregation Number
0.000	0.0146	61
0.100	0.00428	74
0.502	0.00171	90

2.3.4.3 Organic substances

A small amount of organic additives can significantly affect the CMC value of surfactants. Organic compounds with strong polarity including alcohols and amides, show influence on the CMC value particularly at lower concentrations; For other organic substances, larger concentrations are required to reduce the CMC value. They influenced the CMC value by changing the water structure, dielectric constant or dissolution parameters, and improving the interaction of water with surfactants or micelles, such as urea, methylamine, guanidine salts, short-chain alcohols, ethylene glycol, fructose, and xylitol, etc.; Urea, methylamine and guanidine salts can destroy the structure of water, increase hydration of the hydrophilic moiety, prevent gelation, and improve the CMC value of the surfactant aqueous solution, especially for PEG-based non-ionic surfactants. They can increase the CMC value at high concentrations by reducing the cohesive energy density and increasing the concentration of monomeric surfactants. For solutions containing ionic surfactants, the presence of these compounds can reduce the dielectric constant, leading to stronger mutual repulsive force of the ionic end of the micelle.

2.4 Structure and Stability of Polymeric Micelles

PMs are generally prepared from amphiphilic block copolymers or graft copolymers, which consist of a specific sequence of hydrophilic and hydrophobic chain segments, where the hydrophobic moiety of the copolymer joins to form a semisolid core and the hydrophilic part forms a crown with hydrogen bonds between the surrounding water molecules [13]. When the concentration is lower than CMC, the copolymer exists as a single molecule in the whole solution. When the concentration goes above CMC, the polymer obtains micelles by self-assembly. Most polymers have lower CMC (as low as 10^{-6}–10^{-7} M) compared to small molecule surfactants, and are able to form systems with high stabilities [14]. Besides, micelles of some copolymers also have excellent kinetic stability.

Various theories have been developed to predict the structural parameters of micelle (CMC, aggregation number m, nuclear radius R_c, shell thickness L, hydrodynamic radius R_h) based on copolymer characteristics, including functions of molecular weight and chemical composition [15]. In all these theories, the total Gibbs free energy $G^{(m)}$ of micelles can be divided into the sum of several contributions by using different models and mathematical methods, including the Gibbs free energy $G^{(shell)}$ of the core, Gibbs free energy $G^{(shell)}$ of the shell and Gibbs free energy $G^{(interface)}$ of the core/shell interface:

$$G(\text{micelle}) = G(\text{core}) + G(\text{shell}) + G(\text{interface}) \tag{2.9}$$

2.4.1 Thermodynamical Stability

According to the early results of Price and Quintana, micelle formation from block copolymers in organic media is enthalpy-driven [16]. Gibbs free energy ΔG^0:

$$\Delta G^0 = \Delta H^0 = T\Delta S^0 \tag{2.10}$$

The ΔS^0 caused by the loss of the combined entropy is negative, which is not conducive to micellization, because the copolymer chain in micelles has small swelling in the non-associative state. Moreover, the block connection is located at the core/shell interface of micelle, which also leads to a decrease in the number of possible conformations. The negative value of ΔH^0 is due to the exothermic energy exchange caused by the interaction between polymer and solvent replacing the polymer/solvent interaction.

Thus, the formation of micelle nuclei contributes the most to the exothermic process.

This is completely inconsistent with the reported micellization, low molecular weight surfactant and hydrophilic-hydrophobic segmented copolymer in aqueous solutions. A typical example is PEO-*b*-PPO-*b*-PEO block copolymer, whose micellization is an entropy-driven process due to the hydrophobic interaction near the polymer chain and changes in water structure [17, 18].

For block polymer to associate in selective organic solvents, the standard enthalpy and the Gibbs free energy can be approximated at the first level by the following equation:

$$\Delta G^0 = RT\ln(\text{CMC}) \tag{2.11}$$

$$\Delta H^0 = R\left[\frac{\text{d}\ln(\text{CMC})}{\text{d}\left(\frac{1}{T}\right)}\right] \tag{2.12}$$

CMC is the critical micelle concentration expressed in mole fraction, ΔH^0 has nothing to do with temperature, and the integral is:

$$\ln\text{CMC} = \frac{\Delta H^0}{RT} + \text{constant} \tag{2.13}$$

2.4.2 Structural Stability

The scaling theory is developed to describe the relationship between the structural characteristics and properties of a certain segmented copolymer (mainly two segmented copolymer AB) on the basis of a simple model, such as the nuclear radius R_C, corona thickness L and aggregation number Z of a segment in selective solvent. Pierre-Gilles de Gennes first proposed scaling theory in 1978 [18]. There are two most-studied scenarios for AB diblock copolymers with B being the insoluble block; it forms either crew-cut micelles with $R_c \gg L$ or the so-called hairy micelles. For the crew-cut micelles, assuming uniformly stretched nuclear segments and short shell segments, the radius R_C of the core is shown in Equation 2.14:

$$R_C \sim a\gamma^{1/3} N_B^{2/5} \tag{2.14}$$

where N_B is the number of monomer units of the insoluble segment, and γ is the interfacial tension between two chain segments, a is the length of the fragment. The scale P of the aggregation number is shown in Equation 2.15:

$$P \sim N_B\gamma \tag{2.15}$$

In 1982, Daoud and Cotton derived the star copolymer model for describing star-shaped micelles [19]. This model predicts the total radius R of the micelle and the number and aggregation of soluble block monomer units as shown in Equation 2.16:

$$R \sim N_A^{3/5} f^{1/5} \tag{2.16}$$

with f being the number of arms. For block copolymer micelle, the number of arms corresponds to the aggregation number P, it follows that $L \sim P^{1/5} N_A^{3/5}$ with $P \sim N_B^{4/5}$.

The scaling concept describing polymer concentration distributions and free energies is mainly applied to micellular solution containing long polymer chains. In fact, the above scale model cannot include finite chain effects and polymer/solvent interactions. Furthermore, numerical values for micellar properties cannot be obtained directly because scaling laws can only predict trends, such as how a given micellar parameter scales with a given copolymer parameter. Therefore, the scaling model must be complemented by more detailed mean-field calculations and molecular simulations, which are briefly outlined below.

2.4.3 Micelle Structural Tailoring

In 1982, Noolandi and Hong proposed a self-consistent mean-field theory for spherical micellar solutions based on a mixture model of segmented copolymers and homopolymers [20]. It is the theory with the least assumptions at the mean-field level and the most systematic and accurate theory, a very effective theoretical tool for the study of copolymers and their co-mingled systems. They derived the micellar characteristics by minimizing the Gibbs energy of an isolated micelle using numerical values of the Flory–Huggins interaction parameters χ, molecular weight and composition of the copolymer. The theoretical values were in fair agreement with those obtained experimentally. Leibler et al. then improved the theory by minimizing the total Gibbs energy not only for one micelle but for the whole micellar system [21].

The Flory–Huggins solution theory derived the relations between the mixing entropy, the heat of mixing, and the free energy of mixing for polymer solutions from the lattice-like model of liquids using statistical thermodynamics in the form of Equation 2.17:

$$\Delta F_m = \Delta U_m - T\Delta S_m = RT\,[n_1 \ln \Phi_1 + n_2 \ln \Phi_2 + n_1 \Phi_2 \chi_{12}] \quad (2.17)$$

where n_1 and Φ_1 refer to the mole number and volume fraction of the solvent, n_2 and Φ_2, respectively, refer to the mole number and volume fraction of the polymer, χ is called Flory–Huggins interaction parameter, which reflects the variation of the interaction energy between the polymer and the solvent. The value of the polymer solvent action parameter χ can be used as a semi-quantitative criterion for the merits of the solvent. If χ is greater than 0.5, the polymer generally cannot be dissolved; If χ is lower than 0.5, the polymer can be dissolved, and the smaller it is, the better the solvency ability of the solvent is.

The self-assembly of micelles is driven by the difference in solubility of their hydrophilic and hydrophobic moieties in aqueous solution (the determining factor is the solubility parameter of the hydrophilic and hydrophobic moieties in aqueous solution) [22]. The solubility of amphiphilic block copolymers can be described by the Flory–Huggins parameter as shown in Equation 2.18:

$$\chi_{p\text{-}s} = (\delta_p - \delta_s)^2 V_s / KT + 0.34 \quad (2.18)$$

where δ_p and δ_s refer to the solubility parameters of the polymer and the solvent, V_s is the molar volume of the solvent, and 0.34 is the contribution value of entropy. The solubility parameter can be calculated from the Hildebrand–Scatchard equation, ΔE_{vap} is the evaporation energy of the solvent, and V is the molar volume of the solvent used:

$$\delta = \sqrt{\Delta E_{\text{vap}} / V} \tag{2.19}$$

In a polymer solution, the formation of nanoparticles of a polymer containing two different solubilities (such as block polymer) requires the driving force of an external solvent, and two thermodynamic conditions need to be met simultaneously to complete the self-assembly [23]. Firstly, the $\chi_{\text{p-s}}$ of the high hydrophilic chain segment of the solubility should be below 0.5. Secondly, $\chi_{\text{p-s}}$ of the low hydrophobic chain segment of solubility should be above 0.5. The morphology of the aggregates can be realized by adjusting the chemical composition of polymer. Packing parameter β of the system in solution is used to describe this kinetic parameter, and can be obtained by calculating Equation 2.20:

$$\beta = V_{\text{H}}/L_{\text{C}}A_0 \tag{2.20}$$

where V_{H} refers to the volume occupied by the hydrophobic chain, L_{C} refers to the length of the aggregation region of the hydrophobic chain segments, and A_0 refers to the surface area occupied by the hydrophilic chains. Generally speaking, spherical nanomicelles are formed when $0 < \beta < 1/3$, rod-shaped nanomicelles are formed when $1/3 < \beta < 1/2$; vesicles are formed when $1/2 < \beta < 1$; and planar bilayers are formed when $\beta > 1$ (Fig. 2.2). In order to find the specific conditions for the formation of spherical micelles, the range of the parameter β needs to be adjusted to 0–1/3. In other words, the specific self-assembled structures can be regulated by modulating the ratio between hydrophilic and hydrophobic chains according to the packing parameter β defined in Equation 2.20 [24]. For example, He Zhao et al. synthesized amphiphilic segmented copolymers phenyl-PID_{118}-*b*-PLA_{71}, which can self-assemble into spherical structure of nanomicelles in water because its β value is lower than 0.1 [25, 26]. The hydrophilic temperature-sensitive PID shell can stabilize micelles and enhance cell endocytosis, the hydrophobic PLA core can improve the drug loading of hydrophobic drugs.

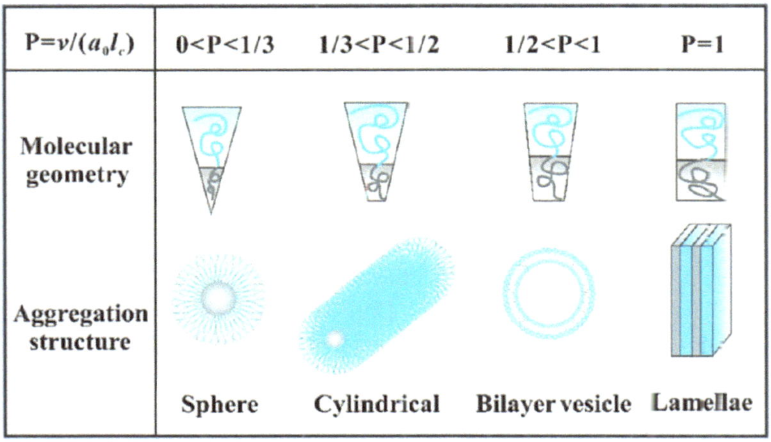

Figure 2.2 Nanostructures formed by amphiphilic block copolymers with different packing parameters. The formula related to the table is shown in 2.18, 2.20.

2.5 NPP of Polymeric Micelles in Drug Delivery

PMs are widely used as nanocarriers for drug delivery and tumor therapy due to their easy preparation and unique physiochemical properties, excellent drug delivery and release, biocompatibility, and tumor targeting [27]. In the early 1980s, Ringsdorf and his colleagues first reported the use of PMs for cancer treatment [28]. After that, PMs gradually became a common carrier for the delivery of anticancer drugs and other bioactive molecules used in cancer diagnosis and treatment.

A large number of studies have shown that the physiochemical properties of nanocarriers vary with their forms, and determine their functions and applications. For example, the *in vivo* stability of micelles is related to the carrier surface charge density (ρ_{charge}), surface chain density ($\rho_{\text{surface chain}}$) defined by us in Equation 2.21, CMC, etc. Drug-loading efficiency can be controlled by particle size (D), core volume (V_{core}) or number of surface chains (N_{chain}), while the drug release behavior is related to parameters such as D, N_{chain}, and ρ_{chain}. D and ρ_{charge} can also affect the endocytosis effect of drug-loaded micelles. Therefore, the controlled release of drugs *in vivo* can be achieved by precise regulation on the physiochemical properties of micelles according to the needs of environmental conditions *in vivo* and related thermodynamic and kinetic theories.

2.5.1 Physiochemical Properties of Micelles

2.5.1.1 Particle size and distribution

The micelle size is mainly characterized by two parameters, D and particle size distribution [29, 30]. The D of micelle can be changed by adjusting the length of hydrophobic chain ($N_{\text{hydrophobic}}$) and hydrophilic chain ($N_{\text{hydrophilic}}$) with the correlation being defined as $D \propto N_{\text{hydrophobic}}{}^{0.16}\, N_{\text{hydrophilic}}{}^{0.6}$ in Equation 2.21. The particle size can be measured by dynamic/static light scattering instruments which usually falls into the range of 10~100 nm. The microstructure of micelles, such as the core and shell dimensions can be further obtained through calculations and theoretical simulations. We have synthesized amphiphilic block copolymers with various chain lengths of poly(*n*-butyl methacrylate) blocks (PBMA) and poly(*N*-acryloylmorpholine) blocks (PAM) to prepare micelles [23]. The relationships between the $N_{\text{BMA}}/N_{\text{AM}}$ ratio and core diameter (R_{core}) as well as corona thickness (T_{corona}) were established, which can be estimated from M_{w} and N_{agg} in the following equations:

$$D \propto N_{\text{hydrophobic}}{}^{0.16}\, N_{\text{hydrophilic}}{}^{0.6} \tag{2.21}$$

$$R_{\text{core}} = \sqrt[3]{\frac{3N_{\text{agg}} N_{\text{BMA}} M_{\text{W(BMA)}}}{4\pi\rho N_{\text{A}}}} \tag{2.22}$$

$$T_{\text{corona}} = (\text{D}_{\text{h}} - 2R_{\text{core}})/2 \tag{2.23}$$

$$\rho_{\text{surface chain}} = S_{\text{surface}}/N_{\text{chain}} \tag{2.24}$$

where N_{agg}, $M_{\text{W(BMA)}}$, ρ, and N_{A} are the polymer aggregation number of block copolymers for each micelle, the molecular weight of BMA, the density of PBMA at the bulk state (approximately 1 mg/mL), and the Avogadro constant, respectively, and micellar diameters

$(D_h) \propto N_{BMA}^{0.16} N_{AM}^{0.6}$. T_{corona} was found to increase from 4 to 9 nm with increasing N_{BMA}/N_{AM} values in the range from 0.104 to 0.775. Both reduced D_h and increased R_{core} as increases in N_{BMA}/N_{AM} values consequently led to the decreases in T_{corona} values from 18 nm to 4 nm. Figure 2.3 shows the fine microstructure tailoring of core, corona, surface properties of micelles.

Nottelet et al. synthesized a series of polyethylene glycol-*g*-poly(ε-caprolactone) (PEG-*g*-PCL) graft copolymers by click chemistry, and prepared micellar solutions with sizes ranging from 30 nm to 80 nm [31]. The size of the micelles decreases with longer hydrophobic PCL chain segments, which is mainly due to decreased hydration from the less hydrophilic outer shell. Moreover, the reduced proportion of PEG trapped in the hydrophobic core leads to less swollen core, hence the smaller size. The size distributions of the micelles are all between 0.21 and 0.40, which are higher than those prepared with conventional linear copolymers. This is possibly due to the higher energy required for the side-arm PEG to phase separation.

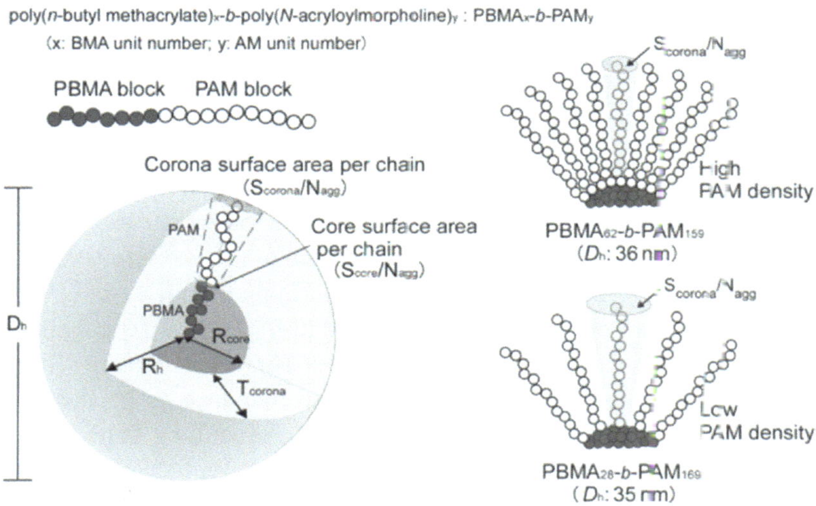

Figure 2.3 The definition and microstructural parameter of micelles by fine tailoring. Reproduced with permission from [23]. The key parameters in the figure are originally defined by us and illustrated in Equations 2.21, 2.22, 2.23, and 2.24.

The micelle size is also related to polymer concentration and segmented copolymer structure. For ABA type three segmented copolymer (A is hydrophobic chain, and B is hydrophilic chain), with the increase of polymer concentration, it is easy to interact with other micelle cores to form larger micelles due to the presence of some hydrophobic chains in the hydrophilic shell, while the micelle size of AB or BAB type three segmented copolymer formed at high concentration is small.

2.5.1.2 Critical micelle concentration

The CMC of segmented copolymers is an important parameter for the thermodynamic stability of micelles, which is closely related to their stability *in vivo* and is generally related to the hydrophobic chain length (N_A) and hydrophilic chain length (N_B). This can be calculated by Equation 2.25:

$$\ln \text{CMC} = \alpha N_\text{A} + \beta N_\text{B} \tag{2.25}$$

where α and β are constants related to external conditions. Pimpha et al. designed and synthesized PLA-*b*-PEG-*b*-PLA with different chain segment ratios [32]. When LA/EG is between 2.43 and 3.60, a micelle with a hydrophobic core of PLA and a hydrophilic shell of PEG could be formed at polymer mass concentrations between 0.5% and 1.0%. When LA/EG decreases to 1.26, only at higher polymer concentrations (>20 wt.%) can aggregates appear, and no micelle structure is found by characterization. In general, CMC gradually decreases with the increase of LA/EG ratio (i.e., the increase of hydrophobicity of segmented copolymers).

A similar trend is found for the PEG-*g*-PCL graft copolymer, with a gradual increase in CMC as the EG/CL ratio increases [31]. In addition, the grafting rate also has an influence on it, and the lower the grafting rate, the lower the corresponding CMC at the same EG/CL ratio. Compared with linear segmented copolymers with similar EG/CL ratios, graft copolymers have a lower CMC, which is mainly due to their high dispersion degree and restricted chain motion caused by side chain steric hindrance.

2.5.1.3 Zeta potential

When charged particles dissociate in solution, they attract oppositely charged ions to the surface, where those close to the surface of the particle are firmly adsorbed, while those further away are loosely combined, forming a so-called diffusion layer. There is a conceptual boundary, that is, when the particle moves in the liquid, the ions inside the boundary move with the particle, while those outside the boundary stay in place. This is called a slipping plane, and the corresponding potential is called zeta potential (ζ). This parameter characterizes the charge carried by the particle surface and is generally measured by electrophoresis. In this way, the electrophoretic mobility (U) of the particle in the applied electric field liquid is first tested, and the zeta potential can be calculated according to Equation 2.25 by combining the known parameters solution viscosity η and dielectric constant ε.

$$U = \epsilon \zeta / \eta \tag{2.26}$$

Vlahos et al. studied the structure, the surface potential and other properties of charged amphiphilic block copolymers (A_5B_{30}, $A_{10}B_{30}$ or $A_{20}B_{30}$) using molecular dynamics simulations with the Langevin algorithm [33]. It was found that when the charged chain segments were short and partially charged, the micelle weight was at least bimodal in distribution, indicating the presence of both small and large particle size micelles within the system. The zeta potential was not a monotone function relative to the length of the charged segment (N_A), and had a more complicated relationship with the charged ratio α. For micelles formed by A_5B_{30} and $A_{10}B_{30}$, the zeta potential decreased as the ratio α decreased, whereas for micelles formed by $A_{20}B_{30}$, the potential when α was 0.4 was higher than that when α was 0.2 or 1. It was found that the zeta potential increased with N_A when the ratio of the confined counterion was lower than 1.11, and an opposite trend was found when the ratio was higher than 1.11.

2.5.2 Stability of Micelles

The stability of a micellar delivery system before it reaches the targeting area is a critical parameter. How to maintain stable can be challenging especially when facing the highly complex *in vivo* environment, including high degree dilution after intravenous administration, high viscosity and high flow rate. We have defined surface areas of respective core-forming and corona-forming chains (S_{core}/N_{agg} and S_{corona}/N_{agg}) as the fundamental parameters (Fig. 2.4) [23],

$$\frac{S_{core}}{N_{agg}} = 4\pi R_{core}^2 \tag{2.27}$$

$$\frac{S_{corona}}{N_{agg}} = 4\pi R_n^2 \tag{2.28}$$

with $R_h = D_h/2$. Decrease in S_{corona}/N_{agg} indicated a strong interaction between neighboring chains inside corona suggesting the loss of mobility. We have also firstly introduced chain unit density inside corona ρ and established their relationship with stability using micelles formed by PBMA-*b*-PMA block copolymer. The result shows that PMA chains inside corona lost their mobilities with increasing ρ_{AM} values leading to increased steric repulsions, which prevented the interaction with proteins. contributing to enhanced stability. The *in vivo* experiments also confirmed that micelles with high S_{corona}/N_{agg} showed much higher tumor accumulation than those with low S_{corona}/N_{agg}, due to the strong entropic repulsion against biomacromolecules in the circulation [34].

The reversible dynamic balance between the free single chains and micelles can also affect the stability of micelles. The free single chains tended to dissociate and insert into the micelles when the concentration of micelles was low, while micelles were easy to fuse and re-split when the concentration of micelles was high [35]. It has been reported that the dissociation rate of micellar cores formed by chain segments with high T_g is much lower than that of micellar cores formed by chain segments with low T_g, which is due to the restricted movement of the high T_g segment in the glassy state [36]. Therefore, micelles with T_g higher than the normal body temperature tend to have limited disintegration during circulation [15].

Interaction of micelles with proteins or cells *in vivo* is another factor contributing to their instability *in vivo*, so avoiding serum protein adsorption of micelles and reducing clearance by the immune system is expected to enhance the stability of micelles *in vivo*. Hydrophilic polymers with N and O atoms in the polymer chain, such as PEG, polyacrylamide, polyvinyl alcohol, poly(diethylenedioxazoline) and poly(vinylpyrrolidone), are able to reduce adsorption by forming hydrogen bonds with water molecules on the surface. The hydration layer helps to reduce the free energy interface and increase the repulsion against biomacromolecules, thus improving the stability of micelles. For instance, a long PEG chain in the shell could significantly prolong the circulation time and prevent protein adsorption with high efficacy [37]. However, the high surface density of PEG will also lead to a strong repulsive interaction of micelles and cells, hampering its endocytosis process.

2.5.3 Drug-Loading Profile of Micelles

Drug-loading capacity is an important index to evaluate the application of micelles *in vitro* and *in vivo*. At present, a variety of methods to encapsulate drugs have been developed, such as dialysis, W/O emulsion, solvent evaporation, co-solvent evaporation and freeze-drying method. In general, the drug-loading capacity and drug-loading efficiency are determined by the interaction between the polymer and the drug, and the driving force of hydrophobic drug binding to the core can also be described by the corrected Flory–Huggins equation. At the same time, other aspects affecting drug-loading effects such as hydrophobicity, electrostatic interaction, dipole-dipole interaction, hydrogen bonding and drug size shall be considered. These effects can be described by the linear solvent-free energy relationship in Equation 2.27:

$$\lg \mathrm{SP} = c + rR_2 + s\pi_2 + a\Sigma\alpha_2 + b\Sigma\beta_2 + vV_x \tag{2.29}$$

where SP is the partition coefficient of the drug in the heterogeneous phase, R_2 is the molar refractive index of the solvent (from London's dispersion force), π_2 is the ratio of depolarization/polarization of the drug, $\Sigma\alpha_2$ is the hydrogen-bond acidity of the drug, $\Sigma\beta_2$ is the hydrogen bond binding alkalinity of the drug, V_x is the McGowan's characteristic volume from the molecular structure, and *c*, *r*, *s*, *a*, *B*, and *v* are regression coefficients.

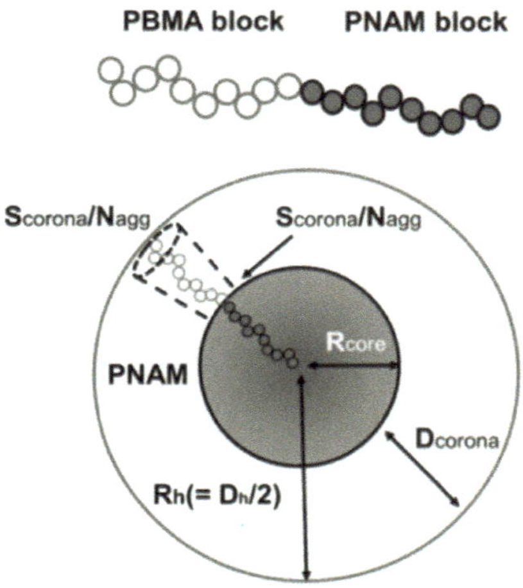

Figure 2.4 Schematic illustration of micellar structure and the parameters related to the microstructure as described in Equations 2.24, 2.27, and 2.28.

Based on the theory of linear solvent-free energy relationship equation, to the core of a micelle, the drug needs to transfer its free energy in water into the polymeric micelle. The efficiency of drug loading is greatly affected by the micelle composition and hydrophobic/hydrophilic ratio, especially the hydrophobicity of the core, the more hydrophobic, the higher drug-loading capacity [23, 38]. We also defined the parameter ζ as [39]:

$$\zeta = \frac{\rho_{\text{shell}}}{V_{\text{core}}} = \frac{3S_{\text{shell}}}{4N_{\text{agg}}\pi R^3} \tag{2.30}$$

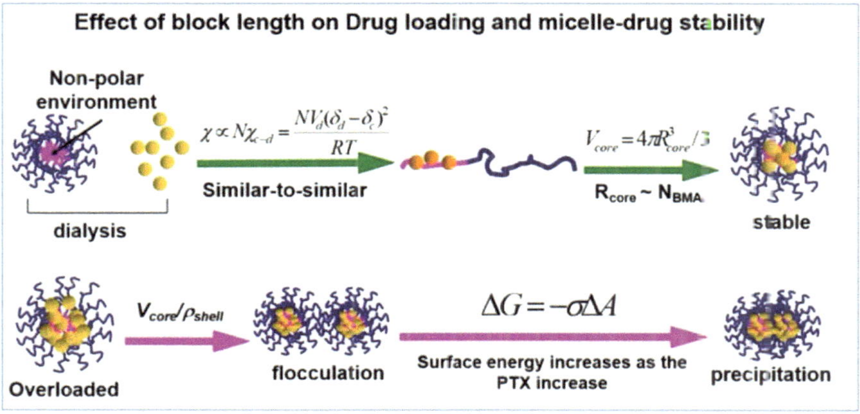

Figure 2.5 The scheme illustrated the mechanism of drug loading and the stability of micelle-drug delivery system in solutions (detailed data not shown here). The equations are defined from Equations 2.27 and 2.30.

High ζ indicates a micellar structure with a relatively small core and thick corona, which usually shows low drug-loading capacity. The detailed process of drug loading and stability of the micelle system is shown in Fig. 2.5. The inserted equations of χ and ΔG are illustrated in Chapter 1. The other parameters and equations are defined and summarized in our studies as shown in Fig 1.16 in Chapter 1 and 2.28. The scheme gives us the basic principles for making a stable and uniform delivery system the experiments.

2.5.4 Endocytosis of Micelles

The endocytosis efficiency of drug-loaded micelles is crucial to the final therapeutic outcome. Taking polybutadiene-*b*-polyethylene glycol as a model block copolymer, Doris et al. investigated the effect of micelle surface charge on cellular endocytosis kinetics by modifying hydroxyl, carboxyl and tertiary amine groups at the terminal, respectively. It was found that micelles with cationic surface charge had the fastest endocytosis, while those with anionic surface charge showed the slowest endocytosis [40]. However, all three micelles entered the cell by cytosolic cell-mediated cytokinesis and had a similar behavior. Liu et al. designed and synthesized a series of poly (2-ethyl-2-oxazoline)-*b*-polylactide segmented copolymer in different proportions, and found that the ratio of two segments can also affect the efficiency of endocytosis. When the ratio of hydrophilic/hydrophobic is between 1.7 and 2.0, the efficiency of endocytosis is the highest [41].

2.5.5 Drug Release Behavior of Micelles

We have found that the drug release behavior is dominated by the thermodynamic detaching parameter (χ) cross-link density (D), and diffusion distance (R) [39], which can be categorized into three steps. In the first step, the drug is detached from the core. A larger R_{core} leads to a higher χ value, indicating a low drug detachment. In the second step, the drug then passes through the cross-linked area at the interface between the core and the corona, which can be explained by a lattice model with the pore size being r. This step is dependent on the ρ_{shell}, and larger ρ_{shell} corresponds to smaller r and smaller D. The third step involves drug diffusion through the corona. The larger R_{shell} means a longer diffusing distance.

Jeong et al. used micelles formed by PEG-*b*-PBLG to encapsulate clonazepam, and the increase in the mass fraction of the drug led to a decrease in the release rate [42]. The differential scanning calorimetry result suggests that the drug tends to crystallize at higher concentrations and the release will thus be affected. However, the drug release curve measured *in vitro* does not take the influence of the *in vivo* environment into account, so it is only partly correct. Generally speaking, the drug release behavior is accelerated *in vivo*. For example, PEG-*b*-PCL micelles encapsulating HCPT can be released *in vitro* within a few days, but it is removed from plasma within a few hours after intravenous injection [43]. Similar results were found in earlier studies, where polymeric micelles showed loss of integrity within 1 hour after intramuscular or subcutaneous injection [44, 45]. This is mainly due to the sudden dilution of the drug-loaded micelles and protein absorption, which promotes rapid drug release. Figure 2.6 shows the detailed factors that affect drug release profile of the micelle system. For hydrophobic drugs like PTX or DOX released from the micelle system, the drug is processed in three steps as shown in Fig. 2.6. The release process is illustrated from the top right panel to the bottom panel in Fig. 2.6 (right): The first step is the thermodynamic disassociation of drug from the hydropobic core, that is, the drug detachment. The second step is the kinetic penetration of drug through the cross section between the core and the shell, where the highly entangled polymer chains result in many small pores with mesh size. The mesh size strongly affects the drug diffusion efficiency (D). The third step is the diffusion distance, which is actually the shell thickness of the micelle. Generally, for a Brown diffusion of drug in solution or in the micelle shell, the time for the drug passing through thickness is too low to be considered. Thus, for the drug release of the micelle system, the accumulated release time and concentration is dominated by the thermodynamic detaching and the mesh size in the core–shell cross section. The function that describes these factors are shown in Fig. 2.6.

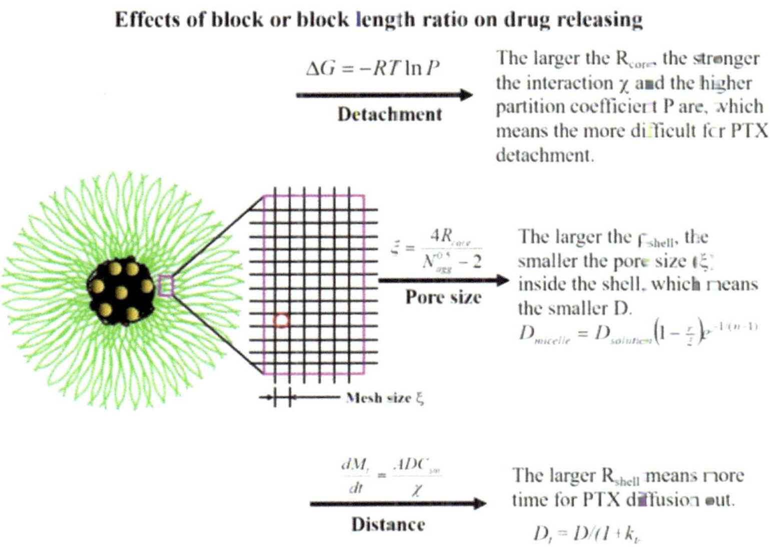

Figure 2.6 The mechanism of drug release in the micelle system. From top right panel to bottom panel: (1) the thermodynamic disassociation of drug (small yellow ball) from the hydropobic core; (2) the kinetically penetration of drug through the cross section between the core and the shell; and (3) the diffusion distance, namely, the thickness of the shell.

For charged drug-loaded micelles, the drug release is not based on diffusion, but through the interaction of charged ions to induce the disintegration of micelles. For example, micelles can be destabilized through ion exchange with compounds such as salts and heparin. Similarly, polymer-metal complex micelles can dissociate by substitution of ions in the medium for metals in the ligands [15].

Based on the series study of different drugs (chemotherapy drug, nucleic acids and proteins) delivered by carriers like micelles, liposomes, nanogels, PLGA microspheres and organic–inorganic complex particles in solutions, we systemically summarize and analyze series of the key physical-chemistry properties, propose and define some of key parameters of microstructure as shown in Fig. 1.16 in Chapter 1. Then the study of NPP is first proposed for covering the *in vitro/vivo* gap of nanoformulations.

2.5.6 Drug Release Behavior of Stimuli-Responsive Micelles

Driven by the aim to develop a smart drug delivery system, various stimuli-responsive polymeric micelles were prepared and investigated. Among them, different chemical and physical stimuli such as temperature, pH and light have been applied to induce the disassembly or morphology changes of micellar aggregates, and hence the on-demand drug release [46, 47].

2.5.6.1 pH-responsive micelles

The pH in normal tissues and blood is maintained between 7.2–7.4, while the mesenchyme cellular environment of tumor tissues is weakly acidic ($pH < 7.0$) due to hypoxia, cell death, improper debris removal and rapid glycolysis leading to lactic acid accumulation, which provides an important theoretical basis for the development and application of pH-responsive nanocarriers [48]. Such drug-loaded micelles are stable under normal physiological environment, and the hydrophobic drugs encapsulated in the core hardly leak after long-term systemic circulation, but they are less stable under low pH environment, in which the structure will be disrupted and the drug will be released specifically to the tumor site.

Currently, pH-responsive micelles are mainly designed through two approaches. The first approach uses polymers containing chain segments with ionizable chemical groups to assemble into micelles that remain deprotonated/deionized at physiological pH, and are protonated and disintegrate at acidic pH with the change in hydrophilic/hydrophobic ratio of the micelle and the release of the drug [49, 50]. Polymers with side chains containing carboxylic acid groups are the most common class of pH-responsive polymers, including polyacrylic acid (PAA) and polymethacrylic acid (PMA) [14]. The second approach uses acid-sensitive chemical bonds to achieve pH response, that is, the acid-sensitive chemical bonds are introduced between the drug and the polymer or inside the polymer, and the drug is released by hydrolysis under acidic conditions. Common acid-sensitive chemical bonds include hydrazone, imine, oxime, acetal, vinyl ether and catechol ester bond [51, 52]. They can be classified into three types based on the location and drug release principles of chemical bonds: (1) chemical bonds located at the connection of two blocks, and the separation of the two after hydrolysis leads to micelle disintegration and drug release; (2) chemical bonds connect hydrophobic chain segments of polymers to hydrophobic groups, and their hydrolysis strips off the hydrophobic groups, leading to a decrease in the hydrophobicity of the inner core and drug release through solubilization; (3) chemical bonds connect drugs to polymers, and achieve drug release after hydrolysis [37].

2.5.6.2 Redox-responsive micelles

Redox reaction is one of the most effective local stimuli. The environment inside the cell is highly reducing, with the concentration of the major reducing agent glutathione (γ-Glamyl-cysteinyl-Glyine, GSH) reaching about 100 times that in the extracellular environment. However, tumor cells have higher GSH concentrations (up to four times the normal cytoplasm) due to their high activity and low metabolism, as well as the structural hypoxia of solid tumors, which provides ideas for the design of redox-responsive systems for tumor therapy [53]. The disulfide bond is a common redox-sensitive chemical bond, which is often used to connect the hydrophobic and hydrophilic segments of amphiphilic block copolymers to synthesize redox-responsive copolymers. Cui et al. have synthesized a high-molecular polymer mPEG-S-S-C16 with redox properties and used it to load the chemotherapy drug DOX. *In vitro* experiments show that under the condition of dithiothreitol reduction, the disulfide bonds in the carrier are reduced to sulfhydryl, the micelle structure is disintegrated, and the wrapped DOX can be quickly released, and effectively kill tumor cells.

2.5.6.3 Temperature-responsive micelles

Temperature is one of the most common sources of stimulation. Temperature-responsive polymeric micelles are assembled from copolymers with temperature-responsive blocks, where the aqueous properties of such copolymers change drastically at a certain temperature and the corresponding micelles release the drug through phase inversion. The temperature at which the transition occurs is referred to as lower critical solution temperature (LCST) or upper critical solution temperature (UCST) of the polymer. When the temperature is below LCST or above UCST, hydrogen bonds are formed between polymer and water and dissolve. When the temperature is above LCST or below UCST, the hydrogen bonds between water and polymer chains are broken and polymer precipitates. Currently, the most widely used temperature-responsive polymer is PNIPAAm when LCST is around 32 °C. In addition, Pluronics and P(HPMAm-Lacn) are often used for drug delivery [47]. Ghasemi et al. synthesized thermo-responsive poly(*N*-isopropylacrylamide-*b*-lauryl acrylate) (PNIPAAm-*b*-PLA) with tailored molecular weight through reversible addition-fragmentation chain-transfer (RAFT) polymerization to prepare drug-loaded smart micelle. The micelle experienced phase transition when the temperature rise up above the LCST and showed a sustained drug release rate [54].

2.5.6.4 Photo-responsive micelles

Photo-responsive micelles usually undergo structural transformations in response to specific light to achieve drug release mainly through photo-induced transition of hydrophobicity–hydrophilicity or photo-cleavage reaction [55, 56]. Photochromic moieties such as coumarin, azobenzene, spiropyran, and 2-diazo-

1,2-naphthoquinone are commonly introduced to construct photo-responsive block polymers. For example, The conformation of azobenzene groups and its derivatives changes from the apolar *trans* form to the polar *cis* form upon light irradiation (340–380 nm), then reforms to the orginal state after being exposed to irradiation at 420–490 nm or put into the dark. Compared with the *cis*-azo bond (dipole moment, μ = 3D), the *trans*-azo bond has a smaller polarity (dipole moment, μ = 0D) and stronger hydrophobicity, which may be utilized to control the assembly and disruption of azo-containing polymeric micelles in repeated cycles and facilitate drug release. Spiropyran is another promising photoisomerization group which can exist in two states, the hydrophilic zwitterionic merocyanine state and the hydrophobic SP state. These two states can transform into each other through a reversible isomerization either upon visible light (620 nm) irradiation (from hydrophilic to hydrophobic) or the UV (365 nm) irradiation (from hydrophobic to hydrophilic). Block copolymer with spiropyran showed improved light responsiveness due to the greater difference between the hydrophilic and hydrophobic state compared with azo moieties.

2.6 Summary and Perspective

Polymeric micelles have shown various advantages as nanocarriers for drug delivery. They can realize the solubilization of water-insoluble drugs through hydrophobic interactions, and achieve enrichment in tumor tissues through enhanced permeability and retention, thus reducing toxic side effects on normal tissues. However, due to the great differences between *in vitro* and *in vivo* environments, and the complex composition of our internal environment (e.g., blood) encountered in practical applications, there are multiple interactions between micelles and biological macromolecules, which makes the precise and rational design of micelles a great challenge. At the same time, the special pathological conditions of tumor microenvironment have stricter requirements on the targeting and nanomedicine. Only when our understanding of disease physiology is matched by advances in materials science, will it be possible to achieve larger breakthroughs in this field.

This chapter attempts to link the physical and chemical parameters of micelles with interactions *in vivo*, for example, how to regulate the particle size and zeta potential of the corresponding micelles, affect their drug loading and release capacity, cellular endocytosis efficiency and the stability of micelles by changing the chemical structure of amphiphilic polymers. It aims to provide theoretical basis and guidance for the precise design of micelles to maximize their advantages and enhance the efficacy of nanotherapy for tumors.

References

1. Torchilin VP, *J Controll Release* **2001**, *73*, 137–172.
2. Sen Gupta A, Cardiovascular nanomedicine: materials and technologies, in *Nanomaterials in Pharmacology* (Lu Z-R, Sakuma S, eds), Springer, New York: New York, NY, 2016, pp. 251–277.
3. Hwang D, Ramsey JD, Kabanov AV, *Adv Drug Deliv Rev* **2020**, *156*, 80–118.
4. Biswas S, Vaze OS, Movassaghian S, Torchilin VP, Polymeric micelles for the delivery of poorly soluble drugs, in *Drug Delivery Strategies for Poorly Water-Soluble Drugs*, 2013, pp. 411–476.
5. Vincent B, *Adv. Colloid Interface Sci.* **2014**, *203*, 51–54.
6. Debye P, Anacker EW, *J Phys Chem* **1951**, *55*, 644–655.
7. Becher P, Arai H, *J Colloid Interface Sci* **1968**, *27*, 634–641.
8. Munk P, Equilibrium and nonequilibrium polymer micelles, in *Solvents and Self-Organization of Polymers* (Webber SE, Munk P, Tuzar Z, eds), Springer, Netherlands: Dordrecht, 1996, pp. 19–32.
9. Jean FG, *Anionic Polymerization*, Springer: Tokyo, **2015**, 925–972.
10. Tanford C, *Proc Natl Acad Sci* **1974**, *71*, 1811–1815.
11. Schott H, Han SK, *J Pharm Sci* **1976**, *65*, 979–981.
12. Mukherjee P, Padhan SK, Dash S, Patel S, Mishra BK, *Adv Colloid Interface Sci* **2011**, *162*, 59–79.
13. Biswas S, *Nanomedicine* **2021**, *16*, 1541–1544.
14. Biswas S, Kumari P, Lakhani PM, Ghosh B, *Eur J Pharm Sci* **2016**, *83*, 184–202.

15. Gohy J-F, Block copolymer micelles, in *Block Copolymers II* (Abetz V, ed), Springer, Berlin Heidelberg: Berlin, Heidelberg, 2005, pp. 65–136.

16. Wright WW, *Br Polymer J* **1984**, *16*, 57–58.

17. Liu T, Liu LZ, Chu B, Formation of amphiphilic block copolymer micelles in nonaqueous solution, in *Amphiphilic Block Copolymers* (Alexandridis P, Lindman B, eds), Elsevier Science B.V., Amsterdam, 2000, pp. 115–149.

18. Linse P, Modelling of the self-assembly of block copolymers in selective solvent, in *Amphiphilic Block Copolymers* (Alexandridis P, Lindman B, eds), Elsevier Science B.V., Amsterdam, 2000, pp. 13–40.

19. Daoud M, Cotton JP, *J Phys France* **1982**, *43*, 531–538.

20. Noolandi J, Hong KM, *Macromolecules* **1983**, *16*, 1443–1448.

21. Leibler L, Orland H, Wheeler JC, *J Chem Phys* **1983**, *79*, 3550–3557.

22. Kabanov AV, Nazarova IR, Astafieva IV, Batrakova EV, Alakhov VY, Yaroslavov AA, Kabanov VA, *Macromolecules* **1996**, *29*, 8999–8999.

23. Li W, Nakayama M, Akimoto J, Okano T, *Polymer* **2011**, *52*, 3783–3790.

24. Nagarajan R, *Langmuir* **2002**, *18*, 31–38.

25. Li W, Li J, Gao J, Li B, Xia Y, Meng Y, Yu Y, Chen H, Dai J, Wang H, Guo Y, *Biomaterials* **2011**, *32*, 3832–3844.

26. Li W, Zhao H, Qian W, Li H, Zhang L, Ye Z, Zhang G, Xia M, Li J, Gao J, Li B, Kou G, Dai J, Wang H, Guo Y, *Biomaterials* **2012**, *33*, 5349–5362.

27. Zhou Q, Zhang L, Yang T, Wu H, *Int J Nanomedicine* **2018**, *13*, 2921–2942.

28. Dorn K, Hoerpel G, Ringsdorf H, *Bioactive Polymeric Systems* **1985**, 531–585.

29. Khougaz K, Zhong XF, Eisenberg A, *Macromolecules* **1996**, *29*, 3937–3949.

30. Kwon GS, Kataoka K, *Adv Drug Deliv Rev* **2012**, *64*, 237–245.

31. Al Samad A, Bethry A, Koziolová E, Netopilík M, Etrych T, Bakkour Y, Coudane J, El Omar F, Nottelet B, *J Mater Chem B* **2016**, *4*, 6228–6239.

32. Chansuna M, Pimpha N, Vao-soongnern V, *J Polym Res* **2014**, *21*, 452.

33. Miliou K, Gergidis LN, Vlahos C, *J Polym SciPart B: Polym Phys* **2018**, *56*, 924–934.

34. Li W, Zhao H, Qian W, Li H, Zhang L, Ye Z, Zhang G, Xia M, Li J, Gao J, Li B, Kou G, Dai J, Wang H, Guo Y, *Biomaterials* **2012**, *33*, 5349–5362.

35. Haliloğlu T, Bahar I, Erman B, Mattice WL, *Macromolecules* **1996**, *29*, 4764–4771.
36. Kataoka K, Harada A, Nagasaki Y, *Adv Drug Deliv Rev* **2001**, *47*, 113–131.
37. Torchilin V, *Adv Drug Deliv Rev* **2010**, *63*, 131–135.
38. Letchford K, Liggins R, Burt H, *J Pharm Sci* **2008**, *97*, 1179–1190.
39. Li W, Masamichi N, Jun A, Teruo O, *Polymer* **2011**, *52*(17), 3783–3790.
40. Gravel E, Thézé B, Jacques I, Anilkumar P, Gombert K, Ducongé F, Doris E, *Nanoscale* **2013**, *5*, 1955–1960.
41. Wang D, Zhou Y, Li X, Qu X, Deng Y, Wang Z, He C, Zou Y, Jin Y, Liu Y, *ACS Appl Mater Interfaces* **2017**, *9*, 6916–6930.
42. Jeong Y-I, Cheon J-B, Kim S-H, Nah J-W, Lee Y-M, Sung Y-K, Akaike T, Cho C-S, *J Control Release* **1998**, *51*, 169–178.
43. Shi B, Fang C, You MX, Zhang Y, Fu S, Pei Y, *Colloid Polym Sci* **2005**, *283*, 954–967.
44. Savić R, Azzam T, Eisenberg A, Maysinger D, *Langmuir* **2006**, *22*, 3570–3578.
45. Burt HM, Zhang X, Toleikis P, Embree L, Hunter WL, *Colloids Surf B: Biointerfaces* **1999**, *16*, 161–171.
46. Kamaly N, Yameen B, Wu J, Farokhzad OC, *Chem Rev* **2016**, *116*, 2602–2663.
47. Rezaei A, Rafieian F, Akbari-Alavijeh S, Kharazmi MS, Jafari SM, *Adv Colloid Interface Sci* **2022**, *307*, 102728.
48. Wang Z, Deng X, Ding J, Zhou W, Zheng X, Tang G, *Int J Pharm* **2018**, *535*, 253–260.
49. Chang G, Yu L, Yang Z, Ding J, *Polymer* **2009**, *50*, 6111–6120.
50. Fleige E, Quadir MA, Haag R, *Adv Drug Deliv Rev* **2012**, *64*, 866–884.
51. Yang YQ, Lin WJ, Zhao B, Wen XF, Guo XD, Zhang LJ, *Langmuir* **2012**, *28*, 8251–8259.
52. Wang C, Wang G, Wang Z, Zhang X, *Chemistry – A Eur J* **2011**, *17*, 3322–3325.
53. Sarisozen C, Joshi U, Mendes LP, Torchilin VP, 10 - Stimuli-responsive polymeric micelles for extracellular and intracellular drug delivery, in *Stimuli Responsive Polymeric Nanocarriers for Drug Delivery Applications* (Makhlouf ASH, Abu-Thabit NY, eds), Woodhead Publishing, 2019, pp. 269–304.

54. Ghasemi S, Ahmadi L, Farjadian, F, *J Mater Sci* **2022**, *57*, 17433–17447.
55. Huang Y, Dong RJ, Zhu XY, Yan DY, *Soft Matter* **2014**, *10*, 6121–6138.
56. Sana B, Finne-Wistrand A, Pappalardo D, *Mater Today Chem* **2022**, *25*, 100963.

Chapter 3

Nano Physical Pharmaceutics of Liposome-Based System

Nian Huang,[a] Mengxin Zhao,[a] Jian Wang,[b] and Wei Li[a]

[a]*Department of Nanomedicine, Naval Medical University, Shanghai 200433, China*
[b]*National Advanced Medical Engineering Research Center, Shanghai 201203, China*

liwei_dds@163.com

3.1 Classification of Liposome

Liposomes are spherical artificial vesicles constituted by cholesterol and natural nontoxic phospholipids [1]. Due to the amphiphilic nature of lipids, their thermodynamic phase properties and self-assembling characteristics influence entropically focused confiscation of their hydrophobic sections into spherical bilayers, where the polar head groups are oriented in the pathway of the interior and exterior aqueous phases. The choice of phospholipids plays an important role in determining the "rigidity" or "fluidity" and the charge of the bilayer. For instance, unsaturated phosphatidylcholine species from natural sources (egg or soybean phosphatidylcholine) construct bilayers with more permeability and less stability, whereas the saturated phospholipids with long

Nano Physical Pharmaceutics
Edited by Wei Li

ISBN 978-981-4968-52-2 (Hardcover), 978-1-003-51393-3 (eBook)
www.jennystanford.com

acyl chains form bilayers with more rigidity and impermeability [2].

Based on the number of bilayers, liposomes can be categorized into unilamellar vesicles and multilamellar vesicles (MLV). The former can also be classified into three subgroups: small unilamellar vesicles (SUV), large unilamellar vesicles (LUV) and giant unilamellar vesicles (GUV) with sizes of 30–100 nm, 100–1000 nm and 1–100 μm, respectively [1, 3]. Unilamellar liposomes are single phospholipid bilayer spheres while multilamellar liposomes have an onion-like structure. Classically, several unilamellar vesicles will be formed inside of the other vesicle, forming a multilamellar structure of concentric phospholipid spheres separated by water molecules (Fig. 3.1). Another classification takes into account the charge present on the liposome surface resulting in the cationic, anionic or neutral types.

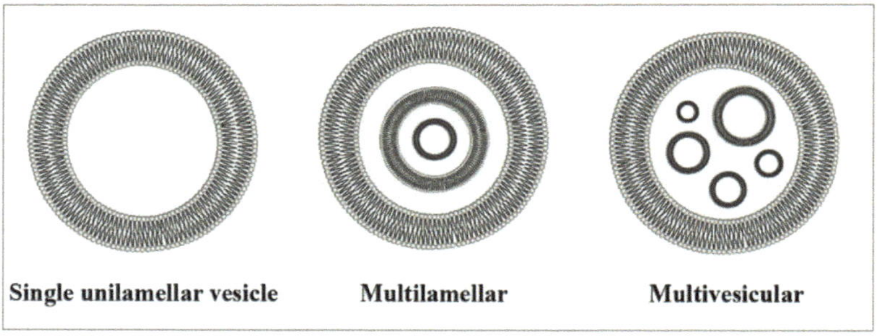

Figure 3.1 Schematic diagram of the structures of liposomes with different structures.

3.2 Preparation of Liposomes

The common synthetic methods of liposomes include thin-film hydration method, reverse-phase evaporation method, injection method, and freeze-drying method [2, 4]. Detailed procedures are described as follows.

3.2.1 Thin-Film Hydration Method

It is the most classic method of liposome preparation requiring no special equipment and skills [4, 5]. Generally, phospholipids and other materials are dissolved in an appropriate amount of chloroform or other organic solvents, then the solvent is removed under reduced pressure to obtain a thin lipidic film adherent to the inner wall of the flask. Then it is dried under vacuum to remove solvents until constant weight of the film is achieved. After adding a buffer solution and vigorous shaking, multiphase liposomes can be prepared with particle sizes ranging from 1 to 5 μm. It can be further dispersed by mechanical methods, such as ultrasonic treatment with a sonicator. Unilamellar liposomes from 0.25 to 1 μm can be prepared, or it may be filtered continuously with polycarbonate fiber membranes with a certain pore size to obtain uniformly distributed liposomes.

3.2.2 Reverse-Phase Evaporation Method

It is one kind of emulsification method, which generally refers to the formation of water emulsion in the organic phase, and is also the most common preparation method [6]. In this method, the mixture is dissolved in organic solvent, and water is added to form a two-phase system. Then, the solution is sonicated or violently vibrated to obtain a uniformly dispersed emulsion. Next, the emulsion is gradually transformed into a liposome suspension by slowly volatilizing the organic solvent. The remaining residual solvent can be eliminated by several methods, such as centrifugation, dialysis or size exclusion chromatography. Notably, the resulting liposomes usually require a final step of extrusion to reduce the size.

3.2.3 Injection Method

This method can be used to produce SUVs in a simple and fast way, which also functions as an alternative to film hydration [7, 8]. The method first dissolves the lipid mixture in a water-miscible organic solvent, and then adds a large amount of water under vigorous stirring. After rapid dilution, lipid molecules first rearrange at the boundary of organic solvent and water to form phospholipid bilayer fragments and then integrate to form vesicles. Many parameters, such as injection rate, stirring rate, lipid concentration and organic solvent/water ratio, have a significant influence on the product characteristics. After the organic solvent is removed by violent stirring, the liposome is obtained by ultrasonic treatment. The monodispersity of SUV produced by this method is better than that of the film water method, and it has unique advantages in terms of speed, low cost and stability.

3.2.4 Detergent Depletion Method

This method has obvious advantages in preparing uniform SUVs of predetermined size [4]. First, the dry film is prepared by the same method as the thin-film hydration method and then dispersed with a detergent solution higher than the critical micelle concentration (CMC). After intense vortex or ultrasound, the mixed micelles consisting of lipid bilayer fragments and detergent molecules are obtained. The detergent molecules are mainly distributed at the hydrophobic edge. Then, as the detergent molecules transfer from the mixed micelles to the solution, the bilayer fragments gradually merge until a closed vesicle is formed, Detergents can be easily removed by dialysis. The liposomes prepared by this method have uniform size distribution and are suitable for the encapsulation of bioactive drugs such as proteins and peptides due to the mild treatment.

3.3 Theory of Liposome Formation

Phospholipid is the main chemical structure that forms a liposome, including phosphatidylcholine, phosphatidylserine, phosphatidylethanolamine, etc. [9]. It is generally composed of two parts, in which the part containing the phosphoric acid group has a strong polarity and hydrophilicity; the two long aliphatic hydrocarbon chains are lipophilic non-polar tail groups. When phospholipid molecules are dispersed in water, according to the lowest energy principle, the hydrophilic heads and lipophilic heads come close together respectively and form bilayers whose inside is lipophilic and outside is hydrophilic, which generally form spherical vesicles due to the lowest energy of the spherical molecules (Fig. 3.2).

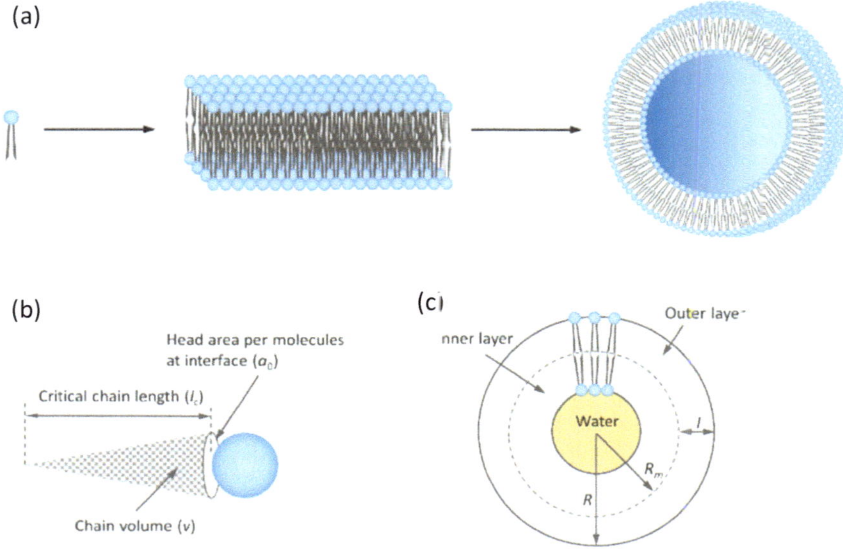

Figure 3.2 The general process and principle of the assembly of liposomes.

The formation mechanism in the lipid preparation process has not been fully studied yet. Lasic group has carried out a series of studies on the formation mechanism of the vesicles [10–13]. After first proposing the theory of detergent depletion method in 1982 [13], it has also been revised and applied to most other preparation methods [12]. The thermodynamic role in the vesicle formation process is the main focus of attention. In this chapter, according to the summary of Pan et al., we have explained the basic theory of lipid self-assembly from the aspect of the free energy of interaction of lipid molecules, the molecular geometry, and the elastic properties of lipid bilayer [9].

When phospholipid molecules are dispersed in water, according to the lowest energy principle, the hydrophilic heads and lipophilic heads come close together respectively and form bilayers whose inside is lipophilic and outside is hydrophilic, which generally form spherical vesicles due to the lowest energy of the spherical molecules (Fig. 3.2). Lasic proposed two main mechanisms: bilayered phospholipid fragment model and the "budd-off" model [10, 11]. Israelachvile further proposed the concept of critical packing parameter P to illustrate the influence of the shape of molecules on the shape of self-assembles [14]. The concept of P is as shown in Equation 3.1:

$$P = V/A_0L_C \tag{3.1}$$

where V is the volume of the hydrophobic part of the surfactant, A_0 is the head area of the molecule, and L_C is the maximum extension length of the hydrophobic chain. Since the critical stacking parameter P of phospholipid is between 0.5 and 1, the shape of its organized assembly is vesicle. Liposome, with the vesicle structure containing both a hydrophilic inner capsule and a hydrophobic interlayer, has natural advantages for realizing the loading of hydrophilic and hydrophobic molecules.

Safinya studied layered cationic liposomes/DNA complexes with different phospholipid structures and different phospholipid/DNA ratios, aiming to clarify the relationship between their structures, charge, and thermodynamic stability [15]. The driving force for the formation of neutral complexes is the release of DNA and lipid counterions. The negatively charged complexes have a higher DNA packing density, while the positively charged complexes have a lower DNA packing density. This indicates that the deviation of the complexes from the isoelectric point is due to the adsorption of excessive DNA or cationic phospholipids. The degree of deviation depends on the membrane surface charge density, or the ratio of neutral and cationic phospholipids. Moreover, charged complexes at a high ion concentration have a tendency to transform into neutral complexes, in which positively charged complexes loss cationic phospholipids and negatively charged complexes loss DNA molecules. By correspondent phase diagrams, the author shows intuitively the influence of its constituent components on its structure (Fig. 3.3).

For this type of complex, when the charges of DNA and the positively charged phospholipid molecule DOTAP are equal (indicated by the dotted line), the left and right charges are opposite, and there is a negative counterionic osmotic pressure, which drives the excess phospholipid molecules or DNA to bind to the complex, causing it to deviate the isoelectric point. At the same time, electrostatic repulsion sets a limit on the amount of DNA or phospholipid molecules entering the complex and causes the width of the single-phase complex area to increase as the phospholipid decreases. It is worth noting that most of the phase diagram features are independent of the chemical structural details of DNA or the phospholipid molecules, and are able to be explained in the nonlinear Poisson–Boltzmann electrostatic model, which is expected to be universally applicable to the non-specific binding of oppositely charged biological macro molecules.

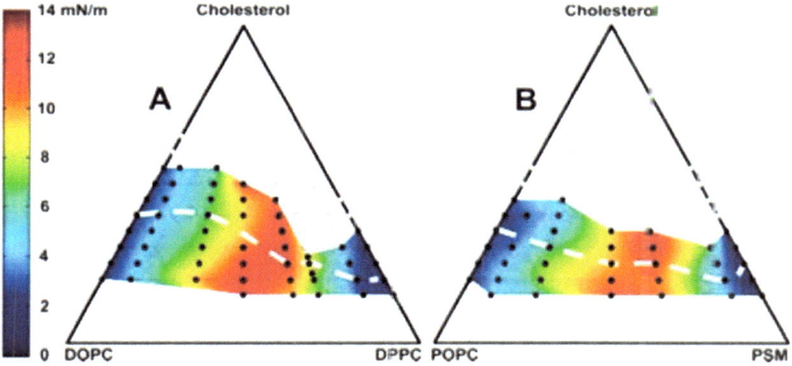

Figure 3.3 Phase diagram of liposome/DNA polyplex formation and the complex stability. Reproduced with permission from [15].

Liposomes have shown great potential in medical applications, and are currently the most widely used non-viral gene delivery system in clinic. In 2018, the FDA approved the RNA interference therapy Onpattro for the first time. It can inhibit the expression of transthyroxine protein after delivering short interfering RNA (siRNA) to the liver, resulting in amyloidosis of transthyroxine (ATTR). The development and commercialization of Onpattro has greatly promoted the development of COVID-19 vaccine. The approved Pfizer BioNTech (BNT162b2, also known as Comirnaty) and Moderna (mRNA1273, also known as Spikevax) have made important contributions to the cause of human health [16]. The current FDA-approved liposome nanoparticles formulations contain four lipids: (1) an ionizable cationic lipid, helper lipids which include (2) 1,2-distearoyl-sn-glycero-3-phosphocholine (DSPC) and

(3) cholesterol, and (4) a polyethylene glycol (PEG)-lipid conjugate. The composition ratio of these lipids has a significant impact on the carrier's performance. A modern nucleic acid delivery system comes from the conventional liposome system for chemotherapy. One example is DOX-loaded liposome, also known as doxycycline, which is the first FDA-approved nano-drug [17]. It contains hydrogenated soybean phosphatidylcholine (mainly DSPC), cholesterol and PEG lipids, with a molar ratio of 56:38:5. This formula is the result of optimization in the past two decades to achieve the required particle characteristics: high encapsulation efficiency by remote loading process, fewer adverse interaction with serum proteins, longer circulation lifetime to utilize enhanced permeability and retention (EPR) effect, higher accumulation at the target site, and lower toxicity. The application of this system to nucleic acid delivery revealed that the large size and high negative charge density of nucleic acid require additional lipid functions. Another formulation containing ionizable lipids, phospholipids, cholesterol and PEG lipids at a ratio of about 50:10:38.5:1.5 mol% was also applied to achieve high nucleic acid encapsulation, cellular uptake, and promote endosomal escape of nucleic acid cargo.

Based on the theories of liposome formation, we have prepared a series of thermosensitive liposomes to realize precise targeting and highly efficient chemotherapy [18]. For example, liposome with a core-shell composition composed of thermosensitive macromolecules was synthesized. It showed rapid drug release with structural changes. Moreover, the surface was modified with HER-2 antibody, which endows it with active targeting to the tumor surface and greatly improves the efficiency of cellular uptake. This provides a reference for the rational design and development of a liposome drug co-delivery system for cancer treatment. In another case, we prepared liposomes to achieve co-loading of hydrophilic melittin and hydrophobic bufotalin [19]. We optimized the delivery system by modulating the ratio of cholesterol, which can improve the rigidity of the phospholipid membrane. The concentration of phospholipid is fixed at the level of 20 μmol/L, and the concentration of cholesterol varies while other conditions remain the same. Liposomes with different molar ratios of phospholipid to cholesterol (1:2, 1:1, 2:1, 4:1, 6:1, 8:1) were prepared. We discovered the composition ratio has a significant impact on the encapsulation efficiency. With increased incorporation of phospholipid, the encapsulation efficiency increases at first and then decreases. When the proportion of phospholipid and cholesterol is 2:1, the encapsulation efficiency reaches the maximum of 60%. This is because liposome with higher cholesterol ratio is physically more stable, which helps to reduce drug leakage caused by ultrasound leading to the improvement of the encapsulation efficiency. On the other hand, cholesterol also enhances the rigidity of the membrane, and reduces the curvature as well as the total surface area, thus leading to lower encapsulated efficiency.

3.4 NPP of Liposomes in Drug Delivery

3.4.1 Physiochemical Properties of Liposome

3.4.1.1 Size of liposome

One of the factors affecting the distribution of liposome is the particle size. Liposomes of a certain size can be easily prepared by squeezing a lipid suspension through a filter with similarly sized pores [20, 21]. The liposomes prepared by this method are slightly larger than the average pore size of the filter (20%–50%). For liposomes of similar compositions, the general trend is that the larger the particle size, the faster they can be absorbed by the reticuloendothelial system (RES). For example, the clearance speed of liposomes extruded through a 400 nm filter is 7.5 times that of a 200 nm filter, and the clearance speed of liposome extruded through a 200 nm filter is five times that of small single-molecule vesicles.

3.4.1.2 Phase transition temperature of liposome

Bilayers composed of different lipids have changes of phase in the aqueous phase depending on various conditions, and the corresponding temperature is called the phase transition temperature (T_c). For example, when the temperature of DPPC liposomes was around 41°C. The arrangement of lipid molecules in the membrane changes significantly from the "rigid" state of the hydrocarbon chain at room temperature to the "flexible" movement, with membrane structure changed from "solid phase" to "mobile phase". All phospholipids have special T_c depending on the nature of the polar group, the length of the acyl chain, and the degree of unsaturation.

3.4.1.3 Membrane permeability of liposome

The membrane permeability is a very important parameter, which affects practical problems and requirements such as the leakage of the package and the targeted release of drugs in liposomes. It is closely related to the composition of membrane lipids and external conditions. For example, the incorporation of cholesterol can reduce the permeability of egg phosphatidylcholine; liposomes composed of high-phase-transition-temperature phospholipids such as DSPC have lower permeability than those of egg phosphatidylcholine. Membranes in the solid phase have lower permeability than that in the mobile phase. Therefore, when applying it *in vivo*, if a lipid whose T_c is lower than body temperature is used, the leakage effect will be great.

3.4.1.4 Membrane charge of liposome

The membrane charge of liposomes is related to the properties of phospholipid molecules and additives. Liposomes containing acidic lipids are negatively charged, liposomes containing bases are positively charged, and liposomes without ions are neutrally charged. The membrane charge can affect the liposome performance through different aspects including the clearance rate, the targeting ability, and the cellular uptake process. For example, liposomes with negative and positive charges on the surface are easier to be uptaken by macrophages than neutral liposomes for the following reasons: (1) The high negative charge density on the cell surface attracts positively charged liposomes easily by electrostatic interaction; (2) The interaction between negatively charged liposomes and non-specific receptors, especially type B scavenger receptors, makes them an easy target for the phagocytosis; (3) Liposomes with high membrane charges tend to bind and activate complement proteins. Chonn et al. reported that in human and guinea pig serum, negatively charged liposomes activate the complement system through the classical C1 pathway, while positively charged liposomes directly activate the complement system through the alternative C3 pathway without C1 activation. In addition, the greater the surface charge density of liposomes has, the more complement proteins are activated [22].

3.4.2 Drug-Loading Behavior of Liposome

Liposomes can be loaded with hydrophilic, hydrophobic and crystalline drugs. According to the principle of "like dissolves like," hydrophilic and crystalline drugs usually exist in the lipid membrane, and hydrophobic drugs are encapsulated in the middle of the phospholipid bilayer. "Passive drug loading" and "active drug loading" are the two commonly used methods.

Passive drug loading is applicable to almost all drugs. Hydrophilic drugs are first dissolved in the aqueous phase, while hydrophobic drugs and lipids are co-dissolved in the organic phase. The following preparation methods include the film dispersion method, solvent injection method, reverse evaporation method and freeze-thawing method. In order to accelerate the formation of liposomes, reduce the particle size, and adjust the uniformity, ultrasonic method and extrusion method can be added on the basis of the membrane dispersion method. The passive drug-loading method is simple and has no special additional steps in the preparation process. However, due to uneven drug dispersion, the drug concentration in the liposome cavity is lower than that in the external solution, resulting in low drug loading and encapsulation efficiency of the liposome. Liposome membrane enhancers such as ethanol, can be used to shift the equilibrium of passive drug loading and effectively improve drug loading and encapsulation efficiency [8].

Active drug loading can improve the encapsulation efficiency of ionizable hydrophilic drugs. The principle is to establish a bilayer gradient across lipids, such as pH gradient, amine ion gradient and calcium acetate gradient. The liposome was hydrated with a buffer solution with known pH, and then the liposome was added to a buffer solution with hydrophilic drugs and different pH to establish the transmembrane pH gradient of the liposome *in vivo* and *in vitro*. Using the different solubility of hydrophilic drugs in different pH buffers, the drug can be transported across the membrane to complete the enrichment of hydrophilic drugs in the aqueous phase of liposomes. With this method, the drug encapsulation rate can reach more than 95% and the drug loading can reach about 10%. Active drug loading improves the loading capacity of ionizable drugs, but it is meaningless for many hydrophobic and non-ionizable drugs. Sur et al. proposed a new method of active drug loading mediated by ionizable cyclodextrin derivatives, that is, hydrophobic drugs are embedded in the cyclodextrin cavity to form ionizable host–guest complexes, and then loaded into liposomes by pH gradient method [23]. Compared with the traditional passive drug-loading method, the encapsulation efficiency can be increased by 20 times.

The drug-loading capacity of liposomes is affected by several factors, including the drug/lipid ratio, the time and temperature of co-cultivation, transmembrane gradient, pH of the outside aqueous phase and buffering ability of the inside phase in terms of the active drug-loading method. The drug/lipid ratio should be as high as possible to maximize the payload of the drug to the tumor. On the other hand, an excessively high drug/lipid ratio may also result in a less stable formulation, possibly due to the dissipation of the pH gradient during drug loading. Tatsuaki et al. prepared thermosensitive doxorubicin (DOX)-loaded liposomes by ammonium sulfate gradient method and pH gradient method (citric acid buffer solution) respectively [24]. When the drug lipid ratio was 1/40 (w/w), the encapsulation efficiency of DOX was close to 100%. When DOX-loaded liposomes were prepared by pH gradient method, when the drug lipid ratio was 1/20 (w/w), the entrapment efficiency of DOX reached the maximum value. When the entrapment efficiency continued to increase to 1/40 (w/w), the encapsulation efficiency decreased to about 80%.

The transmembrane gradient also has an important influence on drug-loading capacity. Generally, the larger the gradient is, the greater the transmembrane driving force, and the more drugs loaded into liposomes, the higher the entrapment efficiency of liposomes. When preparing pixantrone-loaded liposomes, if the transmembrane pH gradient difference is 3 or 4, the entrapment efficiency of liposomes reaches more than 80%. However, when the transmembrane pH gradient difference is smaller than 2, the encapsulation efficiency is less than 40% [25]. DOX-loaded liposomes also have a similar encapsulation effect. When the transmembrane pH gradient difference reaches 3.5, the drug encapsulation efficiency exceeds 98% [26]. Lowering the pH of the internal aqueous buffer solution can increase the pH gradient, but the acidic environment is easy to induce hydrolysis of drugs and phospholipids, leading to the instability of liposomes.

3.4.3 Stability of Drug-Loaded Liposome

In pharmaceutical manufacturing, the stability of drug-loaded liposomes is considered an essential issue. It can be affected by several different factors of liposome pharmaceutical preparations: chemical stability of drug and lipid molecules, colloidal stability and drug retention. For the application of liposomes, liposome-related drugs need to be specifically delivered to solid tumors, and liposomes must retain most of the contents in the circulation. When plasma exists, various factors will affect the relative stability of this preparation.

3.4.3.1 Physical stability of drug-loaded liposome

Regarding the amphipathic drugs that can pass through the membrane easily, there are many factors to affect the stability of liposomes. The existence of cholesterol and phospholipids seems to be the most significant factor that decreases the membrane permeability of these drugs [20]. Besides, the drug-loading method that optimizes liposomes to deliver chemotherapeutic drugs, may cause the internal drug concentration to excel the water solubility of the free drugs, which also acts as a stabilizer. Cholesterol seems to be particularly important in liposomes stabilizing plasma components such as HDL [27]. In addition, some early studies have shown that cholesterol is essential to control the permeability of membranes to ions or small molecules. The addition of cholesterol to the liposome formulations can improve plasma stability and inhibit the transfer of lipid components to plasma lipoproteins [28].

The role of osmotic pressure in the preparation of stable liposomes has been studied in many studies. The study by Allen et al. showed that when trapped under hypertonic conditions, the trapped drug is released faster, and its release is characterized by the initial rapid kinetics, followed by the slow secondary leakage rate [29]. Madden et al. reported that osmotic dissolution merely causes a partial release of liposome content, and after the reseal of liposome membrane, the liposome cavity is still hypertonic [30]. In the study to encapsulate vincristine with sterically stabilized liposomes, it was found that under isotonic conditions, the vincristine loaded with ammonium citrate gradient is relatively stable [31]. In contrast, under hypertonic conditions, 90% of the capsule vincristine is released from DSPC/cholesterol liposomes loaded by the pH gradient method after 24 h. There may be a good balance between the osmotic stability of liposomes, the residual pH gradient, as well as the production of drug precipitation in the liposome cavity. For the pH gradient drug-loading method, it is usually necessary to have a high buffer capacity in the vesicle's internal medium to keep a logical pH gradient and achieve high loading efficiency. High concentrations of DOX usually lead to gel-like precipitates with low osmotic activity. For example, the pH (400 mmol sodium citrate) or ammonium sulfate (250 mmol ammonium sulfate) gradient method is used to stably encapsulate DOX into PEG-DSPE/DSPC/cholesterol or DSPC/cholesterol liposomes. However, after the drug-loading process, the unstable drug complexes or gel may still show an apparent osmotic gradient, which may further increase during the drug-loading process.

3.4.3.2 Chemical stability of drug-loaded liposome

Chemical stability is also another important issue of the drug-loaded liposome [32]. A large number of drugs and lipids are susceptible to alkaline hydrolysis. When applying the calcium acetate gradient method to load amphiphilic drugs, it may cause a very high internal pH, the stability of the drug must be carefully considered to avoid hydrolysis. Besides, fatty acid esters are also sensitive to acid and alkali hydrolysis, certain conditions can cause membrane instability by dissolving lipids. It is common to analyze the lipid components in the newly developed liposome formulations by thin-layer chromatography or high-performance liquid chromatography to determine the chemical stability of the lipids used.

Lipid peroxidation is another common issue in unsaturated lipid components. It can be caused by various factors and can lead to the formation of secondary oxidation products of membrane instability, such as 4-hydroxynonanal and malonaldehyde [32]. Due to the extraction of hydrogen radicals from the diallyl carbon, phospholipids containing diunsaturated fatty acyl chains (such as linoleic acid, linolenic acid or arachidonic acid) are particularly susceptible to lipid peroxidation. Phospholipid, which contains linoleic acid and arachidonic acid, tends to form complex secondary oxidation products, which are particularly harmful to cell membranes [33]. When unsaturated lipids are required, phase transition (PT) is usually an excellent predictor of bilayer fluidity. Increasing the length of the acyl chain will obtain higher T_c, while increasing the amount of unsaturation will reduce T_c. Therefore, the lipid component with the desired T_c can be found by balancing the acyl chain length and the amount of unsaturation found in the specific phospholipid component. EggPC is a widely used liquid phase lipid component, which is liquid crystal at physiological temperature. Unfortunately, it also contains a high proportion of fatty acyl chains and has multiple unsaturation phenomena (18% diolefin, 3% tetraolefin), making it particularly prone to oxidation. The improvement of phospholipid synthesis has led to the increased availability of synthetic lipids (such as POPC) with the cost comparable to natural products. Combined with enhanced chemical stability, POPC is more suitable as the unsaturated lipid component of liposome formulations than EggPC.

3.4.4 Clearance and *in vivo* Circulation of Liposome

3.4.4.1 Influence of liposome size on clearance and *in vivo* circulation

Ahl et al. found that the clearance and size of DSPC liposomes containing a small amount of N-glutaroylphosphatidylethanolamine are similarly dependent [34]. Compared with DSPC/cholesterol liposomes (control group, similar to PEG-DSPE stabilized liposomes), the area under the curve (AUC) of drug blood concentration-time of DSPC liposome also increased. Although the size dependence of these two stabilized preparations of liposomes is relatively low, optimizing liposome size is still important to the delivery system. For neutral liposomes, the optimal size range is much smaller. That is, under the premise of maintaining a reasonable encapsulation rate, the particle size is small enough (preferably less than 100 nm). The composition of liposomes also affects the rate at which it is taken up by RES. For example, adding PEG-DSPE to the liposome components will cause RES to be relatively insensitive to the removal of nanoparticles in the range of 80 nm to 250 nm [35]. Woodle et al. found that aggregation can reduce the circulation time of liposomes. Experimental results also show that when the liposome size increases from 100 nm to 200 nm, the clearance increases by 54% within the same time, indicating that there is a certain relationship between liposome size and circulation time [20].

3.4.4.2 Influence of liposome dose on clearance and *in vivo* circulation

The circulation time of traditional liposomes increases with higher liposome dose, which may be due to the decrease in the phagocytic ability of macrophages in RES after ingestion of higher doses, or the increased circulatory stability of plasma factors by neutralizing liposomes [36]. Among them, liposomes composed of high-phase-transition lipids (such as SM/cholesterol or DSPC/cholesterol) can be neutralized and reabsorbed more easily, but high-phase-transition lipids are difficult to metabolize *in vivo*, suggesting that the neutralization of high-phase-transition lipids disrupts their metabolic pathways. In addition, liposomes can also bind serum proteins in a manner that is inversely proportional to blood clearance. Therefore, it is speculated that the depletion of plasma factors at high lipid doses will lead to an increase in the circulating half-life of liposomes *in vivo*.

The circulation times of traditional liposomes and sterically stabilized liposomes are significantly different when influenced by lipid doses. The steric stability of PEG-DSPE provides a unique advantage for the release of liposomes, and its clearance kinetics is independent of dose [37]. For example, in terms of sterically stabilized liposomes, plasma AUC increases linearly as the liposome dose increases, while the plasma half-life $t_{1/2}$ remains relatively constant. The plasma AUC of traditional liposomes increases slowly at low doses and then increases exponentially with higher dose, and $t_{1/2}$ tends to balance after increasing, indicating that the mechanism responsible for its clearance has reached saturation [37].

3.4.4.3 Influence of liposome charge on clearance and *in vivo* circulation

Early research by Senior et al. showed that because of the presence of negatively charged components, including phosphatidic acid (PA), phosphatidyl serine (PS), and phosphatidyl glycerol (PG), liposomes were absorbed rapidly by RES [21]. However, the relationship between liposome charge and the circulation time is extremely complex, and simple models are not accurate enough to explain. Gabizon et al. did further research on the influence of liposomes containing different phospholipid anions on the efficiency of liposome clearance [38]. Among them, liposomes containing anionic phospholipids such as PG, PA and PS have a shorter circulation time than that of neutral liposomes, while the introduction of ganglioside or phosphatidylinositol will prolong the circulation time. Those liposomes with steric negative charges are cleared more slowly than those without, which sufficiently illustrates the complexity of the relation. Woodle et al. found that the introduction of phosphatidylethanolamine (PE) into distearoylphosphatidylcholine (DSPC) liposomes could significantly increase the circulation time of liposomes *in vivo* [20]. Further studies found that DSPC with a high concentration of PE has a shorter circulation time, which is mainly due to the dense negative charges on the surface of liposomes, and will easily be cleared by RES.

3.4.4.4 Influence of phase transition temperature on clearance and *in vivo* circulation

The introduction of anionic liposome components of high phase transition into liposomes can prolong the circulation life. Research by Gabizon et al. showed that using 1,2-distearoyl-3-sn phosphatidylglycerol (DSPG)/HSPC/cholesterol liposomes (where the source of PG is distearoylphosphatidyl triglyceride, T_c is 53.0 °C) to load and transport DOX, the plasma level of DOX is twice that of HSPC/cholesterol liposomes after 24 h [39]. In addition, the requirement for high-phase-transfer anionic lipid components is also necessary for phosphatidylinositol (PI), among which hydrogenated soybeans are most commonly used as the source of PI in long-circulating liposomes [39].

In the multi-crosslink liposomes with HSPC, DPPC or EggPC as phospholipid components, the correlation among the phase transition of the phospholipid components, the stability of plasma *in vitro* stability test of the formulations, and the plasma level of the drug *in vivo* was observed [40]. The liposomes containing high-phase-transition liposomes form relatively stable preparations, which can retain the drug better, and the circulation lifetime of the drug is significantly prolonged. To be most effective, the phospholipid component must have a phase transition significantly higher than 37 °C. Examination of PT of various phospholipid molecules showed that the T_c of EGPC is lower than 37 °C, while the T_c of DPPC is only a few degrees above 0 °C (with pre-transition at 3 places). However, the T_c values of DSPC and HSPC are both higher than 37 °C, so the liposomes containing HSPC and DSPC have a harder membrane bilayer than that containing EggPC or DPPC, which can resist the permeation of serum opsonins.

3.4.4.5 Influence of steric stability of liposome on clearance rate and *in vivo* circulation

Initial attempts were made to simulate the surface of erythrocytes by adding sterically hindered ganglioside (GM_1) or PE to the liposome preparations, leading to the formation of long-circulating or sterically stabilized liposomes (SSL) [41]. Then, PEG-DSPE was substituted by GM_1 or PE. A common misconception is that the attachment of PEG on the surface of liposomes will prevent RES from uptake of liposomes. On the contrary, it will only reduce the uptake rate. One of the most significant advantages of the SS is the unsaturated log-linear pharmacokinetics, as described earlier. SSL may resist the uptake of high-affinity, low-capacity RES macrophages, leading to a prolonged circulatory lifetime. SSL is mainly distributed in the spleen and liver, similar to the classical liposome (CL) [42]. However, the rate of enrichment in these tissues is much slower than that of CL. In mice, the plasma level of PEG-DSPE liposomes is 3 to 3.5 times as high as that of DSPC/cholesterol (2:1), and 8 to 11 times as high as that of EggPC/cholesterol (2:1) liposome. As mentioned earlier, CL-containing anionic lipid or unsaturated lipid components (e.g., EggPC) are easier to remove from the circulation than CL-containing high-phase transition phospholipid (SMCHOL or DSPC/cholesterol). However, in the presence of PEG-DSPE, some liposomes containing charged lipids or low-phase transition phospholipids are found in plasma after 24 h, and their level is similar to liposomes containing neutral high-phase transition phospholipids. Therefore, the presence of steric stability makes the clearance relatively independent of the remaining lipid compositions of "empty" liposomes [42].

This does not include all phospholipid components. Neither GM_1 nor PEG-DSPE can prevent the rapid clearance of PS-containing liposomes from the circulation, indicating that some membrane components may give a very strong tendency to the types of liposomes that are recognized and absorbed by researchers. In addition, the lipids that make up most of liposomes dominate the circulation. Liposomes may not have a direct effect on the removal of the liposome carriers, and may have an indirect effect on the clearance of capsule drugs. As mentioned earlier, when amphiphilic drugs such as DOX are loaded into these liposomes, if the liposomes do not deliberately prevent leakage through optimization, the rate-limiting step for drug removal from the circulation is the leakage of liposomes. SSL has the advantages of dose independence of liposome clearance, reduced RES clearance, and flexibility of lipid components, and has great application potential.

The mechanism by which steric stability of liposomes extends its circulation lifetime has been widely discussed [43]. The basic concept discussed is that a kind of flexible chain hydrophilic polymer or glycolipid, such as PEG or GM_1, occupies the space near the liposome surface ("exosomal layer") and tends to exclude other macromolecules from this space. Therefore, the process of plasma opsonins entering the surface of liposomes and binding is hindered, thereby inhibiting the interaction of RES macrophages with such liposomes. Excluding the exogenous macromolecules in the exosomal layer will cause steric hindrance ("steric stabilization effect"), which is manifested as an increase in the repulsive force between the two layers and results in a larger separation between the double layers modified by polyethylene glycol than that of the unmodified layer. The effect of GM_1 on the separation between membranes is less than that of PEG. The measurement of the agglutination rate of biotinylated SSL induced by streptavidin shows that as the length of the PEG chains decreases, the steric barrier decreases. At higher PEG lengths, the space barriers are significantly higher than that produced by GM_1 [43].

Several studies believed that the presence of PEG or GM_1 did cause the extent and rate of the binding of plasma protein to liposomes to decrease, although direct experimental evidence was not very clear for liposomes containing PEG-DSPE [44]. In the previous two studies, it was shown that the protein binding or sonic activity of SSL was reduced, and polyethylene glycol stabilized liposomes were incubated with plasma components within 2 to 15 min. Another study showed that after 2 min of incubation, similar results were found *in vivo*, but after 30 min, the binding protein level of CL and PEG stabilized liposome formulations was roughly equal. The study by Chonn et al. showed that the binding level of proteins containing CL and GM_1 liposomes was correlated with *in vivo* circulating lifetime, which was consistent with the results observed by SSL [45]. Therefore, reduced binding of serum opsonins by GM_1-containing liposomes can lead to decreased activation of opsonin, such as complement factor C3, or decreased macrophage uptake. Through the steric hindrance of liposome surfaces by macromolecules or cells, liposomes can obtain a longer circulation lifetime and make it enriched in tumor cells for a longer time.

3.4.5 Targeting Ability of Liposome

Active targeting of drugs is generally considered as a means to increase the access of drugs to diseased sites. However, due to the discontinuous tumor microvascular system, lack of functional lymphatic vessels and high interstitial pressure, the passive capture of liposome drugs in solid tumors leads to rate-limiting enrichment of liposomal drugs. When designing targeting liposomes, the choice of targeting ligand is very important. The ligand should be relatively specific to cancer cells, especially when compared with cells that are easily obtained in the general circulation. Second, exploit epitope-binding that results in liposome internalization. Binding to an internalized receptor didn't require endocytosis, especially in the case of using antibodies or antibody fragments. Ligands of internalized receptors, such as folic acid, usually cause endocytosis, but the binding of proteins or peptides to unrelated parts of the receptor may simply restrict the carrier on the membrane surface. Another problem with the attachment of targeting molecules to the surface of liposomes is that the targeting molecules may pass through tissues other than tumors and thus increase the clearance of liposomes.

Active targeting of solid tumor cell surface proteins that are not internalized is unlikely to provide significant therapeutic effects. When anti-HER2 targeted immunoliposomes were prepared with antibodies that were not internalized, the therapeutic effect did not increase compared with non-targeted liposomes [46]. Vingerhoeds et al. also did not find a higher efficacy of non-replacement immunoliposomes for the OA3 antigen present in 90% of human ovarian cancers [38]. In addition to the advantages related to the reduced immunogenicity of antibody fragments, the decrease in affinity of the fragments to their cell surface targets may help reduce the binding site barrier, allowing the carrier to penetrate deeper into the tumor. Compared with full antibodies, the deep penetration of antibody fragments was previously thought to be due to the reduced size of the molecule and the reduced affinity for its target [47].

The rate and level of drug enrichment in tumors depends on the dose, formulation, and tumor type. Drug enrichment in tumors is most often measured at a single time point, usually after 24 or 48 h. In some studies, since the peak of free drug enrichment is earlier, the free drug is measured at a relatively shorter time (such as 1 hour). Compared with free drugs, the peak DOX in tumors delivered by liposomes is 3–15 times higher. When administered at high doses (20 mg/kg–55 mg/kg), a comparison between CL and SSL-DOX shows that the cumulative amount in the three tumor models are approximately equal. When administered at lower doses (5 mg/kg–10 mg/kg), the accumulation of SSL in the tumor is higher [48]. The effects of dose and formulation are the result of indirect effects on the circulating lifetime and formulation stability of liposomes. The pharmacokinetics and biodistribution of liposomes and liposome drugs, and the kinetics of SSL clearance are logarithmic linear and independent of dosage, which allows liposome drugs to have a relatively higher concentration in circulation, even at low dosages. On the other hand, CL shows saturable dose-dependent kinetics, leading to rapid clearance of liposome at low doses, but the clearance is much slower, so the blood concentration at high doses is higher. It is assumed that at these lower doses, the huge difference in circulating lifetimes between SSL and CL will result in larger liposomes entering into the tumor in the case of SSL. Nonetheless, at higher doses, as the liposome clearance mechanism becomes saturated, the degree of these differences should decrease. Using different lipid components with different circulating lifetimes, a good correlation between increased lifetime and a high liposome level in tumors was observed [49]. This correlation may prove to be too simplistic. Due to factors such as drug-induced RES block and high liposome doses that make the clearance of different formulations closer, the influence of liposome compositions and physical properties (such as size) on the exudation rate may begin to become important. Recent evidence suggests that CL preparations may accumulate in tumors faster than SSL, but due to its shorter circulation lifetime, it tends to have a lower degree of enrichment [48].

3.4.6 Drug Release Behavior of Liposome

It must be emphasized that most of the work described so far is related to drugs that are believed to have membrane activity. They are amphiphilic in nature and can traverse the bilayer at a rate that depends on the physical properties of the membrane and any transmembrane ion or pH gradient. Other drugs, such as ARA-C, are more soluble in water, and after being slowly released from the carriers, they can be absorbed by specific transporters located in the plasma membrane of tumor cells, such as nucleoside transporters. The bioavailability of such compounds depends on how easily they can escape their liposome carriers. We define the bioavailability of a liposome carrier as the amount of free drug that can escape the restrictions of the carrier so as to be redistributed to adjacent tissues and tumors. In order to prevent premature leakage in the circulation and thus avoid non-specific toxicity, a good balance is required while still allowing the drug to be released when it reaches the tumor. For DOX-loaded sustained-release liposomes (PEG-DSPE/HSPC/cholesterol or DSPC/cholesterol), drugs are believed to leak very slowly, so it is similar to the slow infusion of the drug near cancer cells. Scanning confocal fluorescence microscope was used to observe the subcutaneous implants of prostate cancer xenografts. The DOX delivered by SSL-DOX was proved to be adjacent to tumor capillaries and venules in the early stage (1 hour) [50]. After 24 h, DOX was leaked from the liposomes and was found within the tumor in a pattern of diffusion from capillaries to small veins. Free DOX was found in the tumor after 1 h but almost undetectable within 24 h. This may be due to the elimination and metabolism of the drug and the fluorescence quenching after the drug is inserted into the nucleic acid. These results indicate that DOX is indeed bioavailable when it reaches tumors, where it slowly and continuously bombards nearby cancer cells with low levels of cytotoxic agents. So far, the most detailed studies have been conducted using anthracyclines for delivery studies. Although the anti-tumor cytotoxicity of drugs such as anthracyclines and ARA-C is less dependent on the peak level of the drug, the cytotoxicity of other drugs may be more dependent on the peak level of the drug, so the rate the drug released from the carrier is also greater. Therefore, selecting drugs with these characteristics or selectively increasing the release rate at the tumor site is very important for designing an effective carrier.

The mechanism leading to liposome breakdown in tumors and drug release is still unclear. Several possible mechanisms have been proposed, but they are all highly speculative, with little direct evidence provided, mainly due to technical difficulties in monitoring the *in vivo* release of drugs. Some properties of the tumor microenvironment believed to play a role in the instability of liposomes include the slightly acidic pH in the interstitial fluid surrounding the tumor, lipases released by dead tumor cells, and inflammatory cells that respond to tumor-releasing factors, enzymes, and oxidants [51]. In addition, the phagocytes in the tumor can metabolize liposomes and release free DOX, killing neighboring tumor cells through the bystander effect. In an *in vitro* study, the effect of local interstitial medium on the leakage of SSL-DOX is studied [52]. Although plasma leakage is relatively slow ($T_{1/2}$; 100 h), liposomes cultured in malignant pleural effusion leak DOX at a significantly increased rate. Another researcher believes that the ammonium sulfate used for remote loading of DOX can also catalyze the decomposition of liposomes, although the logical basis for its mechanism has not been provided [53]. With SSL, over time, liposomes can release a certain amount of PEG-DSPE, allowing more interaction of liposomes with neighboring cells and or plasma components. Finally, it is possible that DOX passively passes through the liposome membrane, and that drug release accelerates with the progressive destabilization of the sulfate gel with the loss of more and more drug. Finding a method to selectively destabilize liposome pharmaceutical preparations in the tumor area is a major challenge in the field of liposomes. If this is overcome, it may lead to a substantial increase in the bioavailability of the drug at the tumor site, thereby improving the efficacy.

3.5 Summary and Perspective

Liposome has been widely applied to chemistry, medicine, biology, and other fields, providing an ideal strategy for the targeted delivery of high-efficiency drugs. It may effectively overcome treatment barriers and offer a theoretical basis for the development of other novel therapies. Nonetheless, realizing the continuous development of liposome delivery system requires considering the optimization of multi-properties, including the stability of liposomes, loading capacity, *in vivo* circulation time, etc. This chapter provides methodological guidance for the targeted improvement of the relevant performances of liposomes by introducing the physical and chemical properties of liposome and their relations with different properties. Notably, the issues associated with scaling up, stability, and satisfying regulatory still need to be addressed properly to enable sophisticated lipid-based therapeutics in the future.

References

1. Dymek M, Sikora E, *Advances in Colloid and Interface Science* **2022**, *309*, 102757.
2. Akbarzadeh A, Rezaei-Sadabady R, Davaran S, Joo SW, Zarghami N, Hanifehpour Y, Samiei M, Kouhi M, Nejati-Koshki K, *Nanoscale Research Letters* **2013**, *8*, 102.
3. Zununi Vahed S, Salehi R, Davaran S, Sharifi S, *Materials Science and Engineering: C* **2017**, *71*, 1327–1341.
4. De Leo V, Maurelli AM, Giotta L, Catucci L, *Colloids and Surfaces B: Biointerfaces* **2022**, *218*, 112737.
5. Hermann CA, Hofmann C, Duerkop A, Baeumner AJ, *Analytical and Bioanalytical Chemistry* **2020**, *412*, 6295–6305.
6. Shi N-Q, Qi X-R, Preparation of drug liposomes by reverse-phase evaporation, in *Liposome-Based Drug Delivery Systems* (Lu W-L, Qi X-R, eds), Springer, Berlin Heidelberg: Berlin, Heidelberg, 2021, pp. 37–46.
7. Šturm L, Poklar Ulrih N, *International Journal of Molecular Sciences* **2021**, *22*, 6547.
8. Gouda A, Sakr OS, Nasr M, Sammour O, *Journal of Drug Delivery Science and Technology* **2021**, *61*, 102174.
9. Has C, Pan S, *Journal of Liposome Research* **2021**, *31*, 90–111
10. Lasic DD, *Biochemical Journal* **1988**, *256*, 1–11.
11. Lasic DD, *Journal of Liposome Research* **1995**, *5*, 431–441.
12. Lasic DD, *Journal of Theoretical Biology* **1987**, *124*, 35–41.
13. Lasic DD, *Biochimica et Biophysica Acta* **1982**, *692*, 501–502.
14. Israelachvili JN, Mitchell DJ, Ninham BW, *Journal of the Chemical Society, Faraday Transactions 2: Molecular and Chemical Physics* **1976**, *72*, 1525–1568.
15. Stottrup BL, Stevens DS, Keller SL, *Biophysical Journal* **2005**, *88*, 269–276.
16. Kulkarni JA, Cullis PR, Van Der Meel R, *Nucleic Acid Therapeutics* **2018**, *28*, 146–157.
17. Barenholz Y, *Journal of Controlled Release* **2012**, *160*, 117–134.

18. Cao X, Yang C, Zhu X, Zhao M, Yan Y, Huang Z, Cai J, Zhuang J, Li S, Li W, Shen B. *Chinese Chemical Letters* **2023**, 109199.
19. Huang N. Construction of Immunoliposomes Loaded with Melittin and Bufalin and Their Anti-hepatocellular Carcinoma Effects. *PhD Dissertation*, **2019**.
20. Woodle MC, Lasic DD, *Biochimica et Biophysica Acta* **1992**, *1113*, 171–199.
21. Senior JH, *Critical Reviews in Therapeutic Drug Carrier Systems* **1987**, *3*, 123–193.
22. Chonn A, Cullis PR, Devine DV, *Journal of Immunology (Baltimore, Md.: 1950)* **1991**, *146*, 4234–4241.
23. Sur S, Fries AC, Kinzler KW, Zhou SB, Vogelstein B, *Proceedings of the National Academy of Sciences of the United States of America* **2014**, *111*, 2283–2288.
24. Tagami T, Kubota M, Ozeki T, *Journal of Pharmaceutical Sciences* **2015**, *104*, 3824–3832.
25. She ZN, Zhang T, Wang XL, Li X, Song YZ, Cheng XB, Huang ZJ, Deng YH, *Biomaterials* **2014**, *35*, 5216–5225.
26. Mayer LD, Tai LCL, Bally MB, Mitilenes GN, Ginsberg RS, Cullis PR, *Biochimica et Biophysica Acta* **1990**, *1025*, 143–151.
27. Mayhew E, Rustum YM, Szoka F, Papahadjopoulos D, *Cancer Treatment Reports* **1979**, *63*, 1923–1928.
28. Allen TM, *Biochimica et Biophysica Acta* **1981**, *640*, 385–397.
29. Allen TM, Mehra T, Hansen C, Chin YC, *Cancer Research* **1992**, *52*, 2431–2439.
30. Mui BL, Cullis PR, Evans EA, Madden TD, *Biophys J* **1993**, *64*, 443–453.
31. Allen TM, Newman MS, Woodle MC, Mayhew E, Uster PS, *International Journal of Cancer* **1995**, *62*, 199–204.
32. Barenholz Y, Amselem S, Goren D, Cohen R, Gelvan D, Samuni A, Golden EB, Gabizon A, *Medicinal Research Reviews* **1993**, *13*, 449–491.
33. Frankel EN, *Progress in Lipid Research* **1980**, *19*, 1–22.
34. Ahl PL, Bhatia SK, Meers P, Roberts P, Stevens R, Dause R, Perkins WR, Janoff AS, *Biochimica et Biophysica Acta (BBA) - Biomembranes* **1997**, *1329*, 370–382.
35. Liu D, Mori A, Huang L, *Biochimica et Biophysica Acta* **1992**, *1104*, 95–101.
36. Gregoriadis G, Senior J, *FEBS Letters* **1980**, *119*, 43–46.

37. Allen TM, Hansen C, *Biochimica et Biophysica Acta (BBA) - Biomembranes* **1991**, *1068*, 133–141.
38. Vingerhoeds MH, Steerenberg PA, Hendriks JJ, Dekker LC, Van Hoesel QG, Crommelin DJ, Storm G, *British Journal of Cancer* **1996**, *74*, 1023–1029.
39. Gabizon A, Chemla M, Tzemach D, Horowitz AT, Goren D, *Journal of Drug Targeting* **1996**, *3*, 391–398.
40. Bally MB, Nayar R, Masin D, Hope MJ, Cullis PR, Mayer LD, *Biochimica et Biophysica Acta (BBA) - Biomembranes* **1990**, *1023*, 133–139.
41. Klibanov AL, Maruyama K, Beckerleg AM, Torchilin VP, Huang L, *Biochimica et Biophysica Acta (BBA) - Biomembranes* **1991**, *1062*, 142–148.
42. Allen TM, Hansen CB, de Menezes DEL, *Advanced Drug Delivery Reviews* **1995**, *16*, 267–284.
43. Allen TM, *Advanced Drug Delivery Reviews* **1994**, *13*, 285–309.
44. Blume G, Cevc G, *Biochimica et Biophysica Acta (BBA) - Biomembranes* **1993**, *1146*, 157–168.
45. Allen TM, Chonn A, *FEBS Letters* **1987**, *223*, 42–46.
46. Goren D, Horowitz AT, Zalipsky S, Woodle MC, Yarden Y, Gabizon A, *British Journal of Cancer* **1996**, *74*, 1749–1756.
47. Fujimori K, Covell DG, Fletcher JE, Weinstein JN, *Cancer Research* **1989**, *49*, 5656–5663.
48. Unezaki S, Maruyama K, Ishida O, Suginaka A, Hosoda J-I, Iwatsuru M, *International Journal of Pharmaceutics* **1995**, *126*, 41–48.
49. Hwang K. *Liposome Pharmacokinetics*. New York: Marcel Dekker, Inc; **1987**.
50. Vaage JAN, Donovan D, Working P, Uster P, Chapter 4.4 – Cellular distribution of DOXIL® within selected tissues, assessed by confocal laser scanning microscopy, in *Medical Applications of Liposomes* (Lasic DD, Papahadjopoulos D, eds), Elsevier Science B.V., Amsterdam, 1998, pp. 275–282.
51. Martin FJ, Chapter 8.2 – Clinical pharmacology and antitumor efficacy of DOXIL (pegylated liposomal doxorubicin), in *Medical Applications of Liposomes* (Lasic DD, Papahadjopoulos D, eds), Elsevier Science B.V., Amsterdam, 1998, pp. 635–688.
52. Gabizon AA, *Advanced Drug Delivery Reviews* **1995**, *16* 285–294.
53. Lasic D. *Liposomes: From Physics to Applications*. New York: Elsevier; **1998**.

Chapter 4

Nano Physical Pharmaceutics of Inorganic Nanoparticles

Xiaodong Zhu,[a] Beibei Liang,[b] Yong Huang,[a] Wei Yao,[a] and Wei Li[a]

[a]*Department of Nanomedicine, Naval Medical University, Shanghai 200433, China*
[b]*Shanghai Key Laboratory of Molecular Imaging, Shanghai University of Medicine and Health Sciences, Shanghai 201318, China*

liwei_dds@163.com

4.1 Classification of Inorganic Nanoparticles

Nanoparticles, also known as ultra-fine particles, refer to particles with a particle size of 1–100 nm. It is a system composed of a few atoms or molecules between the microscopic system and the macroscopic system. Inorganic nanomaterials refer to nanomaterials whose main body is inorganic substances [1]. They embrace more unique properties relative to the noumenon, including large specific surface area, more active sites, good biocompatibility, excellent electrochemical properties, and four major effects brought about by the nanometer size. The four major effects include:

Nano Physical Pharmaceutics
Edited by Wei Li

ISBN 978-981-4968-52-2 (Hardcover), 978-1-003-51393-3 (eBook)
www.jennystanford.com

1. **Small-size effect.** When the particle size of the nanomaterial is so small that it is equivalent to or smaller than the size of physical features such as the light wavelength, the De Broglie wavelength, and the coherence length or transmission depth of the superconducting state, the periodic boundary conditions of crystals will be destroyed, the atomic density near the surface layer of amorphous nanoparticles will decrease, causing the changes of sound, light, electricity, magnetism, and thermal properties, further leading to the emergence of new features. This phenomenon is called the small-size effect of materials.
2. **Quantum-size effect.** The quantum-size effect refers to the phenomenon that when the particle size in nanomaterials drops to a certain value and is equal to or less than the size of physical characteristics (such as light wavelength or other coherent wavelengths), the electronic energy level near the metal Fermi energy level changes from quasi-continuous to discrete and the energy gap is widened. When the energy level spacing is greater than thermal energy, magnetic energy, and photon energy, the magnetism, light, sound, heat, electricity, and superconductivity of nanomaterials are significantly different from those of macroscopic substances.
3. **Surface/interface effects.** When the size of nanoparticles shrinks to the nanometer size, the number of atoms on the surface of the material is much more than that of atoms on the surface of the non-nanoparticles of the same mass. The surface effect refers to the property change caused by the prominent increase in the ratio of the number of surface atoms to the total number of atoms of nanoparticles with the decrease of the particle size. Due to the decrease in particle size, the number of surface atoms increases sharply, resulting in extremely active surface atoms, which are easily integrated with other atoms to stabilize, thus having high chemical energy and causing surface atomic transport and configuration changes. Meanwhile, it also causes changes in the surface electron spin conformation and electron spectrum.

4. **Macroscopic quantum tunneling effect.** When the total energy of nanoparticles is less than the barrier height, the particles can still pass through the barrier, which is called the tunneling effect. Moreover, the tunneling effect of the macroscopic physical quantities in quantum coherent devices is called the macroscopic tunneling effect. For example, a magnet with ferromagnetism changes from ferromagnetism to paramagnetism or soft magnetism when the particle size reaches the nanometer level.

According to the nature of inorganic nanoparticles (INPs), it can be divided into the following types:

1. **Metal nanoparticles.** Metal nanoparticles refer to metal particles whose component phase is reduced to nanometer level in morphology, which can produce unique photoelectric characteristics through localized surface plasmon resonance (LSPR). Therefore, they are widely used in multiple fields. For example, they can be applied to the sampling of a scanning electron microscope to enhance the electron flow, which is helpful in obtaining high-quality scanning electron microscope images [2];
2. **Magnetic nanoparticles.** Magnetic nanoparticles are mainly formed by magnetic inorganic crystals of small particle size (5 nm to 20 nm) coating on the organic layer in the core–shell structure or embedded in the organic matrix in the form of a multi-core–shell structure. Generally, they have superparamagnetism, ferromagnetism, or ferromagnetic behavior. Among them, iron oxides (such as magnetite and martensite) are widely used in the biomedical field due to their low toxicity and high biocompatibility [3];
3. **Silicon dioxide nanoparticles.** Silicon dioxide nanoparticles are inorganic chemical materials, commonly known as silica, with sizes ranging from 1 to 100 nm. They are featured with multiple unique properties and have a wide range of applications. For example, they can enhance the aging resistance, strength, and chemical resistance of other materials because of their optical properties against ultraviolet rays;

4. **Ceramic nanoparticles.** Ceramic nanoparticles are inorganic non-metallic solid nanoparticles prepared by heating and continuous cooling. They generally exist in amorphous, polycrystalline, dense, porous, or hollow forms [4]. It is mainly composed of oxides, nitrides, or carbides. Due to its heat resistance and chemical inertness, such nanoparticles have attracted more and more attention in catalysis, photo-catalysis, dye photodegradation, imaging, etc. [5];
5. **Semiconductor nanoparticles.** Semiconductor nano-particles are nanomaterials made of silicon, gallium arsenide, and other semiconductor materials, which have many excellent properties. For example, they have both metallic and non-metallic properties, as well as a wide band gap, and their properties can change significantly with the band gap tuning, making them very important materials for photocatalytic, optical, and electronic devices [6]. For example, they can be used for photocatalytic hydrolysis reactions according to their band gaps and band gap positions [7];
6. **Quantum dots.** Quantum dots (QDs) are inorganic semi-conductor nanocrystals with particle sizes between 2 nm and 8 nm, which have the advantages of high brightness, photobleaching resistance, and multicolor fluorescence emission and can be used as novel fluorescent labels [8]. They generally consist of compounds of group II and group VI elements (such as CdSe and CdTe) or group III and group V elements (such as InP and InAs). The structure of QDs has a semiconductor as the core, coated by a layer of the shell to improve optical performance and a cover that allows increased solubility in aqueous buffers [9].

4.2 Preparation of Inorganic Nanoparticles

INPs can be prepared in two ways [10]. One is the "top-down" method, in which large pieces of materials are ground, or pestled into nanoparticles. Etching or photoetching may be the most typical examples in terms of this method. The most critical issue of the "top-down" method is the heterogeneity of synthetic nanoparticles. Particles produced in this way usually have a large size distribution and different shapes. In addition, they may contain a great number of impurities generated during the process of milling media, and considerable defects such as crystal damage may occur. Such materials are often applied for the preparation of ceramic nanoparticles. The above defects may disappear during sintering, and the size and shape distribution of particles are relatively inessential to their application. The other is called the "down-top" method. Generally, a single atom or molecule is compounded into a larger nano-structure to obtain particles with fewer defects, a more uniform chemical composition, and a narrower size and shape distribution. This chapter mainly introduces the "bottom-up" methods of nanomaterials:

1. **Salting out method.** The way that iron oxide particles are synthesized by mixing $Fe^{2+}/^{3+}$ iron solutions with alkali salt has been used since 1852. After that, this synthesis process is used for the preparation of iron oxide nanoparticles applicable to most commercial purposes [11–13]. As for iron oxide nanoparticles, the precipitation or co-precipitation of cations in aqueous solutions is a simple and cheap preparation method. Nanoparticles with a particle size of 5–15 nm can be synthesized by stoichiometry $Fe^{2+}/^{3+}$ = 0.5 [14]. Nanoparticles with a particle size above 20 nm can be prepared through the controllable oxidation of iron hydroxide suspensions [15]. The particle size distribution of nanoparticles is controlled by changing experimental parameters such as the pH value, stirring speed, temperature and ionic strength (Fig. 4.1), and a narrow distribution is achieved. The crystallinity of nanoparticles can also be increased through the dissolution in recrystallization.
2. **Hydrothermal synthesis preparation method.** Under hydrothermal conditions, various nanomaterials can be prepared. In the hydrothermal method, water is used as the solvent, and pressure and temperature are taken into account, which increases the solubility of precursors for preparing nanoparticle precursors and reduces the reaction time. Hydrothermal synthesis can be conducted above or below the supercritical point of water. An appropriate and

sealed reaction vessel is required to control the temperature and pressure during the reaction. In most cases, the synthesis route can produce monodisperse particles with good crystallinity.

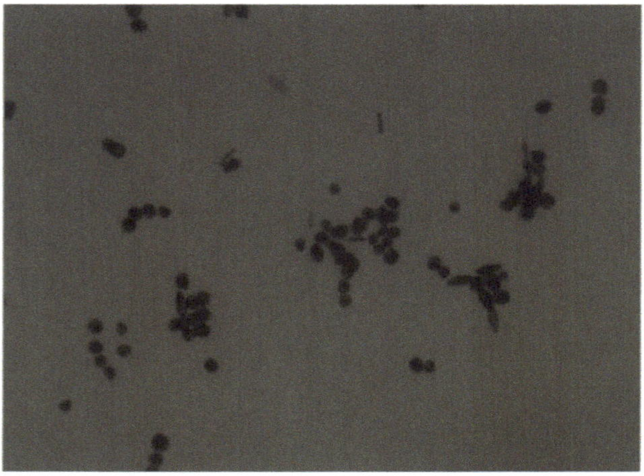

Figure 4.1 Preparation of iron oxide nanoparticles.

3. **Microemulsion preparation method.** Microemulsion is a kind of lotion formed by two immiscible liquids through thermodynamic stable isotropic dispersion, in which microregions of either or both liquids are steadied by an interfacial film of surfactant molecules [16]. In the water-in-oil (W/O) microemulsion, the aqueous phase is dispersed into microdroplets and is covered by a single layer of surfactant molecules in the continuous hydrophobic phase. When two W/O microemulsions that contain necessary reactants are mixed, the microdroplets will constantly crash into each other, agglomerate and break into pieces until precipitation is formed in micelles. These droplets are the nanoreactors for particle formation. Well-crystallized nanoparticles can be achieved by this method, but their versatility in terms of size, shape, properties, and other aspects is relatively restricted. Besides, in comparison with other methods, the yield coefficient of this process is lower, and although its production scale can be enlarged,

it will cost a huge number of solvents to synthesize considerable materials, making this process expensive and adverse to environmental protection.

4. **Polyol preparation method.** Magnetic nanoparticles can be prepared by high-temperature reduction of metal salts dissolved in polyols as solvents, surfactants and reducing agents (through weak reduction of hydroxyl). The average size of magnetic nanoparticles can be controlled by precipitation kinetics or adding seeds to improve heterogeneous nucleation. This process is initially used to produce precious metal particles [17], and its application is then extended to the production of magnetic nanoparticles.

5. **Preparation method via organic medium decomposition.** Monodisperse magnetic nanoparticles with fair crystallinity can be prepared through the thermal (high temperature) disintegration of metallic compounds in organic solvents featuring a high boiling point and stable surfactants. The average particle size can be changed through continuously variable factors such as reaction temperature, concentration and properties of precursor, and the introduction of seeds. In addition, the metal precursor can be added to the preheated mixture of other reagents by hot injection method. The main advantage of this method is to obtain excellent particle crystallinity and controllable particle size distribution. However, nanoparticles are generally hydrophobic, and therefore, Some surface modification methods are needed to improve their dispersion and biocompatibility in aqueous solution, to improve the targeting of nanoparticles and make them better used in the biomedical field [18–20].

6. **Aerosol pyrolysis preparation method.** As an effective method to synthesize magnetic nanoparticles, the aerosol pyrolysis method is similar to spray or laser pyrolysis, and featured advantages such as low costs. In the spray pyrolysis method, the salt solution is sprayed into a series of high-temperature reactors, causing solvent evaporation and decomposition of the salt solution. The solute is precipitated due to supersaturation, thus obtaining nanoparticles.

Particles deposited in this way are subsequently dried and transformed. Small particles (particle size between 2 nm and 10 nm) synthesized through spray pyrolysis coalesce into microporous particles (about 300 nm) and then can be sintered into dense particles.

4.3 Effects on Inorganic Nanoparticle Formation

LaMer and Dinegar published the classical nucleation theory [21] and provided the basic model for precipitation in solutions with the idea that nucleation is the limiting step for this procedure.

In Stage I, the system becomes unstable with the increase in salt concentration. When the nuclear concentration is reached, solute molecules combine to form aggregates. The aggregates are the result of the ion or molecule collision during their self-nucleation processes. The stability of critical nuclei relies on the height of the free energy potential barrier: If it is higher than the heat energy potential barrier, the possibility of critical nucleation is negligible, and particles will not be generated. However, if the free energy potential barrier is low enough, the rate of nuclides approaches infinity. The total free energy is the sum of the nuclear volume free energy and surface tension free energy [22].

$$\Delta G = \Delta G_{\text{surface}} + \Delta G_{\text{bulk}} \tag{4.1}$$

$$\Delta G_{\text{surface}} = \gamma \beta N^{2/3} \tag{4.2}$$

$$\Delta G_{\text{bulk}} = -N k_{\text{B}} T \ln S \tag{4.3}$$

In the above equation, γ is the surface tensile force between the nucleus and external solution, β is the geometry constant (related to the shape of the nucleus), N is the number of molecules, k_{B} is the Boltzmann constant, T is the temperature, and S is the supersaturation ratio. If the critical concentration is not reached, a sponge phase with a featured wavelength will be formed instead of particles [40], which is called spinodal decomposition [23].

In Stage II, the supersaturation of aggregates decreases, and the nucleation rate gradually decreases to zero. When the concentration is decreased to C_{MAX} (critical minimum) from C_{MIN} (critical maximum), the process of nucleation will continue, and particle growth will replace particle nucleation. Therefore, by regulating the initial depression of the precursor, the nucleation time becomes shorter and the monodisperse colloid grows uniformly with the nucleus. When a certain number of molecules gather in the center, smaller particles are formed.

With the decrease of concentration and the remaining supersaturation status of the system (Stage III), the stable nuclear will form disperse particles through the diffusion of substances in the nucleus solution. Only when the aggregated nucleus exceeds a certain volume, the free energy barrier dominates, and the particle size increases with its decrease. The final size and growth rate of particles depends on three variable quantity: the number of nuclei in the growth, the total amount of precursors, and their diffusion coefficient in the media. These variables are in inverse proportion to the viscosity and temperature.

4.3.1 Nucleation Mechanism of Inorganic Nanoparticles

As mentioned above, the nucleation rate is critical for controlling the uniformity of nanoparticles However, because the nucleation rate is very fast in most cases, it is usually impossible to accurately and effectively detect the nucleation time or the formation of aggregates. The nucleation process can be divided into three types: one-time homogeneous nucleation, one-time heterogeneous nucleation, and secondary nucleation.

One-time homogeneous nucleation is the ideal situation described by LaMer and Dinegar, that is, without any solid interface, the molecular solute is completely dissolved in the liquid [24]. Solid interfaces (seeds) in the system induce the nucleation of nascent heterogeneous substances. The surface energy of these external interfaces is lower than those of new particles, so the critical concentration during the nucleation process is relatively low. In this situation, the interface tension of solid/liquid interfaces is replaced with those of solids or liquids separated. Meanwhile, one-time homogeneous and one-time heterogeneous nucleation processes cannot occur (the C_{MIN} of heterogeneous nucleation is greater than that of one-time homogeneous nucleation), so the assembly nucleation rate is the sum of the two nucleation rates.

Secondary nucleation is caused by solute particles in the solution. This nucleation mechanism only occurs when the supersaturation level is higher than the critical level of solute particles in the solution. In most cases, growing particles may touch the wall of the container, the blender, or other particles, and caused residual solute particles, which may lead to the formation of a polydisperse system. Hence, secondary nucleation shall be avoided if a homogeneous hinged nanosystem is intended to be formed [25].

4.3.2 Growth Mechanism of Inorganic Nanoparticles

Through the diffusion of solute molecules to the interface of the crystal nucleus, different growth stages of the crystal nucleus can be separated and identified. The solute molecules are transported from the bulk solution to the particle surface, and these molecules diffuse and attach to the particle surface, and aggregate into crystals. The final process is the release of crystallization heat and the delivery of the particles. In this process, the coordination shells of solvent molecules or counter-ions are far away from the surface diffusion of particles.

In this case, the growth of the crystal nucleus is majorly restricted by the diffusion of solute molecules towards the surface of nanoparticles. Thus, when the supersaturation degree is reduced to a certain extent, nanoparticles will stop growing, which is called restricted diffusion or Lifshitz–Slyosov–Wagner Theory. On the other hand, another theory holds that in the limited reaction growth, the growth of particles is restricted by the reaction between the diffusion solute and the surface of nanoparticles. Both theories are widely accepted [26].

Fortunately, the size distribution during the growth of nanoparticles is self-sharpening in most cases. After the formation of initial particles, the distribution of particle size may vary with time because of cluster-cluster aggregation and Ostwald Ripening (Ostwald Ripening: The smaller crystal particles generated in the solution will gradually dissolve into the surrounding solute because of the large curvature and high energy, and then will be precipitate on the surface of the larger crystal particles, which makes the larger crystal particles further increase. This process is widely used in the preparation of nanoparticles.). In the process of cluster aggregation, a certain number of micro-/nanoparticles gather to form larger and uniform particles. This agglomeration can be random, for example, to produce amorphous or polycrystalline spherical particles, or to form directional, for example, to produce elongated, ellipsoidal, and other morphologies.

During the process of Ostwald Ripening, nanoparticles with a small particle size dissolve and crystallize onto the surface of larger particles. Nanoparticles with a small particle size usually have a higher surface-to-volume ratio. Surface atoms are less steady than volume atoms because fewer adjacent atoms exist, indicating that nanoparticles with a small particle size are more unstable than those with a large particle size. To reduce the total energy, molecules on the surface of small nanoparticles are separated and diffused in the solution. In the solution, these free nanoparticles with a small particle size usually coagulate on the surface of those with a large particle size. After a period of time, all nanoparticles with a small particle size gather and form those with a large particle size, and the particle mass exists in a monodisperse form.

4.3.3 Morphology Control Strategy

During the growth of nanoparticles, the precipitation in supersaturated solution will inevitably produce polydisperse colloids. For our study, we summarize a few general strategies (as shown in Fig. 4.2) and adjust the size and form of inorganic nanoparticles by controlling some parameters related to aggregated nucleation and growth steps.

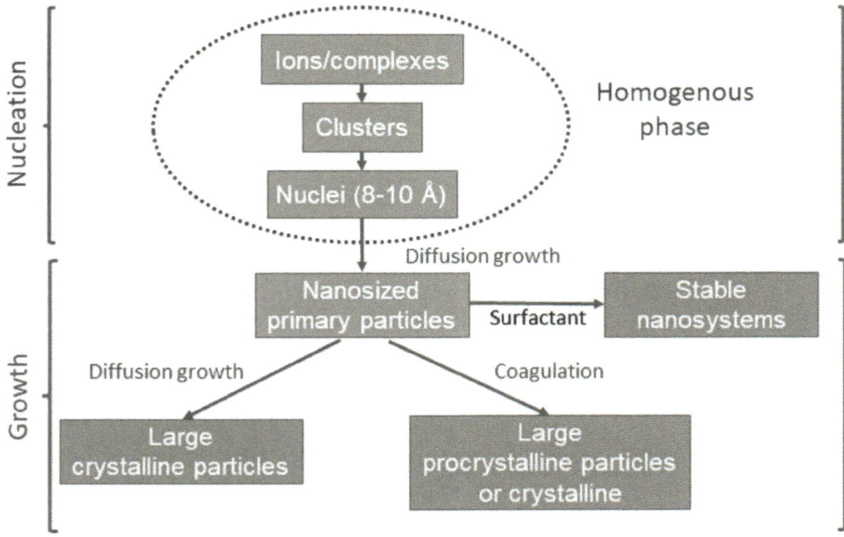

Figure 4.2 Nanoparticles with an evenly distributed particle size generated in different growth phases.

4.3.3.1 Control of nucleation rate

To obtain a monodisperse particle system, it is necessary to ensure that numerous single-crystal nanoparticles are produced in the shortest possible time. These reactions are often controlled by parameters that affect thermodynamics and kinetics and can control the particle size, such as the temperature, the reaction time, and the addition of surfactants. The temperature affects the diffusion coefficient (related to the solubility) of molecules from the solution to aggregated nucleation, and the general free energy, so low temperature will produce fewer and smaller single-crystal nanoparticles. When two different precursors are mixed, smaller particle sizes can be gained through a shorter reaction time. In certain cases, this is achieved by quickly injecting one of the precursors into the diluted solution of the other precursor. Long reaction time is beneficial to the growth of particles as well.

4.3.3.2 Control of growth phases

There are a variety of methods to prepare monodisperse colloidal nanoparticles, all of which are based on the fact that the growth conditions are under control. The capping agent can restrict the growth of nanoparticles by being attached to their surfaces. For instance, in alkaline aqueous media where polyvinyl alcohol (PVA) exists, the size of iron oxide nanoparticles is limited to a few nanometers through the co-precipitation method of iron and molysite [27, 28]. If the capping agent is large enough (e.g., surfactants are added), steric hindrance is provided to prevent the aggregation of mononuclear nanoparticles. Selective binders and surfactants can adsorb and stabilize specific crystal planes, thus inhibiting the growth in the crystal plane with stronger interaction, which can grow more specifically in other planes, forming structural morphology such as wire, rod, or needle [29].

During the process of heterogeneous nucleation where seeds are used, aggregated nucleation and growth steps are separated. Seeds are introduced to the solution of sub-critical concentration to prevent new nucleation. Therefore, it has extensive growth conditions (i.e., different temperatures or solvents) to control the final size and shape of nanoparticles. The quality of initial seeds is also important as seeds with crystal defects tend to produce polycrystal and polydisperse nanoparticles. In some cases, features of certain materials can be used to prepare anisotropic inorganic nanoparticles. For instance, magnetic nanomaterials can be synthesized under the action of the magnetic field, and some metal nanowires can be prepared under the action of the electric field [30].

The stability of solid nanoparticles in aqueous suspension includes two different factors: the durability under the dispersion state (aggregation state), namely the dynamic stability; the durability under the division state (particle size), namely the thermodynamic stability. The dispersion state of nanoparticles is caused by the energy barrier, which is opposite to the Brownian movement direction of particles [31]. This stability is called dynamic stability resulting from the energy barrier caused by the balance between the attraction force (van der Waals) and the repulsion force (static electricity) among surfaces [1]. It determines the chemical reaction rate as activation energy. The classic Dejarguin–Landau–Verwei–Overbeek (DLVO) theory and some recent ones give good explanations for this stability and provide considerably accurate forecasts of the dispersion, aggregation, and flocculation areas of most nanoparticles, colloids, and final microbe systems [2, 8, 32].

On the other hand, thermodynamics is caused by the minimization of surface energy. Nanoparticles in the suspension may have spontaneous changes in size, shape, and/or crystal structure. The cause of such changes is that nanoparticles are generated under the control of kinetics in the initial steps of nucleation and growth involved in the precipitation reaction. Further changes may occur during the maturation to attain the thermodynamic equilibrium of minimal surface energy. Ostwald ripening step led to the secondary growth of nanoparticles so that with time, the average particle size increased overall. This formation process is mainly due to the difference in solubility between nanoparticles of different sizes, which makes the nanoparticles with small solubility recrystallize on larger crystals. The decrease in the total system surface area causes a decrease in free enthalpy. However, the crystal growth is very slow, that is, due to the limitation of surfactant growth, it is possible to obtain a dispersion composed of very fine particles. From the perspective of thermodynamics, the system is not always in a steady state, so generally, the final particle size cannot be reasonably controlled or foreseen.

4.3.4 Dynamic Stability

Dynamic properties mainly refer to the irregular motion of particles in the solution and the diffusion incurred, osmotic pressure and balanced distribution of concentration with height under the gravity field. According to the viewpoint of molecular motion, it is not difficult to understand the Brownian motion of colloids. There are some similarities between colloids and dilute solutions in form, so the kinetic properties of sol can be discussed with the approaches to similar issues in dilute solutions.

4.3.4.1 Brownian motion

After Brown discovered the irregular polygonal movement of pollen in the suspension in 1872 and the observation of the zigzag movement of colloidal particles in 1903, Zsigmondy concluded that the smaller the particle, the stronger the Brownian motion, and the degree of motion does not change with time, but increases with increasing temperature. Afterward, Einstein and Smoluchowski proposed that the Brownian motion is similar to the molecular motion, and proposed that this phenomenon is the result of the impact of constantly hot moving liquid molecules on particles. Then Einstein obtained the Brownian motion equation with the basic concepts and equations of molecular motion theory, assuming that the colloidal particles are spherical:

$$\overline{x} = \sqrt{\frac{RT \cdot t}{L \cdot 3\pi_{\eta} r}} \tag{4.4}$$

In the equation, $\overline{x}$ is the average displacement of the particle moving along the x-axis within the observation time t, R is the radius of the particle, η is the viscosity of the medium, and L is the Avogadro constant. This equation is also called the Einstein-Brown motion equation.

4.3.4.2 Sedimentation and sedimentation equilibrium

In a general dispersion system, the particles in the suspension gradually settle due to gravity, while the case is different with a highly dispersed system. On the one hand, the particles fall by gravity, and on the other hand, Brownian motion tends to promote uniform concentration. When the two forces are equal, the distribution of the particles reaches equilibrium and a certain concentration gradient is formed. This state is called sedimentation equilibrium.

After sedimentation equilibrium has been reached, the distribution of sol concentration with height can be expressed by the law of height distribution. Supposing a container with a cross-sectional area of A is filled with a certain sol, the particle radius is r (suppose the particle is spherical), and the density of particle and medium is respectively, N_1 and N_2 are the numbers of particles per unit volume of the sol at x_1 and x_2, II is the osmotic pressure, and g is the acceleration of gravity, in a layer of sol with a height of dx (suppose the number of particles per unit volume is N), the gravity that causes the particles to fall is:

$$N_A dx \times \frac{4}{3}\pi r^3(\rho_1 - \rho_2)g \tag{4.5}$$

It has been pointed out before that the diffusion force on the particles in this layer is $-Ad_{II}$, and the negative sign indicates that the diffusion force is opposite to the direction of gravity. If the expression II = cRT (c is the concentration of solute) of dilute solution osmotic pressure is quoted, then:

$$-Ad_{II} = -ARTdc = -ART\frac{dN}{L} \tag{4.6}$$

Integrating the above equation, we get:

$$RT\ln\frac{N_2}{N_1} = -\frac{4}{3}\pi r^3(\rho_1 - \rho_2)gL(x_2 - x_1) \tag{4.7}$$

This equation is the equation for the height distribution of the particle, which is identical to the equation for the height distribution of the gas, and it also indicates that the thermal motion of the gas molecules is essentially the same as the Brownian motion of the colloidal particles.

It can be learned from Equation 4.7 that the greater the mass of the particle, the greater the decrease in its equilibrium concentration with height. It should be pointed out that this equation represents the situation after the distribution has reached equilibrium. Systems with larger particles usually have quicker sedimentation and reach equilibrium faster. While particles in the highly dispersed system settle slowly. Therefore, it can explain why many sols can remain for more than a few years without sedimentation in a gravity field. The colloidal particles of small particles can automatically diffuse and make the whole system evenly distributed. This property is called dynamic stability.

4.3.4.3 Interparticle interactions

There is a tendency for particles to coalesce with each other to decease their surface energy, that is, they have instability that makes them easy to coalesce. Therefore, stabilizers are a necessity during the preparation of sols. On the other hand, the small size of particles of sols causes Brownian motion to be violent, which makes it difficult for them to settle in a gravity field, i.e. they have dynamic stability. A stable sol must have both the stability that makes coagulation difficult and dynamic stability. However, the stability that makes coagulation difficult is more important, because while Brownian motion entitles a sol to dynamic stability, it also facilitates continuous collisions among particles. Once the particles lose their stability against coagulation, they will collide with each other and cause coalescence. As a result, the particles become larger and the speed of the Brownian motion decreases, eventually inducing a dynamically unstable system. The process of small particles aggregating into large ones is called aggregation, and the large particle aggregated by colloidal particles is called an aggregate. If the aggregation causes particles to precipitate out of the solution, it is called coagulation. To accelerate the coagulation, other substances can be added as coagulants, such as electrolytes.

When discussing the stability of sols, we must consider the total effect of both the energy that facilitates the mutual attraction among particles that coagulate with each other (V_a) and the energy of mutual repulsion that hinders the coagulation (V_r). The attraction among sol particles is essentially the same as the van der Waals forces among molecules, but it is mutual attraction among particles composed of many molecules, which is the sum of attractions among molecules. It can be proved that this force is not inversely proportional to the sixth power of the intermolecular distance, but inversely proportional to the third power of the distance, so this is a long-range force. The repulsive force among sol particles originates from the structure of the electric double layer on the surface of the colloidal particles. When the distance among particles is large and their electrical double layers are not overlapped, the repulsive force does not work. When the particles are so close that the electrical double layer partially overlaps, the concentration of ions in the overlapping part is denser than that in a normal distribution, and the osmotic pressure of these excess ions will hinder the approach of the particles, which will cause repulsion. The relationship between the distance (d) among the particles and the total action energy (V_a + V_r) is shown in Fig. 4.3. When the distance is large, the electrical double layers are not overlapped, and the attractive force works, so the total potential energy is negative. When the particles are close to a certain distance so that the electrical double layers overlap, the repulsive force plays a major role, and the potential energy increases significantly, but at the same time, the attraction force among the particles also increases with the shortening of the distance. When the distance is shortened to a certain extent, the attraction is dominant and the potential energy decreases. It can be seen from the figure that the particles must overcome the potential barriers to gather together. This is the reason why the particles in a stable sol do not coagulate with each other. In this case, despite the collision among particles aroused by the Brownian motion, they will be repulsive and separated when the particles are so close that the electrical double layers overlap, which will not cause accumulation. However, if the attraction effect is sufficient to offset the repulsion effect due to some reasons, the dispersion of the sol system will decrease first, and finally, all the dispersed phases will separate as precipitates. This process is called coagulation.

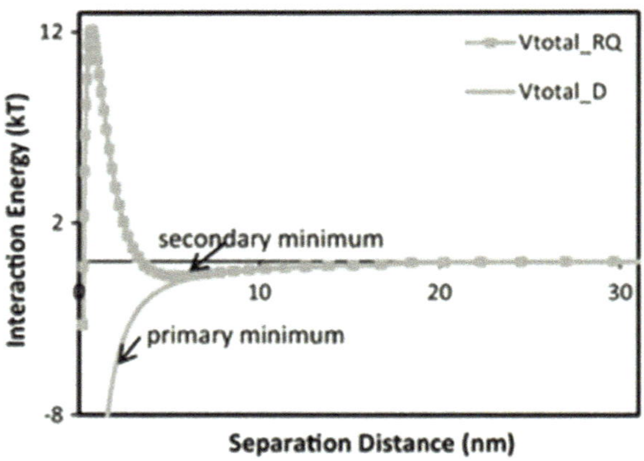

Figure 4.3 The relation curve of the interactive energy and distance between particles in an electrolyte solution as described by DLVO theory. Cited from reference [34] in Chapter 1.

4.3.5 Thermodynamic Stability

At the liquid–solid interface, there are usually two layers of ions with opposite electrical charges of the same amount on the solid surface and in the liquid nearby respectively, thus forming an electrical double layer. The charged ions on the solid surface are called localized ions. In the liquid attached to the solid surface, the ions with the opposite charge to the localized ions are called counter-ions. The cause of localized ions on the solid surface can be summarized as the disequilibrium of adsorption, ionization, isomorphous replacement, and dissolved load. Electrophoresis, electroosmosis, streaming potential, and sedimentation potential are all electrokinetic phenomena.

4.3.5.1 Electrical double layer theory and zeta potential

When a solid is in contact with a liquid, the solid may selectively absorb certain ions from the solution, and the ions may enter the solution due to the ionization of the solid molecules. This makes the solid and liquid phases carry charges of different signs, forming a structure of an electrical double layer on the interface.

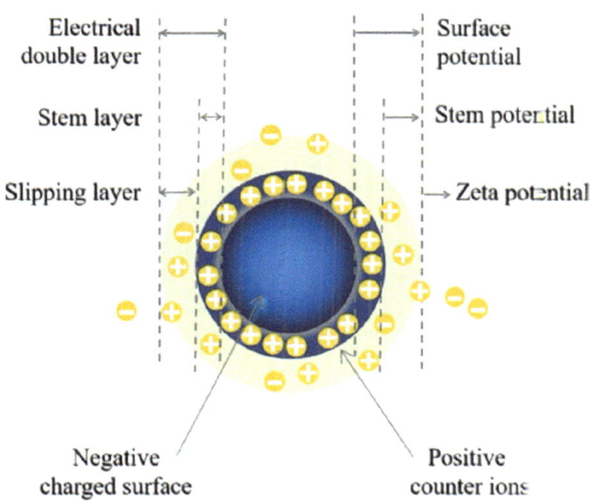

Figure 4.4 Schematic of the electrical double layer formed in aqueous solution.

According to the electrical double layer theory, the electrical double layer can be divided into a compact layer and a diffusion layer, and the compact layer is subdivided into an inner compact layer and an outer compact layer (as shown in Fig. 4.4) [33]. The counter-ions adsorbed by the solid surface in the inner compact layer are non-solvated (at least there is no solvent molecule on the side in contact with the solid surface) because they have tightly adhered to the solid surface. The surface composed of counter-ions on this side is called the inner Helmholtz plane (IHP). The counter-ions outside the IHP are solvated, and the outer Helmholtz plane (OHP) is formed with these as the center. The counter-ions distributed in the outer stern layer are uneven and intermittent, causing the charge discontinuity effect.

4.3.5.2 Electrolyte

The ζ potential is different from the thermodynamic potential φ_0. The value of φ_0 mainly depends on the ion concentration in the solution that is in equilibrium with the solid. The electric potential changes with the concentration of ions in the solvation layer, and a small amount of added electrolyte will have a significant effect on the value of the ζ potential. As the electrolyte concentration increases, the value of the ζ potential decreases, even to an extent that can change its sign [34].

Adding electrolyte to colloid can make colloid agglomerate. The concentration of electrolyte required to make a certain amount of colloidal solution start to agglomerate within a certain time is called the coagulation value. The so-called coagulation value, also known as the critical coagulation concentration, is the minimum concentration of electrolyte required to completely coagulate a certain amount of sol within a certain period, and the coagulation rate is the inverse of the coagulation value. According to a series of experiment results, the following laws can be summarized:

1. The coagulation capacity is mainly determined by the valence of the ions with the opposite charges to that of the colloidal particles. For a given sol, oppositely charged ions are electrolytes with valences of one, two, and three, and the ratio of the coagulation value is about 100:1.6:0.14, which means that the coagulation value is inversely proportional to the sixth power of the valence of oppositely charged ions. This conclusion is called the Schulze–Hardy rule.

2. The coagulation capacities of ions with the same valence are also different. For example, the coagulation capacity of nitrates generated by the monovalent cations of different alkali metals for negatively charged colloidal particles can be arranged in the following order:

   $H^+ > C_S{}^+ > Rb^+ > NH_4{}^+ > K^+ > Na^+ > Li^+$

   Potassium salts formed by different monovalent anions have the following sequence for the coagulation capacities for positively charged Fe_2O_3 sol:

   $F^- > Cl^- > Br^- > NO_3{}^- > I^-$

   This sequence of the coagulation capacities of ions with the same valence is called the lyotropic series. It roughly matches the order of the radii of hydrated ions from small to large because the smaller the radius of the hydrated ion, the easier it may be to get closer to the colloidal particles.
3. The ions of organic compounds have very strong sedimentation capacity, which may be due to their strong aggregation and sedimentation capacity and strong adsorption capacity.
4. The coagulation of the electrolyte is the sum of the action of positive and negative ions. Sometimes ions with the same charges of the colloidal particles (also called ions of the same sign) also have a significant effect. Generally, the higher the valence of the ions of the same electric property, the lower the coagulation capacity of the electrolyte. This may be related to the adsorption of these ions of the same electric property. Therefore, only when the adsorption of the ions with the same charges of the sol is extremely weak can the coagulation of the sol be approximately considered as the result of the independent action of the oppositely charged ions.
5. Irregular coagulation refers to the addition of a small amount of electrolyte in the sol to make the sol agglomerate. When the concentration of the electrolyte is slightly higher, the

precipitation is re-dispersed to form a sol, and the charge carried by the colloidal particles changes the sign. The phenomenon that the newly formed sol can precipitate again with another rise in the electrolyte concentration is called irregular coagulation. Irregular coagulation is the result of the strong adsorption of high-valence ions of the opposite signs by colloidal particles. A small amount of electrolyte can cause the colloid to coagulate, but the absorption of excessive high-valence ions with opposite charges causes the sol particles to have the charges of the ions with opposite electric properties again. Hence, the sol is stabilized again with charges opposite to those of the original colloidal particles. After adding the electrolyte again, the sol coagulates due to the action of the electrolyte ions (such as changes in the ionic strength and the thickness of the diffuse layer). At this time, the concentration of the electrolyte is very high, so even the addition of the electrolyte cannot re-disperse the precipitate.

It can be concluded from the above that the impact of the electrolyte on the coagulation of the sol is very complicated. The common point is that the electrolyte will cause coagulation as long as its concentration reaches a certain value, regardless of its type. Therefore, only when the adsorption of the ions with the same charges of the sol is extremely weak can the coagulation of the sol be approximately considered to be the result of the independent action of the oppositely charged ions.

4.3.5.3 DLVO theory

DLVO theory was published around 1940 by Boris Derjaguin, Lev Landau, the former Soviet scholars, and Evert Verwey, Theo Overbeek, the later Dutch scholars, respectively, on the theoretical analysis of the stability of colloidal particles in dispersion systems. According to DLVO theory, the apparent interaction force of colloidal particles in the dispersion system is determined by the van der Waals force of mutual attraction and the electric double layer force of mutually exclusive colloidal particles.

A simplified expression of DLVO theory: since there are many factors that affect attraction and repulsion, it is complicated to derive the expressions of V_a and V_r, and only the simplified results are quoted here. If two spherical particles are of equal volume, when the distance H between the two spherical surfaces is smaller than the particle radius r, the mutual gravitational potential energy V_a between the two particles can be approximated:

$$V_a = -\frac{Ar}{12H} \tag{4.8}$$

In the equation, A is called the Hamaker constant, which is related to the properties of particles (such as the number of atoms per unit volume, and polarizability), and is the characteristic constant of substances ranging between 10^{-19} and 10^{-20}

The magnitude of the repulsive potential energy v between particles with the same charge depends on the number of particle charges and the distance between them. If the solid surface potential distribution is different, the expression of the repulsion potential energy is also different. The repulsive potential energy between the parallel plates is calculated based on the following equation:

$$V_r = k_1 C^{\frac{1}{2}} \exp\left(-k_2 C^{\frac{1}{2}}\right) \tag{4.9}$$

In the equation, C is the electrolyte concentration, and k_1 and k_2 have a fixed value under certain conditions. It can be seen from Equations 1.8 that V_r decreases exponentially with the increase of electrolyte concentration C. Therefore, the electrolyte concentration has a great influence on the stability of sol. For two spherical particles with a distance of H, assuming that the surface potential is relatively low, the repulsive potential energy can be approximately expressed as:

$$v_r \approx k_\varepsilon r \varphi_0^2 \exp(-kH_0) \tag{4.10}$$

In the equation, k is a constant, ε is the dielectric constant of the medium, φ_0 is the potential on the surface of the colloidal particles and the reciprocal of the radius of the ion atmosphere in the Debye–Huckel equation, and H_0 is the minimum distance between the surfaces of the two spheres, r is the particle radius. It can be seen from Equation 1.7 that the repulsive potential energy increases with the increase of the surface potential φ_0 and the particle radius r and decreases exponentially with the increase of the distance among particles.

The total potential energy between particles is the sum of the attracting potential energy and the repulsive potential energy, i.e. $v = v_a + v_r$. In the potential energy curve, the height of the barrier is an indicator of the sol stability. When the height is zero, the sol will become unstable. Due to the addition of the electrolyte, the potential barrier is reduced to 0:

$$\frac{dv}{dH} = 0,\ v = 0 \tag{4.11}$$

The concentration C of the added electrolyte at this time is the concentration when $v = 0$, which is usually called the coagulation value. After several steps of simplification, a simplified expression of DLVO theory with water as the medium is obtained as follows:

$$C = k \cdot \frac{\gamma^4}{A^2 z^6} \tag{4.12}$$

In the equation, C is the concentration of the electrolyte, k is the constant, A is the Hamaker constant, and γ is a physical quantity related to the surface potential. For a certain sol, A and γ are fixed values, so the coagulation value is inversely proportional to the sixth power of the valence z of the inverse ion, which theoretically reveals the Schulze–Hardy rule, and $C \propto 1/z^6$ can also be obtained from a large number of experimental results, which further proves the correctness of DLVO theory.

4.3.5.4 Stability in aqueous system

A remarkable feature of nanoparticles and molecules similar to them is the persistence of the insoluble part of the nano body, which shows the properties of a rigid lattice structure independent of the dispersion solvent, limiting the freedom of conformation and heat energy. In other words, although nanoparticles can form quite stable aqueous dispersions in water, they are actually insoluble in water. To further understand the hydrochemical properties of nanoparticles, we should also consider the stability of particles in water systems and their progressive solubility. According to the nature of the surrounding environment, the chemically unstable nanoparticle surface can be oxidized, reduced, or dissolved. The solubility of nanoparticles is highly dependent on the solvent, which is the same as other materials. Therefore, the concept applied to colloids is also applicable to nanoparticles. However, based on new considerations of size-dependent properties, the dissolution kinetics and solubility of nanoparticles are expected to be higher. The dissolution rate (k) is in proportion to the surface area (A), as shown in a format 1.4 [11, 24]:

$$k = \frac{AD(C_S - C)}{L} \quad (4.13)$$

In the equation, D is the diffusion coefficient of solute molecules, L is the thickness of the diffusion layer, and C and C_S are the solutions and the solute concentration in the diffusion layer respectively. In addition, the greater the defect ratio of a lattice structure, roughness, and surface curvature, the higher the solubility of nanoparticles in normal solvents. When assessing nanoparticles, another concern is the chemical solutes released when the nanomaterials of the matrix dissolve in the preparation process. These dissolution outcomes may have better fluidity and stability, which have important implications for the preparation environment.

In an aqueous environment, the affinity between the surface of inorganic nanoparticles and water is an important factor that determines their stability. Hydrophilicity depends on the Lewis acids and bases (AB) groups on the surface, which may form hydrogen bonds with water molecules [35]. Surfaces containing these groups are called hydrophilic surfaces, while those lacking these groups are called hydrophobic surfaces. The general hydrophilic and hydrophobic characteristics of solid surfaces can be quantified by various surface-characterizing technologies through experiments, e.g., contact angle analysis or vapor absorption measurement [14].

The stability of nanoparticle dispersions is the key mechanism to determine the properties of these materials in the environment. Because the unique properties of the prepared nanoparticles are mainly related to their size, it is generally believed that the behavior of relatively large nanoaggregates is different from that of well-dispersed particles. Dispersion stability is affected by a variety of parameters, including the intrinsic properties of NPs (especially their surface properties), solution chemistry (such as the ionic strength, ionic composition, pH, and NP concentration), and the active or passive adsorption of macromolecules onto the NP surface [10, 36, 37].

The stability of NP dispersion relies on the frequency of collisions among NPs and the adhesion efficiency of the collisions. In general, these parameters can be used to predict the tendency of NP aggregation. The frequency of collisions among NPs is a result of the combined effect of Brownian motion and local mechanical agitation.

The efficiency of adhesion between two particles is determined by the sum of the interfacial interactions among colliding particles. When the repulsive interaction is dominant, the adhesion efficiency is negligible and the particle dispersion is stable. On the contrary, attractive interactions lead to a comparatively high adhesion efficiency and induce progressive aggregation, i.e., reaching an unstable state.

Many metal oxides will be hydrolyzed in water, forming a hydroxyl layer (M-OH) on the surface. Polar hydroxyl (–OH) groups can help the surface attract and physically adsorb single or multiple additional layers of polar water molecules, thereby imparting hydrophilicity to the oxide. In addition, the amphoteric (M-OH) sites may be charged by reacting with H^+ or OH^- ions. When the pH value is low, the surface of hydroxyl adsorbs protons and has positive charges ($M\text{-}OH^{2+}$) (Equation 1.14), whereas, at a high pH value, the protons are desorbed, leading to negative charges (M^-O^-) on the surface (Equation 1.15).

$$\equiv M - OH + H^+ \Leftrightarrow \equiv M - OH^{2+} \quad (4.14)$$

$$\equiv M - OH + H^- \Leftrightarrow \equiv M - O^- + H_2O \quad (4.15)$$

The density of surface amphoteric points is about 1 mol/m^2, which is determined by the atomic structure of the surface lattice plane of the oxide [23, 25]. The Point of Zero Charge (PZC) of the oxide derived from the respective expressions of all hydroxyl positions is usually defined as the pH value where the surface charge of the particle is neutralized. All amphoteric surface groups are uncharged hydroxyl groups (M-OH). In general, the PZC of a metal oxide is negatively correlated with its oxidation state (i.e., there is no obvious relationship between PZC and the hydration of the compound [26]. The surface neutrality of hetero-metal oxides is determined by the sum of different acid/base pairs (such as aluminum silicate, which involves aluminol and silanol groups). The proton charge on the oxide surface can either be quantified under the above conditions through theoretical modeling or be determined by potentiometric titration. The PZC of effective surface charge density (C/m^2) is distinguished from the isoelectric point (IEP) of the surface, which is the pH value of zero zeta potential measured experimentally on the electrokinetic curve. Under certain circumstances, the effective pH range may be limited due to the relatively strong dependence of the solubility of oxides and hydroxides on the pH value.

In terms of dispersion, the pH-dependent surface charges implied by these amphiphilic sites are mainly based on the assumption related to the electrostatic repulsion (UEL) among particles required for stabilization. Generally, when the pH value of the solution is far from the range of PZC interaction, the highly charged surface forms a stable dispersion due to the high UEL repulsion interaction, whereas when the pH value is within the range of PZC interaction, the low surface charge will lead to the increase in the adhesion efficiency (aggregation) between the surfaces.

4.3.5.5 Impact of polymer compounds on stability

Adding a certain amount of high polymer compound or association colloid to a sol can significantly improve its stability against electrolytes. This phenomenon is called the protection effect, also referred to as steric stability in recent years. People have known for a long time that high polymer compounds can play a stabilizing role. For example, animal glue is used to stabilize the suspension of carbon black in water when ink is made. The cause of such stabilization is that the high polymer compound is adsorbed onto the surface of the sol particles and forms a high polymer protective film, which surrounds the colloidal particles, extends the hydrophilic groups into the water, and has a certain thickness. Therefore, when the colloidal mass points approach each other, the attractive force is greatly weakened, and with this layer of viscous high polymer film, the mutual repulsive force is increased, thus increasing the stability of the colloid. After the sol is protected, some of its physiochemical properties, such as electrophoresis and sensitivity to electrolytes will have significant changes. At this time, the physiochemical properties of the system are similar to those of the added high polymer material. In recent years, it has been discovered that high polymer compounds enable the sols to not only resist the coagulation of electrolytes but also have aging resistance that keeps the particle sizes unchanged for a long time with constant temperature resistance to coagulation in a very wide temperature range. At this time, the sol has lost some of the original lyosol characteristics to show certain properties of hydrophilic sols. For example, the sol can automatically disperse after precipitation to form itself again without any work on the system. Therefore, these sudden changes in properties cannot be generalized with the term "protection." People think that it is more appropriate to adopt the concept of improving the steric stability of the sol, which indicates that these are not limited to the resistance to electrolytes.

4.4 NPP of Inorganic Particles

4.4.1 Properties of Inorganic Nanoparticles

4.4.1.1 Electronic and optical properties

When the light incident on the nanoparticles in the form of precious metals, if the incident photon frequency matches the overall vibration frequency of the precious metal nanoparticles or metal conduction electrons, the nanoparticles or metals will have a force absorption effect on the photon energy, resulting in LSPR phenomenon (Fig. 4.5) [38, 39]. LSPR excitation leads to the selective absorption of a high molar excitation coefficient resonant light scattering wavelength with an efficiency equivalent to the resonant light scattering of 10 fluorophores, which enhances the local electromagnetic field near the nanoparticle surface and improves spectral absorption. The peak wavelength of the LSPR spectrum depends on the size, shape, and interparticle distance of NPs, their dielectric properties, as well as the dielectric properties of their environment, including matrix, solvent, and adsorbent.

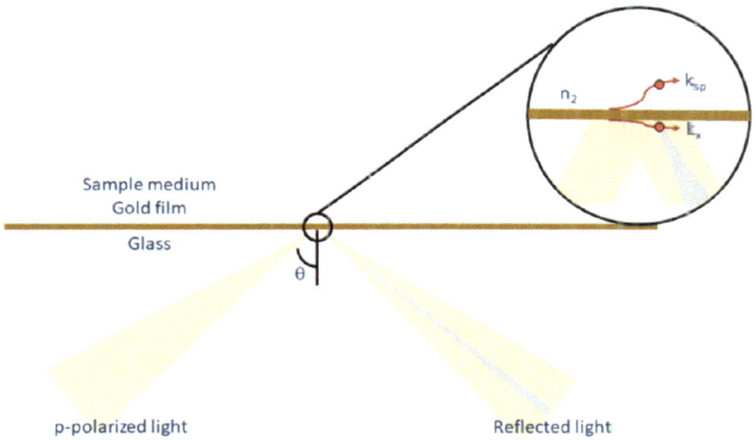

Figure 4.5 Localized Surface Plasma (LSP) on the outer surfaces of NPs.

4.4.1.2 Magnetism

Magnetite nanoparticle (MNP) are very interesting for researchers from different fields, including heterogeneous and homogeneous catalysis, biomedicine, magnetic fluid, data storage magnetic resonance imaging (MRI), and environmental remediation (water purification). Literature shows that NPs have the best magnetic performance when their sizes are below the critical value, i.e., 10 nm–20 nm [40]. In this particle size range, the magnetic performance of NPs can be controlled, which allows these particles to be applied in different fields [41, 42]. The uneven distribution of electrons in NPs entitles them to magnetism, which also depends on their synthesis schemes. MNPs can be prepared with a variety of synthetic methods, such as the solvothermal method

[43], the co-precipitation method, the microemulsion method, the thermal decomposition method, the flame spray synthesis method, etc. [44].

4.4.1.3 Mechanical properties

The unique mechanical properties entitle NPs to new application prospects in various significant fields such as tribology, surface engineering, nano-manufacturing, and nanoscale machinery manufacturing. The mechanical properties of NPs can be accurately understood by measuring different mechanical parameters, such as elastic modulus, hardness, stress-strain, adhesion, and friction. In addition to these parameters, surface coating, solidification and lubrication also contribute to the mechanical performance of NPs [45]. Moreover, in lubrication or lubricated contact, the contrast between the stiffness of the NPs and the outer surface of the contact controls whether the NPs are retracted into the plane or deformed when the contact pressure increases conspicuously. This important characteristic can reveal the performance of NPs under contact. Proper control of the mechanical properties of NPs and their interaction with any surface is essential for improving the surface quality and material properties. To achieve productive results in these fields, it is usually necessary to have a deep understanding of the basic mechanical properties of NPs, such as elastic modulus and hardness, motion laws, friction, and interfacial adhesion, and their size-related properties.

4.4.1.4 Thermal properties

As is known to all, metal NPs have higher thermal conductivities than solid fluids. For instance, the thermal conductivity of copper at room temperature is about 700 times higher than that of water and about 3000 times higher than that of engine oil. Even oxides such as aluminum oxide (Al_2O_3) have higher thermal conductivities than water. Therefore, fluids containing suspended solid NPs have better thermal conductivities compared with traditional heat transfer fluids. Nanofluids are prepared by dispersing nanoscale solid particles in liquids such as water, glycol, or oil. Compared with traditional heat transfer fluids and fluids containing fine particles, nanofluids have better thermal conductivities. Since heat transfer occurs on the surfaces of the particles, particles with larger total surface areas have better thermal conductivities, and a larger total surface area also increases the stability of the nano-suspension [46].

4.4.2 Biological Application of Inorganic Nanoparticles

So far, organic NPs have been extensively studied, including liposomes, polymers, polymer structures, and micelles, all of which are used in imaging or drug and gene delivery technologies (Fig. 4.9). However, inorganic NPs have also attracted people's attention in recent years and can exhibit properties that conventional organic nanoparticles do not possess because of their unique material and size-dependent physicochemical properties [47]. In particular, in addition to inertness, stability, and the tendency of functionalization, optical, magnetic and other physical properties make INPs an attractive alternative. Therefore, magnetic, Au, QDs, carbon nanotubes, and other INPs have broad application prospects in the field of modern biomedicine.

4.4.2.1 Au nanoparticles

Au nanoparticles (AuNPs) are usually called colloidal gold. They have been in use for hundreds of years because of their easy preparation and color changes with particle size. Meanwhile, AuNPs have the same characteristics as other NPs in terms of biological applications (such as size, inertness, ease of synthesis, and biocompatibility). AuNPs have great potential in bio-imaging technology because they can be visualized through the interaction of NPs and light, through which light particles strongly absorb and scatter visible light [48]. When light is absorbed, the light energy excites the free electrons in the gold particles to produce collective oscillations, i.e. the so-called surface plasma. Such absorption, which is located in the visible region of gold, silver, and copper, is visible in the range of 3 nm [49, 50]. The surface plasma resonance (SPR) of AuNPs causes absorption and scattering, the size of which depends on the size of the AuNP. AuNPs smaller than 20nm essentially show absorption, while particles larger than 80nm have increased ratios of scattering to absorption. This characteristic can be used for research on biological imaging, cell delivery vehicles, and biosensors.

Li's group selected the gold nanorod (GNR) to define the density function of the surface molecules of the nanosystem for the first time. They systematically studied the relationship between the surface molecular density of the nanosystem and the related role of the sub-organelles, and the regulation of tumor cell death modes such as apoptosis and necrosis [51]. It is clarified that moderately increasing the density is conducive to promoting the expression of the key protein caspase 8, thereby blocking the formation of the necrosis complex RIP1-RIP3 and promoting the expression of apoptosis-related protein caspase 3, which can inhibit necrosis and promote apoptosis. As a result, it regulates the death of tumor cells through the microstructure of the nanosystem from a physical perspective and reduces the toxicity of nanomaterials. Figure 4.6 shows the molecular mechanism of GNR regulating the transition between necrosis and apoptosis of cancer cells.

After endocytosis, GNRs enter the lysosome, increasing membrane permeability mediated lysosome (LMP). This unstable lysosome is mainly present in cancer cells, while lysosomes remain healthy in non-cancer cells. The destruction of the lysosomal membrane in cancer cells resulted in low expression of LAMP2. In contrast, high expression of cathepsin B and D. When the receptor-interacting protein necrosis complex appears, it will cause necrosis. Cathepsin D is released into the cytoplasm, translocated to mitochondria, and increases MOMP, thereby further promoting the release of Cyt c, and triggering mitochondrial-dependent apoptosis through caspase 9 and caspase 3. In terms of microstructure, the influence of GNR surface characteristics, intracellular molecules related to necrosis and apoptosis on surface molecular density ($\rho_{\text{surf.molecule}}$), and caspase 8, the key protein for the transformation from necrosis to apoptosis, are detailed in the enlarged right figure of Fig. 4.10. Compared with GNR with a large aspect ratio (L/D) and low surface potential, GNR with a small L/D has a higher LMP due to its larger surface molecular density. The key parameter $\rho_{\text{surf.molecule}}$ is thus be defined by us which shows the quotient between the total surface molecules ($N_{\text{surf.molecules}}$) and the surface area ($S_{\text{GNR}}$):

$$\rho_{\text{surf.molecule}} = \frac{N_{\text{surf.molecules}}}{S_{\text{GNR}}} = \frac{N_{\text{surf.molecules}}}{a\pi DL} \tag{4.16}$$

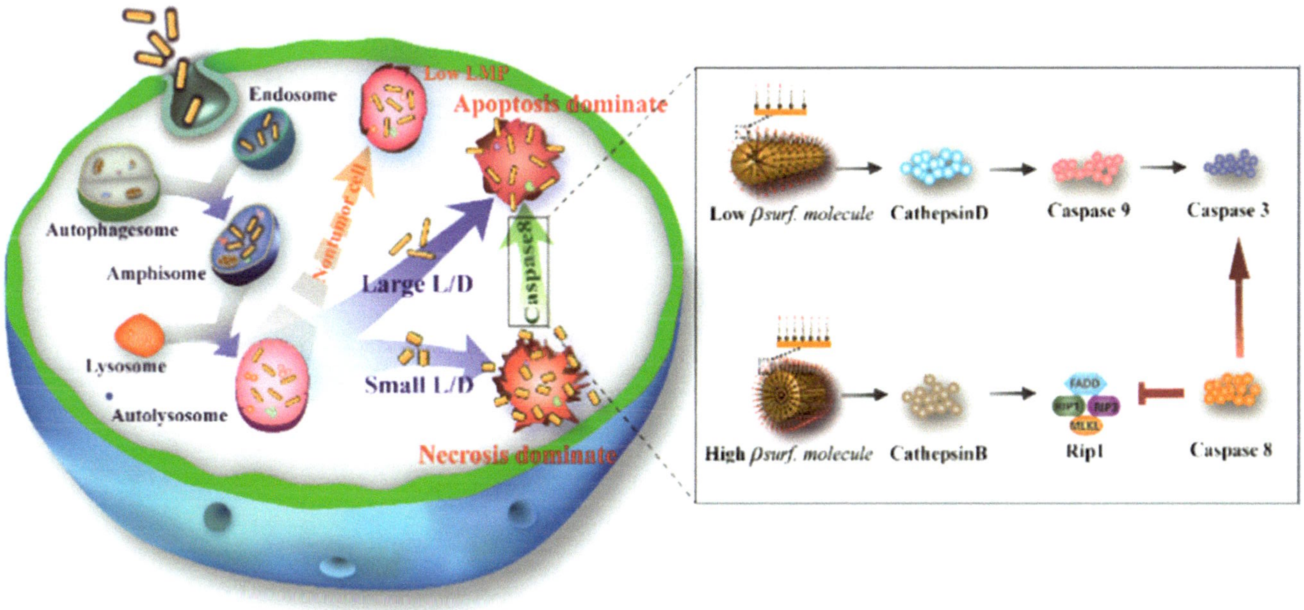

Figure 4.6 Schematic diagram of potential multifunctional methods for disease monitoring and treatment using nanoparticle systems. The formula inserted here is defined in 4.16.

In Formula 4.16, a is a correction factor. If both ends of the gold bar have a corona similar to a hemisphere, a will be 1, $S_{GNR} = \pi DL$; If both ends of the bar have flat sections, it is the cylinder shape of the dotted line in the following figure. As a result, a is less than 1, S_{GNR} can be calculated as $\pi DL + 0.5\,\pi D^2$ (Fig. 4.7).

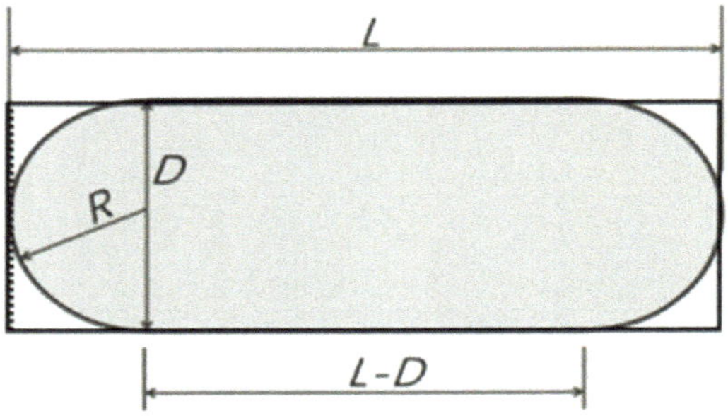

Figure 4.7 Schematic illustrated the model of gold nanorod.

4.4.2.2 Magnetic nanoparticles

Typical MNPs include a magnetic core, which is usually Fe_3O_4 or γ-Fe_2O_3. Other metals such as cobalt and nickel are also used. Nevertheless, they tend to be toxic and easily oxidized [52]. The use of iron oxide MNPs has additional advantages, because in human bodies, iron is mainly stored in the core of ferritin, and the body can process residual iron. Cationic MNPs can effectively enter cells and stay in endosomes for a long time [53]. The iron contained in endosomes and lysosomes (i.e., absorbed by cell nuclei) is metabolized into pure element iron and oxygen by hydrolytic enzymes, where the iron is combined with normal body storage [54]. The steady state of iron is effectively controlled by adsorption, excretion, and storage, so we assume that any excess iron *in vivo* can be processed after the use of iron oxide NPs. However, it should be pointed out that although iron plays an important role in almost all living tissues, its bioavailability is quite limited, and in some cases, it may be toxic to cells in the form of free iron (rather than being related to hemoglobin) [55].

The superparamagnetic iron oxide (SPIO) NP, which has unique mechanical properties, is a new biomedical material with extreme attractiveness. Such superparamagnetism results from the large magnetic moment they obtain in the external magnetic field, whose removal can eliminate the former. This large moment of force causes a relatively big change in the signal of each unit of particles, so only a very small amount is needed to generate good signal feedback [47]. The surface coating can be various substances, including synthetic or natural polymers (such as protein or dextran) [56, 57] and amphipathic molecules (such as fatty acids or phospholipids) [58]. With inorganic iron as the core, biocompatible surface coatings can be designed and prepared, which generally provide stability (i.e., inhibit aggregation) under physiological conditions [58], and then further control surface functionalization, such as coupling with DNA chain fragments, proteins, polypeptides or antibodies. In the past two decades, MNPs have been developed and studied as an efficient delivery vector, which brings chances for MRI contrast enhancement, hyperthermia cancer treatment media, and targeted therapies (such as drug and gene delivery). The above characteristics of MNPs can be used in the research of NMRI, hyperthermia therapy, and magnetic targeted therapy/delivery.

4.4.2.3 Quantum dots

Quantum dots are nanoscale colloidal crystals composed of atoms from the second to sixth groups (e.g., Cd, Zn, Se, Te) or the third to fifth groups (e.g., P, As) in the periodic table. In terms of biological applications, the most typical QDs used are composed of CdSe and wrapped in a ZnS shell (to protect it from high pressure) [59]. By confining electrons to band gaps of different sizes, the corresponding emission wavelength can be tuned from the ultraviolet to the NIR region. Small QDs (2 nm) display blue fluorescence when emitted (380 nm–440 nm), while large particles (5 nm) are emitted in red fluorescence (605 nm–630 nm) [60]. Compared with organic dye molecules, QDs have not only their own advantages but also the advantage of tunable fluorescence. Because of their strong and stable light emission, QDs are not easy to photobleach molecules. Bruchez and other researchers have used QDs with different surface modifications to label entire cells and tissue sections [61, 62]. The light stability makes them very useful in observing cells for long periods, and the above characteristics make them applicable in research fields such as bio-imaging and targeted treatment.

4.4.2.4 Carbon nanotubes

Hollow porous NPs, such as nanotubes, nanoshells, and hollow spheres, have relatively large encapsulation loading capacities, thereby enhancing signals and sensitivity. Carbon nanotubes are cylindrical graphene flakes. Although most of the applications of carbon nanotubes are mainly concentrated in microelectronics, the unique electronic and physical properties enable carbon nanotubes to show some attractive properties when used in biomedicine, including intracellular drug delivery and relatively low toxicity [63, 64]. Two kinds of single-walled carbon nanotubes (SWCNTs; 1 nm–3 nm in diameter and 5 nm–30 nm in length) and multi-walled carbon nanotubes (MWCNTs; 10 nm–150 nm in diameter and 20 nm–800 nm to several microns in length) have been studied for various biological applications. Like QDs, it has been shown that SWCNTs have strong optical absorbance in the highly transparent NIR regions of known biological systems [65]. The above characteristics indicate that they can be used for neural tissue engineering and imaging and cancer treatment research.

4.4.2.5 MXene

MXenes are a large family of two-dimensional metal carbides and nitrides. The structure of MXenes consists of two or more layers of transition metal (M) atoms, which are filled in the honeycomb 2D lattice. The carbon layer and/or nitrogen layer (X atom) occupying the octahedral position between adjacent transition metal layers intervene (Fig. 4.8) [66, 67]. MXenes have some outstanding properties, making them an ideal candidate for various biotechnology applications, including low cytotoxicity to non-malignant cells *in vitro*, some selectivity to cancer cells, low environmental toxicity and biological toxicity, and high antibacterial properties [68]. Their adjustable optics and magnetism also enable MXenes to be used for biosensors and bio-imaging. Moreover, their effective near-infrared absorption and "light heat" conversion imparts them unique application potential in therapeutics, drug delivery, and biological tissue engineering. This high application potential can be further expanded through MXene surface modification and preparation of mixed multi-functional structures. The size, stoichiometric ratio, and chemical composition of MXenes can be optimized by changing the characteristics and ratio of their precursor compounds, reaction temperature and pressure, and ultrasonic conditions. The surface of MXenes can be further customized by using specific biomolecules and nanoparticles. These biomolecules and nanoparticles can develop MRI nanoprobes that can be used for drug transport and imaging at the same time, for example [69, 70].

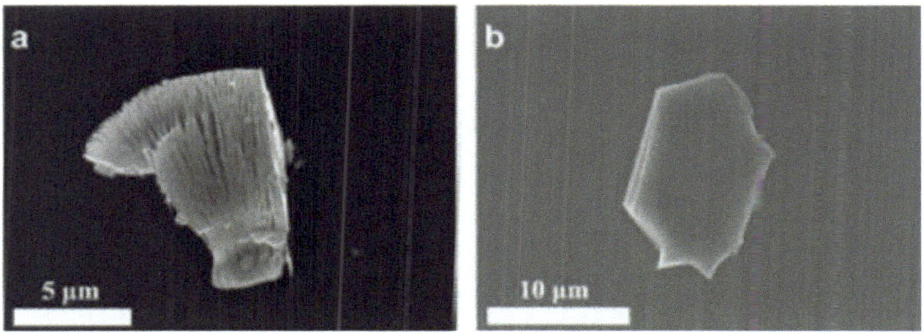

Figure 4.8 SEM images of (a) MXene. (b) delaminated MXene, respectively.

4.5 Summary and Perspective

This chapter introduces the classification and preparation methods of INPs, and the relevant theories of the formation process. It also sheds light on the relationship between the physiochemical properties and biological applications of INPs. The characteristics of these INPs in sound, light, electricity, magnetism etc. provide them with special properties that are different from organic NPs. INPs have great application potential in imaging, clinical diagnosis, and treatment. However, it is difficult to achieve specific biological effects through the reasonable design of INPs. The summary of the physiochemical properties of INPs provides a theoretical basis for the design of INPs and is conducive to the development of new drug delivery systems and further development of nanomedicine.

References

1. Sekhon BS, Kamboj SR, *Nanomedicine* **2010**, *6*, 516–522.
2. Dreaden EC, Alkilany AM, Huang X, Murphy CJ, El-Sayed MA, *Chemical Society Reviews* **2012**, *41*, 2740–2779.

3. Spaldin NA, *Magnetic Materials: Fundamentals and Applications*, 2 edn, Cambridge University Press, Cambridge, 2010.
4. Sigmund W, Yuh J, Park H, Maneeratana V, Pyrgiotakis G, Daga A, Taylor J, Nino JC, *Journal of the American Ceramic Society* **2006**, *89*, 395–407.
5. Thomas SC, Harshita, Mishra PK, Talegaonkar S, *Current Pharmaceutical Design* **2015**, *21*, 6165–6188.
6. Sun S, Murray CB, Weller D, Folks L, Moser A, *Science* **2000**, *287*, 1989–1992.
7. Hisatomi T, Kubota J, Domen K, *Chemical Society Reviews* **2014**, *43*, 7520–7535.
8. Molday RS, Mackenzie D, *Journal of Immunological Methods* **1982**, *52*, 353–367.
9. Tamura H, Matijevic E, *Journal of Colloid and Interface Science* **1982**, *90*, 100–109.
10. Wang Y, Xia Y, *Nano Letters* **2004**, *4*, 2047–2050.
11. Jolivet JP, Tronc E, Chanéac C, *Synthesis of Iron Oxide- and Metal-Based Nanomaterials*, vol. 10, 2000, pp. 167–172.
12. Jolivet JP, Froidfond C, Pottier A, Chanéac C, Cassaignon S, Tronc E, *Size Tailoring of Oxide Nanoparticles by Precipitation in Aqueous Medium. A Semi-Quantitative Modeling*, vol. 224, 2002, pp. U413–U413.
13. Massart R, Dubois E, Cabuil V, Hasmonay E, *Journal of Magnetism and Magnetic Materials* **1995**, *149*, 1–5.
14. Massart R, *IEEE Transactions on Magnetics* **1981**, *17*, 1247–1248.
15. Andrés Vergés M, Costo R, Roca AG, Marco JF, Goya GF, Serna CJ, Morales MP, *Journal of Physics D: Applied Physics* **2008**, *41*, 134003.
16. Langevin D, Micelles and microemulsions, in *Complex Fluids* (Garrido L, ed), Springer, Berlin, Heidelberg, 1993, pp. 327–349.
17. Fievet F, Lagier JP, Blin B, Beaudoin B, Figlarz M, *Solid State Ionics* **1989**, *32–33*, 198–205.
18. Hyeon T, Lee SS, Park J, Chung Y, Na HB, *Journal of the American Chemical Society* **2001**, *123*, 12798–12801.
19. Geng B, Liu X, Ma J, Du Q, *Materials Science and Engineering: B* **2007**, *145*, 17–22.
20. Sun S, Zeng H, *Journal of the American Chemical Society* **2002**, *124*, 8204–8205.

21. LaMer VK, Dinegar RH, *Journal of the American Chemical Society* **1950**, *72*, 4847–4854.
22. Bosholm J, *Acta Hydrochimica et Hydrobiologica* **1990**, *18*, 110–110.
23. Cahn JW, Hilliard JE, *Acta Metallurgica* **1971**, *19*, 151–161.
24. Salas G, Costo R, Morales MDP, Synthesis of inorganic nanoparticles, in *Nanobiotechnology – Inorganic Nanoparticles vs Organic Nanoparticles*, 2012, pp. 35–79.
25. Dirksen JA, Ring TA, *Chemical Engineering Science* **1991**, *46*, 2389–2427.
26. Viswanatha R, Santra PK, Dasgupta C, Sarma D, Growth mechanism of nanocrystals in solution: ZnO, a case study. *Physical Review Letters* **2007**, *98*, 255501.
27. Lee J, Isobe T, Senna M, *Journal of Colloid and Interface Science* **1996**, *177*, 490–494.
28. Morales MP, Veintemillas-Verdaguer S, Montero MI, Serna CJ, Roig A, Casas L, Martínez B, Sandiumenge F, *Chemistry of Materials* **1999**, *11*, 3058–3064.
29. Cheon J, Kang N-J, Lee S-M, Lee J-H, Yoon J-H, Oh SJ, *Journal of the American Chemical Society* **2004**, *126*, 1950–1951.
30. Wang J, Zeng C, *Journal of Crystal Growth* **2004**, *270*, 729–733
31. Tao AR, Habas S, Yang P, *Small* **2008**, *4*, 310–325.
32. Sekhon BS, Kamboj SR, *Nanomedicine* **2010**, *6*, 612–618.
33. Moon JK, Song MW, Pak HK, *Journal of Physics: Condensed Matter* **2015**, *27*, 194102.
34. Outhwaite CW, *Molecular Physics* **1974**, *28*, 217–224.
35. Lesieur S, Grabielle-Madelmont C, Ménager C, Cabuil V, Dadhi D, Pierrot P, Edwards K, Evidence of surfactant-induced formation of transient pores in lipid bilayers by using magnetic-fluid-loaded liposomes. *Journal of the American Chemical Society* **2003**, *125*, 5266–5267.
36. Jolivet JP, Vayssieres L, Chaneac C, Tronc E, *MRS Proceedings* **2011**, *432*, 145.
37. Portehault D, Cassaignon S, Baudrin E, Jolivet J-P, *Journal of Materials Chemistry* **2009**, *19*, 2407–2416.
38. Liu W-C, Tsai DP, *Physical Review B* **2002**, *65*, 155423.
39. Choi BC, Xu H, Hajisalem G, Gordon R, *Applied Physics Letters* **2018**, *112*, 022403.

40. Reiss G, Hütten A, *Nature Materials* **2005**, *4*, 725–726.
41. Faivre D, Bennet M, *Nature* **2016**, *535*, 235.
42. Zhu Y, Goodridge AG, Stapleton SR, *The Biochemical Journal* **1994**, *303(Pt 1)*, 213–216.
43. Qi M, Zhang K, Li S, Wu J, Pham-Huy C, Diao X, Xiao D, He H, *New Journal of Chemistry* **2016**, *40*, 4480–4491.
44. Wu W, He Q, Jiang C, *Nanoscale Research Letters* **2008**, *3*, 397.
45. Guo D, Xie G, Luo J, *Journal of Physics D: Applied Physics* **2013**, *47*, 013001.
46. Lee S, Choi SUS, Li S, Eastman JA, *Journal of Heat Transfer* **1999**, *121*, 280–289.
47. Huang H-C, Barua S, Sharma G, Dey SK, Rege K, *Journal of Controlled Release* **2011**, *155*, 344–357.
48. Popovtzer R, Agrawal A, Kotov NA, Popovtzer A, Balter J, Carey TE, Kopelman R, *Nano Letters* **2008**, *8*, 4593–4596.
49. Boisselier E, Astruc D, *Chemical Society Reviews* **2009**, *38*, 1759–1782.
50. Souza GR, Christianson DR, Staquicini FI, Ozawa MG, Snyder EY, Sidman RL, Miller JH, Arap W, Pasqualini R, *Proceedings of the National Academy of Sciences of the United States of America* **2006**, *103*, 1215–1220.
51. Zhang F, Hou Y, Zhu M, Deng B, Zhao M, Zhu X, Sun Y, Chen D, Jiang C, Wang L, Chen C, Chen H, Chen H, Zheng H, Li W, *Advanced Science* **2021**, *8*, 2102666.
52. Connolly J, Pierre TGS, Rutnakornpituk M, Riffle JS, *Journal of Physics D: Applied Physics* **2004**, *37*, 2475–2482.
53. Sato M, Yamashita T, Ohkura M, Osai Y, Sato A, Takada T, Matsusaka H, Ono I, Tamura Y, Sato N, Sasaki Y, Ito A, Honda H, Wakamatsu K, Ito S, Jimbow K, *Journal of Investigate Dermatology* **2009**, *129*, 2233–2241.
54. Xu C, Sun S, *Dalton Transactions* **2009**, 5583–5591.
55. Dobson J, *FEBS Letters* **2001**, *496*, 1–5.
56. Berry CC, Wells S, Charles S, Curtis ASG, *Biomaterials* **2003**, *24*, 4551–4557.
57. Berry CC, *Journal of Physics D: Applied Physics* **2009**, *42*, 224003.
58. Soenen S, Vercauteren D, Braeckmans K, Noppe W, De Smedt S, De Cuyper M, Stable long-term intracellular labelling with fluorescently tagged cationic magnetoliposomes. *Chembiochem* **2009**, *10*, 257–267.

59. Paul Alivisatos A, Weiwei Gu W, Larabell C, Quantum dots as cellular probes. *Annals of Biomedical Engineering* **2005**, *7*, 55–76.

60. Bruchez M, Moronne M, Gin P, Weiss S, Alivisatos AP, *Science* **1998**, *281*, 2013.

61. Wu X, Liu H-F, Liu D, Haley KN, Treadway J, Larson JP, Ge N, Peale F, Bruchez M, Immunofluorescent labeling of cancer marker Her2 and other cellular targets with semiconductor quantum dots. *Nature Biotechnology* **2003**, *21*, 41–46.

62. J Rosenthal S, Tomlinson I, M Adkins E, Schroeter S, Adams S, Swafford L, McBride J, Wang Y, J DeFelice L, D Blakely R, Targeting cell surface receptors with ligand-conjugated nanocrystals. *Journal of the American Chemical Society* **2002**, *124*, 4586–4594.

63. Kong J, Soh HT, Cassell AM, Quate CF, Dai H, *Nature* **1998**, *395*, 878–881.

64. Bianco A, Kostarelos K, Partidos CD, Prato M, *Chemical Communications* **2005**, 571–577.

65. O'Connell MJ, Bachilo SM, Huffman CB, Moore V, Strano MS, Haroz E, Rialon K, Boul P, Noon WH, Kittrell C, Ma J, Hauge R, Weisman R, Smalley RE, Band gap fluorescence from individual single-walled carbon nanotubes. *Science* **2002**, *297*, 593–596.

66. VahidMohammadi A, Rosen J, Gogotsi Y, *Science* **2021**, *372*, eabf1581.

67. Zhu X, Cao Z, Wang W, Li H, Dong J, Gao S, Xu D, Li L, Shen J, Ye M, *ACS Nano* **2021**, *15*, 2971–2983.

68. Lu B, Zhu Z, Ma B, Wang W, Zhu R, Zhang J, *Small* **2021**, *17*, 2100946.

69. Lin H, Wang Y, Gao S, Chen Y, Shi J, *Advanced Materials* **2018**, *30*, 1703284.

70. Liu Z, Lin H, Zhao M, Dai C, Zhang S, Peng W, Chen Y, *Theranostics* **2018**, *8*, 1648–1664.

Chapter 5

Nano Physical Pharmaceutics of Nanogel Delivery System

Xiaoling Pan,[a] Jing Liu,[b] Xiangling Gu,[b] Na Fan,[b] Wan Sun,[b] and Wei Li[a]

[a]*Department of Nanomedicine, Naval Medical University, Shanghai 200433, China*
[b]*College of Medicine and Nursing, Dezhou University, Shandong 253023, China*

liwei_dds@163.com

5.1 Classification of Nanogels

Nanogel is an intramolecularly cross-linked polymer gel in the form of nanoparticles (1–1000 nm in diameter). And the internal structure of nanogel is a typical network structure, which can be dispersed into nanometer-sized hydrogel particles in an aqueous solution (Fig. 5.1). Nanogels are developed from hydrogels and have the common characteristics of hydrogels and nanoparticles. Like hydrogels, nanogels have a high water content and can shrink or swell according to changes in external conditions; like nanoparticles, nanogels have a small particle size and a large specific surface area. According to the cross-linking forms of the polymer constituting the gel, they can be divided into chemical

Nano Physical Pharmaceutics
Edited by Wei Li

ISBN 978-981-4968-52-2 (Hardcover), 978-1-003-51393-3 (eBook)
www.jennystanford.com

cross-linking and physical cross-linking. The former is a three-dimensional network structure formed by covalent cross-linking between polymer chains. The cross-linking bond is generally strong, and the system can only expand but cannot melt or dissolve. The latter is a three-dimensional network structure formed by non-covalent bonds, including hydrogen bond, Coulomb force, coordination bond, etc. [1]. Most natural gels are cross-linked by hydrogen bonding between polymer segments, but they are easily damaged by heating and other means.

Figure 5.1 Scheme illustrates the 3-D network structure of a nanogel swollen by water.

5.2 Preparation of Nanogels

5.2.1 Non-Covalent Bonding Method

In an aqueous environment and under mild conditions, polymers can be formed into nanogels by physical cross-linking [2]. Various nanogels have been prepared using physical self-assembly between polymers, usually by controlling aggregation between hydrophilic polymers through hydrophobic, electrostatic interactions or hydrogen bonding. Because of its simplicity and mild conditions, this preparation method is mostly used to encapsulate biological macromolecules such as proteins by interactions between hydrophilic polymers for preparing nanogels coated with biological macromolecules. For example, Starch modified with cholesterol can be loaded with insulin to prepare 20–30 nm nanogels through intermolecular hydrophobic interactions [3], and each nanoparticle contains five insulin molecules. Researchers studied the self-assembly of dextran modified with dodecyl and β-cyclodextrin, using their interaction to wrap proteins in water-based media [4]. In this way, self-assembled nanogels of different particle sizes (120–150 nm) were formed and stable for a long time. Similarly, nanogels comprising ovalbumin and lysozyme or chitosan can be prepared from proteins with opposite charges [5, 6]. Such self-assembled proteins are usually ovalbumin, lysozyme or transferrin, and most of the polymers used to prepare nanogels are natural polymers, such as chitosan, glucan, and β-cyclodextrin.

5.2.2 Chemical Cross-Linking Reaction

In normal or reverse-phase microemulsion polymerization, nanogels can be obtained by adding a cross-linking agent to the system to initiate the polymerization of micromolecular monomers, or to initiate the cross-linking reaction of functional groups on the side chains of polymers [7]. This provides a means to modify the structure and properties of nanogels by chemical synthesis under different colloidal conditions. The greatest advantages of this method are that it can control the nanogel structure, has good stability, and has a wide range of applications, which provides conditions for preparing functionalized nanogels.

Some studies have reported that reverse-phase (water-in-oil) microemulsion was used as a medium for polymerization to form nanogels, and monomers with dual functional groups were introduced as cross-linking agents to obtain a stable nanogel network. Unstable bonds are usually introduced into the nanogels. During the preparation of such polymers, unstable bonds are usually introduced into the nanogel in order to release the drug after the unstable bonds are broken. For example, a reversed-phase microemulsion is used to prepare polyacrylamide nanogels by free radical polymerization for encapsulating and transporting proteins and DNA, and an acid-sensitive acetal cross-linking agent is added, the product of which can respond to the acidic environment in the cell to achieve targeted drug release behavior [8, 9]. In addition, the reversed-phase microemulsion system can also be used for synthesizing stable cross-linked nanogels from water-soluble polymers by atom transfer radical polymerization [10]. And disulfide bonds are added to the system to prepare nanogels since the disulfide bonds can be cleaved by glutathione inside the cell, which facilitates the release of the nanogels inside the cell and endows the nanogels with the property of reduction-responsive degradation.

The polymerization reaction of nanogels can also be carried out in oil-in-water microemulsion or water-based suspension. Moreover, the polymerization reaction can be initiated by hydrosoluble monomers in a homogeneous aqueous solution and finally form a slowly growing polygelatinous suspension. For example, researchers synthesized a suspended nanogel by precipitation polymerization using polyethylene glycol (PEG) and polymethacrylic acid (PMA) as main ingredients [11].

In addition to these polymerization methods, covalent cross-linking between polymer chains provides a good solution for producing nanogels with large voids [12] and has been widely used in the preparation of drug-loaded nanogels with different functions. With this method, PEG branch was combined with PEI in the oil-in-water microemulsion environment, the solvents were evaporated through vacuum, and finally the nanogel was dissolved in the aqueous solution. As in the case of the polymerization reactions mentioned earlier, the cross-linking connections between nanogel polymers can be opened by various designs in order to release the load. For example, a segmented PEI cross-linked with disulfide was used to prepare biodegradable ionized nanogels loaded with polynucleotides, which greatly reduced biotoxicity [13]. In another work, hydrosoluble PEI was cross-linked on microglia composed of double-active PEG-*b*-PPG-*b*-PEG triblock copolymers [14] to prepare hydrophobic core of PPO and cross-linked shell of PEI/PEO (PEG-cl-PEI). In addition, Tatiana et al. cross-linked the amphiphilic block copolymers with oppositely charged shortening agents to obtain preliminary particles, then chemical cross-linking of ionic bonds occurred in the inner core, and finally removed the shortening agents to obtain the gel, thereby controlling the spatial distribution of polymer chains in the nanogel [15]. In addition, the spatial distribution of polymer chains in the nanogels was controlled by cross-linking amphiphilic block copolymers with counter-charged condensing agents to obtain preliminary particles, followed by chemical cross-linking of ionic bonds within the core and finally removal of condensing agents to obtain nanogels.

5.2.3 Template Method

Template imprinting or microfluidic technology can obtain nanogels with precise and controllable size and shape (Fig. 5.4). De Simone group proposed "modified photographic template imprinting" for the preparation of nanogels and received much attention. This method can strictly control the size, shape, composition and particle surface function of nanogel particles, allowing the loading of accurate small-molecule active substances or biological macromolecules. For example, taking a fluoropolyether material as the template, on the surface of a non-wetting elastomer, photo-initiated diacrylate-modified oligomers are chemically cross-linked to obtain nanogels with different sizes, shapes, and chemical compositions. Using template imprinting and UV-induced cross-linking, mono-dispersed PEG nanogels with a particle size of 200 nm can also be obtained [16]. Microfluidic technology also provides a method for preparing chemically or physically cross-linked nanogels. The advantage of this method is that it can prepare nanogels with a variety of sizes, shapes and controllable morphologies, and at the same time effectively improve the preparation scale of nanogels.

5.3 Mechanism of Nanogel Formation

The essence of nanogel generation is the moderate cross-linking between polymers. Inorganic substances or metal alkoxides are used as precursors, and these materials are uniformly mixed, hydrolyzed and condensed to form a stable transparent sol system in solution. The solute is aged and the slow polymerization between the colloidal particles gradually forms a three-dimensional spatial network structure, and the network is filled with solvent that loses its mobility to obtain nanogels. The preparation of solid nanogels is relatively simple: it can simply be made by dry gel absorbing liquid and expanding, which is a usually elastic nanogel. However, the preparation of nanogels must meet two basic conditions: (1) reduce solubility, so that the solid matter will precipitate out of the solution in a "colloidal dispersion state"; (2) the formed solid particles neither subside nor move freely, but form a skeleton and continuous network structure.

5.3.1 Cross-Linking of Nanogel

In the cross-linking system of polymers, the conversion rate or reaction time when the first gel molecule appears is called the gel point [17], and gelation is an important feature of the cross-linking system.

5.3.1.1 Gelation theory of nonlinear polycondensation

The viscosity of most nonlinear polycondensation systems increases suddenly and insoluble gel is produced when they reach a certain stage, which is called gelation phenomenon. In production, if gel appears in the reactor, it will bring great trouble to the operation. Therefore, it is important to predict the critical condition of gel theoretically.

Flory [18] introduced the concept of bifurcation point and obtained the critical condition of gelation for a special case for the first time. Tang Aoqing [19, 20] improved the gelation critical condition by using the degree of average polymerization tending to infinity, and treated nearly ten kinds of nonlinear polycondensation systems, thus advancing the theory to a new stage. Take the $A_a - B_b$ reaction as an example, A_a represents a monomers of A group, and B_b represents b monomers of the B group that can react with the A group. N_A and N_B are the corresponding number of molecules, and p_A and p_B are the corresponding degrees of reaction. Then the ratio range of gel in $A_a - B_b$ polycondensation reaction can be expressed by the following formula:

$$\frac{1}{(a-1)(b-1)} \leq r_A \leq (a-1)(b-1) \tag{5.1}$$

Therefore, for A_a – B_b polycondensation, the idea that one of a and b is more than 2 can produce gel can only be realized in a certain ratio range, beyond this range, no matter how much the reaction can only produce sol.

5.3.1.2 Free radical addition polymerization

Compared with condensation polymerization, it is much more difficult to deal with free radical addition polymerization by statistical method. This is due to the fact that during free radical addition polymerisation the number of reactive groups varies with time and the lengths of the reactive polymer chain and the termination reaction differ, resulting in the formation of deactivated polymer molecules.

For the first time, the distribution of addition polymerization was studied by statistical method [21, 22], and the expression of active chain distribution was obtained as follows:

$$P_i = [1-p(t)]\int_0^t \frac{[\alpha(t)-\alpha(\tau)]^{i-1}}{(i-1)!} \cdot \frac{e^{-[\alpha(t)-\alpha(\tau)]}}{1-p(\tau)} V_i(\tau)d\tau \tag{5.2}$$

where P_i is the chain length i is the number of active radicals, V_i is the raising procedure. α represents the average length of active chain past a given time, and $V_i(\tau)d\tau$ denotes the initiation rate in the time $\tau - \tau + d\tau$.

The chain length distribution M_i of terminated polymer can be obtained by the reaction equation of chain termination process:

$$M_i = \int_0^{p(t)} \gamma dp(t) \int_0^t \frac{[\alpha(t)-\alpha(\tau)]^{i-1}}{(i-1)!} \cdot \frac{V_i(\tau)d\tau}{1-p(\tau)} + \frac{1}{2}\int_0^{p(t)} (1-\gamma)dp(t) \frac{f(t)}{\int_0^t V_i(\tau)\frac{d\tau}{[1-p(t)]}} \tag{5.3}$$

where γ denotes a mono-radical termination fraction and $(1-\gamma)$ is a mono-radical termination fraction.

$$f_t = \sum_j \int_0^t \frac{[\alpha(t)-\alpha(\tau)]^{i-1}}{(i-1)!} \cdot \frac{e^{-[\alpha(t)-\alpha(\tau)]}}{1-p(\tau)} \cdot \int_0^t \frac{[\alpha(t)-\alpha(\tau)]^{i-1}}{(i-j-1)!} \cdot \frac{e^{-[\alpha(t)-\alpha(\tau)]}d\tau}{1-p(\tau)} \quad (5.4)$$

5.3.2 Properties of Polyelectrolyte Nanogels

N represents the length of each branch chain in the polyelectrolyte nanogel, *n* is the number of branch chains, *r* is the distance between the end of branch chains, *R* is the radius of gel, *b* is the Kuhn length of the chain, and α is the charged fraction of the chain. Bjerrum length is $l_B = e^2/(4\pi\varepsilon k_B T)$, where ε is the permittivity of the system, k_B is the Boltzmann constant, and *T* is the absolute temperature of the system. Figure 5.2 illustrates the sample of nanogel formed from graft copolymers.

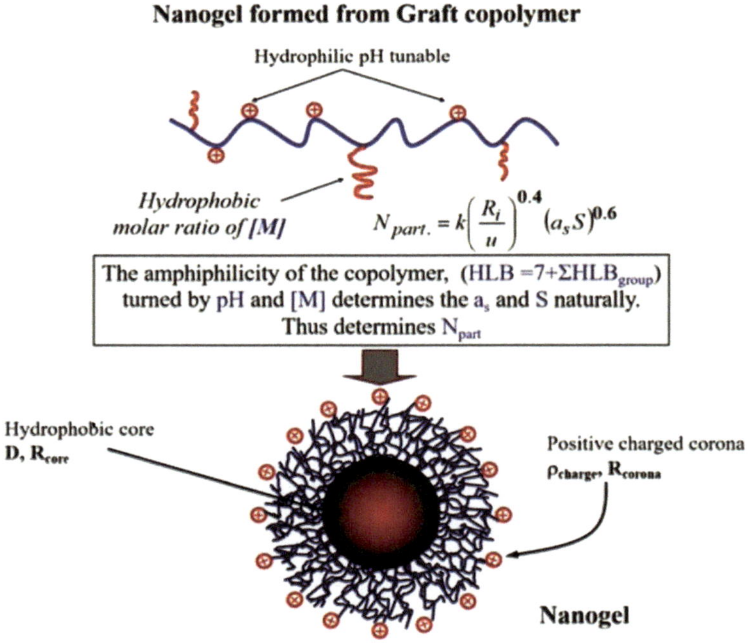

Figure 5.2 The mechanism of structure and properties tunning of nanogel carrier *in-suit* formed from graft copolymers in water. The formula inserted here is shown in Equation 5.7.

In addition, assume that the charged fraction is small ($\alpha \ll 1$), ensuring that the flexibility of the chain is not affected. For strong polyelectrolyte gels, the charged fractionation is constant; for weak polyelectrolytes, the charged fractionation is not fixed but adjustable, for example, by pH. Bjerrum length $l_B = e^2/(4\pi\varepsilon k_B T)$ is in the same order of magnitude as the Kuhn length of the chain, that is, the dimensionless coupling parameter $u = l_B/b \approx 1$ characterizing the strength of the Coulomb interaction, which does not need to consider the effect caused by strong Coulomb interaction such as Manning ion condensation. The case of good solvents is qualitatively similar to that of θ solvents. The volume change of the gel is not taken into account here, so the inferior solvent case is not taken into account for the moment. The volume fraction of gel can be expressed as:

$$\Phi \approx \frac{Nb^3}{r^3} \approx \frac{nNb^3}{R^3} \tag{5.5}$$

The following equation is obtained:

$$r \approx \frac{R}{n^{1/3}} \tag{5.6}$$

5.3.3 Structure and Stability of Nanogel

The assembly behavior of copolymers is described by Equation 1.2 [23]. And the key parameter is the surface charge (also known as zeta potential φ), which can be adjusted by monomer ratio, solvent mass or the pH value of the medium. The diameter of such particles is determined by the number of nanoparticles (N_{part}).

$$N_{part} = k\left(\frac{R_i}{u}\right)^{0.4} (a_s S)^{0.6} \tag{5.7}$$

where k is a constant from 0.37 to 0.53, R_i, u, a_s and S represent the rate of free radical generation, the rate of increase in particle volume, the surface area of the surfactant, and the total amount of surfactants, respectively. The amphiphilicity of a copolymer can be determined by the balance of hydrophilic-hydrophobic relationship (HLB = 7 + ΣHLB$_{group}$). This hydrophilic-hydrophobic balance value (HLB) is also affected by pH and monomer composition [M], and [M] can determine a_s and S, and then N_{part}. A large number of colloidal nanoparticles in nano-microgels can be connected to each other to form a large gel under certain conditions due to the existence of active molecules on the surface. The void is filled with a dispersed medium, and the size of the gel is necessarily related to the number of cross-linked colloidal nanoparticles. At the same time, the cross-linking reaction is also limited by the number of surfactant molecules and surface charge, which affect the size and surface potential of the gel and other physical and chemical properties. For a nano-hybrid system formed by homogeneous polymers, its morphology and structure are mainly affected by electrostatic repulsion (E) [23].

$$E = 32\pi\sigma\alpha\left(\frac{kT}{ze}\right)\tanh^2\left(\frac{ze\varphi^m}{4kT}\right)e^{-hk} \tag{5.8}$$

where σ is the electrostatic constant of the solvent, α is the dielectric constant of the solvent, k is the Boltzmann constant, T is the temperature, z is the number of electrons, e is the solubility of the solvent, φ^m is the double-layer potential in the diffusion layer, κ refers to the thickness of the layer, and h refers to the distance between the two particles. For example, when polyvinyl alcohol (PVA) and polylactic acid (PLA) are mixed in aqueous solution, PLA is insoluble in water, and the electrostatic repulsion force of the carrier mainly comes from the hydrophilic polyvinyl alcohol polymer. This electrostatic repulsion force can adjust the thickness of the double layer of the carrier. The greater the electrostatic repulsive force, the less the thickness of the double-layer film. Therefore, the stability and morphology of the carrier can be adjusted by controlling the content of PVA in the process of carrier synthesis, so as to design the carrier needed.

5.4 NPP of Nanogels in Drug Delivery

5.4.1 Physiochemical Properties of Nanogels

Nanogels have good hydrophilicity and dispersibility, and have good performance to wrap small bioactive agents and biological macromolecules. Swelling, shrinkage and thixotropy are unique properties of nanogels, which can provide many benefits for drug loading and release.

5.4.1.1 Expansion of nanogels

Nanogels are soft nanomaterials, and the swelling effect of nanogels which increases their volume or weight by absorbing liquid or vapor, is a unique property of elastic nanogels. Flory first proposed the theory of gel expansion in 1943. Among many theories, Flory–Rehner theory of gel expansion is one of the most classical theories, which can successfully predict the swelling deformation characteristics of polymer colloids [24]. In recent years, the research on the swelling kinetics of nanogels has received more and more attention and made great progress. It studies the relationship of the swelling ratio of nanogels with time, which is mainly controlled by the weak interaction of the gels. Good solvent molecules swell by penetrating the nanogel network structure. The volume expansion rate depends on the internal osmotic pressure and the conformational transformation ability of the nanogel molecular chain.

Compared with traditional hydrogels, nanogels have a faster response speed [25]. Moreover, the swelling process is divided into two types: infinite swelling and limited swelling. If the nanogels absorb a limited amount of liquid and the network structure of the nanogels only swells without disintegration, it is called limited swelling. If more and more liquid is absorbed and the network structure of the nanogels swells larger and larger, which eventually leads to its rupture, disintegration and complete dissolution, it is called infinite swelling. The occurrence of these two situations are not absolute, and changing conditions can also change the nature of nanogel swelling.

Swelling degree S refers to the limit amount of liquid that can be absorbed by nanogels per unit mass or volume under certain conditions, which is defined as:

$$S = \frac{m_2 - m_1}{m_1} \tag{5.9}$$

or

$$S = \frac{V_2 - V_1}{V_1} \tag{5.10}$$

In the equation, m_1 and m_2 are the mass of the nanogels before and after the swelling of the nanogels, and V_1 and V_2 are the volume of the nanogels before and after the swelling. The swelling degree of the nanogels is related to the structure of the nanogels and the solvent, and increases with the increase of temperature.

The occurrence of swelling is not instantaneous, and there is a process to reach the equilibrium of swelling. The kinetic equation of the first-order reaction of the swelling rate of the nanogels:

$$\frac{\mathrm{d}S}{\mathrm{d}t} = k_{\mathrm{swell}}(S_{\max} - S) \tag{5.11}$$

In the equation, S is the amount of liquid absorbed by the nanogels when the swelling time is t; $S_{\max}$ is the maximum amount of liquid absorbed (in equilibrium), and k_{swell} is the swelling rate constant.

5.4.1.2 Swelling mechanism

The swelling process of hydrogels with cross-linked structure is actually a balance of two opposite trends. On the one hand, the solvent tries to penetrate into the gel network structure to expand its volume, leading to the expansion of the network structure; on the other hand, the extension of the molecular chains between the cross-linking points reduces the conformational entropy of polymers, and the elastic contraction force, generated by the three-dimensional network subjected by stress, causes the molecular network to contract. When these two opposite trends offset, the swelling equilibrium is reached [26]. The swelling ratio of nanogels is related to the temperature, pressure, the degree of cross-linking of polymers, and the properties of the solute and solvent. The quantitative relationship between them is derived from the lattice-like model solution theory and the high elastic statistical theory:

$$P^{\frac{5}{3}} = \frac{\left(\dfrac{i}{2V_u S^{\frac{1}{2}}}\right)^2 + \dfrac{\frac{1}{2} - x_1}{v_1}}{\dfrac{V_e}{V_0}} \tag{5.12}$$

In the equation, $i/2V_u$ is the fixed consumption of the non-swelling network structure, S is the ionic concentration of the external solution of the hydrogel, $(1/2 - x_1)$ is the affinity of the hydrogel to the deionized water or solution, and V_e/V_0 is the cross-linking density of hydrogels.

The swelling process is specifically divided into two stages. The first stage is the formation of solvation layers. Solvent molecules quickly occupy the nanogels and interact with the nanogel macromolecules to form solvation layers. This stage is very short and rapid. It has the following characteristics: (1) The vapor pressure of the liquid is very low. This is because the degree of freedom decreases when the solvent molecules first form solvation layers, so the vapor pressure of the system is very low. This part of the liquid is tightly bound with the gel macromolecules, so it is difficult to completely remove. (2) Volume shrinkage. Although the volume of nanogels increases, as a whole, the increased volume of nanogels is smaller than the volume of the absorbed liquid, thus the volume shrinks. (3) There is a thermal effect. The heat released after dry gel absorbs a certain amount of liquid and swells is called swelling heat, which can be directly measured; (4) Solvent entropy decreases, which is due to the ordered arrangement of liquid molecules in solvation layers.

The second stage is the penetration and absorption of liquid. In this stage, the absorption of liquid is several times or tens of times of that of dry gel plasmids, and there is no obvious thermal effect or volume shrinkage [27]. Because it takes a long time for solvent molecules to penetrate into the core of nanogels, this stage takes a long time. At this time, the pressure exhibited by nanogels is swelling pressure.

5.4.1.3 Affecting factors of nanogel swelling

The swelling of nanogels in aqueous solution is based on the degree of dissociation of functional groups and the hydration of ions, which is affected by multiple factors: (1) the structure of nanogels (chemical structure of polymers, degree of cross-linking, charge density of electrolyte gels, etc.); (2) Environmental parameters, such as pH, ionic strength, temperature, composition of swelling medium, etc. [28].

The pH value of the medium: The pH value of the medium has a great influence on the swelling behavior of the three-dimensional network structure containing ionizable groups such as protein and cellulose. At a certain pH value, the groups ionize, causing a fixed charge to accumulate on polymers, and the resulting electrostatic repulsion can promote the swelling process of gels. For anionic hydrogels, the side chain groups ionize when the pH of the medium is greater than its acidity dissociation constant (pK_a), and gels swell. The cationic hydrogels ionize when the pH is less than pK_a, promoting the swelling process. Using 2-hydroxyethyl methacrylate, methyl methacrylate or *N*-vinyl-2-pyrrolidone as comonomers, the equilibrium and dynamic swelling behaviors of hydrogels containing methacrylic acid or various acrylamide were studied. pH-sensitive gels with a wide range of expansion ratios can be obtained through a series of composition changes [29].

Solution ionic strength: Ionic strength also determines the swelling of polyelectrolyte nanogel. For example, the swelling of ionic nanogel is only affected by the concentration of cross-linker under high ionic strength, but when under low ionic strength, both cross-linker and charge strength will affect the swelling. The general rule is that the cross-linking ratio of the cross-linked gel will decrease when the number of cross-links increases [30].

Temperature: Under certain circumstances, the interactions between the polymer and the solvent chain are affected by temperature, which can lead to swelling or collapse of the gel, so nanogel is temperature responsive. when the system is in equilibrium (limited swelling), the maximum swelling degree decreases with the increase of temperature. This is mainly because the swelling process is exothermic, and the increase in temperature will lead to a decrease in the binding strength of the particles in nanogels, and the maximum swelling degree they can bear decreases.

Others: The aging degree and cross-linking degree of nanogels have an influence on the swelling degree of nanogels. The more aging nanogels are, the higher the degree of cross-linking and the lower the degree of swelling.

5.4.1.4 Thixotropy and desizing effect of nanogels

Subject to an external force, the network structure of nanogels will be destroyed, linear particles will be separated from each other, and the system will show liquidity. If the external force is removed and nanogel is left to stand for a period of time, particles re-cross-link into a network structure, which is called thixotropy. The mutual transformation between sols and nanogels can be repeated.

After sol-gelation, its properties are not completely fixed. During the placement process, the properties of the nanogels are still changing. This phenomenon is called aging. Its manifestation is the desizing phenomenon, also known as dehydration. In other words, under the condition of basically not changing the principle form of hydrogels, it makes the arrangement more orderly through the contraction of the particles, and simultaneously squeezes some liquid out to produce the "sweating' phenomenon [31]. The syneresis of a gel can also be regarded as a process of solubility reduction. The rate of syneresis is a first-order reaction, which is proportional to the concentration of the nanogels.

The swelling behavior of pH-sensitive gels is not only related to the composition of gels, but also related to external changes. For sensitive gels containing weak acid groups, such as polyacrylic acid (PAA) and polymethacrylic acid (PMAA), when the external pH is higher than pK_a, gels swell. And for sensitive gels containing weak base groups, such as polydimethylaminoethyl methacrylate (PDMAEAM), polyvinylpyridine (PVP), etc., there are one or more volume phase change points.

5.4.2 *In vivo* Circulation of Nanogels

Drug-loaded nanogels need to circulate in the body to transport the drug to the target site (Fig. 5.3). Its circulation time and integrity are essential to prevent the drug from being rapidly eliminated by the body or degraded and metabolized. One of the most important barriers to realizing long circulation of nanogels is their opsonization, that is, it is readily cleared by organs of the monocyte macrophage system such as the spleen and liver. Modification of PEG on the surface of the nanogels can enhance their hydrophilicity and reduce their interaction with serum proteins. The final effect is related to the size, shape, molecular weight, and density of the surface PEG of the nanogels. The softness and deformability of nanogels also help to avoid the filtering effect of the spleen. In other words, they can deform under physiological conditions to pass through membrane pores that are many times smaller than their hydrodynamic diameter. The deformability of nanogels can be adjusted by changing the cross-linking density between particles and changing the size of the cross-linked parts, and the introduction of electrolytes into the hydrogel network can increase the gel swelling ratio.

According to the structure of the carrier, targeted behaviors can be divided into three types: The first is active targeting. In other words, the drug-loading system can identify the specific antigen or receptor expressed in the tissue or cell of the lesion site. Generally, it is realized by attaching specific monoclonal antibodies, cytokines, or ligand proteins on the surface of carriers. The second is passive targeting. Carriers with different particle sizes and surface properties can be captured by specific tissues or organs in *in vivo* circulation. For example, the particle sizes of carriers reaching the human brain, bone marrow, and other tissues and cells must be less than 50 nm; nanoparticles of 100 nm–200 nm are easy to be internalized by phagocytes of reticuloendothelial system and accumulate in the liver and the spleen; carriers with a particle size larger than 1000 nm will be absorbed by the first capillary network or mechanically intercepted by the lungs and mainly concentrated in the lungs [32]. The last one is physical targeting, which mainly exploits the differences of the temperature and acidity between the lesion and the normal tissues, or external factors such as electric field, magnetic field, and light on the drug-loading system to achieve targeted movement. Nanogels are usually too large to pass through tight junctions between normal endothelial cells, but they can effectively accumulate in solid tumors or inflammatory tissues. These sites have distinctive structural features, such as vascular leakage defects and loose connections, and enhanced permeability and retention effect (EPR) caused by lymphatic drainage. In addition to the EPR effect, nanogels, through their biological binding with the target ligands, bind specifically to the receptor molecules differentially expressed on the surface of the diseased cells/tissues, prolonging the retention time at target sites and promoting cell uptake. Compared with non-targeted nanogels, ligand-mediated targeted nanogels are mainly distributed in highly expressed tissues, which can avoid excessive accumulation of gel at non-target sites and reduce related side effects.

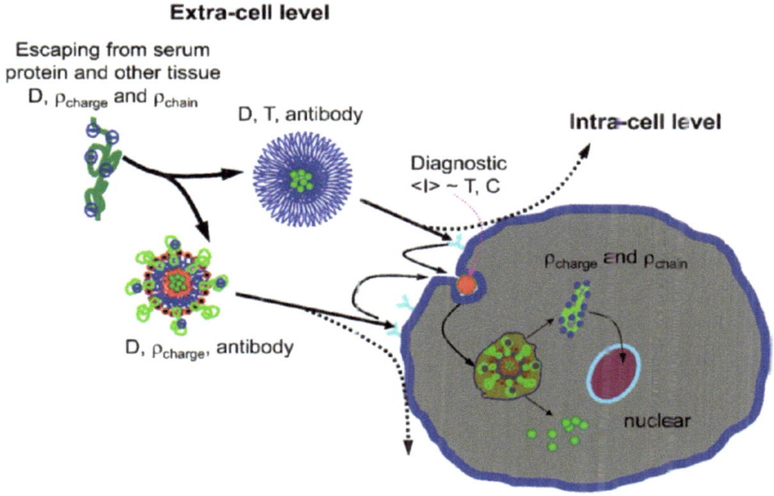

Figure 5.3 *In vitro* carrier–drug complex formation process (micelles and nanogels), microstructural parameters of complex properties, cellular interaction and intracellular fate. The physicochemistry parameters noted here are also shown in Fig. 1.16 in Chapter 1.

After the nanogel is extravasated from blood vessels, it diffuses through the intercellular space and reaches target cells, and can be absorbed by target cells through various endocytosis mechanisms, according to the surface characteristics of nanogels, such as size, softness and charge, as well as the receptor type of target cells. In general, nanogels enter endosomes via endocytosis and then reach lysosomes. The low pH, high degradation enzyme concentration, and reductive environment become the stimulating factors to drug release behavior from gels, giving them the function of targeting specific organelles.

5.4.3 Drug Release Behavior of Nanogels

The loading, transportation, and release process of drugs through nanogels generally involves the following steps: (1) Dissolve nanogels in the drug solution. (2) Because of the concentration gradient, drug molecules diffuse through the barriers and reach the surface of nanogels. (3) The drug is desorbed from polymers at the specific position. (4) The drug diffuses into body fluids or tissues. During the whole process, there are many complex physical and chemical phenomena, and the release characteristics of drugs depend on the interaction of these physical and chemical phenomena [33]. Through the analysis of these phenomena and the study of the kinetic and thermodynamic process involved, a way to control and adjust the release rate of drugs can be found, so that the release of drugs from preparations can meet the expected requirements.

Drugs diffuse and permeate mainly by density changes between reticular polymer chains to achieve drug release behavior. For degradable drug-loaded nanogels, the release of the drug is also related to the degradation rate of the polymer/chemical bonds. Biological agents can often be combined with nanogels through physical embedding and covalent binding. According to the loading method of the drug, this type of gel can be divided into two types. The first one is the drug encapsulated in a biodegradable polymer film or dispersed in a polymer gel, and the drug release behavior is affected and controlled by the degradation rate of the polymer; the second one is to bond the drug on a polymer chain and the release of drug can be regulated by controlling the break of chemical bonds. In different system types and different polymer degradation mechanisms, the drug release behavior is controlled by diffusion, degradation, or both.

The drug release behavior of carriers mainly hinges on the interaction of the drug, the carrier, and the release medium. If the binding force between the carrier and the drug is greater than that between the carrier and medium, then the drug release is mainly decided by the degradation mechanism, and its release rate is affected by the swelling and erosion process of the carrier material; on the contrary, if the drug release behavior is dominated by diffusion, its release rate is affected by the process of drug diffusion. When the drug is bonded to the carrier material in the adsorption manner, the release of the drug is a desorption process and sudden release is easy to occur. For poorly soluble drugs, the release process in some cases is controlled by the solvation process of the drug itself, and is not related to the type of carrier. For a specific drug-loading system, the release of the drug may be based mainly on one mechanism and secondarily on multiple other mechanisms; or there may be different release mechanisms at different release stages. For polymer nanoparticles, the multiphase release pattern often appears, including the initial burst release period and the subsequent sustained release period.

The diffusion of the drug from nanogels is strongly influenced by its microstructure [34], and how to overcome the lack of drug release is an important problem in drug delivery systems. Wei's Li [35] group studied a novel dual-controlled temperature-sensitive nanogel to improve drug release, and drug release kinetics is related to drug diffusion, which is mainly controlled by the microstructure of nanogels. Therefore, establishing the relationship between the drug diffusion coefficient (D_{gel}) inside the gel and the microstructure change during the phase transition of nanogels contributes to the improvement of drug release behavior. This relationship can be explained by the following equation:

$$D_{\text{gel}} = D_{\text{solution}}\left[1-\frac{r}{\zeta}\right]e^{(-1/n-1)} \qquad (5.13)$$

where D_{solution} is the diffusion coefficient of the drug in the solution, r is the size of a drug molecule, ζ is the dynamic correlation length, or the mesh size between two cross-linking points, and

n is the volume swelling rate of the nanogels ($n = V_{swell})/(V_{collapse})$. Diffusion kinetics is mainly affected by ζ and n. As shown in Fig. 5.8, when the temperature causes a phase transition, the uneven collapse will cause a dense surface layer. Due to high chain entanglement or overlap, ζ and n are relatively small and the diameter in the collapse state is about 90 nm. Both ζ and n are able to lead to a decrease in D_{gel}, which indicates that the resistance of drug diffusion is high. However, under ultraviolet irradiation, ζ and n are larger, and D_{gel} is therefore larger (Fig. 5.4), which means a large drug release (Q) and well explains the drug release efficiency of the gel.

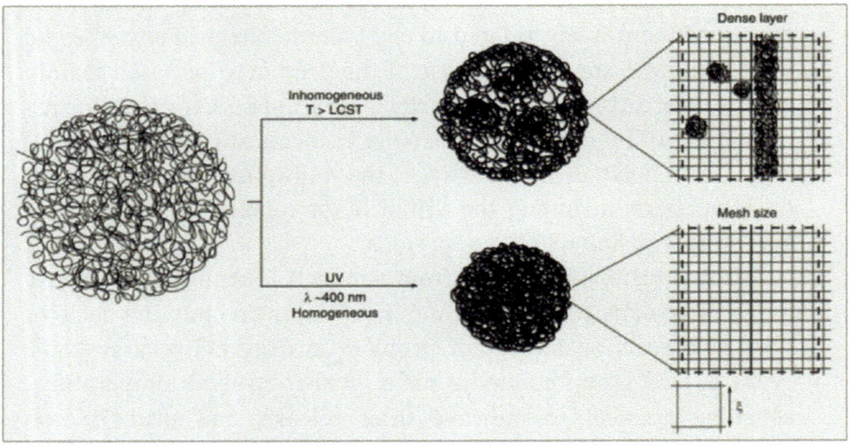

Figure 5.4 The effects of structural collapse and microstructure (mesh size) of the temperature-sensitive nanogels during the phase transition process on the drug release, which is described by the modified drug release function shown in 5.13.

5.4.4 Factors Affecting the Release of Drug-Loaded Nanogels

When the nanogel drug delivery system reaches the diseased tissue or cell, how to control the release and the releasing dose is a key issue. The network of nanogel is sensitive to the environment, can withstand rapid changes in volume, and can be stimulated to control the release of bioactive compounds. In addition, nanogels can be chemically modified to achieve the delivery and release of the targeted drug. According to whether the system responds to the signal or not, the release of nanogels can be divided into the programmed pulsed-release system (PPRS) and the intelligent pulsed-release system (IPRS). The release manner of PPRS is achieved through the pre-designed structure of the carrier, and the lag time and duration of drug release behavior are controlled by the polymer degradation performance. Butun et al. designed a release system similar to the core–shell type. The system consists of alternate drug-containing and drug-free ossein (ossein can gradually degrade in body fluid). The pulse duration, release time, and pulse number of drugs are determined by the composition and layered thickness of ossein [36]. IPRS requires external stimuli and the response of carriers. The intelligent drug pulse release system can be divided into external regulation and self-regulation according to the source of the signals. The former is the release of drugs regulated by external factors such as light, electricity and magnetism, while the latter realizes the drug release behavior through the system itself stimulated by pathological tissues. See the next section for details.

5.4.4.1 Drug-loading methods

The drug-loading process of nanogels requires specific physicochemical properties to encapsulate specific types of drugs. The strategy and the process of drug loading are crucial for the release of the system. The calcium pectate nanogels prepared by the mixing method were faster in drug release behavior than the adsorption method or the swelling method, and the release time was shorter by 50%. The hydrogel prepared by Mackiewicz M. based on poly (*N*-isopropyl acrylamide) has improved in multiple properties [37]. First, cystine(BISS)acrylic derivative containing carboxyl group is used to replace the commonly used *N,N′*-bis(acryloyl)cystamine (BAC) as the cross-linking agent, giving hydrogel particles better stability and ionic strength, and allowing further chemical modification, including drug molecule bonding. Second, the water precipitation polymerization method, by using the semi-batch method to initiate redox, greatly increases the content of BISS and reduces the size of nanoparticles, which is conducive to endocytosis. In addition, the prepared nanogels show a higher loading capacity than dry nanogels (16%), showing better stability and enhanced drug release effect [38].

5.4.4.2 Medium pH

When anion or cationic groups that can be protonated or deprotonated are included in the 3D network of nanogels, their pH response to the medium is more sensitive. If the pH of the environment changes, the proportion of different ionic states of these groups will change, leading to the change in hydrophilicity and finally the change in the structure of the 3D network of the gel. Anionic nanogels generally contain sulfonic or carboxylic acid groups, and if the pKa of the polymer is greater than the pH value of the solvent, its ionic structure results in an increase in electrostatic repulsion within the network and eventually leads to overall expansion. On the other hand, cationic nanogels generally contain amino groups at the end, so when the pH value of the environment around the gel is less than pK_b, the amino groups change from NH_2 to charged NH_3^+ group, thus its electrostatic repulsion, hydrophilicity and expansion rate are increased [39, 40]. The pH of healthy tissue (pH 7.4) and tumor tissue (pH 6.5–7.0) are different. According to the expansion or collapse of gel caused by the protonation or deprotonation of ionic groups in the network, the drug release at specific sites can be controlled.

5.4.4.3 Solvent

Nanogels work by loading hydrophobic drugs into polymer nanonetworks in a solvent, and the details of this process can have a profound impact on their properties. Earlier work [41, 42] showed that when nanoparticles were in an aqueous solvent, hydrophobic effects acted as the driving force, resulting in the hydrophobic core reorganizing into approximately spherical and dense regions, while the hydrophilic parts remained solvated. Thus, in water, not only are hydrophobic drugs difficult to dissolve, but polymer nanoparticles may adopt conformations unfavorable for drug loading. Therefore, in the process of drug loading, nanoparticles are usually first dissolved in water-miscible organic solvents, such as tetrahydrofuran, which is conducive to the formation of hydrophobic cores of nanoparticles [43]. Since the structures of nanoparticles are likely to be strongly influenced by solvents, the particulars of the loading process, especially the selection of drugs and solvent, may lead to differences in loading location and efficiency.

The effects of organic solvents such as dichloromethane, tetrahydrofuran and dimethylsulfoxide on the conformations of two kinds of nanogels were studied by simulating real full-atom polymers in different solvents [44]. When dichloromethane, tetrahydrofuran and toluene are used in the drug-loading step, star polymer nanogels can load large drug molecules inside and the drug-loading rate rises. However, the use of methanol mainly leads to shallow or surface drug loading, so the drug release rate is higher. Ether cannot be used as solvent because it cannot sufficiently swell the nanogels [45].

5.4.4.4 Particle size

The particle size of the nanogels has a great influence on the drug loading and drug release capacity. Researchers have investigated the morphology, particle size and sustained release properties of a drug-loaded nanogel, in which polymethacrylic acid (PMAA) is the raw material, ethylene glycol dimethacrylate (EGDMA) is the cross-linker and amoxicillin (AM) is loaded. All the samples showed uniform spherical shape, and the particle sizes were less than 500 nm. In addition, the particle sizes of nanogels decreased with the increase in the content of the chain terminator (EGDMA). Further, the nanogels can achieve prolonged drug release, which indicates that nanogels can control and continuously release drugs and have long-term antimicrobial activity [46]. However, different release rates were found according to nanogels of different sizes, and the release rate of AM decreased with the decrease in particle size. This may be because smaller nanogel particles form a denser hydrogel network, hindering the release of drugs from the nanogels.

5.4.4.5 Surface charge

The existence of electrolytes in suspension is crucial for the charged nanogel in aqueous medium. In fact, moving ions are distributed in and around the porous nanogel, forming a so-called electric bilayer, which may determine the degree of particle expansion and the efficiency of co-solute absorption. On the other hand, an increase in the concentration of salt leads to a decrease in particle effective charge, which leads to the shrinkage of the nanogel, reducing the absorption of the internal solute with the opposite charge. In addition, the ions and the change in the local concentration of ions may also affect the swelling response of the charged-polymer network, resulting in internal inhomogeneity in polymer density [47].

Arturo et al. explored the charge distribution of charged nanomaterials in electrolyte water suspensions through Monte Carlo simulation and Ornstein–Zernike integral equation theory, and considered the size of ions, the swelling or shrinkage state of the gel, the valence state of the counterions, the type of charge carried by the gel, as well as the interaction between electrostatic and excluded volume effect. When the gel is in a state of contraction, electrostatic repulsion between charged groups will lead to an uneven distribution of polymer and form a low-density pore structure surrounded by dense corona. The excluded volume repulsion strongly inhibits the penetration of ions in the contracting nanogels, and significantly reduces the counterion concentration at the dense structure. In general, the properties and internal structures of nanogels are strongly influenced by the valence of counter ions and the charge of nanogels. On the one hand, with the increase of the valence of counter ions, the charged monomer enters into the inner and outer layers near the coronal structure, forming a shrunken-down nanogel. On the other hand, increasing the charge of particles can promote the swelling curve to move in the direction of increasing the effective radius of nanomaterials [48].

5.4.5 Drug Release Behavior of Stimulus-Responsive Nanogels

Nanogels can integrate drugs, proteins and DNA through physical and chemical methods, and also have adjustable chemical framework, swelling rate, mechanical strength and other properties. Nanogels are also known as smart materials because they can respond to external stimuli (such as pH, temperature, light, etc.) by changing their volume, hydrophilicity, refractive index, etc. [49]. Stimulus-responsive nanogels can be released at a specific dose at a specific time and place. It is a smart drug release system with a feedback mechanism. The drug-loaded nanogels are in a contracted state under normal conditions. After receiving disease information, they can respond and expand the volume to promote the diffusion and release of the drug. When the body recovers, the nanogel shrinks and prevents the continued release of the drug, so as to achieve the intelligent and controlled release of the drug. According to different stimulus sources, temperature-, pH- and glucose-responsive nanomaterials have been developed for applications in drug transport, biological imaging, etc. [50].

5.4.5.1 Temperature-responsive nanogels

Temperature-responsive nanogels are one of the most frequently studied environmentally sensitive systems in drug delivery. The structure of this type of nanogel contains a certain ratio of hydrophilic and hydrophobic groups. Changes in temperature can affect the interaction between groups and solvents, thereby promoting changes in the network structure and volume of the hydrogels. In the hydrogel solvent system, there are generally two kinds of temperature-responsive phase transitions, that is, lower critical solution temperature (LCST) and upper critical solution temperature (UCST). Between them, the LCST-type polymer gel shrinks at a temperature above the critical value, while the UCST-type polymer gel shrinks at a temperature below the critical point. The former is widely used in the temperature-driven drug release strategy.

Poly(*N*-isopropylacrylamide) is currently the most widely used polymer with temperature stimulus responsiveness because its LCST is close to human physiological temperature. While poly (*N,N*-diethylacrylamide) has a slightly lower LCST at 25 °C–32 °C. It can also be used for drug delivery. However, copolymers of acrylamide (AAM) and *N*-isopropylacrylamide (NIPAAm) are used to prepare temperature-responsive photothermally adjustable drug delivery systems. The LCST of copolymers can be adjusted by changing the concentration of AAM. For example, increasing the ratio of AAM/NIPAAm can increase the LCST from 32 °C to 60 °C. The gel is used to encapsulate gold nano sheet, which can cause local temperature rise by absorbing near-infrared radiation (NIR), leading to rapid contraction of the gel and the release of the encapsulated drug [51].

5.4.5.2 pH-responsive nanogels

The swelling properties of pH-responsive nanomaterials mainly originate from the ionizable acidic or alkaline groups on the side groups. However, the electrostatic dissociation effect of the neighboring ionizing groups increases ionization difficulty, and the apparent dissociation constant (K_a) is different from that of the corresponding monobasic or monobasic K_a. The existence of ionizable groups causes electrostatic repulsion between polymer chains, and the swelling degree of the corresponding hydrogel is much higher than that of the hydrogel formed by non-electrolyte polymers. Therefore, the degree of gel swelling can be controlled by influencing the electrostatic repulsive force, such as changing the pH, ionic strength, and type of counter ion of the medium. Moreover, the swelling degree of hydrogels can be regulated by co-polymerization of neutral monomers (e.g., methyl methacrylate), while different comonomers can offer different hydrophobicity, leading to different pH response behaviors.

At present, polyethylene glycol grafted anionic polymethacrylic acid-based nanogels have been developed for oral drug delivery. When the environmental pH is around 2, the pore size of the gel is smaller due to the formation of hydrogen bonds between the carboxylic acid groups and the oxygen atoms on the polymethacrylic and polyethylene glycol respectively. When the pH increases to 7, methacrylic acid is deprotonated, attracting water molecules to enter and causing the hydrogel to swell. This phenomenon is adapted to the conditions of the digestive system, that is, it can protect the drugs in the stomach (low pH environment), while allowing the diffusion and release of the drug in the small intestine [52].

5.4.5.3 Glucose-responsive nanogels

Glucose-sensitive nanogels can swell or collapse in response to changes in glucose concentration, and mainly used in the biosensor field and insulin control release systems. Glucose-sensitive materials are mainly divided into three types: glucose oxidase (GOD), concanavalin (Con A) and phenylboric acid (PBA). The GOD system is inherently pH sensitive. It generates gluconic acid and hydrogen peroxide through the reaction of glucose oxidase and glucose, which leads to a decrease in local pH, causes swelling, and promotes the diffusion of the carried insulin from the gel, reflecting its glucose sensitivity [53]. Con A is a natural agglutinin with four glucose-binding sites. Agglutinins interact with various cells by reacting with glycoproteins and glycolipids on the cell surface. Glucose-sensitive gel prepared by cross-linking with concanavin (Con A) and carboxymethyl dextran, when free glucose competes with the oligosaccharide site of Con A in the external environment, the gel structure is loose and insulin is released [54].

Although glucose-responsive nanogels have excellent performance as an insulin delivery system, they still need improvement in various aspects before clinical application. Their application constraints include slow response to changes in glucose concentration in the environment, and reducing the size of nanogels may be a way to shorten the response time. Besides, they lack a high repeatability and rapid response to the changing glucose concentration, and the ingredients used may induce adverse immune reactions [55].

5.4.5.4 Photoresponsive nanogels

Photoresponsive nanogel is a kind of gel that can change in volume under the condition of light radiation, which is generally divided into two types. The first type of nanogel contains photosensitive groups, such as triphenylmethane and azobenzene, etc., which will undergo photodissociation or isomerization under light irradiation, resulting in swelling of the gel. The other type belongs to hybrid systems, which are nanoparticles with metal and temperature-sensitive groups in the nanogel structure [56]. The principle is that metals such as gold and silver can convert light into heat, so the temperature in the system increases, and temperature-sensitive groups make the nanogel responsive [57]. A light and temperature-sensitive nanogel is shown in Fig. 5.4.

5.4.5.5 Other stimulation-responsive nanogels

With the development of scientific research, more and more stimulus-responsive nanogels have been developed. Magnetic responsive nanogel is a kind of magnetic-sensitive nanogel system containing magnetic materials such as Fe_3O_4, Fe_2O_3. It has good magnetic responsive drug release characteristics and has great application prospects in the diagnosis and treatment of tumors [58, 59]. Nanogels with double or multiple responses have also gradually emerged, such as the magnetic/temperature dual response system [60], temperature/pH dual response system [61], pH/magnetic dual response system [59], and temperature/pH/magnetic multiple-response system [62]. Double or multiple-response nanogels combine the advantages of each response monomer and have more potential, which is worthy of further investigation.

5.4.6 Biological Applications of Nanogel Delivery Systems

With the development of nanotechnology, biomedicine and smart materials, nanogels have shown attractive application prospects due to their excellent drug-loading properties and drug release ability, so they have attracted great attention.

5.4.6.1 Delivery of small-molecule therapeutic drugs

The application of nanogels for drug delivery is an important direction of current research, which mainly has the following three advantages: targeted therapy, controlled release and long circulation. Nanogels are prepared from amphipathic block copolymers, PEG-*b*-polymethacrylic acid (PEG-*b*-PMA) can be synthesized via ionic complexation and cross-linking reaction with a swollen PMA cross-linking network as the core and PEG chain as the shell. This synthesis method can adjust the macro characteristics (e.g., size and swelling degree) of nanogels, by changing the cross-links number or chemical structure [63]. Compared with other nanocarriers, this nanogel shows very low non-specific adhesion, which facilitates specific targeted delivery. Moreover, nanogels with such core and shell can load a large amount of (50%) doxorubicin (DOX) through the electrostatic adsorption of the carboxyl group of the core, which can not only remain stable for a long time but also promote the release of DOX after being protonated in an acidic environment. Meanwhile, it shows good cytotoxicity to tumor cells. The author further used the nanogel to contain the hydrophilic chemotherapeutic drug cisplatin, which not only achieved pH-responsive release but also effectively controlled the drug release time by controlling the degree of cross-linking of the inner core. The released platinum drugs remained active and could form platinum adducts with the DNA in the nucleus of the tumor cell. The experimental results of the mouse model of ovarian cancer indicated that the nanogel drug-carrying system could enhance the anti-tumor effect by increasing the drug concentration at the tumor delivery site and reduce the nephrotoxicity of platinum drugs to improve the safety of treatment. In addition, the tumor growth was more effectively inhibited by modifying the drug-loaded nanogels with targeted ligands (e.g., folic acid, LHRH peptide).

5.4.6.2 Delivery of oligonucleotides

Therapeutic oligonucleotides (e.g., small interfering RNA (siRNAs) and antisense oligodeoxynucleotides (ODNs)), can diagnose and treat tumors, neurodegenerative diseases, lethal virus infections, and other diseases by targeting interference with specific mRNA sequences. Although some oligonucleotide-based treatment methods have achieved initial effects in clinical trials, how to deliver them to target cells is still an urgent concern. Because oligonucleotides are hydrophilic molecules with a negative charge, they cannot penetrate the cell membrane and can be degraded by a variety of endogenous nucleases to stimulate the inherent immune response. Therefore, oligonucleotides need a delivery system to deliver them to the target site without causing side effects. Smart polymer nanogel is a common carrier, mainly composed of polymer polyelectrolytes or copolymers of ionic monomers, giving the gel surface a certain charge. It is able to achieve targeted delivery by adsorbing DNA fragments, bases and other substances, wrapping or forming complexes, and releasing them in response to specific stimuli, with basically the same mechanism of action as that of drug delivery systems.

At present, various cationic nanogels have been modified to deliver nucleic acids such plasmid DNA, siRNA molecules. However, endocytosis or binding of oligonucleotides is not enough to keep specific genes silent, and modification of functional groups with potential buffer function on nanogels may be beneficial to gene therapy. The chemical structure of modified siRNA may also enhance its transfection efficiency and increase its stability. Figure 5.5 shows the general process of DNA-nanogel polyplex formation and release as conducted in our group. It is easily seen that the polyplex formed by a positively charged chain like PEI is not stable because the size and polydispersity (PDI) are large due to the interparticle aggregation. On the other hand, in the polyplex formed by the positively charged nanogel, the negative pDNA is absorbed and trapped in the nanogel resulting in a stable and uniform polyplex.

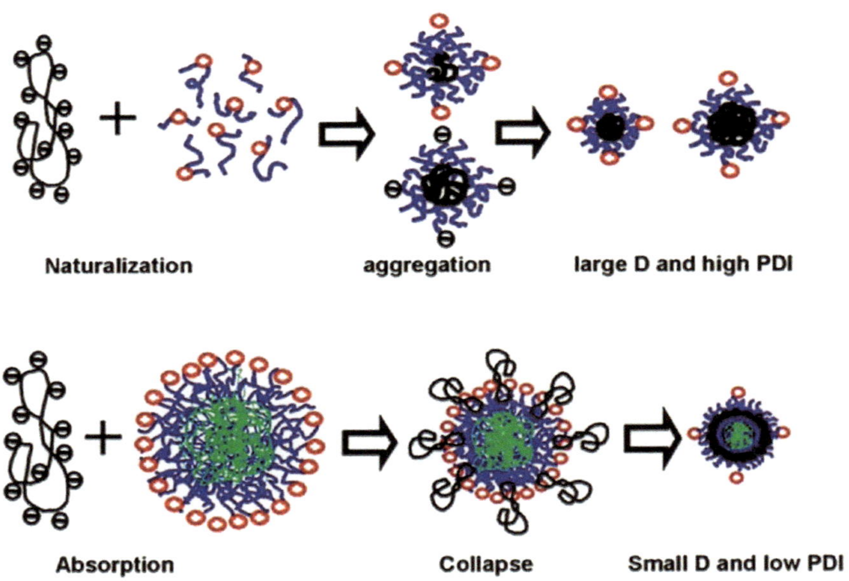

Figure 5.5 The mechanism of DNA complex and polyplex stability as interacted with cationic nanogel (such as PNIPAM-PEI nanogel) and cationic polymers (such as PEI).

5.4.6.3 Delivery of therapeutic proteins

Nanogels can be used to encapsulate biological macromolecules, such as proteins and peptides, to prevent them from being rapidly degraded by the body. For example, cholesterol-modified amylopectin (CHP) gels can form a variety of complex protein complexes through hydrophobic interaction. The complexation can effectively inhibit aggregation and thermal denaturation of proteins and protect them from enzyme degradation. The amount of complexation is related to the hydrophobicity and of molecular weight proteins [64]. However, the protein complexes are unstable under high protein concentration. To improve the long-term stability of the complexes, nanogels with strawberry-like structures were prepared by cross-linking between CHP (modified by acrylate groups) and PEG (modified by thiols), reducing the size of nanogels (40 nm–120 nm). The nanogels had a high encapsulation rate of interleukin-12 (IL-12), and IL-12 level in plasma remained stable after subcutaneous injection for 72 h. Shi et al. synthesized acid-resistant nanogels of *N*-vinylformamide through inverse emulsion polymerization, and the drug loading of lysozyme was up to 60%. About 95% of lysozyme was released at pH 5.8 after 3 h; at pH 7.4; at the same time, only 15% of lysozyme was released, while the released lysozyme retained about 50% of its initial activity. Therefore, although nanomaterials can significantly improve the loading efficiency, the protein activity, structural consistency, and stability after loading are still problems to be considered.

5.5 Summary and Perspective

With the continuous improvement of biomedical technology, the demand for more effective, cheaper, and more tolerant diagnosis, and treatment systems is also increasing. Nanogels show unique advantages in the field of drug delivery due to their high water content and good biocompatibility. With the continuous improvement of nanogel design theory, it is possible to design more functional carrier systems according to the demand by finely optimizing the performance of nanogel carriers, including stability, cell targeting ability, load efficiency, release behavior and so on. However, how to improve the biocompatibility and biodegradability of nanogels is still a huge challenge. With the continuous improvement of synthesis and preparation technologies, it is expected that extensive clinical application of nanogels may be achieved, and new therapies for diseases may be developed.

References

1. Li W, Wang WF, Li RQ, Yao SD, *Nucl Tech* **2002**, *025*, 624–630.
2. Cheng CC, Liang MC, Liao ZS, Huang JJ, Lee DJ, *Macromol Biosci* **2016**, *17*, 1600370.
3. Akiyoshi K, Kobayashi S, Shichibe S, Mix D, Baudys M, Kim SW, Sunamoto J, *J Control Release* **1998**, *54*, 313–320.
4. Daoud-Mahammed S, Couvreur P, Gref R, *Int J Pharm* **2007**, *332*, 185–191.
5. Yu S, Yao P, Jiang M, Zhang G, *Biopolymers* **2006**, *83*, 148–158.
6. Yu S, Hu J, Pan X, Yao P, Ming J, *Langmuir* **2006**, *22*, 2754–2759.
7. Pereira P, Pedrosa SS, Correia A, Lima CF, Olmedo MP, González-Fernández A, Vilanova M, Gama FM, *Toxicol In Vitro* **2015**, *29*, 638–646.
8. Murthy N, Xu M, Schuck S, Fréchet JMJ, *P Natl Acad Sci USA* **2003**, *100*, 4995–5000.
9. Goh SL, Murthy N, Xu M, Fréchet JMJ, *Bioconjug Chem* **2004**, *15*, 467–474.
10. Oh JK, Tang C, Gao H, Tsarevsky NV, Matyjaszewski K, *J Am Chem Soc* **2006**, *128*, 5578–5584.
11. Donini C, Robinson DN, Colombo P, Giordano F, Peppas NA, *Int J Pharmaceut* **2002**, *245*, 83–91.
12. Hennink WE, van Nostrum CF, *Adv Drug Deliver Rev* **2002**, *54*, 13–36.
13. Kohli E, Han HY, Zeman AD, Vinogradov SV, *J Control Release* **2007**, *121*, 19–27.
14. Vinogradov SV, Kohli E, Zeman AD, *Pharm Res* **2006**, *23*, 920–930.
15. Bronich TK, Keifer PA, Shlyakhtenko LS, Kabanov AV, *J Am Chem Soc* **2005**, *127*, 8236–8237.
16. Gratton SEA, Pohlhaus PD, Lee J, Guo J, Cho MJ, DeSimone JM, *J Control Release* **2007**, *121*, 10–18.

17. Carothers WH, *Trans Faraday Soc* **1936**, *32*, 39.
18. Flory PJ, *Principles of Polymer Chemistry*, Cornell University Press, Cornell, 1953.
19. Tang AQ, *Statistical Theory of Polymeric Reactions*, Science Press, Beijing, 1985.
20. Tang AQ, Jaing YS, Wang MJ, *Chinese J Polym Sci* **1963**, *5*, 35.
21. Tang AQ, Shen JC, *Special Issue of Polymer Academic Conference of Chinese Academy of Sciences*, Science Press, Beijing, 1961, p. 153.
22. Tang AQ, *Science China* **1963**, *11*, 605.
23. Li W, Wei H, Li H, Gao J, Feng S-S, Guo Y, *Nanomedicine* **2014**, *9*, 2587–2605.
24. Donnelly C, Tian Y, Potter C, Jones DS, Andrews GP, *Pharm Res* **2015**, *32*, 167–179.
25. Soni G, Yadav KS, *Saudi Pharm J* **2016**, *24*, 133–139.
26. Huang N, Guan Y, Zhu XX, Zhang Y, *Chemphyschem* **2014**, *15*, 1785–1792.
27. Guzman-Sepulveda JR, Deng J, Fang JY, Dogariu A, *Soft Matter* **2016**, *12*, 5986–5994.
28. Kabanov AV, Vinogradov SV, *Angew Chem Int Ed Engl* **2009**, *48*, 5418–5429.
29. Brannon-Peppa L, Peppas NA, *Biomaterials* **1990**, *11*, 635–644.
30. Bontha S, Kabanov AV, Bronich TK, *J Control Release* **2006**, *114*, 163–174.
31. Su DS, Wang SL, *Physical Pharmacy*, Chemical Industry Press, Beijing, 2004.
32. Belmaker RH, Hermoni M, Lerer B, Ebstein RP, Belmaker RH, *J Pharm Pharmacol* **1980**, *32*, 510–511.
33. Kumar V, Prud'homme RK, *J Pharm Sci* **2008**, *97*, 4904–4914.
34. Li Y, Maciel D, Rodrigues J, Shi X, Tomas H, *Chem Rev* **2015**, *115*, 8564–8608.
35. Li W, Guo Q, Zhao H, Zhang L, Li J, Gao J, Qian W, Li B, Chen H, Wang H, Dai J, Guo Y, *Nanomedicine* **2012**, *7*, 383–392.
36. Bütün V, Wang XS, de Paz Báñez MV, Robinson KL, Billingham NC, Armes SP, Tuzar Z, *Macromolecules* **2000**, *33*, 1–3.
37. Mackiewicz M, Romanski J, Drozd E, Gruber-Bzura B, Fiedor P, Stojek Z, Karbarz M, *Int J Pharm* **2017**, *523*, 336–342.

38. He L, Li D, Wang Z, Xu W, Wang J, Guo H, Wang C, Ding J, *Polymers* **2016**, *8*, 36.
39. Vinogradov SV, Bronich TK, Kabanov AV, *Adv Drug Deliv Rev* **2002**, *54*, 135–147.
40. Leber N, Nuhn L, Zentel R, *Macromol Biosci* **2017**, *17*, 1700092.
41. Huynh L, Neale C, Pomès R, Allen C, *Soft Matter* **2010**, 6.
42. Swope WC, Carr AC, Parker AJ, Sly J, Miller RD, Rice JE, *J Chem Theory Comput* **2012**, *8*, 3733–3749.
43. Miller RD, Yusoff RM, Swope WC, Rice JE, Carr AC, Parker AJ, Sly J, Appel EA, Nguyen T, Piunova V, *Polymer* **2015**, *79*, 255–261.
44. Carr AC, Piunova VA, Maarof H, Rice JE, Swope WC, *J Phys Chem B* **2018**, *122*, 5356–5367.
45. Allen TM, Cullis PR, *Adv Drug Deliv Rev* **2013**, *65*, 36–48.
46. Liu T, Liu H, Wu Z, Chen T, Zhou L, Liang Y, Ke B, Huang H, Jiang Z, Xie M, Wu T, *Mater Sci Eng C Mater Biol Appl* **2014**, *43*, 622–629.
47. Adroher-Benítez I, Ahualli S, Martín-Molina A, Quesada-Pérez M, Moncho-Jordá A, *Macromolecules* **2015**, *48*, 4645–4656.
48. Adroher-Benítez I, Martín-Molina A, Ahualli S, Quesada-Pérez M, Odriozola G, Moncho-Jordá A, *Phys Chem Chem Phys* **2017**, *19*, 6838–6848.
49. Cuggino JC, Blanco ERO, Gugliotta LM, Alvarez Igarzabal CI, Calderon M, *J Control Release* **2019**, *307*, 221–246.
50. Vicario-de-la-Torre M, Forcada J, *Gels* **2017**, *3*, 16.
51. Kim YK, Kim EJ, Lim JH, Cho HK, Hong WJ, Jeon HH, Chung BG, *Nanoscale Res Lett* **2019**, *14*, 77.
52. Kim B, Peppas NA, *Int J Pharm* **2003**, *266*, 29–37.
53. Hassan CM, Doyle FJ, Peppas NA, *Macromolecules* **1997**, *30*, 6166–6173.
54. Zhang R, Tang M, Bowyer A, Eisenthal R, Hubble J, *React Funct Polym* **2006**, *66*, 757–767.
55. Suzuki H, Kumagai A, *Biosens Bioelectron* **2003**, *18*, 1289–1297.
56. Alvarez-Lorenzo C, Bromberg L, Concheiro A, *Photochem Photobiol* **2009**, *85*, 848–860.
57. Wu W, Shen J, Banerjee P, Zhou S, *Adv Funct Mater* **2011**, *21*, 2830–2839.
58. Adriane K, Huang J, Ding G, Chen J, Liu Y, *J Drug Target* **2006**, *14*, 243–253.

59. Wu W, Shen J, Gai Z, Hong K, Banerjee P, Zhou S, *Biomaterials* **2011**, *32*, 9876–9887.
60. Purushotham S, Chang PE, Rumpel H, Kee IH, Ng RT, Chow PK, Tan CK, Ramanujan RV, *Nanotechnology* **2009**, *20*, 305101.
61. Quan CY, Wei H, Sun YX, Cheng SX, Shen K, Gu ZW, Zhang XZ, Zhuo RX, *J Nanosci Nanotechnol* **2008**, *8*, 2377–2384.
62. Fan T, Li M, Wu X, Li M, Wu Y, *Colloids Surf B Biointerfaces* **2011**, *88*, 593–600.
63. Liu C, Yang L, Qiao L, Liu C, Zhang M, Jian X, *Polym Int* **2019**, *68*, 1102–1112.
64. Akiyoshi K, Kobayashi S, Shichibe S, Mix D, Baudys M, Kim S W, Junzo S, *J Controlled Release* **1998**, *54*, 313–320.

Chapter 6

Nano Physical Pharmaceutics of Microsphere Delivery System

Zhiwen Qiu,[a] Wanru Tao,[a] Man Wang,[a] Hanwen Sun,[b] and Wei Li[a]

[a]*Department of Nanomedicine, Naval Medical University, Shanghai 200433, China*
[b]*College of Medicine and Nursing, Dezhou University, Shandong 253023, China*

liwei_dds@163.com

6.1 Classification of Microspheres

Microspheres refer to tiny spherical entities formed during the dissolution or dispersion of drugs in a matrix of polymer materials and shall be categorized as matrix-type skeleton particles [1]. In general, the size range of microspheres is 1 μm~500 μm, and those with a particle size of lower than 500 nm are known as nanospheres or nanoparticles, which belong to the colloid category [1]. After being made into microspheres, drugs can be targeted to specific organs and tissues, and slowly released after reaching the organs or tissues, thus prolonging the efficacy, reducing the number of administrations, and improving the therapeutic effect. In addition, they have multiple advantages

Nano Physical Pharmaceutics
Edited by Wei Li

ISBN 978-981-4968-52-2 (Hardcover), 978-1-003-51393-3 (eBook)
www.jennystanford.com

such as improving the stability of the drug, preventing the inactivation of drugs in the stomach or reducing the irritation to the stomach, facilitating the application and storage of liquid drugs, and realizing the sustained release of drugs. With the development of new technologies, the new process and new materials, long-acting biodegradable injection microspheres have become one of the important research areas of new pharmaceutical preparations [2, 3].

According to the synthesized materials, they can be divided into natural polymer microspheres, semi-synthetic polymer microspheres and synthetic polymer microspheres. The natural polymers used include carbohydrates (glucan, starch, chitin and chitosan), acacia gum, alginate, protein (hyaluronic acid, gelatin and collagen, casein, silk fibroin, albumin and zein) and other types of substances, which are generally degradable *in vivo* and safe. The semi-synthetic polymers contain carboxymethyl cellulose salts, methyl cellulose, ethyl cellulose, etc., which have low toxicity, high viscosity and increased solubility after the salt formation. However, due to hydrolyze, they cannot be processed at high temperatures and need to be prepared as they are used. The new biodegradable polymers in synthetic polymers have become an important carrier of microsphere preparations, among which polylactic acid (PLA), polyglycolic acid (PGA), polylactic-co-glycolic acid (PLGA), polylactic acid and polyethylene glycol segmented copolymer (PLA-PEG), ε-caprolactone and lactide segmented copolymer, polycarbonate, etc. are commonly used, which have the advantages of lower toxicity, better stability, better film-forming and sphere-forming properties [4–6].

Based on the structures of microspheres, they can be divided into solid microspheres with the entire microspheres as a dense structure, hollow microspheres with a hollow interior, and porous microspheres with multiple holes on the surface. Compared with the other two, hollow microspheres have unique advantages in the controlled release and sustained release of drugs. However, the preparation is currently only in the experimental research stage due to the complicated preparation method and high cost, but it remains an important research direction in this field for their bright application prospects. Biodegradable polymer porous microspheres can be degraded *in vivo* without toxic side effects, thus having been widely applied in controlled drug release, tissue engineering and other fields [7–9]

6.2 Preparation of Microspheres

The selection of microspheres carriers shall generally put the following factors into consideration: stable in nature, non-toxic and non-irritating properties, certain strength, elasticity and plasticity, the ability to completely wrap drugs and achieve a reasonable release speed, and the fact that after compounding with drugs, they will not affect the pharmacological action and content determination of drugs [10]. The general preparation process of microspheres is exhibited in Fig. 6.1.

According to the specific steps, it can be divided into the following methods: emulsification-chemical cross-linking method, phase separation method, solvent evaporation method, spray drying method, salting-out method, ultrasonic method, supercritical fluid method, etc. These methods may be used in the preparation of microspheres with different structures. The methods commonly used in the preparation of microspheres with various structures are listed below.

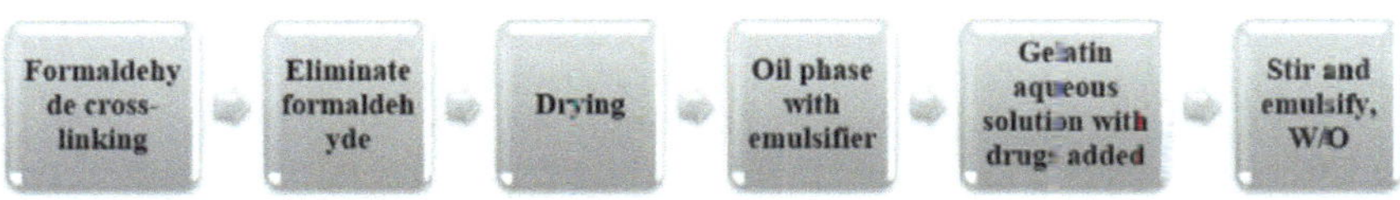

Figure 6.1 Schematic diagram of preparation of drug-loaded microspheres.

6.2.1 Emulsification: Chemical Cross-Linking Method

The emulsification-chemical cross-linking method involves mixing the drug with the carrier solution and dispersing it in an immiscible medium to form a similar water-in-oil or oil-in-water type emulsion, and then forming microspheres by cross-linking. This method involves the cross-linking of groups in polymer materials with the participation of cross-linking agents (such as formaldehyde and glutaraldehyde) to form microspheres, which generally uses the amino groups in the material to condense with the aldehyde group in the cross-linking agent to solidify microspheres [11].

Compared with the traditional solvent evaporation technology and spray drying technology, the microspheres prepared by the emulsification-chemical cross-linking method have a denser structure and a more ideal drug loading capacity. Whereas, microspheres prepared by the traditional method have a low matrix density and are difficult to carry drugs, which leads to the fact that drugs can only be adsorbed on the surface of microspheres, and making it difficult to avoid the dumping effect [11]. The main factors affecting the particle size of microspheres include: the nature of materials, the synthesis concentration, pH value of the solution, the solidification time and the stirring speed, etc. Among them, the concentration ratio of the materials and the drugs is the main factor that determines the drug encapsulation efficiency [12].

6.2.2 Solvent Evaporation

The aqueous drug solution (or buffer) is added to a solvent (such as organic solvents like methylene chloride, chloroform or ethyl acetate) containing carrier material and water insoluble in each other, and emulsified by high-speed stirring or ultrasonic emulsifier under the action of emulsifier to form a water-in-oil (W/O) or water-in-oil-in-water (W/O/W) emulsion. The organic solvents in the emulsion are removed by evaporation using elevated temperature, reduced pressure extraction or continuous stirring, and then prepared into drug microspheres by washing, centrifugation and freeze-drying. Cellulose derivatives, polyesters and polymeric anhydrides can be used as materials for the liquid drying method [13, 14].

The evaporation method is simple to operate and suitable for laboratory operations, but it is difficult to scale up production. This program is usually divided into single the emulsion method and the re-emulsion method, of which the latter is more widely used. The single emulsion method can be divided into O/W and O/O types. The former is suitable for fat-soluble drugs. Under the same circumstances, the encapsulation rate of water-soluble drugs is low, but it can be relatively improved by methods such as pre-emulsifying the water with the drug to saturate the water phase. The re-emulsion method is often used to prepare water-soluble drug microspheres. Generally, polymers are dissolved in methylene chloride, chloroform and other organic solvents, and then an aqueous solution of drugs or active ingredients is added, and stirred or ultrasonically emulsified. After that, they are dispersed in the outer aqueous phase and form the W/O/W type double emulsion to solidify microspheres with the organic solvents removed [15, 16].

6.2.3 Phase Separation

Phase separation is a widely used method for preparing microspheres. Its main principle is to dissolve polymer materials in two immiscible liquids, and use the difference in solubility of the materials in the two solutions to obtain microspheres by filtration, freeze-drying and other methods. The microspheres were optimized by adjusting and controlling the temperature, pH value in the preparation process, as well as adding poor solvents and polymers causing phase separation. The equipment required is simple, together with a wide range of polymer materials, making it suitable for various types of drug microencapsulation. However, problems such as microcapsules or microspheres adhesion, and aggregation are quite common (which is more serious in the gelatin hydrophilic system) [13].

6.2.4 Salting-Out Method

The salting-out effect refers to the decrease of water activity caused by the addition of a large amount of neutral salt, i.e., most or all of the free water in the original solution is transformed into hydrated water of salt ions [13, 17]. At this time, those water molecules forced into contact with the hydrophobic groups on the surface of materials to mask them become free water molecules for further use (not the water molecules involved in the solvation of the polar groups on the protein surface. Instead, they are bound by electrostatic interaction, the link is much stronger than those water molecules in contact with hydrophobic groups), so they are removed to solvate the salt ion, leaving the exposed hydrophobic group. As the salt concentration increases, the hydrophobic surface of the material is further exposed, and the material aggregates and precipitates due to hydrophobic interaction. The method of salting out uses the principle of reversible expansion of polymer materials in solvents [18]. When the polymer materials expand, drugs are added, then the pH is adjusted, and the salting-out agent is added to precipitate the polymer materials and drugs. After centrifugation and drying, microspheres that carry drugs can be obtained. This method of preparing drug-loading microspheres is relatively simple and does not require special equipment. However, the disadvantage of this method is that the concentration of the carrier material during the preparation of microspheres cannot be too high; otherwise, the salting-out agent cannot be evenly distributed, which will easily cause the material to aggregate and form large clusters. Since the drug is bound to the surface of microspheres by adsorption, the extent of sustained drug release is also very limited. In addition, the surface charge balance of microspheres is easy to break when the pH changes *in vivo*, which changes the solubility of microspheres and further accelerates the release of drugs.

6.2.5 Spray Drying

Spray drying involves dispersing drugs in the solution of the materials, and then spraying the mixture into the hot air stream so that the generated droplets are dried and solidified under a short hot air impact to obtain microspheres [19, 20]. This method uses material drug suspension rheology and the nozzle design to control the size and shape of microspheres. Compared with other methods, this method requires short time, and is featured with simple production process, easy operation and control, high product purity, and good dispersibility. However, special equipment is required, and the factors affecting the preparation of microspheres are more complicated, including: the viscosity of the mixed liquid, the uniformity, the concentration of drugs and materials, the spray rate, the spray method and the hot air temperature [13, 21, 22]. Spray drying has been applied to natural polymer materials such as albumin and chitosan.

Compared with the solvent evaporation method, this method avoids the high-speed stirring in the emulsification process and the long-term contact between the drug and the organic solvent, which can effectively protect the activity of biological samples, and opens up a new way for the preparation of unstable drug microspheres. In addition, due to the rapid volatilization of the solvent during the drying process, the temperature of the droplets can be kept lower than that of the drying air, thus facilitating the preparation of temperature-sensitive drugs.

6.2.6 Ultrasound Method

Air albumin microspheres for acoustic contrast can be prepared by using ultrasonic energy to heat the albumin solution to a temperature close to denaturation, which will cause "cavitation" of the micro air originally present in the solution and the air brought in from outside. Under the action of sound waves, the existing bubbles in the liquid expand and then burst to generate a large number of microbubbles [13, 21]. These microbubbles undergo further cavitation under the action of sound waves, thereby splitting to generate more microbubbles. The cavitation process is controlled by controlling the ultrasonic energy and sound vibration time, and finally the microbubbles are made into microspheres with good uniformity to meet the needs [23].

6.2.7 Supercritical Fluid Method

In the foregoing methods, a large amount of organic solvents are used, and post-treatments such as washing and drying are required, which are likely to cause residual organic solvents and affect product quality [13, 24]. At present, the phase separation technology based on the supercritical fluid-atomized solvent extraction system has been used in the preparation of microspheres. The system mainly includes rapid expansion of supercritical solution (RESS), gas anti-solvent recrystallization (GAS) and precipitation with a compressed fluid anti-solvent (PCA) [25]. This process can meet the requirements of light-proof, air-isolated, waterproof, sterile, etc., thus enabling production expansion [13].

RESS: The supercritical fluid containing drugs and polymers is rapidly expanded in the expansion chamber to reduce the solubility rapidly so that the solute can precipitate to form microspheres. This method does not require the use of other solvents, which is conducive to environmental protection. However, CO_2 is relatively weak in polarity and can dissolve fewer substances, so it is mainly applied to fat-soluble drugs.

GAS: Drugs and polymers are dissolved in an organic solvent, and then a supercritical fluid (anti-solvent) that is miscible with the organic solvent but has poor solubility for the drug and polymer is used to precipitate to obtain microspheres. It is applicable to compounds that are slightly soluble or insoluble in supercritical fluids. It is favorable for the expansion of organic solvents to increase the content of supercritical fluid in the system, which will ensure a low level of the residual amount of the solvents, as well as a narrow particle size distribution.

PCA: This method is similar to spray drying. The drug-containing polymer solution is sprayed into the supercritical fluid vapor phase to form fog droplets which will then be penetrated by the supercritical fluid under high pressure to reduce the dissolving capacity of the solvent, thereby causing the drug and polymer to precipitate to form particles. In this way, relatively monodisperse microspheres with small particle sizes can be prepared [26].

6.3 Mechanism of Microsphere Formation

6.3.1 Stability of Nano-Microspheres and DLVO Theory

Microspheres are thermodynamically unstable systems. There is a tendency for particles to coalesce with each other to reduce their surface energy, that is, microspheres are featured with instability which makes them easy to coagulate. Therefore, a stabilizer is necessary when preparing microspheres. In addition, due to the small particles of microspheres, the Brownian motion (which refers to the never-ending irregular motion of particles suspended in liquid or gas) is violent, so they are not easy to settle in the gravitational field, which means they have dynamic stability. Stable microspheres must have both stability and dynamic stability that are not prone to coagulation. However, the stability that is not prone to coagulation is more important, because although Brown motion makes the microspheres have dynamic stability, it also promotes the continuous collision among particles. If the particles lose their stability against coagulation, the mutual collision will cause agglomeration, and the result is that the particles are enlarged, with the speed of Brownian motion decreased, and eventually the microspheres will become a dynamically unstable system. The process of particle aggregation from small to large is called aggregation, and the large particles formed by the aggregation of particles are called aggregate [27]. If the final result of aggregation causes particles to precipitate out of the solution, the aggregation process is called coagulation. In order to accelerate the aggregation, other substances can be added as a coagulant, such as electrolytes [13]. In addition, certain physical factors may also cause the coagulation of microspheres, such as light, electricity, and heat [13, 28].

When discussing the stability problem, we must consider the force between particles (Fig. 6.2), which generally has the following five kinds: (1) van der Waals gravity or electromagnetic gravity; (2) Electrostatic repulsion; (3) Born force or short distance repulsion force; (4) Space force; (5) Solvability. Derjaguin, Landau, Verwey and Overbeek put forward DLVO theory based on the mutual attraction and mutual repulsion force between particles, which can better explain the influence of electrolyte on the stability of particle multiphase dispersion system. The following mainly discusses the calculation of attraction and repulsion force between particles [29].

The DLVO theory provides a method to calculate the repulsive energy and attracting energy among colloidal particles, and based on this, the stability of colloids is quantitatively processed, the relationship between the coagulation value and the electrovalence of counterions can be concluded, and the Schulze–Hardy rule is theoretically clarified. The classical DLVO theory studies the stability and coagulation of colloids from the perspective of the interaction between the repulsive potential energy and the attracting potential energy among colloidal particles. It is believed that there are two interaction forces among charged colloidal particles, that is, the electrostatic repulsion when the electrical double layer overlaps, i.e., the long-range Van der Waals attracting force among the particles, and their interaction determines the stability of colloids [30]. The theory believes that the total potential energy V_{TD} among particles in the dispersed system is the sum of the repulsive potential energy (electrostatic repulsive potential energy) V_{ER} and the attracting potential energy (long-range Van der Waals potential energy) V_{WA}, namely:

$$V_{TD} = V_{ER} + V_{WA} \tag{6.1}$$

6.3.2 Factors Affect the Potential Energy

The potential energy in the dispersion system is affected by the particle size and shape of colloids, the distance among particles, Hamaker constant A, the particle surface potential ψ, and the electrolyte concentration n_0 [31]. For a long time in the past, the classic DLVO theory has been used to explain the cohesion and dispersion behavior between colloidal particles. However, research in the past 10 years has found that there are many special interaction forces between hydrophilic or hydrophobic particles, which play a decisive role in the stability of the colloidal dispersion system. The classic DLVO theory cannot explain the cohesion and the dispersion of these systems, and an extended DLVO theory is thus proposed as shown in Equation 6.2:

$$V = V_{ER} + V_{WA} + V_{SR} + V_{HR} + V_{HA} + V_{MA} \tag{6.2}$$

Here V_{SR} refers to the space-repulsive potential energy, V_{HR} refers to the hydration interaction repulsive energy, V_{HA} refers to the hydrophobic interaction attracting energy, and V_{MA} refers to the magnetic attraction potential energy. If the distance (H) between the surfaces of two spherical particles with equal volumes is much smaller than the particle radius (r), the mutual gravitational potential energy (A_α) between the two particles can be approximated based on Equation 6.3:

$$A_\alpha = -\frac{Ar}{12H} \tag{6.3}$$

where A_α is the gravitational potential energy, A is the Hamaker constant, r is the particle radius, and H is the surface spacing of spherical micelle particles. The Hamaker constant is related to the properties of particles (such as the number of atoms per unit volume, polarizability and solution pH), and is the characteristic constant of substances, which are approximately between 10^{-19}~10^{-20} J [32].

The magnitude of the repulsion potential energy (V) between particles with the same charge depends on the number of particle charges and the distance between each other. If the surface potential distribution of a solid is different, the expression of the repulsive potential energy will be different too. The repulsive potential energy between parallel plates is:

$$V_r = K_1 c^{\frac{1}{2}} \exp\left(-K_2 c^{\frac{1}{2}}\right) \tag{6.4}$$

Here c is the electrolyte concentration, K_1 and K_2 have fixed values under certain conditions, and V decreases exponentially with the increase of electrolyte concentration c, therefore, the concentration of electrolyte has a great influence on the stability of microspheres. For two spherical particles with a distance of H, assuming that the surface potential is low, the repulsive potential energy can be approximately expressed as:

$$V_r \approx K\varepsilon\varphi_0^2 \exp(-kH_0) \tag{6.5}$$

Here K is a constant, ε is the dielectric constant of the medium, φ_0 is the electric potential of the colloidal particle surface, K is the reciprocal of the ion atmosphere radius in the Debye-Hickel equation, H_0 is the minimum distance between the surfaces of the two spheres, and r is the particle radius. It can be seen from the equation that the repulsive potential energy increases with the increase of the surface potential φ_0 and the particle radius r, and decreases exponentially with the increase of the distance among particles.

The total potential energy among particles should be the sum of the attracting potential energy and the repulsive potential energy, that is, $V = V_a + V_r$, and the value change is roughly. In the potential energy curve, the height of the potential barrier is a sign of the stability of microspheres. When the height is zero, microspheres will become unstable. Find the point where the potential barrier is reduced to zero due to the addition of electrolyte, namely:

$$\frac{dV}{dH} = 0,\ V = 0 \tag{6.6}$$

The concentration c of electrolyte added at this time is the coagulation concentration when $V = 0$, which is usually called the coagulation value. After several steps of simplification, a simplified expression of the DLVO theory with water as the medium is obtained:

$$c = K\frac{r^4}{A^2Z^6} \tag{6.7}$$

6.3.3 Factors Affect the Stability of Microspheres

6.3.3.1 Properties of polymers

Aggregation or association is a process in which primary particles collide with each other to form a specific structure (aggregates) under the action of force. The study of particle aggregation in polymer solutions and dispersions has attracted extensive attention of scientists. The aggregation of polymer chains can form a variety of structures such as spheres, core–shell, columns and vesicles with uniform density. Because the aggregation of polymer chains has many important applications in industry, such as the stability of lotion particles, the preparation of dyes and pigments in the printing industry, etc. The aggregation of a polymer single chain is more complex [33]. Wu et al. studied the thin layer phase separation behavior on the surface of a homopolymer dilute solution. When the collision time is not much different from the time scale of permanent interpenetration of two adjacent aggregates, aggregation does not play a leading role [34]. Our group's research shows that the expressions of τ_C and τ_E are as follows:

$$\frac{l_0}{\langle v\rangle} < \tau_{\mathrm{C}} < \frac{l_0^2}{\langle D\rangle} \tag{6.8}$$

$$\tau_{\mathrm{E}} \sim \frac{\alpha^2 N^3 \phi_{\mathrm{p}}^{3/2}}{D_1} \tag{6.9}$$

where l_0 is the range of interaction; $\langle v\rangle$ is the average thermal velocity, $\langle D\rangle$ is the diffusion coefficient of a particle, D_1 is the diffusion coefficient of a monomer, a is the length of a unit monomer, N is the degree of polymerization, and ϕ_{P} is the concentration of polymer-rich phase [35].

When particles collide with each other, the polymer chains on adjacent microspheres will entangle with each other, and the time required for this process is τ_{E}. The time when particles collide with each other is τ_{C}. If $\tau_{\mathrm{C}} \ll \tau_{\mathrm{E}}$, the two colliding particles cannot be bonded together. They are like two elastic "glass balls" in the collision process, which is determined by the viscoelastic properties of long high molecular chains [34]. Our previous studies have shown that even if the polymer concentration is very low (10^{-6} g/mL), in the collapsed state, the ϕ_{P} is still as high as 30 wt.%. Therefore, for long chains, the relaxation in the aggregation of polymer is very slow. If there is no hydrophilic group on the surface of such polymer particles to play a stabilizing role ($\tau_{\mathrm{C}} \gg \tau_{\mathrm{E}}$) [36, 37], they are thermo-dynamically unstable, as shown in Fig. 6.2.

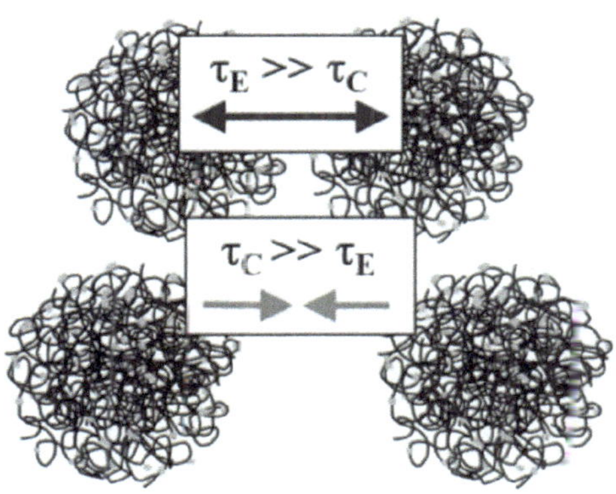

Figure 6.2 Schematic illustrates the inter-collision mechanism between viscoelastic particles.

6.3.3.2 Surface charge of microspheres

Based on the type of surfactants, monomers or initiators used, the surface of microspheres is positively or negatively charged. The greater the surface charge, the greater the repulsive force among microspheres, and the more stable the microspheres in an aqueous solution. Therefore, the electric potential is one of the important indicators to characterize the stability of the microsphere dispersion (see Fig. 6.3).

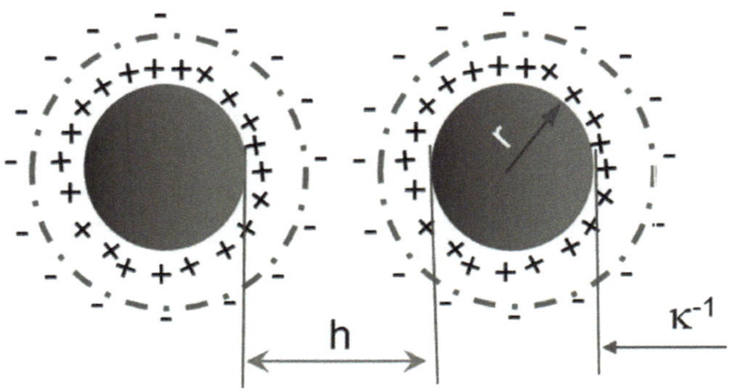

Figure 6.3 The micro-electronic properties and force among charged particles in solution.

Long-distance (~100 nm) attracting interaction is usually called the Van der Waals force (or London dispersion force). The sum of the potential energy of these two interactions is negatively correlated to the center-to-center distance. For the highly charged microspheres suspended in a solution containing stray electrolytes, the experiment is usually affected by various effects and becomes complicated, so the effective calculation of the potential is still controversial. Generally speaking, the potential energy must be added with the third term: the stronger long *r*.

$$U(r)=\frac{(Ze)^2}{\varepsilon}\left(\frac{e^{kr}}{1+kr}\right)^2\frac{e^{-kh}}{h} \tag{6.10}$$

Here h is the center-to-center distance of the two charged microspheres, Z is the number of charges of each charged microspheres, r is the radius of the charged microspheres, and e is the dielectric constant of the dispersive medium. The magnitude of this electrostatic repulsion is significantly reduced with the increase of the stray electrolyte concentration, which is due to the shielding effect of the counterions in the electrical double layer.

6.4 NPP of Microspheres

6.4.1 Physicochemical Properties of Microspheres

6.4.1.1 Particle size and apparent structure of microspheres

The shape, size and pore structure of microsphere particles depend on a lot of experimental variables, such as temperature, stirring speed, pH, stirring type, porogen, polymer and their concentration. Therefore, it is very challenging to synthesize porous particles with predetermined porosity.

The microspheres were polymerized under different conditions. In the copolymer system of GMA and DVB or EGDA (Route 1), polymers with a higher content of GMA and PGMA are separated from polydimethylsiloxane during the polymerization process. The polydimethylsiloxane migrates inward, and monomers and polymers migrate to the interfaces of the dispersed phase and the continuous phase. When the porosity is sufficient, a capsule structure will be formed. Due to the relatively low thermal gravity, micelles that migrate to the interface will deform, making the surface of microspheres smooth. In the cross-linking system St (Route 2), the phase separation occurs later, and monomers have a better swelling effect on micelles. Therefore, the prepared micelles is larger and have a higher thermal weight. The "eggshell" can be supported by micelles, and the surface of microspheres has a spherical convex structure. In the F (GMA-St-EGDA) system (Route 3), the phase separation occurs earlier due to the poor compatibility of monomers and porogen. At the same time, monomers show poor swelling capacity towards micelles, so micelles prepared in this way are smaller. However, the T_g is relatively moderate, which can not only form a support frame, but also cause shrinkage and wrinkle deformation of the "eggshell" [38].

The volume shrinkage of the "eggshell" shows the shape of wrinkles, which should be directly related to two factors. On the one hand, the micelle skeleton formed by phase separation acts as a support, which determines the basic roughness. The results show that the surface of P(St-EGDA) and P(St-DVB) microspheres are spherical protrusions, and the compatibility between St and dimethylsiloxane is better than that between GMA and dimethylsiloxane. On the other hand, the swelling effect of toluene and the promoting effect of phase separation are beneficial to the formation of folds [39].

The performance of microspheres can be further improved by coating thin shells of different chemical compositions and different thicknesses. The structure, size and composition of these mixed particles can be changed within a controllable range to adjust their optical, electrical, thermal, mechanical, electro-optical, magnetic and catalytic properties [40]. This modification also facilitates the adjustment of the interaction between microspheres, and stabilizes the dispersion of these balls in a given medium. In addition, in the subsequent steps, solvent extraction, high-temperature calcination and other ways can also be used to remove the inner core, thereby generating a hollow sphere of the coating material. In many cases, these hollow spheres may show fundamentally different optical properties from solid spheres. When used as a filler (or pigment), hollow spheres definitely have an advantage over solid spheres because hollow spheres have a lower density. Hollow spheres have nothing inside, which also makes them particularly useful in other niche applications. For example, it can be used as a container to carry various substances and is widely used in research fields such as catalytic reactions and drug delivery [40].

6.4.1.2 Factors affecting the particle size of microspheres

Drug concentration. There are two ways to add drugs into microspheres: one is to add drugs into microspheres during the formation of microspheres. The other is to prepare blank microspheres first and then adsorb drugs, so as to add drugs into the microspheres. With the increase of drug concentration and drug loading of microspheres, the particle size will continue to increase.

Effect of additives. By reducing the interfacial tension between the dispersed phase and the dispersion medium the surfactant can vary the size of the emulsion drop in the preparation process, thus affecting the size of the particle. Microspheres prepared by different surfactants are not necessarily the same. Different dispersion media will greatly affect the particle size of microspheres.

Preparation method. The size of the microspheres depends greatly on the preparation method, and the diameters of the microspheres obtained by different preparation methods are not necessarily the same. The same preparation method adopts different treatment processes, and the obtained microspheres have different diameters.

Mixing speed and emulsification time. Generally, the faster the stirring speed is, the smaller the particle size is. The ultrasonic treatment is smaller than the stirring method. The longer the emulsification time is, the smaller the microsphere particles are and the more uniform the particle size distribution is. In addition, the size of the microspheres was affected by the intensity and time of irradiation, curing time and temperature, cross-linking agent, amount and type of catalyst.

6.4.2 Drug Loading and Encapsulation Efficiency of Microspheres

The determination of drug content generally adopts the solvent extraction method, in which the structure of microspheres is broken to release drugs. The principle of solvent selection: maximize the dissolution of drugs and minimize that of carrier materials; ensure that solvents will not interfere with the determination. Generally, an ultraviolet spectrophotometer or high-performance liquid phase is used to detect the drug concentration [13, 41].

$$\text{Loading efficiency (\%)} = \frac{L_a}{L_t} \times 100\% \tag{6.11}$$

where L_t is the total weight of the drug-loading microspheres, and L_a is the actual measured content of the drug in the drug-loading microspheres.

$$\text{Drug encapsulation efficiency (\%)} = \frac{M_a}{M_o} \times 100\% \quad (6.12)$$

where M_o is the total amount of the drug administered, and M_a refers to the actual measured content of the drug in the drug-loading microspheres [42].

6.4.3 Drug Release Behavior of Microspheres

The drug release from the microspheres can be simply described as the following processes: the drug adsorbed on the surface of the microspheres first dissolves into the aqueous phase. As the water molecules infiltrate into the matrix, the drugs in the pores dissolve and diffuse out. The polymer degrades due to hydrolysis, and the drug encapsulated in it is released slowly. Studying the drug release behavior in the microspheres is of guiding significance for the development of microsphere preparations.

6.4.3.1 Mechanism of drug release by microspheres

The mechanisms of drug controlled release can generally be divided into five types: (1) diffusion control, (2) degradation corrosion control, (3) solvent activation control, (4) ion exchange control, and (5) magnetic control. In diffusion control, the drug is released through diffusion in the polymer. In the degradation dissolution control, drugs are released due to the degradation of polymers or the dissolution of the whole carrier. The drug release from the microspheres can be simply described as the following processes. The drug adsorbed on the surface of the microspheres first dissolves into the aqueous phase. As the water molecules infiltrate into the matrix, the drugs in the pores dissolve and diffuse out. The polymer is degraded due to hydrolysis, and the drug encapsulated in it is released slowly shown in Fig. 6.4. For most drug controlled release systems, diffusion control and degradation dissolution control are the two main mechanisms. At present, polymer materials as carriers are mainly combined with drugs in three forms: cumulative, homogeneous and polymer drugs. Cumulative drug controlled release system is to embed drugs in polymer membranes by physical methods, and drugs are released through diffusion and degradation of polymer membranes.

Homogeneous drug controlled release system is to evenly mix the drug in the polymer matrix, and the drug will be released through diffusion in coordination with the degradation, dissolution and release of the polymer. Polymer drugs refer to drugs and polymers connected by chemical bonds, and drugs will also be released due to the chemical bond-breaking reaction [43, 44].

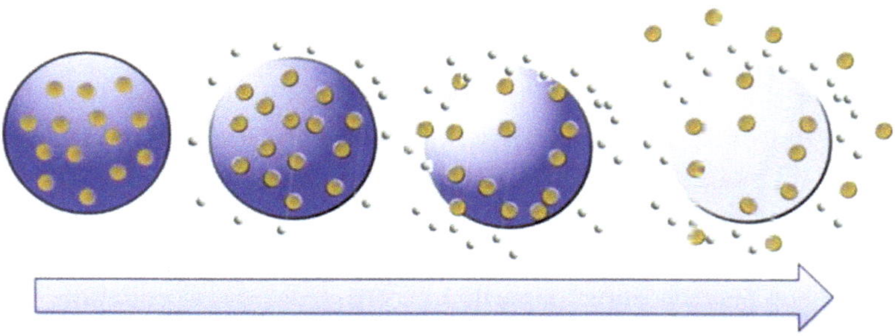

Figure 6.4 Scheme illustrating the process of microsphere surface corrosion for controllable drugs release.

If the drug is uniformly distributed or molecularly distributed in the microsphere matrix, the release process can be expressed by Equation 6.13:

$$F = 1 - \frac{6^{n=8}}{\pi^2}\sum_{n-1}\frac{1}{n}e^{Q2K_r t} \tag{6.13}$$

Here F is the number of drugs released, K_T is the release rate constant, D is the diffusion coefficient of the drug in microspheres, r is the radius of microspheres, and Q is the release quantity at the time of t.

If the drug is suspended in the microsphere matrix, it is an inhomogeneous dispersion, and its release process is usually expressed by Higuchi's Equation (6.14):

$$\frac{3}{2}\left[1-(1-F)^{\frac{2}{3}}\right]-F = \frac{3C_s D_t}{Ar^2} \tag{6.14}$$

Here C_S is the dissolution degree of the drug in the medium, and A is the drug-loading rate per unit volume.

If microspheres are used as a drug carrier, the concentration in microspheres is almost constant during the release process, which can be expressed by Equation 6.15:

$$\frac{d(Q_{\mathrm{n}}-Q)}{dt}=\frac{DS}{n}(C_{\mathrm{m}}-C) \tag{6.15}$$

Here Q_{n} is the drug content of microspheres, S is the drug release area of microspheres, n is the diffusion thickness of microspheres, C_{m} is the concentration of the drug in microspheres, and C is the concentration of the drug in the release medium at the time of t. When $C_{\mathrm{m}} > C$, the parameters on the right side of the equation are all constants, expressed by K, and integrated to obtain the equation: $Q = K_{\mathrm{t}}$.

This is the most simplified zero-order rate equation, and some people also use $Q = SC_0DK_1ItrD_1$ to denote the zero-order release of microspheres. Here C_0 is the initial concentration of microspheres (C – Cs), K_1 is the interface velocity constant, I is the interface thickness, and D_1 is the interface diffusion coefficient of drugs. The actual drug release method is more complicated, so usually, it is the result of several methods conducted alternately, which may not be consistent with the theoretically derived equation. Sometimes different drug release equations are met at different stages of release, and there is also a relationship between group and individual drug release characteristics.

The factors controlling the drug release rate mainly include the concentration difference of the drug in the polymer matrix, the pore size distribution of the polymer, the interaction between the polymer and drug molecules, the solubility of the drug in body fluid or medium, the size and aggregation morphology of drug molecules, and the swelling and degradation of the polymer in the medium [45, 46].

6.4.3.2 PLA microspheres delivery system

The sustained-release system of polylactic acid microspheres is an ideal drug delivery system. However, this kind of microsphere preparation is not commonly used in clinical practice. The main technical problem causing this situation is that the preparation will quickly release a large amount of drugs before and after the first day of entering the body. This phenomenon is called "burst release," the degree of which is generally measured through the cumulative drug release (%) within 24 h. Since the burst release in the early stage of administration may cause the blood concentration to approach or exceed the toxic level, resulting in obvious toxic side effects, the burst release phenomenon has become a problem faced by the researchers of the microspheres controlled release system [47].

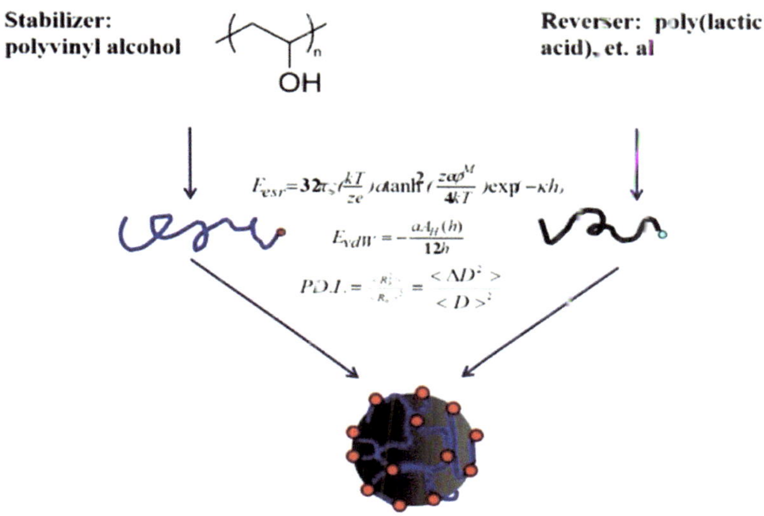

Figure 6.5 The process and key parameters for structural tunning of microsphere (such as PLA microsphere). Here, the inserted equations are also detailed in Sections 1.3 and 1.4 in Chapter 1.

Figure 6.5 describes a general formation process of a biodegradable PLA microsphere, which is a well-known useful carrier for drug delivery. The size, size distribution and stability of such a system are mainly regulated by the chain length, the ratio of hydrophilic chain to hydrophobic chain, and the mixing methods used in the experiments. And the regulation mechanism is shown in the inserted formula, which is discussed in Sections 1.3 and 1.4 in Chapter 1. Drugs are uniformly dispersed in the degradable microspheres, and the drugs adsorbed on the surface or blended with PLA of the microspheres are released first. As the carrier gradually degrades, more release pores are formed, so that the drug dissolves little by little. The drug release is slow in the initial stage, and the drug release rate continues to increase with the rapid degradation of the carrier in the later stage. It is generally believed that the main reason for the burst release of a microsphere controlled release system is that part of the drug exists on or near the surface of microspheres, or there are holes inside or on microspheres. Therefore, in the initial release stage, the drug can be released directly or quickly through the holes to form a burst release. The specific factors that affect the degree of burst release include the molecular weight of polymer, the purity of the polymer, the physical and chemical properties of the main drug, the preparation method and parameters of microspheres, and the drug loading rate of microspheres. According to the preliminary research on the causes and factors affecting the burst release, the current methods to hinder burst release or reduce its degree mainly include directly removing the drug on the surface of microspheres by physical means, such as extraction and washing, changing the nature of the microspheres matrix, changing main drugs' physical and chemical properties, using appropriate microspheres preparation methods or adding additives [48].

The PLA/PEG microspheres were prepared by O/W solvent evaporation method. The dried microspheres were immersed in dilute gelatin solution at room temperature, collected and dried. On the one hand, because of the high hydrophilicity of polyethylene glycol (PEG), the introduction of PEG fragments into PLA polymers can increase the diffusion ability of water or drugs in PLA/PEG microspheres. On the other hand, the microspheres were dipped in the diluted gelatin solution to form a gelatin film on the surface of the microspheres, which can effectively inhibit the burst release (it can reduce the burst release by 98%). The microspheres coated with gelatin film can also control the drug release rate, which is much lower than the release rate from a single gel film. This indicates that a hydrogen bond may be formed between PEG and gel molecules. In addition, the higher the gel concentration, the better the controlled release of drugs. Therefore, the release rate can be controlled by impregnating the microspheres with gelatin solutions of different concentrations. The application of PLA microspheres in the biomedical field is increasingly determined by size, shape and internal structure. The release curve is generally determined by the porosity of the microspheres. The results provide not only a new strategy for the preparation of shape-controllable PLA microspheres but also a new possibility for the application of PLA products in the field of biomedicine [49].

Microspheres have unique physical and chemical properties, such as low density, large specific surface area, and large adsorption capacity. They are not only widely used in biopharmaceuticals, but also play an important role in food safety detection, disease diagnosis, and environmental monitoring.

6.4.4 Route of Administration of Microspheres

6.4.4.1 Cavity administration

As early as the 1950s, people began to use intra-articular injection of liquid anti-inflammatory drugs to treat rheumatoid arthritis, but the solution was eliminated very quickly from the intra-articular cavity. Only by using high concentration and repeated administration of drugs can we obtain better efficacy, but at the same time, it brings greater side effects and pain to patients. Microspheres can effectively prolong the action time of drugs in the articular cavity [50, 51].

Inject microsphere drug directly into the joint cavity, and control the particle size of the microsphere to reduce the degree of drug leakage to the tissues or blood outside the joint cavity. The larger the microsphere, the easier it will be swallowed by the phagocytes in the intraluminal synovial fluid after intracavitary injection to delay the drug clearance time. Osteoarthritis (OA) is the most common musculoskeletal disease in arthritis. Usually, drugs are directly transported to the diseased joints to change the distribution of drugs in the body [52]. The clinical flurbiprofen axetil lipid globular carrier injection can reduce adverse reactions and increase the therapeutic effect, indicating that the injection of drugs into microspheres in the joint cavity can reduce the number of treatments and provide a new choice for the treatment of patients in the future.

6.4.4.2 Injection administration

After the drug-loaded microspheres are administered in this way, the administration frequency can be reduced, and the toxic and side effects caused by multiple injections can be reduced. In the clinical treatment of tumors in the liver, spleen, kidney and breast, the size of microspheres is controlled by intravenous injection [53, 54]. The microsphere suspension injected into the vein is transported with the blood flow. The microspheres with a particle size of lower than 3 μm will be quickly cleared by the macrophages of the reticuloendothelial system, so it is mainly concentrated in the liver, the spleen and other tissues rich in reticuloendothelial systems, and finally reaches the lysosomes of Kupffer's cells in the liver. Most of the microspheres with a particle size of 2 to 12 μm are ingested by the liver and spleen capillary networks. Microspheres with a particle size of 7 to 12 μm will be mechanically filtered by the lung (the diameter of the pulmonary capillaries is 3 to 11 μm). Microspheres with a particle size of above 12 μm can be temporarily or permanently blocked in the capillary bed. And microspheres with a particle size smaller than 0.1 μm can leave the blood circulation through the gaps among blood vessel cells (the width of cell membrane gaps of pancreas, intestine and kidney is 50 to 60 nm, while that of liver, spleen and bone marrow is 100 nm). In addition, microspheres can be combined with tablets, capsules and other dosage forms to further improve the bioavailability during administration.

6.4.4.3 Administration by arterial embolism

Arterial embolism therapy refers to the direct injection of microspheres with a particle size of 50 to 250 μm from the artery of lesion (canceration) parts to embolize the small arteries there, cut off the nutrition for tumor cells, slow down or inhibit the growth of tumor cells and, through the continuous release of drugs from the blocked area in microspheres, keep the treatment concentration in tumor tissue for a long time and reduce the distribution of drugs in other tissues as much as possible. In this way, it can greatly improve the efficacy of chemotherapy drugs and reduce toxic and side effects. The good performance of drug microspheres makes them an ideal embolic dosage form in transcatheter arterial embolization. Albumin, gelatin and starch are commonly used natural base materials [55].

Arterial embolization treatment refers to the selective injection of pharmaceutical preparations into the artery to embolize certain parts, making specific parts lack oxygen, die and shrink, so as to achieve the therapeutic effect. Wei et al. conducted clinical observation on 43 patients with advanced bladder cancer and hematuria and treated them with microsphere embolization. In the later stage, bladder hemorrhage was stopped after the first TACE operation. MRI examination of 27 patients found that the bladder tumor shrank by more than 50% after 30 days from the operation, indicating that arterial embolism can be used for tumor treatment and inhibit tumor growth [56].

6.4.4.4 Magnetic microsphere administration

Magnetic polymer microspheres began to emerge at the end of the 1970s. It refers to polymer microspheres with magnetic response containing ultrafine powders of magnetic metals or metal oxides (such as Fe, Co, Ni and their oxides). Because of its magnetism, microsphere can move to a specific position directionally under the action of magnetic field, or quickly separate from the surrounding medium. These properties make it have a very broad application prospect, so it has broad application prospect in cell separation, protein separation and purification, immobilized enzyme, immunoassay, targeted drugs, DNA separation and nucleic acid hybridization and other fields. Wang et al. improved the inverse micro lotion polymerization method (point polymerization method) to develop a monodisperse polyacrylamide microsphere with visible and near-infrared light prepared by glutaraldehyde cross-linking. Furthermore, polyacrylamide Fe_3O_4 composite microspheres were successfully prepared by one-pot method (PAM@FeNPs). The multi-component microspheres have the characteristics of good magnetism and stable fluorescence, which has huge potential application in biomedicine imaging and drug delivery system, and more importantly, provides an insight into preparation of multifunctional composite microspheres [57].

6.4.4.5 Oral administration

Generally, due to the acidic environment in the stomach and the presence of degrading enzymes, drugs are absorbed by the stomach with bioavailability reduced. Especially for polypeptide and protein drugs, it is very important to choose the appropriate carrier for administration. Oral administration of drug microspheres can avoid the degradation of active drug molecules by gastric acid and enzymes, improve the stability of drugs and reduce the frequency of administration. Oral vaccine microspheres under extensive research, such as the antigen-alginate microspheres, have been prepared to protect the stability and activity of the antigen so as to significantly increase the G level of immunoglobulin in blood [58, 59].

6.4.4.6 Mucosal administration

Mucosal drug delivery system (MDDS) is a new type of drug delivery system. It mainly refers to the drug delivery mode in which drugs and appropriate carrier materials are applied to the cavity mucosa to play a local role or absorb into the systemic circulation to trigger systemic treatment. Mucosal administration has the following advantages: (1) Compared with oral administration, drugs enter the systemic circulation through mucosal absorption, and can effectively avoid the first pass effect of the liver without passing through the liver; (2) The types and quantities of enzymes in mucosa are limited, and the activity of enzymatic degradation of drugs is relatively low; (3) Local administration can increase the drug concentration at the active site and improve the bioavailability; (4) Local application of medicine is convenient for patients to operate independently and improve patients' compliance. Guided by optimizing the clinical administration mode and improving the comfort of patients in the treatment process, based on the dry cross-linking mechanism and the drug diffusion finite element model, researchers designed a gel patch that can quickly and stably adhere to the wet oral mucosa and load photosensitizers for local PDT, providing a new drug delivery system that can replace the traditional local mucosal administration mode [60, 61].

6.4.4.7 Ocular administration

The eye is one of the most sophisticated organs in the human body. It has unique anatomical and physiological characteristics. Generally, the eye is divided into anterior segment and posterior segment by the lens plane. The ophthalmic drug delivery preparation refers to the drug preparation that is directly used in the eye to play a local therapeutic role or enters the body circulation through the eye to play a therapeutic role [62]. Ocular administration is the main route of administration to treat ocular local diseases. Compared with systemic administration, it is more conducive to drug enrichment at the target site and can reduce systemic adverse reactions. At present, more than 90% of eye preparations sold in the market are traditional dosage forms such as eye drops and ointment. The main defects are low bioavailability, large adverse reactions and poor patient compliance. In recent years, great progress has been made in the clinical application of ophthalmic drug delivery systems. Many new dosage forms aimed at prolonging the retention time of drugs on the ocular surface, increasing the permeability of ocular surface tissues and improving the bioavailability of ocular target tissues have been marketed. The researchers used the spray drying method to prepare levofloxacin chitosan/gelatin microspheres. The encapsulation efficiency, drug loading and stability were all good, which increased the absorption efficiency of drugs in the nasolacrimal duct and the retention time on the ocular surface, providing a theoretical basis for the study of new ocular drug delivery systems [63].

6.4.5 Biological Application of Microspheres

6.4.5.1 Sustained-release microsphere formulation

Peptide drugs. During the preparation of polypeptide drug-loaded microspheres, due to its small molecular weight and strong. Hydrophilic compounds tend to escape to the outer aqueous phase during the emulsification and solvent evaporation steps, resulting in low embedding rate. Therefore, researchers studied the characteristics of polypeptide drugs and prepared drug-carrying microspheres with high embedding rate and high drug activity. Using a nearly monodisperse high molar mass poly(N-isopropylacrylamide) (PNIPAM) sample, Wu et al. successfully changed the conformation of a single PNIPAM chain from curl to a fully collapsed stable single-chain globule in extremely dilute aqueous solution, which enabled us to study the transition from globule to curl of a single homopolymer chain in solution for the first time. This also laid a foundation for the later study of the stability of microspheres in solution. Qi et al. [64] prepared uniform-sized PLGA microspheres by Shirasu porous glass premix membrane emulsification and used to encapsulate exenatide, a drug for treating Type 2 diabetes. The experimental results showed that the hypoglycemic rate of the exenatide solution group increased significantly in the first 14 days, but the blood glucose returned to the level of the control group (normal saline group) after the 14th day of drug administration was stopped. Yang et al. [65] took thymosin as the model drug, optimized the concentration of internal aqueous phase, oil phase emulsifier, PLGA concentration, molecular weight and external aqueous phase pH, and prepared thymosin loaded microspheres with an average particle size of 24 μm and uniform particle size (Span value of 0.6) by using the rapid membrane emulsification combined with solvent evaporation method, and the embedding rate was more than 80%. On the basis of controllable particle size, the effect of PLGA molecular weight on drug release was investigated. The results showed that with the increase of the molecular weight of PLGA, the burst rate of microspheres decreased, and the whole release process was closer to zero-order release. And because the preparation process of membrane emulsification technology is simple and mild, it overcomes the influence of traditional methods on drug molecular structure and significantly improves the actual use efficiency of drugs.

Protein drug sustained-release microspheres. Protein drugs have high molecular weight, complex structure, easy denaturation and inactivation under severe preparation conditions, and poor stability, thus reducing the efficacy and even toxic side effects on the body. The membrane emulsification process is mild, which can solve this problem. Zhang et al. [66] used insulin as a model drug and prepared an average particle size of about 7.5 μm by the rapid membrane emulsification technology and the two-step curing process. The oral calcium alginate chitosan microspheres are narrow (span value is 0.731) and spherical. The method can mildly prepare lotion with uniform size, so as to retain the immune activity of insulin and effectively absorb it in the gastrointestinal tract. In the experiment of diabetes rats, the prepared insulin-loaded oral microspheres have a continuous hypoglycemic effect.

Small molecule drug sustained-release microspheres. Different from protein peptide drugs, small molecule drugs are easy to escape quickly during the preparation of microspheres due to their small molecular weight, which leads to a decrease in the entrapment rate. To solve this problem, the researchers obtained emulsion droplets with uniform particle size through rapid membrane emulsification technology, avoiding Oswald ripening in the curing process, and ensuring high embedding rate of drug-loaded microspheres. Li et al. [67, 68] took ropivacaine (RVC) as the model drug and prepared a uniform drug-loaded microsphere with an average particle size of 17 µm, a Span value of less than 0.6, and an embedding rate of more than 80% by using emulsion method and membrane emulsification technology. In average particle size, the effects of PLGA with different terminal groups (hydroxyl, carboxyl, ester) on drug loading were investigated on a controlled basis. The hydroxyl-terminated ropivacaine microspheres exhibit the best long-term analgesic effect.

6.4.5.2 Application of uniform drug-loaded microspheres in vaccine preparation

Vaccines are one of the most significant medical interventions in the fight against infectious diseases. An ideal vaccine should satisfy this same necessity of effectiveness, affordability, protection in application and charge. The aspect of protection and avoidance of severe effects is complicated. The aspect of safeness and the extent of the manufacturing of antibody responses are intently linked to mode of application. Biodegradable delivery technology for vaccines which are administered by the intravenous path may resolve the shortcomings of the same conventional vaccines.

Liu et al. [69] prepared water/oil (W/O) vaccine preparations with uniform particle size through rapid membrane emulsification technology. The water phase is an antigen solution and the oil phase is an oil adjuvant. Studies have confirmed that lotion preparations with uniform particle size have a stronger immune response than preparations with wide particle size distribution.

6.4.5.3 Application of uniform drug-loaded microspheres in the treatment of malignant tumors

At present, most traditional anti-tumor drugs have low specificity and bioavailability in the human body. Therefore, large doses are required to increase the probability of reaching the tumor site, as shown in Fig. 6.6. Based on the enhanced capability and retention effect of tumor blood vessels, drugs can be passively targeted. Therefore, nanosphere (NP) preparations are considered to be one of the most promising ways to effectively deliver drugs to tumor tissues [70]. However, uneven particle size and low drug-loading rate will affect the clinical treatment effect [71, 72]. Lv et al. [73] also used oil/water/oil (O/W/O) double emulsion method combined with rapid membrane emulsification technology to prepare nanospheres with uniform particle size. On the basis of this research, Fu et al. [72] prepared homogeneous chitosan nanospheres (CNPs) with a particle size of about 100 nm by combining rapid membrane emulsification technology, and realized co-encapsulation of doxorubicin (DOX), PTX and Fe_3O_4 nanocrystals by combining O/W/O double emulsion method and temperature programmed solidification. The CNPs with uniform particle size are loaded with hydrophobic taxol, doxorubicin and iron oxide nanocrystals Fe_3O_4 at the same time. Under the magnetic targeting effect mediated by Fe_3O_4 and the active targeting effect mediated by RGD, anti-cancer drugs are efficiently delivered to the vicinity of tumor tissue to play a killing role. The actual results showed that the three-week survival rate of mice treated with RHCNP: Fe/P/D reached 100%, which was far higher than other control groups, showing a good anti-tumor effect. The carboxyl group on the surface of the nanospheres was used to introduce a polyethylene glycol

(PEG) chain with stealth effect and a tripeptide sequence (RGD) targeting tumor cells to prepare a paclitaxel nano drug delivery carrier (RGDPEG-CNP: PTX) with both stealth and targeting capabilities. The experimental results show that the drug delivery preparation with uniform size nanospheres can improve the drug bioavailability and anti-tumor effect, and has low side effects.

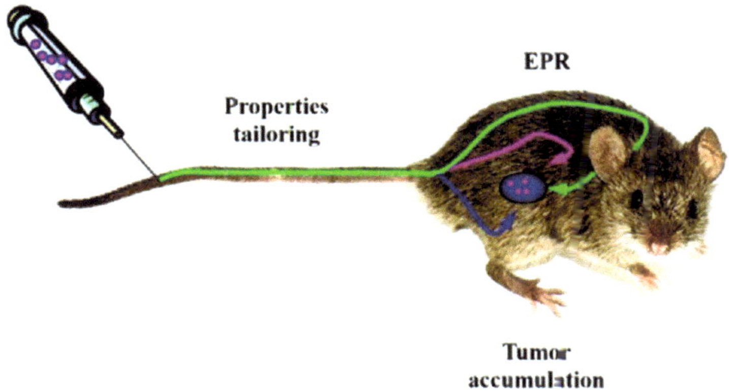

Figure 6.6 Schematic of the general *in vivo* fate of the microsphere from injection site to tumor.

6.5 Summary and Perspective

In the past 30 years, under the influence of various factors such as changes in the living environment, the aging society, the development of medicine, cognitive concepts and so on, the disease spectrum in China has undergone significant changes. Malignant tumors, cardiovascular diseases, diabetes and mental diseases have increased rapidly, which will greatly increase the consumption of medical resources. Microspheres have the advantages of prolonging the drug action time, enhancing efficacy, reducing toxic and side effects, and have significant clinical therapeutic effects on chronic diseases such as malignant tumors. The market demand will be on the rise.

One of the important directions for the development of drug dosage forms is to improve the targeting of drugs. At present, there are new and gratifying achievements in the research of drug microspheres targeting, such as nano-microspheres, magnetic microspheres, bioadhesion microspheres and antibody-mediated microspheres. Nano-microspheres are easier to pass through the gastric mucosa, intestinal mucosa, nasal mucosa and even the stratum corneum of skin. Due to this, significant attention is being paid to them at present. The magnetic microspheres prepared by magnetic properties have magnetic responsiveness, can induce specific aggregation and can be used for MRI. Bioadhesive microspheres prepared by using adhesive materials such as deacetylated chitosan and polycarboxyethylene can improve the absorption of drugs at the biological mucosa. Absorption accelerators such as microspheres co-carrier lysophospholipid or protease inhibitors are used to improve the drug absorption of gastrointestinal mucosa. The special affinity of antigen-antibody is used to prepare antibody microspheres, which can improve the targeting of drugs. In addition, the development and application of new natural polymer materials is also imperative. This paper briefly describes the research progress of microsphere preparations and provides corresponding theoretical guidance and basis for related work in the future. Hopefully, with a deeper understanding of microspheres, new approaches will be developed to conquer the obstacles in the preparation and application process.

References

1. Kim HK, Park TG, *J Control Release* **2004**, *98*, 115–125.
2. Sun Y, Zhang C, Mao Y, Pan D, Qi D, Di N, *New Journal of Chemistry* **2019**, *43*, 11164–11170.
3. Tuttolomondo MV, Villanueva ME, Alvarez GS, Desimone MF, Díaz LE, *Biotechnology Letters* **2013**, *35*, 1571–1577.
4. Nicholson, Linda K, Tetsuo A, Makoto D, Timothy AC, *Biopolymers CTJ*, **1993**, *33*, 847–861.
5. Jeong B, Kim SW, Bae YH, *Advanced Drug Delivery Reviews* **2012**, *64*, 154–162.
6. Dupin D, Fujii S, Armes SP, Reeve P, Baxter SM, *Langmuir* **2006**, *22*, 3381–3387.
7. Qing S, Lyu C, Zhu L, Pan C, Wang S, Li F, Wang J, Yue H, Gao X, Jia R, *Advanced Materials* **2020**, *32*, 2002085.

8. Phillippi JA, *Science Advances* **2022**, *8*, eabl6364.
9. Tu Z, Zhong Y, Hu H, Shao D, Haag R, Schirner M, Lee J, Sullenger B, Leong KW, *Nature Reviews Materials* **2022**, 1–18.
10. Yang H, Xie Y, Hao G, Cai W, Guo X, *New Journal of Chemistry* **2016**, *40*, 589–595.
11. Wu XS, *Artificial Cells, Blood Substitutes, and Biotechnology* **2004**, *32*, 575–591.
12. Dastidar DG, Saha S, Chowdhury M, *International Journal of Pharmaceutics* **2018**, *548*, 34–48.
13. Wu XS, *Artificial Cells Blood Substitutes and Biotechnology* **2004**, *32*, 575–591.
14. Takai C, Hotta T, Shiozaki S, Matsumoto S, Fukui T, *Colloids and Surfaces A: Physicochemical and Engineering Aspects* **2011**, *373*, 152–157.
15. Jayakrishnan J, Thanoo BC, *Journal of Biomedical Materials Research* **1990**, *24*, 913–927.
16. Mandal T, Bostanian L, Graves R, Chapman S, Idodo T, *European Journal of Pharmaceutics and Biopharmaceutics* **2001**, *52*, 91–96.
17. Lammel A, Schwab M, Slotta U, Winter G, Scheibel T, *ChemSusChem* **2008**, *1*, 413–416.
18. Juluri BK, Chaturvedi N, Hao Q, Lu M, Velegol D, Jensen L, Huang TJ, *ACS Nano* **2011**, *5*, 5838–5847.
19. Desai KGH, Park HJ, *Drug Development Research* **2005**, *64*, 114–128.
20. Yu F, Zhang J, Yang Y, Song G, *Journal of Power Sources* **2009**, *189*, 794–797.
21. Dibbern EM, Toublan FJ-J, Suslick KS, *Journal of the American Chemical Society* **2006**, *128*, 6540–6541.
22. O'Donnell PB, McGinity JW, *Advanced Drug Delivery Reviews* **1997**, *28*, 25–42.
23. Sivakumar M, Gedanken A, *Ultrasonics Sonochemistry* **2004**, *11*, 373–378.
24. Yuan X, Lin S, Zhao K, Han Y, *Materials Letters* **2022**, *309*, 131461.
25. Suslick KS, *Science* **1990**, *247*, 1439–1445.
26. Maqbool F, Moyle PM, Thurecht KJ, Falconer JR, *International Journal of Pharmaceutics* **2019**, *563*, 174–183.
27. Yang C, Kizhakkedathu JN, Brooks DE, Jin F, Wu C, *The Journal of Physical Chemistry B* **2004**, *108*, 18479–18484.

28. Li X, Wang B, Liu Q-J, Zhao R, Song D-P, Li Y, *Langmuir* **2021**, *37*, 6744–6753.
29. Shaw GS, Samavedi S, *ACS Biomaterials Science & Engineering* **2021**, *8*, 1–15.
30. Trefalt G, Michal B, University of Geneva, Switzerland, **2014**, *304*, 1–10.
31. Boström M, Williams D, Ninham B, *Physical Review Letters* **2001**, *87*, 168103.
32. Muneer R, Hashmet MR, Pourafshary P, *ACS Omega* **2020**, *5*, 31624–31639.
33. Qiu X, Kwan CMS, Wu C, *Macromolecules* **1997**, *30*, 6090–6094.
34. Wu C, Wang X, *Physical Review Letters* **1998**, *80*, 4092.
35. Wu C, Zhou S, *Physical Review Letters* **1996**, *77*, 3053.
36. Xu X, Zhang X, Zhang L, Wu C, *Biomacromolecules* **2004**, *5*, 1893–1898.
37. Qi J, Lai X, Wang J, Tang H, Ren H, Yang Y, Jin Q, Zhang L, Yu R, Ma G, *Chemical Society Reviews* **2015**, *44*, 6749–6773.
38. Wang, J, Yang Z, Xu J, Ahmad M, Zhang H, Zhang A, Zhang Q, Kou X, Zhang B, *Macromolecular Rapid Communications*, **2019**, *40*, 1800768.
39. Stanger BZ, Leder P, Lee TH, Kim E, Seed B, *Cell* **1995**, *81*, 513–523.
40. Xia Y, Gates B, Yin YD, Lu Y, *Advanced Materials* **2000**, *12(10)*, 693–713.
41. Abdekhodaie, Mohammad J, Wu XY, *Biomaterials*, **2006**, *27*, 3652–3662.
42. Dhakar RC, Maurya SD, Sagar BP, Bhagat S, Prajapati SK, Jain CP, *Der Pharmacia Lettre* **2010**, *2*, 102–116.
43. Arifin DY, Lee LY, Wang C-H, *Advanced Drug Delivery Reviews* **2006**, *58*, 1274–1325.
44. Versypt ANF, Pack DW, Braatz RD, *Journal of Controlled Release* **2013**, *165*, 29–37.
45. Sa'adon S, Abd Razak SI, Fakhruddin K, *Procedia Computer Science* **2019**, *158*, 436–442.
46. Son G-H, Lee B-J, Cho C-W, *Journal of Pharmaceutical Investigation* **2017**, *47*, 287–296.
47. Qi F, Wu J, Li H, Ma G, *Frontiers of Chemical Science and Engineering* **2019**, *13*, 14–27.

48. Lee BK, Yun Y, Park K, *Advanced Drug Delivery Reviews* **2016**, *107*, 176–191.

49. Yu B, Lu M, Fu S, Zhao Z, Liu Y, Wang K, Fu Q, *Colloids and Surfaces B: Biointerfaces*, **2018**, *172*, 105–112.

50. Walenkamp G, *Journal of Chemotherapy* **2001**, *13*, 66–72.

51. O'Brien DAR, Kaye SM, Poppas PJ, Mahase SS, An A, Christos PJ, Liechty B, Pisapia D, Ramakrishna R, Wernicke AG, *Journal of Neurosurgery* **2021**, *135*, 1695–1705.

52. Wieland HA, Michaelis M, Kirschbaum BJ, Rudolphi KA, *Nature Reviews Drug Discovery* **2005**, *4*, 331–344.

53. Burger J, Tomlinson E, Mulder E, McVie J, *International Journal of Pharmaceutics* **1985**, *23*, 333–344.

54. Agarwal A, Raza M, Dhiraaj S, Pandey R, Gupta D, Pandey CK, Singh PK, Singh U, *Anesthesia & Analgesia* **2004**, *99*, 117–119.

55. Gammel JA, *Journal of the American Medical Association* **1927**, *88*, 998–999.

56. Wang N, Wu XS, Hannia L-U, Donahue E, Siddiqui A, *Journal of Biomaterials Science, Polymer Edition* **1997**, *8*, 905–917.

57. Wang J, Wang J, One-pot Synthesis of Multifunctional PAM@FeNPs Composite Microspheres, *IOP Conference Series: Materials Science and Engineering* **2020**, *768*, 022004.

58. Trentham DE, Dynesius-Trentham RA, Orav EJ, Combitchi D, Lorenzo C, Sewell KL, Hafler DA, Weiner HL, *Science* **1993**, *261*, 1727–1730.

59. Schüller J, Cassidy J, Dumont E, Roos B, Durston S, Banken L, Utoh M, Mori K, Weidekamm E, Reigner B, *Cancer Chemotherapy and Pharmacology* **2000**, *45*, 291–297.

60. Robertson S, Taylor P, Sear J, *Veterinary Record* **2003**, *152*, 675–678.

61. Neutra MR, Kozlowski PA, *Nature Reviews Immunology* **2006**, *6*, 148–158.

62. Fialho SL, Da Silva-Cunha A, *Clinical & Experimental Ophthalmology* **2004**, *32*, 626–632.

63. Varela-Garcia A, Gomez-Amoza JL, Concheiro A, Alvarez-Lorenzo C, *Polymers* **2020**, *12*, 2026.

64. Qi F, Wu J, Hao D, Yang T, Ren Y, Ma G, Su Z, *Pharmaceutical Research* **2014**, *31*, 1566–1574.

65. Yi W, Guanghui MA, *CIESC Journal* **2021**, *72*, 6176–6187.

66. Zhang Y, Wei W, Lv P, Wang L, Ma G, *European Journal of Pharmaceutics and Biopharmaceutics* **2011**, *77*, 11–19.
67. Li X, Wei Y, Lv P, Wu Y, Ogino K, Ma G, *Colloids and Surfaces A: Physicochemical and Engineering Aspects* **2019**, *562*, 237–246.
68. Li X, Wei Y, Wen K, Han Q, Ogino K, Ma G, *European Journal of Pharmaceutics and Biopharmaceutics* **2021**, *160*, 143–151.
69. Liu Y, Wang L, Zhang Y, Zhang W, Chen X, Yang T, Wang Z, Ma G, *International Immunopharmacology* **2014**, *23*, 603–608.
70. Longo F, Mansueto G, Lapadula V, Stumbo L, Del Bene G, Adua D, De Filippis L, Bonizzoni E, Quadrini S, *International Journal of Clinical Practice* **2012**, *66*, 753–757.
71. Wei W, Lv P-P, Chen X-M, Yue Z-G, Fu Q, Liu S-Y, Yue H, Ma G-H, *Biomaterials* **2013**, *34*, 3912–3923.
72. Fu Q, Lv P, Chen Z, Ni D, Zhang L, Yue H, Yue Z, Wei W, Ma G, *Nanoscale* **2015**, *7*, 4020–4030.
73. Lv P-P, Ma Y-F, Yu R, Yue H, Ni D-Z, Wei W, Ma G-H, *Molecular Pharmaceutics* **2012**, *9*, 1736–1747.

Index

Howard Clinebell

Ecotherapy: Healing Ourselves, Healing the Earth

Pre-publication
REVIEWS,
COMMENTARIES,
EVALUATIONS. . .

"Ecotherapy bridges disciplines to call for a necessary shift in consciousness. This visionary, yet grounded and much-needed book presents a spirituality that involves deepening our relationship with the natural world, appreciating the sacredness of our planetary world, and taking responsibility for what happens to our Earth. It combines theoretical formulations and experiential exercises with action-oriented suggestions that can be readily implemented by teachers, therapists, and students."

Lesley Irene Shore, PhD
Author, *Healing the Feminine: Reclaiming Woman's Voice*
and *Tending Inner Gardens: The Healing Art of Feminist Psychotherapy*

More pre-publication
REVIEWS, COMMENTARIES, EVALUATIONS...

"Such a delightful book! Finally we begin to understand that the loss of the primordial forests, our inability to see the stars at night, our suspicion of the water we drink and the air we breathe—all this is something more than an economic or recreational or aesthetic loss. It is a soul loss. We lose ourselves. It is difficult to see how we can possibly be saved without the beauty and wonder and numinous presence that come to us in our experience of the natural world about us. Only there do we come to ourselves in our integral being. To have this articulated so clearly in this book by Howard Clinebell is a healing gift to us all in this closing decade of the twentieth century."

Thomas Berry
Author, *The Dream of the Earth*

"Psychological health and the health of the earth are deeply intertwined. The healing of persons and of the biosphere are mutually dependent. In this book psychotherapy is vastly expanded, and environmentalism is greatly deepened. The boundaries between therapy, education, spirituality, and ecojustice melt away in Clinebell's revisioning of our common calling."

John B. Cobb, Jr.
Coauthor, *For the Common Good*

"On a stage already filled with creative contributions to the field of pastoral care and counseling, Howard Clinebell has drawn back a new curtain with *Ecotherapy: Healing Ourselves, Healing the Earth.* His insightful, informative, integrative, and evocative synthesis of the inner psychospiritual environment of the person and the outer natural environment of the earth, for the purposes of mutual healing, represents a high point in a superb writing career, the book for which Howard Clinebell may be most remembered. Read it. It will nourish your soul and expand your therapeutic sensitivities."

C. Roy Woodruff, PhD
Executive Director, American Academy of Pastoral Counselors

Ecotherapy

Ecotherapy

Healing Ourselves, Healing the Earth

Howard Clinebell

Routledge
Taylor & Francis Group
LONDON AND NEW YORK

First published 1996 by The Haworth Press

2 Park Square, Milton Park, Abingdon, Oxon OX14 4RN
711 Third Avenue, New York, NY 10017, USA

Routledge is an imprint of the Taylor & Francis Group, an informa business

First issued in hardback 2016

Acknowledgements
"Earthdance" and "Two Wolves" by Richard W. Voss, copyright © Richard W. Voss. Used by permission. "A thousand plastic flowers. . ." by Frederick Perls from *Gestalt Therapy Verbatim,* copyright © *The Gestalt Journal.* Used by permission. "Wild Geese" from *Dream Work* by Mary Oliver, copyright © Grove/Atlantic, Inc. Used by permission. "Native Hawaiian Prayer" from *Hawaiian Antiquities* by David Malo, copyright © Bishop Museum Press. Used by permission. "Prayer of Healing" from *Only One Earth,* copyright © United Nations Environment Programme. Used by permission.

Cover and interior designed by Joseph Bonyata.

Cover photo: Cecropia leaf, copyright © Gerry Ellis/Ellis Nature Photography. Used by permission.

Library of Congress Cataloging-in-Publication Data

Clinebell, Howard John, 1922-
Ecotherapy : healing ourselves, healing the earth / Howard Clinebell.
p. cm.
Includes bibliographical references and index.
1. Nature—Psychological aspects. 2. Environmental psychology. 3. Environmentalism—Psychological aspects. I. Title.
155.9'1–dc20 95-46959
CIP

ISBN 13: 978-0-7890-6009-9 (pbk)
ISBN 13: 978-1-138-15045-4 (hbk)

Contents

—Drawing by Kelly Frick Richards. Used by permission.

"Care for the earth.
It was loaned to you by your children."

–Proverb from Kenya, East Africa
Shared by Peter Rukungah

Dedicated to the healing and wholeness of children and grandchildren—yours, mine, and the world's!

If the earth were only a few feet in diameter, floating a few feet above a field somewhere, people would marvel at it. People would walk around it marveling at its big pools of water, its little pools, and the water floating between the pools. People would marvel at the bumps on it, and holes in it, and they would marvel at the very thin layer of gas surrounding it and the water suspended in the gas. The people would marvel at all the creatures walking around the surface of the ball, at all the creatures in the water. The people would declare it precious because it was the one, and they would protect it so that it would not be hurt. The ball would be the greatest wonder known, and people would come to behold it, to be healed, to gain knowledge, to know beauty and to wonder how it could be. People would love it and defend it with their lives, because they would somehow know that their lives, their own roundness could be nothing without it. If the earth were only a few feet in diameter.

"Earth Ball" by Olaf Skarsholt

Appreciation

My heartfelt thanks to the countless folks who enriched the process of my researching and writing this book over the last fifteen years. Some contributed in indirect ways by intellectual osmosis of which neither they nor I am aware. The direct contributors include a variety of ecological theoreticians whose writings challenged and stimulated my thinking. Their contributions are identified in footnotes. Books by persons whose thought is strikingly relevant to the theory and/or practice of ecotherapy are marked with double asterisks in the bibliography.

Unnamed contributors include numerous clients who must remain anonymous to honor the confidentiality of those therapeutic relationships. Together we used many of the methods described herein as resources on their journeys of healing and growth. Other unnamed contributors include graduate students and workshop participants with whom the core concepts of this book were discussed vigorously and refined repeatedly. Among these were students in an ecotherapy course at Vanderbilt University Divinity School who gave critical feedback on a draft of the book. I am also grateful to Dean Joseph Hough of that seminary for the guest professorship that gave me the creative writing time to complete the book near a fine university library.

I am glad to express special thanks to others who contributed in direct ways. Peter Rukungah, a Claremont Ph.D. graduate hailing from Kenya, did extensive book research and conducted the sixteen cross-cultural interviews reported herein. Dorothy Owens, Ph.D. candidate at Vanderbilt University Graduate School, provided in-depth help in researching and polishing the book. Ecotherapist and pastoral counselor Richard W. Voss provided invaluable feedback on the typescript, shared rich experiences of doing wilderness ecotherapy, and gave permission to use his expressive poems. Pastoral psychologist Jim Ashbrook gave me invaluable input on the brain's functioning in ecobonding and ecoalienation, as well as feedback on the typescript. Psychologist Jeffrey Martin shared helpful insights about issues in chapter 2. Psychologist Mel Blanchette suggested the chapter title "Finding the Lost Self and the Lost Earth." Theologian John Cobb gave his guidance and support at many stages of the book. Artist Kelly Frick Richards gladly gave permission to use her "Children ask the world of us" drawings. Physician Art Madorsky gave me feedback from a medical perspective and helped line up survey respondents in the health professions. The editorial expertise of Roland Seboldt, Tim Staveteig, and Marshall Johnson guided my work at many points.

Each of the sixty-two counselors, therapists, clergy, teachers, and medical professionals who responded to the exploratory survey have my thanks for sharing their insights. Several tribal members at the Charles Cook Native American Theological School in Tempe, Arizona, introduced me to the earthy healing of the sweat lodge. Michael Rice (Red Hawk) illuminated Native Americans' "spirit quest" experience in nature.

It is a joy to offer loving thanks to my children and other family members who shared from their areas of expertise and gave the perspectives of a younger generation. Susan Clinebell provided extensive and invaluable critical feedback on early drafts of the book. Don Clinebell provided guidance on ecological music. Stephanie Clinebell affirmed my love of nature in various thoughtful ways. John Bell opened the door to insights from two members of the Puyallup tribe of Native Americans. Anita Bell shared insights from Eastern therapies and her experiences as a health professional. (Dear Anita's death in her mid-years, while I was completing this book, makes me painfully aware of both the vulnerability and preciousness of life and the people we love.) Jeff Clinebell gave his uncle feedback on animal therapy from a veterinarian's viewpoint, and Debbie Clinebell did the same on ecoeducation from her teaching experience. My brother, Paul Clinebell, gave me insights about environmental engineering. Nevan, Brennan, Tessa, Andrew, and Jamie have their granddad's gratitude for enhancing, by their presence in his life, awareness of the preciousness of what is at stake for *all* children of the earth in the ecojustice crisis!

My loving thanks for a wide variety of gifts to earth-caring Charlotte Ellen, my partner for five decades. Her gifts include insights from her clinical experiences in ecofeminist therapy and her profound concern for healing the earth. These gifts also include sharing exciting encounters with nature through the years, most recently a white-water rafting adventure through the Grand Canyon. I thank her too for keeping our home and yard alive with a wondrous variety of plants and flowers, together with birds, butterflies, and music, here in Santa Barbara, where nature's nurturance is so bountiful.

I must, of course, take final responsibility for what you will find in this book, especially its limitations. But these friends, colleagues, and family members enriched and deepened immeasurably the creative pool in which the cognitive and clinical theories in these pages were birthed over the years.

Prelude

Earth, which has seemed so large must now be seen in its smallness.
We live in a closed system, absolutely dependent on Earth
and on each other for our lives and those of succeeding generations.
The many things that divide us are therefore of infinitely less importance
than the interdependence and danger that unite us.

—From a message to the world by six
biologists at an international meeting[1]

From the author's perspective, this book will:

- Affirm what you are already doing to help save the planet for future generations. The book says a ringing Yes! to your present earth-caring concern and action, with the expectation that it will enhance both of these.
- Shed light on humankind's most serious health challenge ever: how to save our precious planet as a clean, viable habitat for all the children of the human family and all other species.
- Describe how psychotherapists, pastoral counselors, teachers, medical healers, parents, and other earth-caring persons can use our strategic opportunities to cope constructively, even creatively, with this unprecedented challenge.
- Explore an expanded, ecologically grounded theory of personality development, the widely ignored earthy dimensions in understanding human identity formation. Finding this lost sense of our groundedness is the theoretical foundation for the educational and therapeutic approaches described.
- Outline basic principles of a model for doing ecologically oriented psychotherapy, counseling, medical healing, teaching, parenting, and community action.
- Offer a variety of methodologies for helping yourself and your clients, patients, students, or family become more nurtured by nature, thereby strengthening both motivation and energy for nurturing nature more caringly.

• Describe six life-saving perspectives on human-earth relatedness, orientations that can help people make their lifestyle more earth-caring as well as earth-nurtured.

• Demonstrate why the keys to effective ecotherapy and ecoeducation include hope, humor, and, most important, wise love. This includes love of ourselves, other people, the place where we live, the divine Spirit, the earth, and the biosphere. In short, all these loves are a part of the *love of life*.

• Suggest ways to use these earthy approaches in a variety of social contexts and cultures.

You may be wondering whether this book is about the well-being of persons, the well-being of the earth, or an earth-grounded understanding of human personality. It is about all three and their dynamic interrelationships.

When I began researching this book a decade and a half ago, I planned to write only for counselors, psychotherapists (including pastoral psychotherapists), and personality theorists, all of whom have crucial roles in creating a greener, cleaner planet. As my own thinking and teaching evolved in earth-grounded directions and I talked with ecologically aware teachers, the parameters of this book enlarged. It became clear that teachers have crucial roles in preventing earth alienation and facilitating earth bonding in learners. Ecoeducation emerged as complementary to ecotherapy. Subsequently, as I engaged in dialogue with various health professionals, I saw that they also have strategic roles in resolving their planet's environmental health crisis. They possess the expertise and status to motivate and guide community action to clean up health-damaging toxicity in our air, water, soil, workplaces, and homes. Health professionals also need to be aware of the interdependence of body alienation and nature alienation, interacting factors that often prevent patients from caring for their bodies as well as for the earth.

Therapeutic insights and educational methods are not the exclusive property of professionals (fortunately). They are vital human resources that belong to everyone. Thus the light dawned that I had neglected the most influential group of teachers: parents, grandparents, and other adult family members. They have the most vital and profound opportunities to shape earth-caring attitudes, values, priorities, and lifestyles in the younger generations. Ecoparenting can enhance family life and transmit earth literacy via everyday experiences. Fortunately, many of the principles and methods of ecoeducation and ecotherapy can be utilized by nonprofessionals in their personal life and with their families.

Readers accustomed to more "objective" books on scientific issues, including health, may be puzzled by the personal and passionate tone they will encounter at places in these pages. This tone is intentional and seems appropriate. If saving a viable planet is truly a life-and-death issue, it is essential to integrate the passions we feel with solid scientific data that shed light on the crisis.

The Author's Roots—A Personal Earth Story

Knowing something about an author's life journey often sheds light on a book's perspectives. And, as you will discover later, telling your earth story is an essential part of both ecotherapy and ecoeducation. For these reasons I will share some crucial parts of my earth story. I do this not because my story is in any sense normative, but because I know it well and it is a deep part of myself. Your earth story undoubtedly includes many comparably vivid but very different nature memories. These experiences helped shape your sense of self-in-the-world and your degree of bonding or alienation with nature. As you read my autobiographical accounts, I hope you will be reminded of your own earthy memories, whether you were raised in a city, a village, or on a farm. I hope that these earth memories will bring to life for you the theories being discussed throughout this book.

I invite you now to meet Junior, the little boy inside Howard. Junior embodies the old adage, "You can take the boy out of the farm, but you can't take the farm out of the boy." I grew up as a city boy in the midwestern community of Springfield, Illinois, with a population of about 80,000, but my parents were roots-in-the-soil people, both having been raised on Illinois farms. My dad's university degree was in agriculture. He worked in farm-related jobs most of his adult life. My folks transformed the four city lots on which our old frame home was located into a veritable mini-farm. Combining know-how and hard work, they grew a wonderful variety of delicious vegetables, beautiful flowers, and succulent fruit. During the economic disaster of the Great Depression, the garden was much more than an avocation. It was a major source of food, without which our family undoubtedly would have been hungry at times. I can still savor the juicy red raspberries and the large kettles of tomatoes and peaches being cooked for canning. I have memory pictures of shelf after shelf of multicolored glass jars in the cool basement at summer's end, preserving some of the earth bounty for the cold winters.

Mom and Dad made sure that Junior and his siblings spent time, in spite of our protestations, working alongside them in the large garden. In retrospect, I am aware that these experiences gave me precious gifts. These included an appreciation for the importance of cooperating with the powerful growth energies of nature and the ongoing earth cycles—preparing the soil, planting seeds, cultivating and harvesting the food, and then waiting while the often-frozen soil rested under a blanket of snow. In spring's rebirthing, I remember an aching back from working the rich, black prairie soil and spading under the plants left there to regenerate the soil for the new crop.

I learned how generous nature can be when she is respected and cared for. But any temptation to wax romantic about the earth's beauty and productivity is curbed by the occasional grief that accompanies depending on nature for sustenance. I recall my parents' grief when a winter freeze killed the peach trees. I remember the never-ending fight to deter an army of voracious insects, as well as the endless struggle to control the ubiquitous weeds that always seemed to grow much faster than the vegetables and flowers.

I have poignant and painful memories from one summer during my early adolescence. The Depression was fierce for our family, and I could find no job even though I desperately needed one. So Dad thoughtfully arranged with a farmer friend for me to plant several acres of popcorn just outside the city. I sweltered in the heat and humidity all that Illinois summer, faithfully cultivating and weeding the corn. Then, just as the ears were beginning to fill with kernels, tragedy struck. Early one August morning, I rode my bike to my popcorn field full of hope. I was devastated to discover that a "twister" (a small tornado that occasionally accompanies midwestern thunderstorms) had leveled most of my precious crop. My dreams of selling a bountiful harvest and using the money for a more functional bicycle for use in selling magazines had been obliterated by what probably was less than a minute of violent wind.

During most summers in my childhood and early teens, I spent considerable time on my grandparents' farm five miles from Kinmundy, a tiny southern Illinois farmers' shopping village of a few hundred souls. Grandma's loving care of her grandson and Grandpa's simple but powerful religious faith had a powerful impact on my early life. His faith was a gift to me, in spite of his nightly reading of a long, boring Bible chapter by the flickering light of a coal-oil lamp, followed by uncomfortable kneeling on the wood floor during his fervent but interminable prayers.

Only since the environmental awakening began in the 1990s have I sensed how deeply I was shaped by learning to garden and farm when horses were the primary power. In my olfactory memory bank, I can still smell freshly plowed fields and the sweat of the horses pulling the corn cultivator as I rode behind them back and forth along endless rows. I can smell the newly mown redtop hay as I drove a team pulling the mower and then the rake back and forth. I still can feel aching muscles from pitching endless wheat and oat bundles onto a hayrack, followed by the exhilaration of riding high on the load as the horses pulled it for storage in the haymow.

I also can relive the imprinted experiences of the cycles of growing animals—the sexual stimulation of watching chickens and ducks and large farm animals copulate. This was my first, less-than-adequate sex education. I can recall the excitement of riding bareback on a huge galloping draft horse down the muddy dirt road, followed by the humiliation of falling off and having to walk home coated with mud from my matted hair to my bare feet. I remember the oral satisfactions of the wild blackberries picked in fencerows, followed by the agony of scratching the bites in unmentionable places from the chiggers that infested those berry patches. Feeling-laden memories include playing in the rushing, rain-swollen "crick," excitement mixed with fear of the large snapping turtles who lived in it. My bodily memories include the discomfort of the frigid outhouse seat on frosty mornings and the pain of sitting on a bee there in the summer, as well as the year-round discomfort of using old Sears and Roebuck catalog pages in lieu of toilet tissue.

I can still savor the boyish triumph of finally mastering the art of milking cows by hand. My memories include the butchering of hogs, a bloody whole-family ritual, and the anticipated enjoyment of later eating delectable ham cured in brine and the smokehouse.[2] Delicious food memories include the aroma of the bread and breakfast biscuits baking in the wood-burning oven, and the feasts prepared for holidays and on days when male neighbors came to help Grandpa with the threshing while the women contributed their favorite dishes. I remember walking with Grandpa in the watermelon patch with a large knife. He would thump melons until he found one he liked. Then we would sit on the ground and enjoy the delicious melon heart. The rest was thrown over the fence to be consumed by the pigs, accompanied by their noisy, competitive grunting and jostling.

My parents loved to fish and my dad was an avid hunter.[3] Our summer vacations usually were fishing trips to rivers and lakes in Illinois, Minnesota, and Wisconsin. I have memories of choice family experiences on fishing excursions. These include the excitement of landing a large northern pike or a plump crappie, together with the scary rush of adrenaline of being in the middle of a huge lake when a violent wind from a thunderstorm arose unexpectedly. Dad also took me hunting for ducks, rabbits, and quail—hoping, I am sure, that I would acquire his passion. To his disappointment, hunting never "took" with me because I tended to identify with the animals. But I cherish the memories of precious times spent with him in the woods, fields, and river bottoms.

In my teens, a high school buddy and I took several float trips down the Sangamon River, the stream up which Abraham Lincoln had tried in vain to navigate a small steamboat to New Salem, his young-adult hometown. Once we were carelessly floating with the current when suddenly, in a narrow, swift place in the river, our double-bowed skiff hit a log that dumped us into the swirling water. We struggled mightily trying to catch the boat and save the oars and the few pieces of our camping gear that floated. So obsessed was I that I continued fighting the current until I was far beyond exhaustion. Just when my strength was so spent that I could barely stay afloat, one of the oars came within my reach in the turbulent stream, probably saving my life. This near-death experience gave me a scary awareness of the deadly potential of the natural world when it is treated carelessly.

Connectedness with nature was enhanced in my adult years. The miraculous struggles of human birth and growth came alive as wonderful gifts as our three children joined us. During their growing-up years, our family often camped and hiked in spectacular mountain ranges—first in New England's White Mountains, later in the Colorado Rockies, and then in California's magnificent Sierra Nevada. As I write, I am reliving cherished memories of backpacking in the high country and car camping with our family. In my later mid-years, the small men's sharing and support group I was in purchased a sailboat. The challenge of learning to sail on the ocean was rewarded bountifully during three sailing tours of the five wild, wondrous islands of the Channel Islands National Park (I can now see on clear days from our home). All these are parts of a continuing love affair with mountains and the sea.

I feel lucky as well as blessed that my connections with nature in the early years and beyond had such earthy intensity. Spiritual highs during summer youth camps in inspiring natural settings (mingled with sexual surges) had a profound impact in helping to shape my spiritual development,[4] my life orientation, and my vocational choice. When the current environmental movement was dawning, I wondered why I was resonating like an old-fashioned tuning fork. It felt so right and so important. Gradually it dawned on me how profoundly nature experiences had imprinted my sense of self, including feelings about our species and the earth, life and death, and many other things. I also saw why I feel so ungrounded when I do not make space in my schedule for reconnecting with the natural world. During these all-too-frequent times, I experience ecoalienation, the dim awareness that I have cut myself off from being nurtured by nature. When I succumb to this part of my male programming, I overfocus on *doing*, controlling and succeeding. I neglect just *being* by slowing down enough to experience nature and people in more depth.

In spite of my struggles with distancing from nature through the years (or could it be *because* of them?) I have gravitated toward growth and wholeness in developing cognitive maps to guide practice. During the pressure-cooker years of graduate education and personal psychotherapy in young adulthood, nature images came spontaneously to my mind as I worked therapeutically with myself and then with growth-blocked clients. As a revised model for pastoral care and counseling began to develop in my practice, it was no accident that the label "growth counseling" came to mind. Through several decades of graduate teaching, I have found that for numerous counseling students, struggling to develop their own professional therapeutic skills, the growth orientation is empowering. It was exciting and heart-warming to witness these bright young people from some twenty-five diverse cultures awaken to their own growth potentialities. To walk alongside them as they learned to facilitate growth toward wholeness in others, brought rewards that far outweighed the frustrating times in teaching.

Clarification of Terms

A variety of terms are being utilized currently to refer to healing interrelationships of humans and the rest of the biosphere. Let me clarify some words used in these pages. *Ecotherapy* refers to both the healing and the growth that is nurtured by healthy interaction with the earth. *Ecoeducation* is this growth-stimulating process. Terms that are comparable to ecotherapy include *green therapy*, *earth-centered therapy*, and *global therapy*. *Ecopsychology* and *psychoecology* refer to what is called the "greening of psychology" and the psychologizing of ecology respectively. I have used *ecotherapy* rather than a word like *ecopsychology* for two reasons. The focus of this healing and growth work encompasses the total mind-body-spirit-relationship organism, not just the psyche. Furthermore, my approach is on the application of ecopsychology but also ecobiology and ecospirituality in therapy and education. The term *biophilia* means, as defined by Harvard zoologist Edward O. Wilson, the innate, genetically rooted affiliation of humans to other animals and living organisms. *Biophobia* is fear of nature. Ecotherapy aims at incorporating biophilia into healing and growth practices and thus at utilizing the healing energies of nature. I use the terms *wholeness*, *wellness*, *well-being*, and *holistic health* interchangeably, and biophilia is regarded as an essential dimension of all such health.

An Invitation to Network

The concepts and methods in these pages are rooted in vivid experiences of nature. They have evolved over the years and continue to do so. As I experiment with earth-based methods in my teaching and therapy, and savor enlivening interaction with nature myself, critical reflections generate continuing changes in both theory and methods. It is encouraging that numerous other clinicians and teachers have described their experimentation with earth-grounded methods. As I had completed all but the final revisions of this volume, a significant book appeared—*Ecopsychology: Restoring the Earth, Healing the Mind*, edited by Theodore Roszak, Mary E. Gomes, and Allen D. Kanner. I am pleased that several papers in that volume affirm the healing efficacy of earthy methods in the clinical work of mental health professionals, mainly psychologists.[5] As I read these papers eagerly, I felt less lonely, less like a voice crying in the ecological wilderness. The validity of much of what you will find in these pages is undergirded by my own clinical work. It is heartening that clinical and experimental testing by several authors of those papers seems to confirm many of the theories found in these pages.

From a scientific perspective, however, the testing by myself and others provides mainly anecdotal evidence. Theories often go beyond the empirical experiences that stimulated their creation. Although empirical support is increasing with encouraging alacrity, many of the ecotherapeutic formulations in this book must be regarded as tentative and speculative until they are tested by "hard" controlled research as well as by much more practice-oriented applied research.

In any area so inviting of continuing research, it is vital to share learnings as widely as possible. I hope this book will stimulate such professional dialogue. To raise issues, share comments and criticism about how the ecotherapy-ecoeducation model relates to your personal or professional experiences, write me at the address below.[6] Tell me if your work as a therapist, educator, health professional, or parent contradicts, confirms, or causes you to modify what is presented here. Such dialogue will expedite the needed refining of approaches to this crucial aspect of our lives and work and world.

Researching and writing this book gave me an unexpected but very welcome gift for which I feel deeply grateful. It changed my life in significant ways, more than anything I had written before. It did this by deepening my awareness of who I am and what is most important in my remaining days or weeks or years. It helped me rediscover the earthy roots of myself and gain increased passion for loving this good earth in personal as well as professional ways.

Any book has a life of its own quite beyond the author's control after it is written. As this one is launched, my passionate hope is that you and others will find in this book fresh ways to open more fully to the joys of being intentionally nurtured by nature, while you further the wholeness of persons by furthering the healing of Mother-Father Earth and its wondrous network of living creatures.[7] As you read, may the health of our earth home be enhanced as you create new ways to be a responsive and responsible member of the blossoming biosphere. May this book be an instrument to help you heal persons by healing the earth—and vice-versa!

Notes

1. Meeting in Menton, France, in May 1970.

2. All this was long before anyone was aware of the dietary hazards of high salt and cholesterol intake.

3. On a farm near the Illinois River, these activities combined both favorite recreation and a welcome source of a variety of protein in the diet.

4. Hymns celebrating the wonders of nature were favorites in the pietistic religious life of my childhood family and in the youth camps.

5. San Francisco: Sierra Club Books, 1995.

6. 2990 Kenmore Place, Santa Barbara, CA 93105, USA; fax 805-682-2816.

7. I have chosen the term *Mother-Father Earth* for two reasons. The obvious reason is that creating new life usually takes two. I also want to avoid the age-old dualistic sexist put-downs of women resulting from associating their gender with "Mother Earth," bodies, sexuality, and emotions whereas men often have been associated with the sky and with other "elevated" things like the mind, rationality, philosophy, and science.

1

Introduction: Using the Ecological Circle for Self-Care, Earth-Care, and Soul-Care

Teach your children what we have taught our children, the earth is our mother. . . . This we know. The earth does not belong to humans; humans belong to the earth. This we know. All things are connected like the blood which unites one family. . . . Whatever befalls the earth befalls the children of the earth. People did not weave the web of life, they are merely a strand of it. Whatever they do to the web, they do to themselves.

—attributed to Native American Chief Sealth (Seattle), 1854[1]

The most serious, most dangerous health challenge all of us in the human family face is to reverse the planet's continuing ecological deterioration. It is the most profound health issue of *all* times, from a historical perspective. Why? For the first time in the long human story, our species faces a health challenge that if not resolved will foreclose opportunities to solve humankind's countless other problems, including a multiplicity of health problems. The human species now must be included on the endangered species list. This is the bottom-line health challenge we all face.

For the first time ever, one species—with the questionable self-label the "wise humans"—has the awesome power to threaten the health, perhaps even the survival, of all species.[2] Using our superb intellectual endowment, this species has created and misused technology, squandered limited natural resources, and multiplied in unrestrained ways so that the earth's biosphere is being depleted more rapidly than it can repair itself by natural processes. In a real sense, the earth's autoimmune system is threatened with irreparable damage. Unless we humans devise effective, nonviolent ways to resolve these planetwide threats to the life-nurturing environment, the whole human family and all the animals, birds, and plants have a very problematic future. Therefore, there is no more urgent health issue for all of us to learn to live in earth-friendly ways as earth-carers and peacemakers. Only in this way can we help save the planet as a healthy place for ourselves and for all other living creatures, today and tomorrow.

But, before we are overcome with gloom and doom, remember that this environmental health crisis is also an unprecedented opportunity. Now, as never before in history, the whole human family has the most urgent demand to cooperate across the plethora of social, cultural, political, and language barriers that divide us. The challenge is to turn this global crisis into what philosopher-psychologist William James once called "a moral equivalent for war." He used this phrase in discussing the remarkable ways that a struggle to defend against a common external enemy unifies and empowers internally conflicted nations. He declared that humanity needs a constructive equivalent for the intense but genocidal commitment fostered by the planet-destroying war system.

Today the human family has unprecedented, potentially unifying or divisive common enemies—global violence, overpopulation, economic injustice, suppression of freedom, and the deadly destruction of the environment. All these problems transcend national, ethnic, cultural, religious, linguistic, and racial boundaries. They can only be solved by international collaboration. In our postmodern world, how these problems impact any country ultimately influences all other countries on our shrinking planet. As the Earth Summit and the follow-up meetings in Cairo and Beijing illustrate so hopefully, it *is* possible for the leaders of most of the nations as well as countless NGOs (nongovernmental people's organizations) to join forces in devising and implementing effective solutions to the root causes of the planet's pain.

Why is it crucial that those of us in the healing, teaching, and helping professions, along with parents, understand the complex interrelationships of personal health and sickness with the wholeness and brokenness of the biosphere, and all the people-serving institutions that impact our personal well-being day by day? Doing everything we can to maximize individual and family health obviously is very important. But to focus only on maintaining personal health while ignoring the social causes of much illness in today's world, is increasingly inadequate. Maintaining high levels of individual health already is a precarious, privatized goal—the luxury of a shrinking minority of those of us who live in affluent countries. It will become increasingly so unless we approach health problems socially, even globally, and work "like miners under a landslide"[3] to heal both the individuals and the socioeconomic and cultural causes that breed pandemic sickness around the world.[4]

It is now prudent for all of us to recognize that as Brian Swimme and Thomas Berry declare:

> the well being of the ecosystem of the planet is a prior condition for the well being of humans. We cannot have well being on a sick planet, not even with our medical science. So long as we continue to generate more toxics than the planet can absorb and transform, the members of the Earth community will become ill. Human health is derivative. Planetary health is primary.[5]

How can we motivate people to make the difficult lifestyle changes that will be needed to save the biosphere? One key is to awaken widespread awareness of the heavy price our lifestyles are costing us in terms of personal health as well as the environment. These costs will continue to soar until increasing numbers of us change our lifestyles drastically in ways that are simultaneously self-caring and planet-caring. An ecologically aware physician and an environmentalist highlight this point in their book *Well Body, Well Earth:*

> The best measure we have for designing our future technologies is human health. There is nothing that seems more immediate and important than personal health, our own and that of our loved ones. It is here that we feel the greatest urgency to solve problems of environmental pollution, and it is here that the consequences of our actions are the most dramatically demonstrated. . . . Like radio buoys guiding a ship at sea, disease and health can guide our lives. Steering our course by these signals not only leads to a life relatively free of disease. It can also guide us to the upper limits of personal fulfillment. Again, because of our interconnectedness, the fulfillment of any single individual or system ultimately benefits all systems around it. Guided by disease and health, Gaia, the living Earth, benefits from every person's human fulfillment.[6]

We humans display blatantly contradictory attitudes and behaviors in our individual health practices and in the ways we treat the environment. In many circles, interest in these interdependent dimensions of health is high, but we continue to send mixed messages by our behavior. For example, Americans are reported to jog twenty-seven million miles each day for their health, but eat three billion gallons of ice cream (mainly fat and sugar), and produce one and a half million tons of toxic waste. Around the world, parents long passionately for children's well-being, yet collectively we have made wellness impossible for millions of children by doubling the planet's population twice in the twentieth century.

Awakening to the Ecological Circle

An effective methodology in both ecotherapy and ecoeducation is to offer participants opportunities to *tell their ecological story*. This means inviting them to recall and verbalize imprinting experiences with the earth, from their childhoods as well as more recent years. This usually involves experiencing and expressing their thoughts, images, and feelings about the earth, including any awareness they have of the perils confronting it. I told a little of my earth story in the preface to stimulate your own recall. Here is an experience I had a few years ago that intensified my awareness of the earth's pain.

A few years ago, while giving lectures in five South American countries, I arranged to spend a few days at two different places along the Amazon River in northern Brazil. I planned this to provide a relaxed mini-vacation in a hectic schedule as well as to provide a firsthand look at the troubled rainforests. Quite unexpectedly the experience shook me to my depths and had an impact on me as a person searching for some light in the ominous shadows of the global environmental crisis.

Before the visit, I had read extensively about the perilous plight of rainforests around the equator. My "head-level" information came to life vividly during this visit. It was a rude but invaluable awakening to the interdependent agonies of ravished nature and oppressed peoples. Gestalt therapy often uses the technique of asking people to describe incidents and dreams in the present tense; I will use this approach in describing these experiences in Amazonia.

My plane is circling the airport of a city some fifty miles from the Amazon's mouth. I watch eagerly for my first glimpse of the river and the jungle. It is early evening, but so near the equator, there is plenty of light for a good look. As the plane descends, I feel a wave of disappointment. I can barely see the forest or the river. Everything is shrouded by a pall of smoke from the burning forests.

The next day I am having my first close encounter with the majestic river. As the medium-sized paddlewheel boat traverses the enormous channel, I recall hearing that this river carries eleven times the water that flows in the Mississippi, my country's largest river. The lush vegetation is nearer and nearer to our boat as we now chug up a tributary into the heart of the rainforest. I fear for the small, brown-skinned boys who are risking their lives swimming to our slow-moving boat and climbing aboard. They try to sell us trinkets they have made from forest materials and ask for handouts. I buy a little box made of palm leaves from a boy who looks about seven or eight.

The crew is tying up our boat at a small dock made from rough wooden planks. We get into a sizable canoe propelled by an outboard motor. Our group of seven, plus a trilingual guide, motors up a small, narrowing creek. We must duck the hanging vines. The beauty and aliveness of the forest is breathtaking. I hear strange birdcalls. The canoe pulls in at a tiny crude dock. Our guide says he will lead us along a path through the rainforest, joking that we are lucky that this is the dry season so we can walk rather than swim.

My excitement rises as the guide takes us deeper into the forest, commenting on exotic plants, flowers, vines, trees, and birds along the trail. After a half-hour walk, we reach a clearing where the trees have been slashed and burned to make a soccer field near a tiny school. We walk to the middle of the field. The guide suggests that we have a close look at the "soil" under our feet. I am shocked as I pick up a handful of dirty sand with almost no topsoil in it.

Our guide is aware of the ecological plight of his rainforest. He comments matter-of-factly: "When the trees and other plants are stripped away and their roots lost, the shallow rainforest topsoil is quickly leeched away by the torrential rains that fall almost nonstop for several months each year." The result? Within four years the topsoil is too depleted to grow any type of food crops. In about seven years it is reduced to almost pure sand like that under our feet. It can no longer grow even enough grass to pasture cattle.

The suffering of the raped earth is real. What I know in my head comes alive in my whole being, as I hear the sound of the earth crying. My male eyes are dry but I feel inner tears—tears of anger, shame, and grief. Something I have read comes back: The desperate poor from the slums of the cities of northeastern Brazil are pushed into the rainforest to slash and burn them. Why? To produce cheap beef for export to the fast food chains in affluent Northern Hemisphere countries like mine. An unwelcome memory picture intrudes: I see myself buying a hamburger at a fast-food chain that profits from using inexpensive imported beef. With each such hamburger, I see myself causing a magnificent, towering tree to come crashing down. I feel a tightening in my chest as I remember that 40 percent of the oxygen produced on the earth comes from plants in the rainforests. Cutting down the rainforest is like amputating 40 percent of the "lungs of the planet" on which all other living things, including myself and my loved ones, depend for survival.

Now we are returning along the forest trail to our canoe. The jabbering tourists in our party are strangely silent. I can hear the subtle sounds of the forest—the wind swaying the trees and the music from the canopy a hundred feet above our heads, unfamiliar, haunting bird songs. The relative quiet creates space in my consciousness for reflecting on what I have just experienced. I remember why rainforests around the globe are being decimated at the rate of a football-field-sized area every minute of every day. It is not because the people who live there are evil or stupid. They are not. It is because they are very, very poor and trying desperately to raise the level of their families' well-being.

Now we are back in our canoe going down the widening forest stream. I see the rickety houses on stilts along the banks where the half-Portuguese/half-Indian residents live. I realize why the little boys risk swimming out to the paddlewheel boat. They are from these homes where everyone must do all he or she can to help the family survive. It is no wonder the average child only makes it to the fourth grade in the little schools the government provides.

As my mind struggles to process what I am experiencing, I remember more about why the poor are pushed from the slums into the rainforests. Their government is unwilling to bite the bullet of land reform, redistributing to the urban superpoor some of the good farming land that is held in huge tracts by a tiny oligarchy of wealthy families. The government is also motivated by their "debt trap"—a gigantic debt to the World Bank and other transnational Northern Hemisphere banking institutions. The income the government derives from exporting cheap beef is a vain effort to make a dent in this astronomical debt. But the debt is an impossible trap, so huge that the payments cannot even pay the spiraling interest. If all the vast rainforests of Brazil, the largest in the world, are turned into deserts, their national debt will probably be at least as overwhelming as it is today. An insightful Brazilian leader of laboring people is said to have observed that, if the Amazon is the lungs of the world, then the debt is its pneumonia.

This human and ecological tragedy is compounded by the economic insanity of this exercise in global frustration. When the rainforests in all the countries along the Amazon are gone, the ex-slum dwellers and native peoples who manage to survive will be even more destitute. Why? Because they will no longer be able to harvest nature's bounty—food crops as well as rubber, Brazil nuts, and coconuts that bring some income to forest dwellers. Through many centuries the native peoples learned to live well in the heart of the rainforests. The average per acre income from harvesting the living forest's rich, renewable supply of products is around seventy-five dollars (U.S.) a year. In contrast, the average yearly income from an acre of stripped rainforest land is only around thirty-five or forty dollars—during the short time the topsoil lasts.

As our boat passes numerous families in little clearings or on the narrow decks around the front of their stilt homes, I remember that when Europeans first invaded the Amazon basin some four centuries ago, an estimated eight million native people lived here in harmony with nature. Now, as a result of Western diseases and the destruction of their forest habitat, fewer than 300,000 are left. Furthermore, the well-being of many of these tribal peoples is threatened in unprecedented ways by Northern Hemisphere lumbering, oil, and mining companies. The fish on which native peoples depend for food have become dangerous to eat because of the massive pollution of the streams from gold mining and oil exploitation. It is becoming clear why native peoples around the world refuse to celebrate their "discovery" by European invaders (like Columbus) who stole their land and its abundant resources while "Christianizing them to save their souls."

We are back on the larger paddlewheel boat bucking the Amazon's mighty flow as we recross it. As I sit alone near the bow, my mind makes a leap to my home. I see the smog damage to millions of trees in the forests above the Los Angeles basin. And I know that, if I had been more aware, I could have had a comparable awakening in many less exotic places closer to home—even in my own back yard. The problems of the Brazilian rainforests are, in different ways, also the problem of the Los Angeles basin where I lived for three decades. I see the Sequoia Gigantia, probably grown from seeds of one of the 2,000-year-old giants in Yosemite or Sequoia National Park. I remember planting it as a four-foot "child" in our front yard, after it had served as our live Christmas tree for several years. As the years passed, it grew to more than forty feet. To adorn it with colored lights for the holidays required stretching our extension ladder each year. It became a member of our family. As it gradually turned brown and died in spite of our efforts to revive it, we experienced growing grief. I knew that its demise probably resulted from the yellowish smog that often poisoned the air.[7] I wonder whether the dirty air back home had discolored the lungs of all our family. Then I think of the poor ethnic minorities who can only afford to live in the most polluted inner city. And I remember the people in the terribly poisoned air enveloping the poverty-wracked megacities I have visited in developing countries, the victims of poverty-driven ecological oppression. As the boat docks in its home port, my heart is heavy with the experiences of an unforgettable day.

My awakening during those days by the Amazon was actually a dramatic reawakening of an awareness that had been developing over the years, but often forgotten. I knew again that this awareness must influence the way I function in my personal lifestyle as well as my professional world. The conviction was reinforced that privatized counseling, therapy, and education that ignores the social and environmental contexts of people's lives will no longer fly in my professional life.[8]

Exploring the Ecological Circle's Meaning

Those unforgettable days in the rainforest made me aware of the need to implement in my life and work what I came to call the ecological circle. Simultaneously, I got in touch with three realities that together constitute the ecological circle. I experienced the awesome aliveness of nature and the delicious taste of being nurtured by this vital life energy. I got in touch with the transcending source of nature's healing energy, the divine Spirit. These two experiences triggered enhanced motivation to nurture nature caringly in whatever setting I am in. A closer look at each of these is now in order. (It will be helpful to keep in mind the diagram of the ecological circle above.)

Few people are aware of how utterly dependent our lives are on being continually nurtured by nature. Every breath we take, each bite of food we eat, every drop of water we drink is a silent, usually unrecognized expression of this dependence. We usually take these abundant gifts for granted. But if we stop to reflect, it is obvious that we cannot be healthy, or even survive, without the continuing, minute-by-minute renewal of these bountiful gifts from nature. Being "nurtured by nature," as the phrase is used in this book, means flinging wide our inner windows of grateful awareness of these gifts of life and deepening our intimate interaction with the natural world in ways that are both healing and enlivening. Intentionally deepening this openness to this life-sustaining nurturance by nature is the foundational experience of implementing the ecological circle.

The second experiential dimension of the circle occurs for many people when they pause, while being intimate with nature, to enjoy the experience of the creative source of all life that many of us call God. The creation mystics in various religious traditions through the centuries have tuned in on this spiritual reality in nature. For many people, it is like cool, refreshing water from an underground spring on a hot, dry day. Whether people think of these experiences by using religious concepts or think of them as marvels of nature explicated by science, is not the heart of the matter. What *is* important for empowering the ecological circle is to open oneself to the self-transcending reality immanent in the awesome creative power in nature. Whether one understands the transcending reality in nature in religious or secular terms or both, connecting with this reality can bring multiple benefits. Experiencing this enlivening energy can enhance people's love for the natural world, deepen positive bonding with the earth, and add an earthy grounding to their spirituality. In other words, it can bring all of life down to earth.

Concerning the spiritual reality in nature, Martin Buber observed:

> Creation is not a hurdle on the road to God, it is the road itself. We are created along with one another and directed to a life with one another. . . To look away

> from the world, or to stare at it, does not help a man to reach God; but he who sees the world in Him, stands in His presence. . . If you hallow this life, you meet the living God.[9]

The third experiential dimension of the ecological circle flows from one or both of the first two. Feeling a loving connection with nature can energize motivation to respond by nurturing the earth more caringly. The circle is completed only when people reach out in earth-nurturing action. For many people, including some who describe themselves as nonreligious, the motivation to nurture nature is enhanced when the experience of being nurtured by nature is deepened by a transcending, spiritual reality in that experience.

Whether or not people sense a spiritual dimension in nature, informed caring for the earth is really enlightened self-interest. Simultaneously, it can heal persons and heal the living environment that is in them as they are in it. By completing the circle, people participate in re-creating nature, including themselves.

Each of the three interdependent dimensions of the ecological circle experience is valuable in itself as part of ecotherapy and ecoeducation. Each is needed if the human family is to learn to become more earth-caring by learning to live more lightly on the earth. All three can contribute in different ways to the goal of learning to live healthier lifestyles.

To summarize, the three dimensions could be called *inreach*, *upreach*, and *outreach*. In relation to the environment, inreach means opening ourselves to be more intentionally nurtured by nature. Upreach refers to the energizing spiritual awareness that can motivate and empower us to engage in the often difficult tasks of outreach. Outreach means participating with others in action that will help save the planet, action that is an essential aspect of the holistic, contextualized therapy and education described in these pages. Furthermore, earth-caring and people-caring are two sides of the same process in that together they may enhance wellness in ourselves, in other people, and in our living environment.

Case Illustrations of the Ecological Circle

The following cases illuminate the three dimensions of the ecological circle. The childhood experiences of Gloria Johnson, a pastoral psychologist and counselor, helped equip her to implement the ecological circle in her adult life and work. She recalls: "As an African American born and raised in the South, I was taught that the earth will take care of you but you must respect it, listen to it, and relate to it. As a child my entertainment, my toys came from nature. We braided plugs of tall grass in the earth. This is how we learned to braid hair. We had dolls that we made from corncobs and corn silk. The medicines that my parents used often were from the herbs and roots my father collected during his hunting trips. We played in the light rain. It was exhilarating, warm and nurturing. Emulating our parents, we children had small garden plots of our own to have "hands-on," personal relations with nature." These nurturing experiences brought her what she describes as a deep respect, deep love, and appreciation for the wonders and beauty of nature. She recalls: "I have cultivated the ability to communicate with nature, to experience its aliveness, and to realize how essential nature is to my well-being. The most profound memory I have from my childhood is lying on the grass in our front yard, talking to God, taking in the beauty of the trees, the clouds, the sun, while I talked. This memory is a place of safety and security for me. The memory is where I still go now for security."

Johnson reflects on how her professional functioning has been influenced by relating lovingly with nature: "My sense of being related to nature has helped me be more accepting of people in counseling and education, less judgmental, because I now am able to see myself in others. My ability to be more with people has been facilitated by having been so communal with things in nature. I am more respectful and inclusive in my teaching and counseling. I am being taught and counseled as I teach and counsel. I participate in life's cycle."[10]

The counseling case that follows also illustrates the three dimensions of the ecological circle.[11] This family's experience raised my consciousness about the importance of social context issues in counseling and therapy. Several years ago, at the height of the cold war, a school psychologist referred a student I will call Bill (not his real name) to the pastoral counseling and growth center where I was the clinical codirector. Bill's presenting problem was his parents' extreme frustration and dismay that he was failing miserably in his high school courses. He had been a good student until a couple of years before. The crisis had produced a painful breakdown of communication within the family system.

The pastoral psychotherapist who was to see Bill wisely asked the school psychologist for Bill's test results. She was amazed to discover that his IQ was around 150. Being trained in family systems theory and therapy, she knew that when one member of a family is suffering self-sabotaging dysfunction like Bill's, the whole family is feeling pain. So her second wise move therapeutically was to invite the young man's parents and two younger siblings to come with him to the first session.

When they arrived, Bill took a chair as far from the rest of the family as he could get and sat looking at the floor dejectedly. Suspecting that he was depressed but also angry about being brought by his parents, the therapist asked each family member, "How do you feel about being here?" Bill: "It's a bummer." Therapist (in a nonjudgmental tone): "It sounds like you'd really rather not be here. I can understand how you might feel upset about having to come here." This brief interchange let Bill express some of his angry resistance and, equally important, let him know that the therapist had some awareness of his negative feelings.

After further interaction to strengthen rapport with the family, the therapist said, "Bill, you must have some reasons why you're not cracking your books. I'd be interested to know what your thinking is on that." Bill shrugged his shoulders dejectedly and shook his head. Therapist: "What do you mean by this [shrugging her shoulders]?" Bill: blurted out with pained intensity: "Why bother when the world's going to hell in a basket?' Therapist: "Things look terribly hopeless to you." Bill: "Yeah, it sucks!"

Bill's despair about the world situation (psychiatrist Robert Lifton calls this "radical futurelessness") came as a total surprise to his family, who had never discussed their feelings about this.[12] Toward the end of the second or third session, a turning point came in the family's communication when Bill's father leveled with him: "The mess in the world worries me, too, Bill.'

During the next few months, Bill and his family did hard work therapeutically in conjoint family sessions, interspersed with individual sessions with Bill and the therapist. Opening up about their previously undisclosed fears about the future proved to be a major key to unlocking their communication. Their frozen emotional climate warmed up as they communicated their fears about the world situation. As this gradually happened, Bill's angst-driven depression lifted. He began to study and in short order pulled himself out of his self-defeating emotional tailspin of despair. Not surprisingly, his grades improved.

As conjoint family therapy continued, realistic concerns about the world situation continued to come up. After about three months, the therapist made what turned out to be a particularly strategic intervention. She asked if they had considered getting involved in some group working in the area of their shared worries about the world. Bill's mother responded that their church sponsored a group called Pax Christi, which did peacemaking work. The therapist asked how each person felt about getting involved to see if it was for them. One of the most therapeutic actions that Bill and his family took was to become involved in that peace-with-justice study/action group. Encouraging such involvement often proves, as it did for Bill and his family, to be among the most valuable channels of self-healing of feelings of hopelessness and helplessness about society's huge problems.

Clinical and teaching experience has shown that encouraging clients and students to become involved in peacemaking and environmental action often is a no-lose strategy. Putting some time, energy, and funds into working for ecological sanity can pay important dividends. Engaging in such volunteer activities often enhances individuals' sense of purpose in life and thus increases their mental health. It may also result in some degree of constructive changes in their community and world. Thus, earth-caring outreach can be an investment in one's own future and perhaps also in the future of the world.

Therapists' and Teachers' Crucial Roles

Why should professionals in the healing and teaching fields implement the ecological circle in their work? Why is it important for us in these professions to help our clients, patients, and students become more open to be nurtured by nature and thereby to increase their caring concern for the environment? The evidence is mounting that what is at stake in this crisis is the survival of the planet. As the caption of the drawing early in this book puts it, "Children ask the world of us." Therefore, persons in all academic disciplines and professional groups have an obvious responsibility to use their expertise to do all they can to give future generations a viable planet home. Persons in the "people professions" have special contributions to make to help resolve this unprecedented crisis facing the human family. And those of us trained in both the psychosocial sciences and the therapeutic disciplines have the capacity to make unique and much-needed contributions to both earth-caring and peacemaking.

Brian Swimme and Thomas Berry frame their call for professional involvement in terms of the scientific story of the universe's origins and continuing unfolding: "The human professions all need to recognize their prototype and their primary resource in the integral functioning of the Earth community. The natural world itself is the primary economic reality, the primary educator, the primary governance, the primary technologist, the primary healer, the primary presence of the sacred, the primary moral value."[13]

Counselors, psychotherapists, health professionals, teachers, clergy, and parents share a common concern for developing whole persons and enhancing the good life. Any definition of the good life that makes sense in our world must include protecting the good earth on which our wellness depends. Awareness of the rootage of the human mind-body-spirit organism in the natural world impacts all aspects of our work with people in subtle but important ways. However the good life is defined in particular ethical or religious systems, there is no doubt that the ethical challenge of the ecological crises towers like a moral Mount Everest over our differences and over the lofty peaks of other critical ethical issues facing our species.

Healers (including therapists, health professionals, and clergy) and teachers (including parents, the most important teachers) have strategic opportunities to help reflect and shape the attitudes and values of their culture. These values guide individuals in their lifestyles and institutions in their policy decisions. These persons are therefore key agents of transformation in influencing change toward earth-friendly lifestyles and policies. Forming and transforming values, attitudes, and behaviors to make them more ecologically constructive is an essential aspect of all truly holistic education and therapy.

Consider the common shared interests of therapists and teachers in preventing and healing human alienations. Healing alienation from our bodies and their self-healing immune systems is an objective of both health professionals and body therapists. Healing inner alienations from repressed parts of the mind is the objective of intrapsychically focused psychotherapists. Overcoming alienations from other persons is the objective of relationship-oriented therapists. Preventing alienations from parts of our personalities and bodies, as well as from other persons, is a key objective of wholeness-oriented teaching. Preventing or healing inner alienation from the divine Spirit is a central objective of religious education and pastoral counseling.

To all these interdependent alienations must now be added the twofold alienation from nature—from our own inner "wildness" and from organic bonding with nature. This alienation is a bottom-line cause of violent behavior toward nature, toward our bodies, and toward others perceived as "wild." Healing and preventing this violence involves healing the earth-alienation that is at their roots. Helping people learn to open themselves to be nurtured more deeply and often by nature is one crucial focus of holistic healing, teaching, and parenting.

We healers and teachers should utilize our skills to help people implement the ecological circle in their everyday lives because the brokenness and toxicity of the planet affect our clients, patients, and students (whether or not they are aware of it), as well as ourselves and our children. Most of us Westerners suffer from some degree of alienation from our deep rootedness in nature. This alienation impacts the total body-mind-spirit organism in wholeness-diminishing ways.

The brokenness and toxicity of the planet diminish our relationships with the people we serve in our professions. The health of relating with family and friends also is diminished by earth-distancing lifestyles. Emotional need deprivation in intimate relationships is exacerbated by ecological deprivation and toxicity. Conversely, relationships often are enriched when people share experiences in natural settings. Earth bonding and people bonding are complementary needs that, when satisfied, are mutually reinforcing. Enjoying intimate connectedness with the natural world sometimes opens people's whole organism to deepening emotional and bodily intimacy, including sexual intimacy with, others.[14]

For those of us whose major interest is mental and emotional health, it is important to recognize that many of those we serve are suffering from conscious or subconscious ecological angst. This often includes anticipatory anxiety and grief about what they fear is the impending fate of the natural world. Our methods must deal with this anxiety along with other anxieties and worries that constrict living. In my experience, anxiety about the biosphere is more likely to surface in teaching-learning contexts than in therapy. It is often present in psychotherapy or medical contexts but not articulated because persons do not perceive these as appropriate settings to raise such issues. Therefore, it is important for counselors, therapists, nurses, and physicians to invite discussion by asking an appropriate question, such as "Is your problem related in any way to the situation in the world?"

Therapists and other healers deal with both the light and the dark sides of life. The latter includes the intrusions of pain, frustration, disappointments, heartache, injustice, oppression, and death that come, sooner or later. By encouraging clients, students, or patients to open themselves to experiencing the healing power of nature, we can enable them to discover an invaluable source of both healing and growth. When people connect with the healing energies of the cycles of life, death, and rebirth inherent in nature, they often discover that they view death from a different perspective. It is still profoundly sad but not devastating.

Those in the counseling and psychotherapy professions are trained to respond actively when we identify or suspect the threat of suicide in our work. In the wider context the issue is now environmental suicide prevention. The challenge confronting all healers is to make preventing ecosuicide an integral part of both ecotherapy and ecoeducation.

John Seed, a pioneer Australian ecologist, has written a moving essay entitled "To Hear within Ourselves the Sound of the Earth Crying." Persons trained to do counseling, psychotherapy, or creative teaching have developed special sensitivities and skills as listeners. We know that real listening involves hearing what people say but also what is "between the lines," unspoken because it is too painful. We are also aware that such listening is tremendously important to the healing, growth-nurturing relationships we seek to develop with our clients, patients, or students. Many teachers and therapists also have discovered that to be most healing, listening needs to have a quality aptly described by theologian Paul Tillich as "loving listening." This means listening with caring as well as understanding, listening responsively and without judging. To make the vital, unique contributions needed to help resolve the ecological crisis, counselors, therapists, teachers, parents, and health professionals must learn to practice a new dimension of listening—*responsive and loving listening to the earth*. Unless more and more of our species learn to listen with love and caring to the anguished cries of our earth as well as its shouts of joy when it celebrates its creativity, it is doubtful that our planet will survive as a healthy place for unborn generations to come.

But there is a catch in the admonition to listen to the earth. To the degree that people suffer inner alienation from their inherent bonding with nature, the earth's crying will be muted or in a strange language they will not understand. Pioneer feminist theologian Nelle Morton discusses communication issues among women. She observes that women's voices have been ignored so long that many no longer dare to speak what they really think and feel. Nelle declares that what women need, therefore, is to be "heard into speech" by other women. Ecotherapy and ecoeducation often involve teaching nature-alienated persons to hear themselves and the earth into speech so that they can hear the earth's often subtle messages of both joy and suffering. For all the above reasons, healing and enhancing relationships with the natural as well as the human environment are essential goals in our counseling and therapy, our healing and teaching.

The unprecedented ground swell of concern around the globe for saving the planet is an expression of a profound rediscovery: that our personal and species survival depends on learning to love and respect the wonderful web of living things. At some level, more and more people are becoming aware that preserving a livable planet for all the children of the human family, and for all other species, is a paramount survival issue facing humankind.

Exploratory Survey of Therapists and Teachers

As part of the preparation for writing this book, a survey was conducted to sample the ecological views and experiences of a variety of psychotherapists and teachers. A total of sixty-two usable survey forms were returned. Because this survey was designed to be only exploratory, the sample was not selected randomly.[15] Its findings can only be suggestive, not definitive. It is my hope that the validity of the findings will be checked out by the use of random survey samples of the professional groups represented. I must emphasize, however, that when viewed in the context of the early stages of knowledge in the fields of ecotherapy and ecoeducation, the findings of this survey are illuminating as preliminary indications of potentially useful data.

The primary professional identities of the survey respondents:

Pastoral psychologists/ psychotherapists 17
Clergy generalists who do counseling in congregations 13
Psychotherapists (secular) 12
Teachers 10
Medical professionals 8

It is noteworthy that many counselors and psychotherapists, both clergy and secular, equated "environment" with the human environment. For those in medical professions, "environmental issues" tended to be limited to toxicity in the home and work settings.

Following are the survey questions with a summary of the responses (not everyone responded to all questions):

In what setting were you raised? Large city—23; Town—24; Farm—15.

Were your childhood experiences with nature such that you learned to enjoy and respect the natural world and let yourself be nurtured by it? Yes—54; No—5 (included one person raised in an urban ghetto).

Do you have a close, nurturing relationship with nature now? Yes—40; No—13. Several respondents expressed regret that their busy schedules prevented them from spending more time enjoying natural settings.

Does your relationship with nature (in childhood and now) influence the ways you do counseling, therapy, or education? Yes—45; No—11. Here are samples of the "Yes" responders' descriptions of how their professional functioning is influenced:

• "The images I will pick up on when clients speak."

• "Illustrations from nature—seeds, balance in life, mutual dependence on our environment."

• "The bountifulness of nature, the goodness, beauty and infinite variety . . . provide images of hope and trust, of a seed growing in dark earth, a huge eggplant that will feed the hungry (this came from a dream I once had)."

• "I believe in nature's thrust to heal what is broken."

• "I use nature to demonstrate relaxation and re-creation, as opposed to wreck-creation."

• "I'm convinced there is a healing power in nature. I cherish it for my counselees."

• "I use nature metaphors, for example, that mature plants take three years to develop adequate root systems after being transplanted."

• "I wonder if the cycles of nature and the experience (on the farm) that things can, do, and will change, help me be a little more relaxed when counseling people who can't see any way out of their current predicaments."

• "My relationship with nature includes an acceptance of death as a part of the life cycle" (this from an oncology nurse who herself had a life-threatening struggle with cancer).

Do your goals as a therapist or educator include healing or enhancing relationships with the natural environment? Yes—35; No—12. A physician respondent added this comment after his "No": "A bizarre question." One of the "Yes" responders commented: "A person's well-being must include the relationship with non-human nature. Otherwise we are only working with a part of who one is."

Is it appropriate to ask clients or patients (who do not bring this up) whether environmental issues are contributing factors to the problems that bring them to counseling or therapy? Yes—20; Yes, but I haven't done so—1; No—0.

Do you understand environmental, peace, and justice problems in our world as contributing to the problems of living of some of your clients, students, or patients? Yes—35; No—10.

If yes, what were the prevalent feelings expressed about these social context issues? Among feelings mentioned, many of them repeatedly, were: powerlessness, helplessness, apathy, resignation, outrage, betrayal, despair, anxiety, fear, guilt, grief, fear about drinking the tap water, concern for the health of themselves and/or their children, and hope (mentioned only twice).

Do you routinely inquire of clients, patients, or students about the health or toxicity of their home and work environment? Yes—23; No—11. Several therapists who said "No" added that they thought this was a good idea and they would implement it.

Do you regard it as potentially healing or empowering to encourage clients, patients, or students to become involved in environmental, peace, and justice work? Yes—49; No—1. This almost unanimous affirmation of the therapeutic and educational values of outreach in action was unexpected. One psychologist commented: "I believe that it is good for individuals to become pro-active, to 'empower themselves,' to believe that 'I can make a difference, as an individual, on this planet, as well as in my family and work situation.'"

Do you regard issues of meaning, value, or spirituality as important factors in determining people's lifestyles including how they treat the environment? Yes—42; No—0. (This perspective will be explored in chapter 4.)

Do you have evidence that, in forming their identity, people internalize the natural environments—toxic or healthy—within which they grew up? Yes—15; No—8. (This theory will be explored in chapter 2.) One pastoral psychotherapist commented: "Most people speak of family and extended family as formative. The absence of the natural environment is itself striking."

Experiencing the Ecological Circle

The purpose of this right-brain exercise is to experience the three aspects of the ecological circle. Put in other terms, the experience may highlight the joy of being nurtured by nature, the urgency of the ecological crisis, and the energy-for-action of a transformed future. Reflection on the experience may then encourage envisioning actions required to move toward the future visualized. I am indebted to Joanna Macy for the heart of this exercise.[16] The greatly modified version below has evolved in my practice as I have experimented with it in numerous ecology and peacemaking workshops over the years. If you teach, you may decide to use it in your work.

Instructions: When you come to this mark / stop and do what has been suggested.

Sit in a quiet place with a piece of paper within easy reach. / Tightening all your muscles, hold them in tension for five seconds and then relax them for five seconds. Repeat this several times, continuing to breathe deeply through the cycles. Do this (or other full-body relaxation methods that work for you) until your body-mind organism is deeply relaxed but very alert. /

Using one of your many creative abilities, form a mental picture of the most beautiful place in nature you know, on a warm sunny day. / Be in that place now. / Let your body, mind, and spirit—your whole self—enjoy its beauty and its serenity. Feel its pulsing aliveness in your body. Let its life energy touch every cell from the top of your head to the bottom of your feet. / Focus this healing energy on those places in your body that are tense or feeling some discomfort. / Continue to experience being warmly nurtured by nature as long as you wish. / As you enjoy all this, discover how what you are experiencing touches the heart of your spiritual life. In your own way, be open to experiencing the loving Spirit present as the creative source of the beauty and order, the peace and power, of this lovely setting. /

Now, imagine that it is thirty years in the future and you are in the same place outdoors in nature. Imagine that the destructive environmental forces and population explosion of the 1990s continued to escalate over the next thirty years. Pollution has increased tremendously as the world's population soared from just under six billion to nearly ten billion. Literally thousands of species have been wiped out forever. Look around you and see the devastating impact of all this on the natural setting. Stay in touch with your feelings. /

Now, shift your mental gears and imagine a radically different scenario. It is thirty years hence, and you are again sitting in your favorite place outdoors. But imagine that sometime back in the mid-1990s people like you in your community and all over your country, as well as in many other countries, decided to really protect and love the earth. As a long-term result of the Earth Summit, they decided to make caring for the global environment a top priority both personally and politically. They did this to make sure their children and grandchildren would have a clean, healthy planet on which to live. They organized politically to make sure their own governments and the United Nations enforced wise environmental and population policies and ended the insanely wasteful arms race, destroying all the chemical, biological, and nuclear weapons of mass destruction. Let yourself enjoy how it feels to be in a sane, healthy, safe world with an international peacekeeping and earth-caring force, a world where children everywhere are insured enough food, housing, health care, and education to develop fully their unique capacities. / Look around you and enjoy this place that now is much cleaner and even more beautiful and alive than it was thirty years ago.

You notice that a child is approaching you, a little girl of eight or nine. She seems shy but obviously very curious. She has read something in school about the way the world was thirty years ago. She comes up to you and begins to ask questions. Carry on a conversation now, being aware of your feelings as the two of you chat. Listen to your responses to her questions about the old days as she asks, "What was it like back then? / Were there really bombs so big that they could blow up the whole world? / Was there really pollution that poisoned the air and the water so people and animals were getting sick? / Is it true that there were millions of children like me who were poor and hungry and sick, because governments were spending so much on things to kill people? / Wasn't it scary then? / What did you do to get through such a scary and discouraging time? / How did you keep feeling hope so you could help make the world like it is today? / What did you do to help give us children our beautiful, peaceful world?" /

The little girl spontaneously flings her arms around your neck and gives you a hug of gratitude. You feel her tears of joy as she whispers with deep feeling, "Thank you! Thank you! Thank you!" Then, a little embarrassed, she leaves you. You close your eyes and think about the whole experience. / What did you feel when you realized you were in a safe, sane world without massive weapons and environmental degradation? Perhaps your heart felt like a heavy weight had been lifted. Or did you find it so hard to believe that you could not let yourself really experience such a world? / How did you feel when she asked you what you did to help give children like her such a wonderful world? / Continue to process your trip into a healthy future. You may want to ask yourself questions; for example, Do I need to do more ecology, peace, and justice work? What do I need to do to be more involved in helping heal the planet for my own children or for children like the little girl I met in the future? How can I use my professional or parenting skills in this healing process? / When you are ready, open your eyes and make notes to remind yourself of important things you discovered and what you plan to do as a result of imagining a transformed world. /

If you were surprised at how much better you felt when you imagined that wonderful changes had occurred, you now have some valuable awareness of the heavy load of chronic ecological anxiety that environmentally aware people are carrying like subconscious packs on their psychological backs, every hour of every day. How useful this awareness proves to be, of course, depends on what you do with it. If you felt some discomfort when the little girl asked you what you had done (as I often feel when I do the exercise), commend yourself for awakening some healthy feelings of personal responsibility that can motivate you to get more involved in loving and saving the wonderful living network that makes all life possible.

Notes

1. This statement recently was discovered to have been written in its present form in the last few years by a media writer. It may well reflect the spirit of what Chief Seattle said, however, and it certainly rings true to the Native American world view and reverence for nature.

2. Some opposition to environmental concerns stems from the view that the very assumption that our species can damage the earth irreparably in a few years or even centuries is arrogant and fallacious, in light of the earth's profound wisdom developed over some five million years. Actually this assumption is supported by mounting evidence from the earth sciences that our age of high-tech and chemical revolutions has given humans the dangerous capacity to damage the earth in unprecedented and irreparable ways.

3. A phrase by a well-known mid-twentieth-century preacher and teacher, Halford Luccock.

4. The runaway AIDS epidemic provides a deadly illustration of this point. It confronts the whole human family with massive, unprecedented medical, economic, educational, sexual, ethical, and public policy problems. This epidemic is potentially a threat to everyone's well-being, even survival. This global health crisis illustrates the necessity of developing innovative, transnational solutions to an escalating epidemic.

5. Brian Swimme and Thomas Berry, *The Universe Story: From the Primordial Flaring Forth to the Ecozoic Era, A Celebration of the Unfolding Cosmos* (San Francisco: Harper San Francisco, 1992), 237.

6. Mike Samuels and Hal Zina Bennett, *Well Body, Well Earth*, (San Francisco: Sierra Club Books, 1983), 92, 94.

7. Our giant sequoia was but one representative of the beautiful pine forests on the slopes of the mountains towering above Claremont. When we hiked there, it was obvious that the health of millions of trees was being damaged by air pollution. This aerial poison comes mainly from millions of cars driven by nine million people in this overpopulated area where a temperature inversion traps the garbage in the air. The crisis of a poisoned environment is reducing health around the entire planet—including, in all probability, the place you call home and the place where you practice your healing, teaching, or parenting skills.

8. As I reflect on this Amazon experience, I remember several previous experiences in which a similar awareness hit me around different problems in society. One was the day more than two decades ago when I marched with thousands of others into Montgomery, Alabama, on the final day of Martin Luther King, Jr.'s, march from Selma. On that muggy day we walked, protected by rows of heavily armed national guard troops, through the streets singing. We repeated "We Shall Overcome" again and again as we walked through the African American ghetto of Montgomery. A few of the less intimidated black residents cheered from their porches. Angry whites watched with taunts and obvious rage behind the rows of troops lining the streets. We knew who the likely targets would be of the violence the troops were there to prevent. Exhausted by the time we reached the open square in front of the capitol, we sat on the pavement for an hour or more as King delivered a powerful, inspiring message from the steps of the capitol with the confederate flag above it. Five-gallon apple juice jugs were passed along the rows of parched marchers sitting in the square. The jugs were filled with tap water from the spigot in the little Baptist church across from the capitol, the church once pastored by King. No experience of communion ever approached the depth of meaning for me of drinking from that common jug. I knew that day that I could no longer practice privatized counseling or teaching that ignored the social malignancy of white racism.

9. Quoted on a plaque at the Star of the North Retreat Center, St. Albert, Canada. About the animals, Buber says, "Creatures are placed in my way so that I, their fellow creature, by means of them and with them, find the way to God. A God reached by their exclusion would not be the God of lives, in whom all life is fulfilled."

10. I appreciate Gloria Johnson's willingness to share her earth story with us.

11. This case also was described briefly in Howard Clinebell, *Well Being: A Personal Plan for Exploring and Enriching the Seven Dimensions of Life* (San Francisco: Harper San Francisco, 1992), 186–87. It is presented here in somewhat more detail for professional readers.

12. The nuclear issue was part of their "family secret," things a family subconsciously agrees not to talk about.

13. *The Universe Story*, 255. To respond to the challenge, professional horizons must be expanded and professional identities made more inclusive. We must come to include the healing and wholeness of the biosphere as an essential part of healing individuals and families in the context of a world on the brink of geosuicide. There is a double reward from stretching our horizons of concern and identity. It can increase

the healing-helping-growthing power of our professional skills. Equally important, it can also enable us to make small but valuable contributions to saving the planet.

14. What about the self-isolated hermitlike people who choose to live alone in secluded, often wild natural settings? Those who have been deeply hurt in early relationships with humans often feel safer and more at home in the organic world of nature than with people. For them, nature is a haven from the fear of having old, unhealed wounds deepened by further hurts in human relations. In the same vein, I remember talking with a nature-aware occupational therapist who had long experience in working in mental hospitals. She reported that some of her patients, in the process of treatment, do gardening and other earthy experiences long before they will risk relating to people or talking about feelings and interpersonal issues.

15. Many respondents were individuals known to the author, which increased the possibility that respondents had a pro-environment bias. The sample was heavily weighted by counselors, therapists, and teachers whose primary professional identity is as clergy. This factor makes it likely that respondents who linked the ecological crisis with spiritual and ethical issues were considerably overrepresented. It is encouraging that most of the clergy responses reflected considerable insight and informed concern about ecological issues.

16. Macy's original version is found in *Despair and Personal Power in the Nuclear Age* (Philadelphia: New Society Publishers, 1983), Exercises 32 and 33, pp. 140–41. A version of this exercise, as I have modified it, is in my book *Well Being*, 203–4.

Part One
A Grounded Model of Human Development and Healing

2

Humankind's Earthy Roots: Finding the Lost Self and the Lost Earth

Perhaps most important, we need to assess our own relationship to the natural world and renew, at the deepest level of personal integrity, a connection to it.

—Al Gore, *Earth in the Balance, Ecology and the Human Spirit.*[1]

The inner life of the human depends immediately on the outer world. Only if the human imagination is activated by the flight of the great soaring birds in the heavens, by the blossoming flowers of Earth, by the awesome sight of the sea, by the lightning and thunder of the great storms that break through the heat of summer; only then will the deep inner experiences be evoked within the human soul.

—Ecotheologian Thomas Berry[2]

When one reviews the major guiding paradigms in therapy and education today, in light of the global environmental crisis, a clear need comes into focus. Nothing short of an earth-grounded but also transcending model of human beings and their development is required to provide a conceptual foundation for doing healing and teaching in the world of today and tomorrow. What is needed is an understanding of human growth and healing that includes the deep earth-rootedness of all aspects of our species. This chapter describes such an enlarged guiding paradigm and cites evidence from many sources that supports the model. The central motif of the model is that rediscovering and befriending the lost dimension of self—the earthy core of our being—and finding the lost earth, the missing or diminished bonding with the earth, are mutually reinforcing processes. In our earth-alienated world, helping people to discover these interdependent living processes may enliven humans to help save the biosphere.

The Greening of the Embodied Self: The Nature of Ecobonding

The conceptual paradigm that is the theoretical foundation of ecotherapy and ecoeducation is an ecologically based personality model but much more. It also is an earth-grounded conception of our species' total being—body, mind, spirit, and relationships. It thus goes beyond the focus of ecopsychology. Discovering, befriending, and intentionally developing one's profound rootedness in the life-giving biosphere is the process that produces what is called healthy biophilia and ecobonding. Ignoring, denying, or rejecting this inherent earth-rootedness is called ecophobia and ecoalienation. Ecobonding involves claiming and enjoying one's nurturing, energizing, life-enhancing connectedness with nature. Ecophilia is the love of life associated with this bonding with the earth. Ecoalienation involves seeking to distance oneself from our inescapable life-giving dependence on nature. Ecophobia is the fear of claiming one's dependence and bonding intimately with nature.[3]

Nearly two hundred years ago, William Blake, the English poet, painter, and engraver, described vividly the contrasting responses of two groups of people to a tree: "The tree that moves some to tears of joy is in the eyes of others only a green thing that stands in the way. Some see nature all ridicule and deformity . . . and some scarce see nature at all. But to the eyes of the man of imagination, nature is imagination itself."[4] These contrasting groups are extreme forms of what I am describing as ecoalienated and ecobonded, respectively.

Healthy bonding with nature is not merging with nature or finding oneness with nature in the way that other animals seem to do. Although we have much in common with them—more than 95 percent of our genes are like those of chimpanzees—the other animals' degree of undifferentiated oneness with nature is both undesirable and impossible for our species. This is because we seem to be distinguished from other animals by our transcending consciousness, choicefulness, intelligence, and spirituality.

Humans have been described as a strange blend of animal and angel, of dust and destiny. Such terms describe our species' paradoxical condition of being in nature and yet also set apart from nature. Creative bonding is accepting the sometimes painful paradox of our lack of animal unity with nature but also accepting the deeply rooted earthiness of our bodies, minds, and spirits.[5] Such bonding means interacting intimately with the natural world in creative closeness frequently and intentionally. This is mutually beneficial to our well-being and to our unique responsibility as a species in determining the biosphere's well-being. The bonding involves valuing both this earth nurturance and the transcendence of nature that is our distinctive humanness. It is allowing each of these to balance and enrich the other.

Western personality theories and therapies have had primarily an intrapsychic and, to a lesser degree, an interpersonal orientation. (Social work has had a long tradition of emphasizing social context issues of people's lives and problems.) But the formative interaction between persons and their natural environment has been virtually ignored in most personality theory and therapy until recently. It is as though the crucial interactions within and among humans occur in a natural vacuum, without any relation to a particular place on the earth or to the life-generating and sustaining earth. If this aspect of human identity is so crucial, why have Western developmental theories and psychotherapies tended to ignore it? Consider three possible reasons:

- If a fish were able to develop a theory of fish existence, it probably would not focus attention on how being continually immersed in water influences everything else about "fishness." In a similar way, the profound, lifelong human rootedness in the earth is so basic that people usually do not examine this crucial aspect of their lives critically.
- Western developmental theories and therapies have been generated, for the most part, by urban males who could ignore the earth in their theory-building more easily in that context than in settings where close bonds with the natural world are assumed to be essential for survival.
- There is a profound chasm in the dualistic cosmologies that has dominated Western thought for many centuries. Matter and mind have been kept in separate compartments by most scientists and religionists. The physical world and the body have been understood as machinelike and characterized by fixed mathematical principles called laws. Mind and spirit, in contrast, often have been understood as fuzzy realms of amorphous feelings, ideas, and beliefs. In contrast, the conceptual foundation of ecotherapy and ecoeducation is a unified understanding of humans as holistic living organisms interacting with the world understood as a living organism

The central premise of ecotherapy is that our early relationships with the natural world have a profound shaping impact on the development of a grounded sense of identity for our whole body-mind-spirit organism. Our identity formation is influenced at a deep, pre-verbal level, by our early experiences in nature, but also by our culture's views of the natural world as these are experienced by us directly and also through our parents' feelings, attitudes and ways of relating with nature. Furthermore, the quality of our continuing relationship with nature is a major dynamic in our sense of being firmly grounded versus. feeling "up in the air" or unrooted. Thus, our relationship with the earth, mother-father of all living things, is an often-ignored but foundational factor influencing our overall wellness and the wholeness of our identity. This view affirms the wisdom of philosopher Simone Weil's observation that "to be rooted is perhaps the most important and least recognized need of the human soul."[6]

The core ecotherapeutic concept of personality development and human well-being is an extrapolation and integrations of four conceptual models:

• An expanded understanding of mind-body development involving an extension of objective relations;

• A social systems based understanding of personhood extended to include interaction with the community of all living things (the biosphere or earth community) as well as with the interpersonal community.

• To these two theoretical frameworks of understanding is added an extended spiritual perspective derived from the discipline of psychology of religion. This is the understanding that a central and essential dimension of being human are the capacities for self-transcending, valuing, spiritual insights and experience. To this view is added the ecological perspective that sees our total being, including our spiritual and valuing life, as deeply rooted in the biosphere. This ecological spirituality is intertwined with all the interacting dimensions of our humanity. Let's examine each of these three ingredients:

Object relations theory and the therapies based on it are among the most creative developments in current psychoanalytic or psychodynamic thought. The central theme of this theoretical model is that human personality and identity are formed as children take into themselves (internalize) their most important "objects," an unfortunate usage referring to the primary need-satisfying adults in their early childhood on whom they depended for their very survival. D. W. Winnicott, a pioneer object relations theoretician, refers to the continuing state of internal relatedness that characterizes what he calls the "true self." He holds that the most fundamental need of persons, at the very core of their selves, is the need for a "perfect environment." He writes: "The facilitating environment is necessary, and without it being good enough the maturational process weakens or wilts."[7] "Environment" is used here and elsewhere in objective relations thought to refer to the *interpersonal* environment with the primary caregiver, especially the mother, and not the natural environment.[8]

The extension of object relations and social systems theory proposed herein can be summarized as follows: In early childhood, humans internalize both the human environment and the natural environment. We internalize our positive experiences with need-satisfying adults and our negative experiences with those who are need-depriving. But we also internalize our positive and negative experiences in nature, including the dominant feelings and attitudes of our culture toward nature, as these are screened to us through our parents' relationships with the natural world. Our crucial internalized relationships with our primary caretakers—our biological or adoptive mother and father—are paralleled by our internalized relationship with Mother-Father Earth. The internalized unloving, undependable "bad mother" or loving, dependable "good mother" (as expressed in the formation of the senses of self in objective relations thought) interacts with the internalized fearful and threatening "bad earth" experiences and the security-giving, nurturing experiences with the "good earth." To traditional object relations theory, I would add this: Paralleling the good and bad mother experiences are the good and bad father experiences, all of which are internalized in our identity formation. These earth-oriented internalizations become a core dimension of everyone's identity, for better or for worse, the usually unexamined center of identity formation. Furthermore the quality of our current relationships with the natural world is deeply influenced by whether our internalized natural objects are primarily positive and nurturing or threatening and toxic.

Winnicott discusses infants' earliest personality development in relating to external reality by interaction with the mothering person and the mother's breast. He introduces the concept of transitional objects. Infants invest such security-giving objects with magical, protective power over which they fantasize they can exercise control. Examples are a particular blanket (like Linus's security blanket in the Charlie Brown cartoons) or a stuffed animal. Transitional objects give the infant a sense of having something in external reality to provide security when the mothering person (upon whom the child is dependent for survival) is not present. Winnicott observes that relationships with transitional objects continue to be crucial throughout life. "Out of these transitional phenomena develops much of what we variously allow and greatly value under the headings of religion and art and also the little madnesses which we legitimate at the moment, according to the prevailing cultural pattern."[9]

In parallel ways, it is my view that transitional objects from nature are very important throughout life. This throws light on the intense bonding that many children and adults sustain with their pets and their plants, their flowering window boxes and their gardens. Enormous energies are invested in them because they are security-giving transitional objects. They give us a sense of grounding that may substitute for our initial intimate bonding with nature—in our remote genetic programming as well as in our intrauterine life of being bonded with the mother's body. They are ways of maintaining an inner-outer connection with wildness in ourselves and also in the external world.[10]

Our alienation from our unconscious inner wildness, and therefore from our bodies as the closest manifestation of nature, is a basic cause of fear and defensiveness about intimate bonding with nature. This inner wildness is a potentially invaluable but largely repressed residue from our genetic forebears and their eon-spanning experiences of learning to survive in the wilderness. Our forgotten wildness might be described as the deep dimension of the unconscious mind that Freud speculated about and that Jung developed more fully, calling it the collective unconscious. This designation, however, is partial in that the inner wildness is rooted in our total selves, including the physiological as well as the psychological, spiritual, and interpersonal dimensions.

Like all repressed memories, repressed wildness continues to haunt our "civilized" lives. As these energies accumulate, they may eventually produce wild, irrational, and often violently destructive mental processes or behavior. This destructiveness may be turned inward on ourselves in masochism, irrational (perhaps psychotic) ideation, or potentially suicidal depression. Or the repressed wild energies may be directed outward at civilized society, at those perceived as wild or simply different, or directed at wilderness preservation.

If we get to know and befriend this wild side of our mind-body-spirit organism and integrate it into our overall self-identity, this wasted repressed energy becomes available for playfulness, adventure, constructive relating, and creativity in all areas of our lives. It can also help to encourage us to be more intentionally nurtured by nature in all of its expressions from unspoiled wilderness to the more domesticated setting of a loved houseplant, a tree or garden, or a favorite pet.

This wild dimension of our deeper body-mind-spirit is probably a part of our genetic legacy. It is a potentially valuable residual from the eons of prehuman evolution that still function in our unconscious mind and in our bodies. It is a part of our genetic heritage from the many millennia when our remote ancestors learned to survive in wildernesses long enough to pass along their genes, eventually to us. In many "civilized" people, this potentially enriching and empowering aspect of inner life is deeply repressed. They suffer awareness-blocking anxiety when the wild man or wild woman within each of us threatens to rise into consciousness. The threat of this unconscious energy erupting reinforces anxiety about wildness, causing the wild person within to be more deeply repressed rather than befriended and integrated with the "civilized" side.

The popularity of recent films and books that explore this topic suggests the power of the need for positive bonding with inner wildness. An example is the best-selling book by Jungian Clarissa Pinkola Estes entitled *Women Who Run with Wolves: Myths and Stories of the Wild Woman Archetype*,[11] a collection of stories, ancient myths, and fairy tales that the author hopes will enable women to own and channel their inherent wild strength. Commenting on her book's popularity, the author says:

> Many women are unmothered and unguided. They don't have what should have been passed down from mother to daughter, and so they're feeling half-alive. . . . We're born with inner strength and by virtue of inadequate training, we begin to move farther and farther from it, and we become weaker, and the weaker we become, the weaker our insight becomes. I'm talking about ancient advice, the kind that helps you make life choices.[12]

It is clear that what Estes means by "wild woman" is not a carousing, reckless, or irresponsible one, but a woman who has claimed her wild inner female strength and is using it to live creatively and courageously.

The socialization process by which both women and men grow up often tends to alienate them from their wildness. Feminist psychologists have critiqued this process insightfully. In a male-dominated sexist society, this process occurs in traditional families when little girls are trained to be "nice" and little boys are trained to express their wildness in macho aggression. Both genders are disempowered by this early learning.

Social Systems Theory and Earth Rootedness

The theory of personality development undergirding ecotherapy and eco-education integrates an enlarged social systems theory with an enlarged object relations theory. The latter theory, as described above, provides a conceptual framework that can link intrapsychic, interpersonal (parental-family), sociocultural, and nature-based dimensions of the identity formation process. Wholly intrapsychic understandings of personality development are inadequate as a basis for conceptualizing either the essential social nature of humans or their essential groundedness in the earth. Humans are essentially relational beings. Our personalities are formed in and by a sociocultural context of primary relationships. Our species does not just *have* formative family relationships and cultural contexts that are screened and transmitted through early family relationships. In a profound sense, the heart of human identity *is* these relationships and their socioculture but also their natural contexts.

The widespread current breakdown of interpersonal communities is exacerbated by the loss of a sense of support from the community of living things—the biosphere that is all around us and within us. Families that together strive to create islands of environmental mutual healing in their own lives, perhaps reaching out to their local neighborhood, often find that bonding within their own family is enhanced as they reconnect with the earth community in caring ways. Likewise, local neighborhoods or congregations that work together for healthier natural surroundings often find that the sense of community within these social systems is enriched.

The Impact of Ecological Angst

As teachers, therapists, or parents, we need to ask ourselves a crucial though difficult question: What is the impact on the development of a sense of self in children and youth of the knowledge that they may have a relatively short healthy future on a healthy planet? For the first time in the millennia-spanning human story, many young people are aware that they may have no long-range, viable future. Countless teenagers cope with this deadly awareness, as do some of their elders, with denial. They go on with business as usual, numbing their awareness by all manner of escapes and addictions. But, their vain attempts to escape into this self-protective womb of denial does not really protect them from the spiritual virus that psychiatrist Robert Lifton calls radical futurelessness. Young persons without a deep, reality-based feeling that they have a long-range healthy future are infected with futurelessness and its attending hopelessness. Furthermore, many parents and grandparents sense, usually at subconscious levels, that the self-repairing, life-nurturing systems of our living planet are increasingly threatened. They know that the biological immortality adults have always found in their children and grandchildren may indeed be in jeopardy.

Death, Contingency, and Ecoalienation

Reflecting on the ecological crisis, Jungian analyst Lynda Wheelwright Schmidt declares:

> The death of wilderness would be an incomprehensible experience beyond cycles and the rhythms of birth and death. It would be a sterilization, a one-sidedness as shocking as prison. . . . The reason a human-made, human-sized phenomenon [like a garden or a park] can work for us is that it is a reference to something greater, something infinite. We may not actually have to be in, or even see, the wilderness for it to reconnect us to the Self, the feeling of completeness. But we do need to know that it exists . . . A garden without a wilderness to refer to would no longer connect us to the infinite. The call to save the wilderness is a call to save us all. . . . Because we emerged from the wilderness, we need to reemerge with it to heal our feeling of abandonment.[13]

The rejection and repression of our earthy roots frequently is a defense against existential anxiety or fear of death. It is a way of denying human finitude, contingency, vulnerability, and eventual death. The dominant urban, scientific, death-denying mentality tends to be used to support illusions of control over the uncertainties, contingencies, and vulnerabilities of the many experiences that really are beyond human control. Destructive natural catastrophes such as earthquakes, floods, and hurricanes are traumatic reminders of the illusory nature of our defenses. Everyday life is full of less traumatic reminders. Chapter 4 will explore the ways in which healthy spirituality and bonding with nature together can provide constructive alternatives for dealing with death anxiety.

What are the characteristics of healthy, reality-based relationships between humans and other animal species in the biosphere? This is a complex issue. At one extreme, the human grandiosity called "species-ism" is the attitude that other animals exist mainly or solely for human benefit. This belief system is untenable because it ignores our own animal vulnerability, interspecies genetic interconnectedness and interdependence. (In terms of the food chain, all species "use" each other for sustenance.) A humane survival ethic for the entire biosphere must include the recognition that the seemingly unique human capacities for abstract thought, creativity, and choice, do not give us a right to exploit and misuse other species. (Examples of misuse of other animals are abundant; for example, cruel factory poultry farming methods and livestock raising and slaughtering practices.) On the other hand, attempts to completely level differences between humans and at least the "higher" animals do not reflect the choicefulness, power, and responsibilities accruing from our special human capacities. Humans' unique capacities carry with them the profound responsibility to use other species in ways that respect their value and protect their future. Native American ritual practices surrounding hunting and farming recognize this responsibility to the rest of the biosphere. Constructive relations must somehow be characterized by an inclusive love of life and respect for all living things. The biblical emphasis on responsible human stewardship of God's world is an ancient articulation of such an ethical position.

Intentionally Deepening Ecobonding

Ecoeducation and ecotherapy first seek to enable people to accept their continual dependency on nature as crucial for the well-being of themselves and the earth. Our dependence is undeniable; we must breathe from the biosphere's shared oxygen pool and eat the food of the earth's bounty. But ecotherapy and ecoeducation should go far beyond this acceptance to teach people how to deepen and enrich experiences of being nurtured by intimate nature bonding, more fully and regularly.

This deepening ecobonding has many rewards. Internal bonding with nature and external bonding with the living community of the biosphere influences the quality of all other aspects of our identity and life. As is recognized in most psychological theories, healthy identity is characterized by a robust sense of self-worth and inner strength. And, as Freud and other developmental theorists hold, our self-image is rooted in our body image. But a missing dimension of most theories is that healthy identity includes a strong sense of being firmly grounded. This means discovering the reality of our body-mind-spirit self being deeply, securely rooted in the biosphere. Such groundedness tends to enliven inner feelings of security and strength. It also can serve as a bridge to integrating awareness of the interconnectedness of all aspects of the self—mind, body, spirit—and interactive connectedness of these with the external world of relationships, culture, society, and nature. Such grounded identity has an anchored awareness of organic relatedness with one's body, with the earth, and with the other living creatures that share the biosphere with us. This ecobonding influences, if not determines, how open we are as adults to intimate interaction with nature. Having a solidly grounded identity enables us to become maximally receptive to daily experiences of being nurtured by nature.

To apply Martin Buber's interpersonal philosophy, healthy eco-identity involves relating to plants and animals in an I-You way and on occasion an

I-Thou way, rather than an I-It way. (I-It relationships involve objectification that often results in exploitation of the It. I-You relating involves reciprocity and mutual respect. I-Thou relating involves treating the other as somehow having spiritual value.) In his classic volume *I and Thou*, Buber describes how this intimate, caring bonding can happen: "I contemplate a tree . . . and if will and grace are joined as I contemplate the tree I am drawn into a relation, and *the tree ceases to be an it*."[14] Thomas Berry makes a similar point when he observes, "The outer world (of nature) activates the inner world. The most important thing is to recognize that this is not a subject-object relationship but a subject-subject one, and inter-communion process." [15]

For some people, one of the unexpected rewards of being able to enjoy intimate bonding with nature is celebrating those exciting times when our own wildness is awakened by the wildness of nature. Occasionally I have felt a strange exhilaration while sailing on the Pacific Ocean during a storm or huddled under a rocky ledge high on a mountainside during a thunderstorm. These feelings were complete enigmas to me for a long time. They came to make sense only as I began to let go of the fear by beginning to own and gradually enjoy the wildness in the deep recesses of my earthy being. When I mention such experiences in workshops or classes, persons sometimes tell of similar encounters with their own inner wildness, triggered by wildness in nature.

The Nature and Consequences of Ecoalienation

Ecoalienation is a dual distancing, a splitting of humans from connectedness with the natural world that they are in and that is in them. On the one hand, this disconnectedness has a fundamental inward dimension that needs to be attended— inner alienation from nature in people's whole mind-body-spirit organisms. This inner alienation usually is paralleled by a disconnectedness from "real" nature as experienced in the outer world. There seems to be a resonance between the inner and outer dimensions of ecoalienation, each of which tends to reinforce the other. Furthermore, ecoalienation and bonding, as well as interpersonal alienation and bonding, are mutually reinforcing. Psychiatric social worker Terrance O'Connor points out that the very patterns of denial, control, and projection that sabotage intimate human relationships are the patterns that endanger the natural world. She tells about crying as she sat in the wilderness and read about the ecocrisis: "It was like reading the details of one's mother's cancer." [16]

Paul Shephard holds that our species' ecocidal habits go back to the invention of agriculture, which made possible a false sense of separation from and dominance of the natural habitat. As a result, in what he calls collective madness, humans evolved profound alienation that reflects a breaking of a harmonious latent sense of self and world that is intrinsic to our very natures.[17]

Alienation from a healthy inner-outer connecting with nature (and from the inner strength and stable identity derived from this) is epidemic in Western society. The foundational grounding from the inner and outer connectedness with nature is weak or missing in many people today, especially those who live most of their lives distanced from nature's power and nurturance in industrialized, high-tech, polluted mega-cities. Preventing this alienation by enhancing ecobonding is an essential but often-neglected goal of holistic education, including parenting. Healing this alienation should be one of the essential goals of whole-person counseling, psychotherapy, and other forms of healing practice. Many, if not most, Western healers still need to become aware of this vital need.

Over three decades ago, Alan Watts anticipated the outer-inner nature of ecoalienation in his insightful volume *Nature, Man, and Woman:* "Our view of nature is largely a matter of changing intellectual and literary fashions, for it has become a world strangely alien from us." He then identified how this alienation is connected dynamically with our alienation from vital aspects of ourselves: "It is our ignorance of and, indeed, estrangement from ourselves which explains our feelings of isolation from nature."[18]

Erich Fromm, philosophically oriented psychotherapist and theoretician, sheds valuable light on this ecoalienation as a basic cause of much of humankind's anxiety. He declares: "Man—of all ages and cultures—is confronted with the solution of one and the same question of how to overcome separateness, how to achieve union, how to transcend one's own individual life and find at-onement."[19] Fromm rejects pseudo-answers as abortive attempts to overcome our existential separation. These pseudo-answers include seeking to return to Eden by merging with nature (for example, through animal and nature worship); controlling, others by violence; obsessive work; luxurious indulgence; orgasmic states (such as those induced by drugs and sexual orgies). He finds constructive answers to our species' alienation from nature in artistic creativity and genuine love of humans and of God.

Fromm also identifies separation from nature as a significant cause of the existential anxiety that is inherent in human life. He declares: "What is essential in the existence of man is the fact that he has emerged from the animal kingdom, from instinctual adaptation, that he has transcended nature—although he never leaves it; he is a part of it—and yet once torn away from nature, he cannot return to it."[20] This emphasis is similar to Paul Tillich's view that existential anxiety flows from the human awareness of transcendence of nature and yet being ever trapped in nature with the omnipresence of death.

Fromm rightly emphasized that there is an irreversible loss of prehuman oneness with nature, and that the more the human family "separates from the natural world, the more intense becomes the need to find new ways of escaping separateness." [21] What he, and many other psychoanalytic theoreticians did not see, was that there is another major dimension of people's broken connection with nature. This dimension is a reversible alienation from healthy bonding that has been widespread only in recent centuries as a result of urbanization, industrialization, and the dominance of mechanistic science. Healing this side of ecoalienation is both possible and therapeutically desirable. Fromm did not grasp the answer that ecotherapy gives to the separation-from-nature anxiety: to intentionally diminish the anxiety by deepening our healthy bonding with the natural world. This is very different from merging with nature, which is neither possible nor desirable because of our species' self-awareness, intelligence, and spiritual capacities. Our bodies and deep unconscious minds remind us that we are a part of the natural processes, that we are kindred to the other animals. But our intelligence and our souls cause us to transcend the rest of nature. Instead of a merging with nature, what ecotherapy seeks to facilitate is intimate nurturing interaction with nature. In this process humans can prize their transcending capacities, choose to enhance being nurtured by nature, and to respond by nurturing nature caringly. These responses provide much more than a balm in Gilead for our longing to return to Eden. They bring multiple benefits to our total well-being.

Theodore Roszak explores what he regards as the dangerous and painful consequences of the fact that the ecological unconscious is deeply repressed in many people: "We discover a repression that weighs upon our inherent sense of loyalty to the planet that mothered the human mind into existence. If psychosis is the attempt to live a lie, the epidemic psychosis of our time is the lie of believing that we have no ethical obligation to our planetary home."[22] He sees this repression as the deepest root of our collective, societal madness and opening access to it as the path to recovering our collective sanity. Extrapolating from Freudian oedipal theory, Roszak asks:

> Lust for the mother, hatred for the father . . . these guilty secrets have long since been laid bare. But what of the guilt that comes of annihilating whole species of our fellow creatures, not because we must do so to survive, but in ignorance and for the sake of nothing better than ephemeral amusement, petty pleasures, quick riches? We are, after all, in ways that may even be part of our innermost genetic inheritance, tied to the beasts from whom we evolve. At what risk of madness do we break faith with them?[23]

Psychiatrist Walter R. Christie, a brilliant ecological theory-builder, has generated an insightful understanding of the evolution of consciousness. He traces the development of consciousness from its earliest undifferentiated rootedness in nature called "Ground Unconsciousness" (before 200,000 B.C.E.); through magic-centered consciousness (200,000 to 15,000 B.C.E.); to deity-centered consciousness (15,000 to 1500 B.C.E.); then gradually into science and technology-centered consciousness (from 1500 B.C.E. to the present).

Like Fromm and others (including myself), Christie holds that the anxiety that caused alienation from nature is the emergence of self-awareness and the awareness of death. Experiencing the life-and-death power of nature and seeing death as the inevitable end of life made nature terrifying. Consequently, consciousness erected an illusory wall by which people try to convince themselves that they are really separate from nature. The games of superiority and separation from nature support the illusion, but they never work in the long run.

In the present earth crisis, Christie holds that humans must move toward a planetary consciousness in which they experience themselves connected to the whole—to nature and the planet. As the reasoning ego created the marvels of culture and technology that we enjoy today, "the mystical-physical-mythological (intimate) relationship with nature was sacrificed," and an unnatural separation occurred. With this came a loss of imagination, spirituality, creativity, and intuition as a mode of knowing. Planetary consciousness, which Christie sees emerging today around the world, will allow us to experience our spontaneous feelings (from childhood); the sacredness of our bodies and their unity with the earth body; the depth of our unconscious mind, with its mythic power and intuitive spiritual wisdom; identification with the "lower" forms of life; appreciation for ancient myths; and our awareness of "no boundaries" with humankind and the biosphere. He declares:

> Nature is our teacher, because much of who we are is indistinguishable from her, although mystics who have gone before us say that in the higher realms of consciousness is revealed a world of pure light and energy that permeates all natural forms. However, we still have much to learn; we are gifted creatures, but we are dangerous creatures too.[24]

He then quotes astronaut Edgar Mitchell. After seeing the "beautiful, harmonious, peaceful-looking planet . . . that gave you a sense of home," Mitchell was hit with this counter image: "Beneath that blue and white atmosphere was a growing chaos that the inhabitants of planet Earth were breeding among themselves—the population and technology were growing out of control." Christie concludes a series of articles on his theories with this ringing challenge: "To see nature is to see ourselves . . . To preserve nature is to preserve the matrix through which we can experience our souls and the soul of the planet Earth. The choice is clear, because there really is no choice at all."

In my view, two of many indications of the contemporary "greening" of consciousness are the planetwide ecology movement and the convergence of insights from many sources regarding the earthy roots of human beings. But such insights have been available throughout the centuries, long before the crisis-stimulated greening of much thinking today. Christie cites as examples the American Transcendentalists, including Ralph Waldo Emerson. The remainder of this chapter examines a variety of earth-grounded views that, like Christie's, parallel and illuminate many of the basic concepts undergirding ecotherapy and ecoeducation.

Insights from Therapists and Teachers in Survey

Responses from some of the professionals surveyed for this book affirm the extension of object relations theory proposed above. Consider these responses by some people who said "Yes" to the survey question: *Do you have evidence that people internalize in forming their identity, the natural environments—toxic or healthy—within which they grew up?*

- "People have talked to me about how they carry the effects of early environments, often a subtle influence, a feeling of at-homeness in certain environments or an aversion to others associated with painful experiences. I believe that the depth of our relation to the natural world has a great impact on our identity."
- "People internalize everything that happens to them. Every piece plays a part in the complex process of forming an identity. I see firsthand how the natural environment, its nurturing bountifulness and goodness, is a part of my own identity."
- "Influences 'basic trust' in the universe. Those who lack this tend to feel betrayed by the environment."
- "I think there is a difference in people who were raised in the city as compared with the country—for example, whether one is comfortable with solitude or needs outside entertainment."
- "A person's level of awareness is dependent on their upbringing. An [unfortunate] example locally [among rural people] is the acceptance of chemical fertilizers and pesticides."
- "Rural persons remember the purity of the setting, animals, growing crops, being directed by the seasons. City persons try to direct the seasons."
- "People raised in urban ghettos [as this respondent was] possess a different outlook than those from bucolic settings."
- "Even very fragile clients who grew up with a sense of nature can image [and get strength from] that calm place."

Ecological Insights from Artists, Poets, and Philosophers

Earth scientists and biologists have made crucial contributions to illuminating the genetic connections of our species with other animals and the earth. Some of their insights enhance the conceptual groundwork for understanding ecobonding and ecoalienation. A major historical example is Charles Darwin's monumental *Origin of Species*, a volume that opened wide the doors to biological human-animal connectedness. His revelations were and still are deeply threatening to those whose feelings of security and self-worth, and their belief systems are based on the illusion of total separation between humans and other animals.

Many dynamic early writings about the earthy roots of humans came, not from scientists, but from poets, artists, philosophers, and nature mystics in the major world religions. Just before the middle of the nineteenth century, nature mystic and poet Walt Whitman depicted the hoped-for recovery of bonding with the lost earth in his *Leaves of Grass:*

> We two, how long we were fool'd,
> Now transmuted, we swiftly escape as Nature escapes,
> We are Nature, long have we been silent, but now we return,
> We become plants, trunks, foliage, roots, bark. . .

Whitman continues this poem using sensuous and beautiful, as well as violent nature images:

> We are two fishes swimming in the sea together,
> We are what locust blossoms are, we drop scent around lanes morning and evening. . .
> We are two predatory hawks, we soar above and look down. . .
> We prowl fang'd and four-footed in the woods, we spring on prey. . .
> We are two clouds forenoons and afternoons driving overhead,
> We are seas mingling, we are two of those cheerful waves rolling over each other and interwetting each other. . .
> We have circled and circled till we have arrived home again. . .
> We have voided all but freedom and all but our own joy.[25]

In a society that was investing enormous energy in denying humankind's roots in nature, as well as the power of humanity's earthy sexuality, it is not surprising that Whitman's sensuous, earth-grounded poetry caused vigorous controversy. Today, in some ultra-conservative religious circles, the same emotionally charged investment in denying our continuity with the ongoing, creating, destroying, and rebirthing processes of nature, operates with intense passion.

Henry David Thoreau gave an illuminating description of powerful experiences of his inner wildness. During his time of learning what nature had to teach at Walden Pond, he wrote:

> Once or twice, I found myself ranging the woods, like a half-starved hound, with a strange abandonment, seeking some kind of venison which I might devour, and no morsel could have been too savage for me. The wildest scenes had become unaccountably familiar. I found in myself, and still find, an instinct toward a higher, or, as it is named, spiritual life, as do most men, and another toward a primitive rank and savage one, *and I reverence them both*. I love the wild not less than the good. The wildness and adventure that are in fishing still recommend it to me. I like sometimes to take rank hold on life and spend my day more as animals do.[26]

Thoreau's ambivalence in contrasting his wildness with his "good" feelings reflects the dominant beliefs of his society, beliefs that are unfortunately still common today.

John Muir, self-trained pioneer environmentalist and grandparent of the Sierra Club, described his own intense ecobonding at times when he was immersed in nature in Yosemite and elsewhere in the High Sierra ranges:

> Here is a calm so deep, grasses cease waving . . . . wonderful how completely everything in wild nature fits into us, as if truly part and parents of us. The sun shines not on us, but in us. The rivers flow not past us, but through us, thrilling, tingling, vibrating every fiber and cell of the substance of our bodies, making them glide and sing.[27]

If you have related intimately with a magnificent, awesome mountain, Muir's words may awaken memories of comparable enlivening bonding with nature. Such experiences provide self-validating knowledge that nature is truly "a part and parent" of all living creatures, including ourselves.

Author John Burroughs, a friend of Walt Whitman, was a pioneer naturalist who wrote powerful ecological prose in the early decades of the twentieth century. He was aware of the birthing process of the universe and wrote: "I saw afar down the huge first Nothing, and I know that I was there." He saw how knowledge of the eon-spanning journey of continuing creation can provide one path to enlivening our experience of nature and ourselves in nature:

> When our minds have expanded sufficiently to take in and accept the theory of evolution, with what different feelings we look upon the visible universe. . . . Evolution makes the universe alive. In its light we see that mysterious potency of matter itself, that something in the clod under foot that justifies Emerson's audacious line of the "worm striving to be man." We are no longer the adopted children of the earth, but her real off-spring. Evolution puts astronomy and geology in our blood and authenticates us and gives us the backing of the whole solar system. This is the redemption of the earth: it is the spiritualization of matter. . . .
>
> Revert to the time when life was not, when the globe was a half-incandescent ball, or when it was a seething, weltering waste of heated water, before the land had emerged from the waves, and yet you and I were there in the latent potencies of the chemically and dynamically warring elements. We were there, the same as the heat and flame are in the coal and wood. . . . The creative cosmic chemistry in due time brought us forth, and started us on the long road that led from amoebae up to man. . . . Creation has been a continuous process, and the creator has been this principle of evolution inherent in all matter. Man himself was born of this principle. His genealogy finally runs back to the clod under his feet. One has no trouble accepting the old Biblical account of his origins from the dust of the earth when one views that dust in the light of modern science.[28]

Building an ecological world view, on the awesome story of the universe's continual unfolding, Brian Swimme and Thomas Berry point to the interconnection between finding a multidimensional sense of the self and finding the lost connection with the earth and the universe: "Our individual self finds its most complete realization within our family self, our community self, our species self, our earthly self, and eventually our universe self."[29]

Humankind's Genetic Earth-Rootedness

The eminent Harvard biologist and ant expert Edward O. Wilson, author of *The Diversity of Life*, offers evidence that is in harmony with the ecotherapeutic understanding of humans' earthy roots. He holds that our love of and connectedness with the natural world is rooted in our genes—a deep inherited but forgotten connectedness with the world of nature. He uses the word *biophilia* to mean "the inherent [genetically based] human need to affiliate deeply and closely with the natural environment, particularly its living organisms." He gives evidence that "biophilia is part of our mental and emotional apparatus—as much a product of our history as love and bonding and having children." Our love of pets and parks, fondness for houseplants and gardening, bird watching, and nature documentary films are expressions of this genetically programmed longing for connections with the rest of the living world. He says, "It's a remarkable fact that Manhattan penthouse dwellers tend to turn their terraces into tropical rainforests. The first thing they do is fill everything up with palms, tropical plants and flowers."[30]

In exploring these genetic strivings for bio-connecting, Wilson points out, as suggested above, that "only in the last moment of human history has the delusion arisen that people can flourish apart from the rest of the living world." During his extensive scientific field research in rainforests around the world, he has made remarkable discoveries. He has found native peoples who have names for more than a thousand species of plants. Believing that our ancestors had such comprehensive botanical knowledge for thousands of years, he concludes: "If our survival once depended on our ability to identify, to have a feeling for these organisms, to understand their life cycles, to live with and around them, this has got to have affected the evolution of our sensibilities and our emotions, and it is certainly worthy of an investigation to see what innate responses human beings have toward biological diversity."[31]

It is noteworthy that, like many other biological and earth scientists, Wilson is a committed activist in earth-saving efforts. He estimates that an alarming 75,000 species are becoming extinct each year. His sense of urgency motivates passionate efforts to inspire people to become "biophiles" to help save the earth's wonderfully diverse web of living creatures.

Wilson's view is that the human species has had some three million years of survival programming in how to interact constructively with nature. At a deep body-mind level, we know, however dimly, that if we continue to reject this programming and do not establish a respectful interaction with nature, we will lose not only a vital dimension of our humanness, but eventually our planet home as a self-renewing, life-nurturing organism. It is noteworthy that Wilson's views are paralleled by those of Paul Shepard discussed below.

Psychologist and pastoral therapist James B. Ashbrook illuminates the neurological roots of ecobonding and ecoalienation, using his dual expertise in neuropsychology and psychology of religion.[32] If we are to claim humankind's true place in the biosphere and the universe, it is essential that we accept the fact that the whole mind "comes out of nature and does not function apart from nature." A whole-brain understanding of earth bonding and alienation includes functions of both hemispheres of the new brain or neocortex—the computer-like observing, objectifying, symbolic functioning of the left hemisphere and the imaging, patterning, meaning-making, feeling functions of the right hemisphere. But in addition there is the subsymbolic, nonconscious activity of the old brain with its somatic and its environmental roots. Furthermore, the cultural context powerfully influences all aspects of whole-brain, mind-body bonding with nature. As Ashbrook puts it, ecobonding "comes from the whole mind, cortical and subcortical, new brain emergence and old brain empathy."

Because the neocortex is involved, being more fully at home in nature involves choices—avoiding cognitive dualisms that divide, befriending our bonding, and caring for our natural home. Our ecological precariousness calls for the "integration of our cultural patterns into our biological and ecological universe." We must "move from the new brain's prominence and domination back into the old brain's primacy and purpose, the best evolutionary adaptation under the circumstances. Our destiny lies in the recovery of our relatedness to the whole of creation, not in getting beyond that origin."

The new brain or neocortex, particularly its left hemisphere functions, distinguishes humans from other animals. It represents an evolutionary development often seen as humankind's crowning glory. But when the cognitive functioning and rationality of the left hemisphere of the neocortex are regarded as the essence of the human mind and therefore of life's fullness, what Ashbrook calls the "new brain illusion" results. This obsolete view of the mind reinforces a sense of radical separation from the other animals and deep alienation from awareness of our organic relatedness with the evolutionary process and the rest of the biosphere.

The new brain is "an extension and elaboration of the old brain—the reptilian and mammalian brains"—and the whole mind's functioning involves the constant interaction of the old brain and the new. The limbic system of the old brain is at the biological root of human bonding with nature—a bonding that obviously was present in the responses of our reptilian and mammalian predecessors that enabled them to survive and pass on their genes. But for humans, the old-brain experience of ecobonding is intermingled with and enriched by cognitive processes in the new brain that add unique dimensions (including the spiritual) that probably transcend the experience of other animals. Although the brain chemistry of this bonding process is not yet understood, it seems clear that there is a biochemical foundation of it in the mind-body.

Ashbrook understands intimate bonding with the natural world as involving:

> the undefended, curiosity and attending engagement of the young child—and the childlike—in primary sensory processing. One is oriented to the world in an expectant and open way. The result is parasympathetic relaxation in the autonomic nervous system with its own restorative immune-strengthening processing. One is freed of having to defend oneself in terms of survival and self-esteem. All one's energy goes into engaging the world instead of defending against the world.[33]

The trustful response during such non-fight-or-flight relating to the mothering earth is comparable to an infant's relaxed body-mind response when bonding empathically with a loving, trustful mothering person. "Similarly, response to the natural world puts the arousal of the sympathetic system at the service of the relationality of the parasympathetic system, not the other way around. The courage to be as oneself is derived from and contributes to the courage to be part of the whole, in this case nature."[34]

As the human brain and family evolved together, "empathic caring marks the most striking change in evolutionary adaptation." Now, Ashbrook holds, as we rediscover our "oneness with the whole created order," we must invest this capacity for empathic caring in caring for our earth home. He then provides a clue concerning how all this relates to the spiritual dimension of the ecological circle: "The hidden wisdom which permeates our universe (the older cortex) is being revealed through the image and likeness of the creative Spirit in us (the whole cortex). And that Spirit explores everything, even the depths of our own creative imaginative Spirit."[35]

Contributions from Ecopsychology and Various Psychotherapies

Philosopher-historian Theodore Roszak, along with a variety of mental health professionals, has made major contributions to the development of an earth-grounded view of personality. The twenty-seven authors whose papers constitute *Ecopsychology: Restoring the Earth, Healing the Mind* have given green psychology a wealth of insights, many of which have affinities with ecotherapy. Roszak posits an ecological unconscious and suggests that the ways humans relate to nature are projections of unconscious needs and desires. They can be used to learn about people's deeper motivations and fears. He proposes an extension of object relations thought similar to the ecotherapeutic theory described above:

> Granted the need for a "perfect environment," we might let it be the real environment of all living things: the planetary biosphere, which is everybody's "primary care-giver.". . . Let us imagine parenting that is responsible for making that environment as perfect as possible. What, after all, do parents owe their young that is more important than a warm and trusting connection to the Earth that accounts for our evolutionary history?"[36]

Roszak also has done a comprehensive survey of earth-rooted understandings of personality in classical psychotherapeutic theories. Sigmund Freud, he observes, had a "dour, alienated vision of nature," even though he was convinced that the psyche is biologically grounded. But, because he also held to the social Darwinism that was prevalent in intellectual circles in his time, Freud saw the "primitive" grounding of the psyche as something to be rigidly controlled. He regarded the id as a primitive, pleasure-driven beast, a wild horse that the ego is trying to tame to ride in a socially conforming manner. His view of nature was essentially negative because it was based on a deep mind-matter dualism. Freud saw human consciousness as a strange anomaly that does not really belong in materialistic, mechanistic nature. He dismissed the "oceanic feeling" in which people feel a "oneness with the world" as a primitive, psychic state expressing the desire to return to the womb, a regressive attraction that must be outgrown. It is essential for the ego to sever this original intimate bonding with nature in order to be healthy.[37]

As Roszak points out, however, Freud eventually "felt compelled to grant that our infantile sense of oneness with the world plays one major role in adult life. From it, he believed, arise the fires of Eros: the emotional force that binds the self to others."[38] And, in spite of Freud's belief in the inherent and necessary gap between humans and nature, in his practice of psychoanalysis he occasionally prescribed patients to experience the healing power of nature.

One Freudian theoretician who brought the natural environment into the psychoanalytic mainstream is psychiatrist Harold Searles. In his 1960 book *The Nonhuman Environment*, he explored the relation of the biosphere to child development and neurosis, and developed clinical methods using settings in nature for treating chronic psychotics. But he concurred with Freud's thinking that the "oceanic feeling" of union with nature is a regression of the primitive infantile ego into the "chaos" of nature. Separation from the nonhuman environment is a goal of maturing and therapy. Interestingly enough, he used Buber's thought, but held that only as the separation from the nonhuman environment is achieved can one establish an I-Thou relatedness with nature.[39]

Carl Jung was raised in a rural setting, and his writings demonstrate a sense of intimate relatedness to nature not present in Freud. In his autobiography, *Memories, Dreams and Reflections*, he describes the sense of wonder he found in nature as a boy. During this period he drew away from the human world and immersed himself in the wonders of nature. But, as Roszak makes clear, although Jung rejected Freud's materialistic reductionism and his view of nature, his psychology retained little of Freud's connectedness with the body and the physical world. He focused so exclusively on inner "psychic reality" as to deepen rather than heal the chasm between the mind and nature. With most other post-Freudians, he regarded psychological alienation from the natural world, including its other animal members, as normative and irreversible.

Several contemporary Jungians (in contrast to Jung) have moved to an ecopsychological understanding of personality. James Hillman proposes "prescribing nature" as one part of therapy. Jungian ecopsychologist Stephen Aizenstat believes that "human behavior is rooted most deeply in nature's intentions . . . The rhythms of nature underlie all human interaction: religious tradition, economic systems, cultural and political organization. When these human forms betray the natural psychic pulse, people and societies get sick, nature is exploited, and entire species are threatened."[40]

Jane Hollister Wheelwright and Lynda Wheelwright Schmidt, mother and daughter Jungian analysts, have an explicit ecopsychological view of persons. Lynda grew up on a huge family ranch that stretches along a wild expanse of the Pacific Ocean near Santa Barbara, California. Their autobiographical book, *The Long Shore*, includes a chapter by Lynda on "The Wilderness as Healing Power." Musing on the healing of wild nature, she points out that much of Western psychology today focuses on the pain and anxiety of abandonment caused by children being separated too early from the "safe place" with their mothering person (female or male):

> As soon as we realize ourselves, our initial harmony with the [interpersonal] environment is lost. We are thrown out of Eden. . . . If there has been an idyllic merger with a central, protective adult, male or female, it is soon disrupted. . . . No longer does that adult provide a safe place for us to nestle in. . . . Psychotherapy and other forms of psychic healing have moved into this breach, with their methods of providing a safe place, a re-creation of the original "safe place" with the mother-person (male or female). From this safe place, clients can explore and find healing for not only the wound particular to their own life but also the abandonment wound everyone shares.[41]

Unlike most psychotherapists, Schmidt does not stop with the interpersonal environment. She sees profound healing of the sense of abandonment in bonding with wild nature and points to the need of city-dwellers to find healing in wilderness:

> Alone in the wilderness we also experience fear and isolation, but we have a history of millions of years of relating to wilderness literally and bodily. . . . Entering the wilderness and its microcosms—gardens and parks—gives us an opportunity to reconnect with that instinct and rests our fragile psyches from the exhaustion of trying to stay intact in the civilized world, which is so alien to many of us. . . . For those of us who divide our lives between civilization and wilderness, even a brief experience of the silence that hums while we sit still or move quietly through it will bring us to ourselves, will bring us refreshment and renewal of enthusiasm for life. Merger with a therapist can heal our abandonment wound, but merger with nature can reconnect us to the ancient roots of the Self as well. . . . The healing that comes to me from being in the wilderness is the opportunity to leave ordinary consciousness and return to my animal, instinctual way of being.[42]

In my own personal as well as clinical experience, an unexpected psychological benefit from intimate bonding with nature is the awakening of creative mental processes. Somehow the energies of nature enliven these. When I do my early morning aerobic walks nearly every day, up and down the foothills surrounding my home, I often experience a spontaneous flow of fresh thoughts and images. Connecting with the beauty of the sky, the mountains, the birds, and the distant ocean, together with big-muscle exercise in the natural setting, seems to release this flow. It enlivens the dance between the functions of the right and left hemispheres of the brain. This playful dance is the essential generator of creativity.[43]

Schmidt points to the power of wilderness to stimulate not only the creative process but also the search for meaning: "In order to protect our creativity, we need now to honor the great resource that was our first mother, our wilderness. Attention paid to the wilderness will give us the heart and soul we need to continue our quest for the meaning of our presence in the vast universe." [44]

More than a century ago, Henry David Thoreau, in an illuminating passage in his journal, discussed the way in which what I am calling alienation from nature and the wilderness within stifles creative imagination:

> The intellect of most men is barren. They neither fertilize or are fertilized. It is the marriage of the soul with nature that makes the intellect fruitful, that gives birth to imagination. . . . Men have circumnavigated this globe of land and water, but how few have sailed out of sight of common sense over the ocean of knowledge.[45]

He observes that without nature-awakened imagination, most persons do not really *live* in the world. They merely pass through it as they endure dull lives of quiet desperation. If I were asked for a fruitful image of a main message of this book, I could not think of a more appropriate one than Thoreau's phrase "the marriage of the soul with nature."

A major psychotherapy system founder who took significant steps toward an ecological understanding of personality is Fritz Perls. In his *Gestalt Therapy Verbatim*, he stated: "Even a cow can't live in Los Angeles." He used nature images in discussing human growth and aliveness:

> A thousand plastic flowers
> Don't make a desert bloom
> A thousand empty faces
> Don't fill an empty room.[46]

Perls indicates, but only in passing, that his holistic, organismic understanding of persons includes their relation to the natural environment:

> We exist as an organism—an organism like a clam, like an animal, and so on, and we relate to the external world just like any other organism of nature. . . . You cannot separate the organism and the environment. A plant taken out of its environment can't survive, and neither can a human being if you take him out of his environment, deprive him of oxygen and food, and so on. So we have to consider always the segment of the world in which we live as part of ourselves. Wherever we go, we take a kind of world with us.[47]

Perls focuses on the notion of ego boundaries between the person and the environment:

> The very moment we breathe, is the air that comes in still part of the outside world, or is it already our own? If we eat food, we ingest it, but can still vomit it up, so where is the place where the self begins, and the otherness of the environment ends? So the ego boundary is not a fixed thing. If it is fixed, then it again becomes . . . an armor, like in the turtle.[48]

But, in spite of this ecological orientation, elsewhere in Perls's writings the "environment" usually refers to the human or social environment. His primary therapeutic focus is helping people detach from symbiotic relationships with others and reclaim their own center and inner strength. In this context the environment becomes a negative concept for Perls. He uses the Spanish word *maduro*, which means "ripe," in discussing maturing, the goal of therapy, and he sees neurosis as a growth disorder. He declares:

> Maturing is the transcendence from environmental support to self-support. . . . What we are after is the maturation of the person, removing the blocks that prevent a person from standing on his own feet. We try to help him make the transition from environmental support to self-support. . . . Being grounded in one's self, is about the highest state a human being can achieve.[49]

As far as I can discover, Perls never considered healthy *interdependence* nor focused on the healing possibilities of such interdependence with either the social or the natural environments. Gestalt therapist William Cahalan transcends Perls. He describes several techniques he uses in therapy, including silent walks outdoors, "finding oneself" within one's home neighborhood, and asking clients what animal they are most like. He reports that work with the nonhuman dimension tends to bring up clients' relation to ultimate reality "that some call the spiritual," and gives a sense of thankfulness that "often naturally leads to a desire to give back, to live less as a consumer and more in balance with the Earth, which in a sense is our true body, our real self."[50]

Ecofeminist Theories of Personality

Some of the most insightful understandings of human personality's earthy roots are in the thought of the ecofeminists. (The term *ecofeminism* refers to both the diverse spectrum of efforts by women to save the earth, and the transformed feminist self-understanding that reflects an earth-centered view of the relation of women and nature.) Ecofeminists Irene Diamond and Gloria Feman Orenstein identify three philosophical streams in the diverse ecological movement of ecofeminism. One position emphasizes that the earth is sacred, that her rivers and forests and all her living creatures have intrinsic value. Another strain focuses on the fate of human life and the earth as being deeply intertwined because human life is fully dependent on the earth. For this reason justice in society cannot be achieved if separated from the earth's well-being. The third ecofeminist perspective is shared by many native peoples whose connection with their land is at the center of their identity. This strain emphasizes that humans must learn to walk the narrow line between respecting the earth's cycles, needs, energies, and ecosystems, and using the earth respectfully as a resource for human beings.[51]

Ecofeminist poet and playwright Susan Griffin opened many eyes (including mine) with her lyrically confronting *Woman and Nature: The Roaring Inside Her.*[52] Writing from what she calls her intuitive or uncivilized self, she lifts up the dynamic connections between men's attitudes toward and treatment of women and nature, throughout all of the patriarchal period. She pictures the male vision of the universe that equates women with nature and defines men as rational creatures who are above and superior to both nature and women. Then she paints passionate verbal pictures of how women claim their wisdom about the earth and what is still wild in them, and thus enable the deeply divided in them to come together again. Griffin writes:

> This earth is my sister; I love her daily grace, her silent daring, and how loved I am. How we admire this strength in each other, all that we have lost, all that we have suffered, all that we know: we are stunned by this beauty, and I do not forget: what she is to me, what I am to her.[53]

Griffin concludes with this soaring affirmation of the bonding between nature and herself as a woman:

> When I let this bird fly to her own purpose . . . the light from this bird enters my body, and when I see the beauty of her flight, I love this bird. . . . I fly with her . . . live in this bird whom I cannot live without, as part of the body of the bird will enter my daughter's body, because I know I am made from this earth, as my mother's hands were made from this earth, as her dreams came from this earth. . . . all that I know speaks to me through this earth and I long to tell you, you who are earth too, and listen as we speak to each other of what we know: the light is in us.[54]

Another classic ecofeminist contribution is by Carolyn Merchant, a feminist historian of science and author of *The Death of Nature: Women, Ecology and the Scientific Revolution*. She observes: "The ancient naming nature nurturing mother links women's history with the history of the environment and ecological change."[55] The contemporary women's movement and the ecology movement, which arose simultaneously in the 1960s and early 1970s, share an egalitarian ethic. Merchant calls for reversing the interrelated subjugation of both nature and women, and recovering the core of the organismic, holistic world view that can help liberate women and men and, with them, liberate the environment to their fullness.[56]

Ecofeminist Charlene Spretnak challenges the patriarchal socioeconomic and political structures, and the obsolete values that guide them. She holds that Hiroshima, DDT, Bhopal, and Chernobyl require a radical rethinking and revaluing of the present policies, rather than simply a fine-tuning of policies based on ignorance, fear, and greed that are leading humankind to ecocide and species suicide.[57]

Speaking from her expertise in Buddhism, peacemaker Joanna Macy holds that people today need to "awaken to the ecological self." This means shedding their old skin-encapsulated egos and moving to an expanded, encompassing self with a deep sense of interconnectedness with all life. Macy calls the ecologically grounded understanding of persons the "eco-self" as contrasted with the "ego self." Survival as a human species depends on learning to live in harmony with an earthy reality that the environment is in us just as we are in it. Reflecting Buddhist wisdom, she holds that moving from the dominant but erroneous Western view of the ego-self to the eco-self is an expansion of our self-identity that is essential for saving the planet. The eco-self is what I am calling the "lost self."

Macy affirms an important methodological point: that one health-giving result of greening our sense of self is that this eliminates the need to engage in sermonizing or moralizing about ecology. She regards both of these, as do I, as boring and unproductive of constructive change. By enlarging our identity to the eco-self position, our understanding of what constitutes real self-interest is expanded so as to include the well-being of the earth.[58]

Macy shares a moving account of a college student who heard her talk on the Chipko (tree-hugging) movement in north India, in which courageous villagers risk their lives to defend their surviving forests. They do this by literally hugging the remaining tree trunks to block the chain saws of the loggers. Following her talk, a male student wrote: "I think of the tree-huggers hugging my trunk blocking the chain saws with their bodies. I feel their fingers digging into my bark to stop the steel and let me breathe. . . . I give thanks for your life and mine, and for life itself. I give thanks for realizing that I too have the powers of the tree-huggers." [59]

Ecofeminist psychologist Lesley Irene Shore has a perspective on human development that is directly relevant to the heart of ecotherapy's developmental theory foundation. Her book *Tending Inner Gardens: The Healing Art of Feminist Psychotherapy*[60] is a goldmine of insights and methods for doing ecotherapy from an ecofeminist perspective. She observes that current psychological theories were formulated in a male-oriented social climate and reflect the social structures and cultural metaphors of that context. They overvalue moving toward autonomy, separation, and individuation, while devaluing the feminine orientation and experience of the central value of moving toward interpersonal relatedness. As a psychologist she declares:

> Our profession struggles to grow, to break free of chains that shackle our minds and blind our thoughts to ideas that may be right in front of our eyes. I believe that we're trapped inside western psychology. . . . The blinders include our assumption of linearity and either/or quality of our thought. These characteristics of our mechanistic view of the universe have been all too apparent in our thinking about development.[61]

Earth-Healing Wisdom from Native Peoples

As I write these words I am at home in the foothills of Santa Barbara on a gorgeous spring day. The unusually abundant winter rains have produced a profusion of wildflowers in our yard. The air is perfumed by the blossoms of our lemon, orange, and grapefruit trees. Seeing the ocean and the wonderfully wild offshore islands through the morning mist awakens fond memories of sailing among them. As the nurturing energies of the living world surround and fill me, I feel the presence of the Chumash—the cultured, peaceful, and earth-bonding Native Americans who lived here and undoubtedly loved this as *their* place for many centuries.

Anthropologist Jack Forbes describes early California Indians:

> They perceived themselves as being deeply bound together with other people (and with the surrounding non-human forms of life) in a complex interconnected web of life, that is to say, a true community. . . . All creatures and all things were . . . brothers and sisters. From this idea came the basic principle of non-exploitation, of respect and reverence for all creatures.[62]

Therapist Richard Voss contrasts the understanding of personality in Western personality theory with indigenous understandings: secular mythology vs. sacred mythology; emphasis on progress vs. emphasis on process; dichotomous vs. unified; structures of mind vs. mind-body-earth identity; linear motion vs. circular motion; exclusivism vs. inclusivism; emphasis on time vs. emphasis on place; written theory vs. oral tradition.[63]

Jeanette Armstrong, a leader in the international political movement of indigenous people, declares from her Native Canadian (Okanagan) background: "We are told that we are responsible for the Earth. We are keepers of the Earth because we are Earth. . . . The Okanagan word for 'Earth' uses the same root syllable as the word for our spirit-self. . . . It is the whole-Earth part of us that contains immense knowledge."[64]

The Education of Little Tree is a tender, touching story by Forrest Carter.[65] With the power of fiction, the story communicates authentic awarenesses of Native Americans' bonding with nature. The author writes:

> Following the spring branch up the hollow was how I found the secret place. It was a little ways up the side of the mountain and hemmed in with laurel. It was not very big, a grass knoll with an old sweet gum tree bending down. When I saw it, I knew it was my secret place, and so I went there a whole lot. Ol' Maud [his dog] taken to going with me. She liked it too, and we would sit under the sweet gum and listen—and watch. Ol' Maud never made a sound in the secret place. She knew it was secret.[66]

Being "too young to keep a secret," he told his Granma about his secret place in nature.

> She wasn't surprised—which surprised me. Granma said all Cherokees had a secret place. She told me she had one and Granpa had one. She said she had never asked, but she believed Granpa's was on top of the mountain, on the high trail. She said she reckined most everybody had a secret place, but she couldn't be certain, as she had never made inquiries of it. Granma said it was necessary. Which made me feel right good about having one.

Forrest remembers:

> Granma said I would come to know the old sweet gum tree in my secret place had a spirit too. Not a spirit of humans, but a tree spirit. She said her Pa had taught her all about it. Granma's Pa was called Brown Hawk. She said his understanding was deep. He could feel the tree-thought.

Granma told Forrest that one time when she was a little girl, her Pa was worried because the beautiful, tall white oaks on the mountain where he often walked were excited and scared. He knew something was wrong. Then one morning early, just as the sun broke over the ridge of the mountain, Brown Hawk saw the lumbermen marking the white oaks and figuring how they would cut them down. When they left, he said the trees began to cry. After that he could not sleep. So he talked to the Cherokees and they decided to try to save their white oaks. Each night, after the lumbermen left, they would dig up the road built for the wagons to haul away the oaks. So the lumbermen posted armed guards on the road. But at night the Cherokees would dig up the road where there were no guards. Then one day a healthy oak fell on the road, killing two mules and smashing a wagon. The lumbermen left and never came back.

> Granma said the moon waxed full, and they held a celebration in the great stand of white oaks. They danced in the full yellow moon, and the white oaks sang and touched their branches together, and touched the Cherokee. Granma said they sang a death chant for the white oak who had given his life to save others, and she said the feeling was so strong that it almost picked her up off the mountain.

Like the Cherokees, native peoples in many countries take for granted that a deep connection with nature that leads to living cooperatively with nature is essential for human well-being, even human survival. Al Gore quotes this moving prayer attributed to the Onondaga tribe of Native Americans in upstate New York:

> O Great Spirit, whose breath gives life to the world and whose voice is heard in the soft breeze . . . make us wise so that we understand what you have taught us, help us learn the lessons you have hidden in every leaf and rock, make us always ready to come to you with clean hands and straight eyes, so when life fades, as the fading sunset, our spirits may come to you without shame.[67]

A vivid though tragic illustration of the deep rootedness of native peoples in their place in nature is the great Apache shaman Goyahkla, or Geronimo, as he is usually remembered. He was born around 1823 in what was then

Mexican territory and is now western New Mexico near the headwaters of the Gila River. Geronimo was not a chief but a charismatic spiritual leader to whom chiefs came for his wisdom and spiritual power. In a bloody, unprovoked raid while he was on a peaceful trading mission to their country, Mexican soldiers slaughtered his wife, mother, and three children, along with twenty other women and children, and took sixty others to be sold as slaves.

He probably acquired the name Geronimo when Mexican soldiers invoked Saint Jerome in their fear when he attacked them with his knife in spite of their rain of bullets. Geronimo became the last powerful Native American leader to lead his people to stand against "the tide of Manifest Destiny that pushed the United States land claims steadily west."[68]

In 1881, the great Native American leaders such as Sitting Bull, Chief Joseph, and Crazy Horse had either been "pacified" or killed. The Chiricahua Apache, with Geronimo as their inspirational leader, were the only Native Americans still free even though their four great chiefs were dead. For five years, after the murder of his family and repeated violence and betrayals by both Mexicans and Americans, Geronimo led an utterly hopeless resistance—the longest that any Native Americans were able to hold out against impossible odds. Geronimo eventually surrendered when promised that he would be reunited within five days with the new family he had gained and his people allowed to return to their homeland. But his captors had lied again. He and his surviving tribal members were held captive at Fort Sill in Oklahoma, far from their "place" on the land.

As a prisoner in 1904, he wrote to President Theodore Roosevelt pleading to return his people to Arizona: "It is my land, my home, my father's land, to which I now ask to be allowed to return. I want to spend my last days there, and be buried among those mountains. If this could be I might die in peace, feeling that my people, placed in their native homes, would increase in numbers, rather than diminish as at present." Roosevelt refused on the grounds that outrage by whites against the Apache was still too intense. Geronimo met the army general who had lied to him when he surrendered. Trembling with anger, he demanded that the general explain his lies: "I have been away . . . now twelve years. The acorns and pinon nuts, the quail and wild turkey, the giant cactus and the palo verdes—they all miss me. They wonder where I've gone. They want me to come back." The general refused to explain or to help the tribe return to their homeland. Geronimo died at the age of 85, never seeing his beloved Sierra Madres again. He died separated from the place where he and his people were at home in a special way, the place where the earth loved and nurtured them.

What were the sources of Geronimo's enormous power, inspiration, and respect? In addition to his fearlessness, he had a razor-sharp intellect, could divine distant events, and give wise counsel. But his dynamic spiritual power undoubtedly came from his relationship with the earth. The centerpiece of the rich ceremonial life of his band, the Chiricahua Apache, was and is still today the Dance of the Mountain Spirits. In this empowering ritual, Geronimo and his band drew strength from the earth. For him it must have been healing strength after the devastating tragedy of his multiple losses. As it was for all native peoples, his place of birth was vitally important to Geronimo. The Chiricahua believed that their god, Ussen, had given them their homeland—now southeastern Arizona, southwestern New Mexico, and a large tract of what was northern Mexico along the crest of the Sierra Madre Mountains. It is said that after his wanderings, when he returned to his birthplace near the hot springs by the fork of the Gila River, "he would roll on the ground in four directions." For the Apache, four is a sacred number. Anyone familiar with the medicine wheel of Southwestern Native American tribes will recognize that ritual acts in the four directions are meant to draw healing energy from the earth. During the years that Geronimo led his people, "the mountains in particular were congenial terrain, for among their cliffs and gorges an Apache felt all but invulnerable. Here too dwelt the Mountain Spirits, divine beings who cured illness and protected the Chiricahua from their enemies."

Native American Marie Wilson, a spokesperson for the Gitksan-Wet'-suwet'en Tribal Council, is one whose wise views show how our society can benefit by learning from Native American women's insights about humans and the earth. The central theme of ecofeminism—that sexist exploitation of women is at the root of exploitation of the earth—is expressed well when she states:

> A North American Indian philosopher has likened the relationship between women and men to the eagle, which soars to unbelievable heights and has tremendous power on two equal wings—one female, one male—carrying the body of life between them. The moment one is fractured or harmed in any way, then the powerful bird is doomed to remain on the earth and cannot reach those heights.[69]

It is important not to overidealize the past or present ecological behavior of native peoples. Humans have always had some negative impact on their natural environments. Certainly today, the demoralized condition of many oppressed indigenous peoples—living in what has been called the Fourth World—causes them to live in earth-unfriendly ways. But the huge difference between traditional indigenous attitudes and behavior and those of industrialized, urbanized societies is that native peoples tend to be earth-bonded with earth-reverencing cosmologies, whereas non-natives tend to be earth-alienated with cosmologies that support noncaring, nonrespectful earth attitudes.

The Role of Wild Animals in Human Development

Paul Shepard, professor of human ecology at the Claremont (California) Graduate School, has generated a provocative theory about the profound roots of humans' relationship with animals, supporting his view by multicultural evidence. He reviews the evolution of humans as "thinking animals," showing how the development of our minds and the characteristics that make us human have been and continue to be intimately involved with interaction with animals. Animals play a prominent role in the learning processes of children as well as in our social relations as adults, including the core of our imagination as expressed in our languages, dreams, images, art, speech, and play. When humans' relationships with other species are diminished, their caring about and for all life, including their own species, is lessened. In spite of the implied human hubris in keeping animals as pets, childhood pet care may be a first step by which growing persons move in the direction of a sense of stewardship toward all nonhuman life. Shephard declares:

> What the teacher, like the parent, needs to remember is that all childhood nature study is fundamental to the child's competence as a person and to elements of the personality, especially with the formation of attitudes, definitions, and attention structure that define consciousness uniquely for the individual. . . animals have a large claim on the maturing of the individual and his capacity to think and feel.[71]

But, in a discussion entitled "Alone on a Domesticated Planet," Shepard describes domesticated animals, including pets, as "minimalized" animals with whom humans have a master-slave relationship. They are a part of the anthropocentric, urbanized, industrialized, technological world where

> the intricate symbioses of wild ecosystems, the kingdoms of otherness half hidden in secret lives, and private purposes of migration, hibernation, and metamorphosis in the water or in the air come to an end and are unknown. That world cannot reveal hidden purposes of life where we have annexed and tamed its purposes. Deep in his heart each of us knows that the mystery of the self is as great as that of the universe—perhaps is the same mystery.[70]

Bioregional Understandings of Place

Bioregionalism is a growing contemporary intellectual movement with direct relevance to an earth-based understanding of human personality. It emphasizes that each earth region has its own unique, interdependent regional culture and identity, both of which are deeply rooted in the natural environment of that area. Bioregionalism means learning to become native to one's place by living in harmony with both its special gifts and its limits. Two pioneers in bioregionalism, Peter Berg and Raymond Dasmann, summarize that it

> involves becoming native to a place through becoming aware of particular ecological relationships that operate within and around it. It means understanding activities and evolving social behavior that will enrich the life of that place, restore its life-supporting systems, and establish an ecologically and socially sustainable pattern of existence within it. Simply stated it involves becoming fully alive in and with a place. It involves applying for membership in a biotic community and ceasing to be its exploiter.[72]

Bioregionalism is one of several contemporary movements that seek to bring together ecology, the science that studies the interdependence of all living systems, with social ecology, which seeks to develop social behavior in which we humans can meet our requirements for our life from nature in sustainable ways in harmony with the well-being of the environment. Kirkpatrick Sale's *Dwellers in the Land: the Bioregional Vision* discusses the meaning and importance of this vision.

> The crucial and all-encompassing task is to understand *place*, the immediate specific place where we live . . . The limits of its resources; the carrying capacities of its lands and waters; the places where it must not be stressed; the places where its bounties can best be developed; the treasures it holds and the treasures it withholds—these are the things that must be understood.[73]

The well-being of the biotic (living) community in a region must be the ultimate criterion guiding social and economic development of sustainable human life there.

Sale shows the radical contrast of the bioregional paradigm's guiding principles and those of the industrial-scientific paradigm that dominates so-called modern society. The bioregional paradigm emphasizes decentralization of power, governance, and economies, complementarity, and diversity rather than centralization, hierarchy, and uniformity, the norms in the industrial-scientific paradigm. Bioregionalism emphasizes conservation, stability, self-sufficiency by biotic regions, and cooperation rather than exploitation, "progress" (meaning growth), world economy, and competition.[74]

Ecofeminism and bioregionalism have many common objectives. Judith Plant observes that if one extrapolates from the bioregional idea of the decentralization of power and self-governing forms of social organization, one comes to an emphasis on the home as the place where new values and behaviors are created in the next generation. The word *ecology* itself points to this: the Greek root *oikos* (from which *eco* is derived) means home. She declares, "Home is the theater of our human ecology." But traditional family models (created relatively recently by the industrial revolution) are not conducive to the urgent task of creating a new identity, a new vision of human relatedness to each other and to nature. The feminist and the bioregional values and understandings complement each other, and the two movements are essential allies in the task of saving the planet.[75]

Exploring This Chapter's Issues Further

For fuller information about these books, see the annotated bibliography at the end of this book.

Gregory Bateson, *Mind and Nature, a Necessary Unity*
Peter Bishop, *The Greening of Psychology*
Ruth Fletcher, *Teaching Peace: Skills for Living in a Global Society*
Willis Harman, *Global Mind Change*
James Lovelock, *Gaia, A New Look at Life on Earth*
David W. Orr, *Earth in Mind*
Theodore Roszak, *The Voice of the Earth*
Theodore Roszak, Mary E. Gomes, and Allen D. Kanner, eds., *Ecopsychology*
Paul Shepard, *Nature and Madness*
Edward O. Wilson, *Biophilia*
E. O. Wilson and Stephen R. Kellert, eds., *The Biophilia Hypothesis*

Notes

1. New York: Penguin Books, 1992, 366.

2. Unpublished paper, "The Wild and the Sacred," presented to an artist group, October 1993, 6

3. Arne Naess, father of deep ecology, describes the "ecological self," a broadening of the self by identification with the biosphere as a whole.

4. Letter dated 23 August 1799, *The Columbia Dictionary of Quotations*, N.Y.: Columbia University Press, 1993, 622.

5. The interdependence of humans with the biosphere is represented concretely by facts such as these. All living beings drink from the planet's precious pools of oxygen and water, recycling these as they use them, so that they can be reused by other organisms. Earth scientists estimate that the entire supply of water is recycled every three years.

6. Kirkpatrick Sale, *Dwellers in the Land: The Bioregional Vision*. Philadelphia: New Society Publishers, 1991, 47.

7. D. W. Winnicott, "The Value of Depression," *British Journal of Psychiatric Social Work*, no. 3 (1964): 123–27.

8. I concur with feminist psychological theorists, such as Dorothy Dinnerstein, who hold that co-parenting involving fathers as well as mothers in intimate infant

care would produce a variety of salutary consequences in addition to reducing the overloading and scapegoating of "bad mothers." It would cut the mythological connection of "Mother Nature" with human mothers, which has tended to give male encounters with nature overtones of sexual conquest (Theodore Roszak, *The Voice of the Earth* [New York: Simon & Schuster, 1992], 292–93)

9. D. W. Winnicott, *Human Nature* (New York: Schocken Books, 1988), 107.

10. Jungian Lynda Wheelwright Schmidt's discussion of the psychological impact of the death of wilderness (see below) is in harmony with this theory of pets and gardens as transitional objects that assist us in keeping in touch with wild nature within ourselves.

11. New York: Ballantine Books, 1992.

12. From an interview with Estes reported by Linda Heiser, "Women Devouring Surprise Hit 'Wolves,'" *Santa Barbara News Press*, 10 October 1993, D3.

13. *The Long Shore, A Psychological Experience of the Wilderness.* (San Francisco: Sierra Club Books, 1991, 202.

14. *I and Thou*, translated by Walter Kaufmann (New York: Charles Scribner's Sons, 1970). Emphasis added.

15. Berry, unpublished manuscript.

16. "Therapy for a Dying Planet," in *Ecopsychology: Restoring the Earth, Healing the Mind*, edited by Theodore Roszak, Mary E. Gomes, and Allen D. Kanner (San Francisco, Sierra Club Books, 1995), 154.

17. Paul Shephard, "Nature and Madness," in *Ecopsychology*, 56.

18. New York: Pantheon, 1958, 2.

19. *The Art of Loving* (New York: Harper & Row, 1956), 9.

20. Ibid., 7.

21. Ibid., 11.

22. *The Voice of the Earth* (New York: Simon & Schuster, 1992), 14.

23. Ibid., 50. Ecopsychologist Ralph Metzger illuminates the nature of ecoalienation. He employs standard psychopathological categories such as addiction, dissociation, autism, and amnesia as "diagnostic metaphors" to shed light on the pathological ecoalienation between human consciousness and the remainder of the biosphere ("The Psychopathology of the Human-Nature Relationship," in *Ecopsychology*, 55ff.).

24. All quotes from Christie are from a series of brilliant articles on the evolution of consciousness and its relation to nature, published in *Habitat*, the journal of the Maine Audubon Society, from 1984 to 1986. It is noteworthy that Christie does not "mix" his earth theory with his psychiatric practice because "this isn't therapy in the ordinary sense and because it is professionally dangerous for a well known psychiatrist to open such work to patients with major transference problems" (Walter R. Christie, letter to author, 1994).

25. "We Two, How Long We Were Fool'd," from *Leaves of Grass*, edited by Harold W. Blodgett and Sculley Bradley (New York: New York University Press, 1965), 108.

26. *Reflections at Walden*, edited by David E. Scherman (Kansas City, Mo.: Hallmark Edition, 1968), 29. (Emphasis added.) Thoreau had the courage of his convictions and was involved in protesting problems in the social as well as the natural environment. He went to jail rather than pay taxes to support what he believed to be a grossly unjust war—the Mexican American War. He also said that a government that allowed slavery was not his government, and helped at least one slave to freedom via the Underground Railroad. However, his emphasis on self-reliance was

somewhat compromised along with most young men at his time and up to this day, by the fact that he took his laundry home for his mother to do for him.

27. Quoted by Joseph Cornell, *Listening to Nature, How to Deepen Your Awareness of Nature* (Nevada City, Calif.: Dawn Publications, 1987), 42.

28. John Burroughs, *Time and Change*, vol. 15 in *The Complete Works of John Burroughs* (New York: Wm. H. Wise & Co., 1924), 193–94, 187 188.

29. *The Universe Story: From the Primordial Flaring Forth to the Ecozoic Era, A Celebration of the Unfolding Cosmos* (San Francisco: Harper San Francisco, 1992), 268.

30. Anne Wingfield Semmes, "Modern Day Shaman Explores Biophilia," *E Magazine* 4, no. 2 (March/April 1993), 15.

31. Ibid.

32. The substance of this section, unless otherwise noted, is from Ashbrook's paper, "The Human Brain and Human Destiny: A Pattern for Old Brain Empathy with the Emergence of Mind," *Zygon* 24, no. 3 (September 1989).

33. James B. Ashbrook, letter to author, 30 May 1995.

34. Ibid.

35. Ibid.

36. Roszak, *Voice of the Earth*, 293, 294.

37. Ibid.

38. "When Psyche Meets Gaia," in *Ecopsychology*, 16.

39. Roszak, *Voice of the Earth*, 294–96.

40. "Jungian Psychology and the World Unconscious," in *Ecopsychology*, 93.

41. *The Long Shore*, 194.

42. Ibid., 198–201.

43. For a discussion of creativity, see Howard Clinebell, *Well Being: A Personal Plan for Exploring and Enriching the Seven Dimensions of Life* (San Francisco: Harper San Francisco, 1992), 57.

44. *The Long Shore*, 202.

45. *Reflections at Walden*, 42–43.

46. Lafayette, Calif.: Real People Press, 1969, 2.

47. Ibid., 6.

48. Ibid., 7.

49. Ibid., 28, 37.

50. "Ecological Groundedness in Gestalt Therapy," in *Ecopsychology*, 223.

51. Diamond and Orenstein, eds., *Reweaving the World: The Emergence of Ecofeminism* (San Francisco: Sierra Club Books, 1990), xi–xii.

52. New York: Harper and Row, 1978.

53. Ibid., 219.

54. Ibid., 227, xvi.

55. San Francisco: Harper and Row, 1980, xvi.

56. Merchant points out that nature in most Western and non-Western cultures had been traditionally understood as female. She then shows how the "fathers" of the sixteenth- and seventeenth-century Scientific Revolution, including Francis Bacon, René Descartes, and Isaac Newton, destroyed both humanity's ancient view of nature as a living organism and the Renaissance image of the nurturing earth, both of which views produced social orders based on cooperation between humans and nature. All this was replaced by a view of nature as a machine that men (I use the term advisedly) must control and exploit. The new mechanistic world view was used

in the modern period to justify the exploitation of nature and unrestrained commercial expansion, in the name of progress, and to sanction a new socioeconomic order that subordinated women. Ibid., chapters 7, 8, 9.

57. Spretnak, "Ecofeminism: Our Roots and Flowering," in *Reweaving the World*, 9.

58. Macy, "The Greening of the Self," *Common Boundary*, July/August 1990, 22–25. Macy's eco-mentor, Norwegian philosopher Arne Naess, coined the term *deep ecology*. She quotes him in clear agreement: "Unfortunately, the extensive moralizing within the ecological movement has given the public the false impression that they are being asked to make a sacrifice—to show more responsibility, more concern, and a nicer moral standard. But all of that would flow naturally and easily if the self were widened and deepened so that the protection of nature were felt and perceived as protection of our very selves." Commenting on this viewpoint, Macy says, "It would not occur to me to plead with you not to saw off your leg. It wouldn't occur to me because your leg is a part of your body. Well, so are the trees in the Amazon rain forest. They are our external lungs. We are beginning to realize that the world is our body." Ibid., 22.

59. Ibid. See also Macy's "Awakening to the Ecological Self," in *Healing the Wounds: The Promise of Ecofeminism*, ed. Judith Plant (Philadelphia: New Society Publishers, 1989), 201–11.

60. New York: Harrington Park Press, 1995.

61. Ibid., 22–23.

62. Quoted by Kirkpatrick Sale, *Dwellers in the Land: The Bioregional Vision* (Philadelphia: New Society Publishers, 1991), 7.

63. Personal communication with the author, October 1994.

64. "Keepers of the Earth," in *Ecopsychology*, 324.

65. Albuquerque: University of New Mexico Press, 1976. When published it was thought to be remembrances of Carter's life with his Eastern Cherokee hill country grandparents during the Great Depression years. It is now known to be an insightful work of fiction.

66. This and the following quotes are from *The Education of Little Tree*, 58, 59, 61, 62.

67. *Earth in the Balance*, 259.

68. This account is from David Roberts "Geronimo," *National Geographic* 182, no. 4 (October 1992): 59, 71.

69. "Wings of the Eagle, A Conversation with Marie Wilson," in *Healing the Wounds*, 212.

70. Paul Shepard, *Thinking Animals: Animals and the Development of Human Intelligence* (New York: Viking Press, 1978), 210.

71. Ibid., 253.

72. See Judith Plant, "Searching for Common Ground: Ecofeminism and Bioregionalism," in *Reweaving the World*, 158.

73. Philadelphia: New Society Publishers, 1991, 42.

74. Ibid., 50.

75. Plant, "Searching for Common Ground: Ecofeminism and Bioregionalism," in *Reweaving the World*, 160–61.

3

Principles of Healing Persons by Healing the Earth: The Ecotherapy-Ecoeducation Model

Nothing is more practical than a good theory.
—The author's first philosophy professor

The tools of psychotherapy are similar to those of gardening. Fertilizing and weeding, raking and hoeing, digging and mulching can't make people grow, but they can assist nature's growing and healing process. . . By tending inner gardens we help people nourish the soil within their selves and grow into whole person.
—Leslie Irene Shore, *Tending Inner Gardens, The Healing Art of Feminist Psychotherapy*

The purpose of this chapter is to spell out the objectives and working principles of the ecotherapy-ecoeducation model. This model implements the ecological circle (chapter 1) and the earth-grounded understanding of personality (chapter 2), its theoretical foundation. The chapter will also describe six transforming perspectives on human-earth relations that ecotherapists and ecoeducators should keep in mind and help others adopt.

The Model's Objectives

The ecotherapy-ecoeducational model aims at enabling people to gain the insights, motivation, and skills to experience the transforming energy of all three dimensions of the ecological circle by:

1. facilitating the healing of alienation from the earth and enhancing

openness to being nurtured by nature more intentionally and often, in loving, respectful interaction with the natural world;

2. enabling people to become more aware of the self-transcending or spiritual dimension in their experiences of nature; and

3. motivating them to learn how to adopt more earth-caring lifestyles and behaviors that will help save the biosphere.

Psychologist Sarah Conn states well the objectives of ecologically aware therapies, including ecotherapy:

> The challenge of ecologically responsible psychotherapy is to develop ways to work with the "purely personal" problems brought by clients so that they can be seen not only as unique expressions but also as microcosms of the larger whole, of what is happening in the world. The goals of therapy then include not only the ability to find joy in the world, but also to hear the Earth speaking in one's own suffering, to participate in and contribute to the healing of the planet by finding one's niche in the Earth's living system and occupying it actively.[1]

Ecotherapy and ecoeducation are complementary healing and growth processes that occupy partially overlapping positions on the same illness-wellness continuum, a continuum representing degrees of alienation from or positive bonding with nature. Ecotherapy is essentially a healing process, taking place somewhere within the half of the continuum representing degrees of illness, dysfunction, and alienation, including alienation from the earth. Ecoeducation, a whole-person growth process, functions at some point within the other half of the continuum with persons whose ecobonding outweighs their ecoalienation. The same principles and many of the same methods are useful all along the continuum. Healing and growth are not essentially different processes but simply applications in the lives of persons with different ratios of inner bonding and alienation with nature.

Often both are used at different points in the same healing-teaching processes, whatever the primary agenda. Healing alienation from nature by ecotherapy usually includes the use of what I have called educative counseling. This means providing personalized information that is directly relevant to an individual's particular ecological issues. Conversely, in learning settings of many types, ecoeducation usually includes the use of ecotherapeutic healing methods at points in the learning process where there is evidence of ecoalienation.

Of course, the ecological dimension of therapy and education is only one focus of concern for healing and growth, but one that is important and often neglected. This dimension must be integrated in appropriate ways with the multifaceted dimensions of holistic therapy and education.

Working Principles of Ecotherapy and Ecoeducation

The following basic principles guide the process of enabling clients or students to move toward the above objectives. Ecotherapy and ecoeducation utilize a variety of interventions designed to facilitate healing and growth in each of the three interdependent dimensions of the ecological circle. These interventions seek to enable persons to enlarge and enhance their body-mind-spirit selves in three ways: (1) by becoming more fully, intentionally, and regularly nurtured by nature; (2) by becoming more aware of the larger meanings of their place in nature and the universe (ecological spirituality); and (3) by becoming more involved in nurturing nature by active earth-caring. This triple focus can bring reciprocal healing benefits to persons and the quality of their human relationships, and to the natural environments around them.

A frequent sequence of interventions begins by focusing on helping people open themselves to the healing, creativity-enhancing energies of the natural world, and in the process, encouraging their awareness of the wider meaning, mystery, and power in all this, whatever their religious background. This energy and awareness become sources of motivation and power to help people reach out in ways that make small but crucial contributions to the healing of our wounded world. A closer lock at these three interrelated processes is in order:

Enhancing creative bonding with nature. In both the ecotherapeutic side of therapies and the ecoeducational dimension of teaching, interventions are designed to increase people's sense of caring connectedness with nature, and to awaken empathic awareness of the pain of the biosphere. Whereas ecotherapy uses methods designed to overcome internal and external alienation from nature, ecoeducation's methods are designed to strengthen self-care by deepening the nurturing bond with nature. All such methodologies seek to make more available to people's total body-mind-spirit selves the creative energies that flow when we deepen our nurturing connectedness with nature.

Fortunately, enabling people to deepen their sense of connectedness with nature may also help them overcome other dimensions of alienation in their lives. A space industry engineer sought counseling, as he put it, "to get my wife off my back about what she claims is my lack of feelings." With the client's consent, the counselor wisely invited the man's wife to the next session. She expressed deep frustration at the husband's super-rational, low-affect way of relating. She declared: "He spends most of his day at his damn computer terminal. Sometimes I feel like I'm talking to a computer when he comes home! The only live feeling he expresses is anger."

It emerged that the man's painful emotional blunting was rooted in his macho male programming as a little boy, teaching him that only certain "strong" feelings and behaviors are accepted by "real men." These childhood learnings seemed to have been both reflected in and reinforced by his choice of a thing-oriented (as compared with a people-oriented) vocation. A major turning-toward-healing in his therapy occurred when the counselor helped him explore his alienation from the natural world, from his body (the natural world *in* him), and from most of his intense emotions. As he did a deep breathing exercise in one session, long-suppressed tears began to trickle down his cheeks. His bottled-up feelings of painful alienation from his body flowed. These feelings were intertwined dynamically with his feelings that the natural world is something to be tamed, civilized, or bulldozed into its "place." It became clinically evident that his feelings about "mother earth" were paralleled by his feelings that women are threatening and need to be controlled or kept at an emotional distance. The latter feelings impacted his behavior with his wife and other women in his life in negative ways—ways that began to change as his earth feelings changed. As he worked through these painful feelings in therapy, his inner emotional channels gradually became more open. What flowed was grief and anger from his lonely cut-off-ness from himself, people, and the living world, followed by the quiet joy of feeling more alive. Not surprisingly, his sense of positive connection with his wife increased, enhancing both their spiritual and sexual intimacy.

Discovering for oneself the self-transcending or spiritual reality available in nature. If people show any openness to increasing their awareness of the self-transcending meaning of their place in the continuing re-creation of the natural world, encouraging them to explore this further is appropriate. In making such interventions it is crucial to have both awareness of and respect for the individual's idiosyncratic world view and spiritual orientation. The benefits that often accrue when people open themselves to greater ecospiritual awareness include putting their experiences of the natural world into a self-transcending and therefore self-enlarging spiritual context. Doing this enables their experiences of nature to enrich their spiritual lives.

Spiritual, ethical, and meaning-of-life issues often are key considerations in enabling people to develop both more creative connectedness with nature and more earth-caring lifestyles. It is my experience that passionate commitments to saving the earth flow from religious or spiritual sources in many people. And, because spiritual and value pathologies usually are among the root causes of alienation from nature (as well as alienation from people), some focus on these is an essential aspect of ecoeducation and ecotherapy.

Reaching out to nurture nature more caringly. It is valuable for people's wellness for them to learn ways to be nurtured by nature more fully and frequently. The healing, growth-enabling effects of this are increased if people take a valuable next step: channeling some of the energy derived from nature's nurturance, including spiritual nurture, into action aimed at helping heal the interrelated human and natural environments.

It is crucial, therefore, to encourage students, family members, and therapy clients to complete the ecological healing circle by respecting, protecting, and nurturing nature actively in love, and to do so with other people. This will enhance their experiences in the circle's other two dimensions and may also enhance relationships with friends and family. It may also bring the satisfactions of knowing that together they are making small but significant contributions to people-caring through earth-caring. Nurturing nature includes working to heal and prevent further degradation of the natural world, beginning with living a more ecologically caring lifestyle. Of course, the nature and timing of all the above therapeutic or educational intervention depend on the dynamics and flow of particular relationships.

Ecological peacemaking innovator Joanna Macy describes the synergistic interaction of receiving care from the earth and giving care back to the earth:

> Because the relationship between self and world is reciprocal, it is not a question of first getting enlightened . . . and then acting. As we work to heal the Earth, the Earth heals us. No need to wait. As we care enough to take risks we loosen the grip of ego and begin to come home to our true nature. For, in the co-arising nature of things, the world itself, if we are bold to love it, acts through us.[2]

Ecophilosopher Theodore Roszak articulates a comparable theme: "Ecopsychology holds that there is a synergistic interplay between planetary and personal well being. . . . The needs of the planet are the needs of the person, the rights of the person are the rights of the planet." [3]

Because of this synergism, encouraging people to become involved in working with others to heal the planet is an essential part of ecological therapy and education. Such action is really enlightened self-interest because it helps to heal ourselves and those we love. Such outreach may be difficult at times, but it is definitely good for one's health. As the holistic artist Sister Corita once affirmed, "To be fully alive is to work for the common good!"

Essayist-novelist-poet Alice Walker has shared this astute observation about her philosophy: "My activism pays the rent on being alive and being here on the planet. If I weren't active politically, I would feel as if I were sitting back eating at the banquet without washing the dishes or preparing the food." [4]

Awakening to the Social Context of Personal Pain

A closely related working principle is that awakening people to the societal and environmental causes of their individual or family pain may contribute to their growth and healing. The personal issues and problems that motivate people to seek therapy or remedial education usually have important societal roots in the political, economic, institutional, spiritual, or environmental contexts in which they live. Occasions when it is appropriate to raise contextual issues occur much more frequently in education than in therapy, of course. But a therapist's asking about a relevant contextual issue may be illuminating and helpful at times in that healing process. In any case, ecotherapy and ecoeducation must avoid the cul-de-sac of etiological privatism derived from the fallacy that alienation from nature is exclusively an intrapsychic problem. Intrapsychic problems usually are the places to begin but not to end in psychotherapy, especially when interpersonal, social, and environmental issues clearly are involved.

It is important for therapists and teachers to know, though they should mention this in their work only when it is relevant, that the most vulnerable targets of violence against nature as well as themselves are the persons socially defined as less valuable, or less powerful. Such targets include children, the poor, ethnic minorities, women, the disabled, and homosexuals—and also nature itself. As people become motivated to adopt wellness lifestyles, they should be encouraged to do whatever they can to confront and change the social causes of the violence that spawns ecological as well as human brokenness. These social pathologies include poverty and economic oppression, extreme wealth, sexism, classism, militarism, consumerism, age-ism, pathogenic religious and ethical movements, racism (including environmental racism),[5] and species-ism (the belief that nonhuman species exist only to be used to satisfy human needs and desires). All these institutionalized belief structures and societal malignancies feed the global trashing of the planet.

Those who are oppressed and in inferior social and power positions in any society are much more likely to be victims of environmental pollution than are those in a position of superior power and prestige. For example, children who grow up in the toxic slums that surround many of the planet's megacities, are victims of sociosystemic ecodeprivation. Compared to those living in relative affluence, they usually must surmount greater obstacles to achieve a healthy relationship with nature. The vast majority of persons in the underclasses of poorer nations are highly vulnerable to both ecological poisoning and earth alienation. This is because their whole social context suffers economic exploitation and injustice at the hands of both the economically privileged and the power privileged class, often including the military. Furthermore, poor nations often suffer from economic violence inflicted by affluent nations that maintain their advantage by exploiting the poorer nations. The debt trap in which Brazil is caught (described in chapter 1) is an example of a socioeconomic and political context issue that impacts the ecological health of millions of people in that nation. The crucial importance of correcting injustices in socioeconomic and political systems is also evident in the discussion of ecofeminist and Native American issues in later chapters.

Here is a striking example of how feelings about social and political context issues impact our health. It also illustrates the debilitating impact of violence and the threat of violence on our health. Norman Cousins, the late, great editor and author, performed a fascinating experiment on himself to demonstrate that earth-caring, peacemaking, and holistic health care are interdependent. He did this during the height of the frighteningly insecure "security" provided by the cold war's MAD (Mutual Assured Destruction) policies of the two superpowers. At that time, he was adjunct professor on the medical school faculty at UCLA. Cousins asked the lab there to take a blood sample and test it to establish a base line for how effectively his immune system was functioning. Then, for a few minutes, he allowed himself the luxury of imagining that one of the superpowers (the USA) had revised its basic policies toward the other to make these truly rational and constructive. A second blood sample was taken and tested after his imaging. The lab reported that the various types of body-protecting cells of his immune system had increased an average of 53 percent. Even this brief imaging of a desirable fantasy had produced dramatic strengthening of his immune system.[6]

To experience this experiment, put it in the current global context. Close your eyes and imagine that the nations of the world have honored the hopeful commitments to saving a viable planet they made at the Earth Summit in Rio, the subsequent Cairo conference on sustainable development and population, and the Beijing conference on the full equality of women. Picture the nations implementing mutual commitments to eradicate international economic injustices and thereby reducing disease and poverty to a minimum, and to provide adequate education, housing, and medical care. They have been cooperating in an effective global program to control the devastating population explosion and end the oppression of women. They also have been working together to clean up the tragic pollution in both the developing and the developed countries. To accomplish all this, the nations decided to stop wasting billions on preparations for war. Instead, for five years they have invested these huge peace dividends in worldwide programs to meet a wide range of human and environmental needs. This effort has been coordinated by a greatly strengthened United Nations with a genuine international police force to maintain real "national security" for all countries. Let yourself enjoy picturing in your mind the wonderfully healthy condition for all the world's children and the flowering of the biosphere. Experience how it feels in your body-mind-spirit to be living in a sparkling and clean healthy, nonviolent world where basic human needs are well met, in your own country and all around the world. Remember that converting the billions spent on weapons of killing could much more than cover the costs of creating such ecosocial environments for the whole human family.

If ecotherapy and ecoeducation are to generate whole-earth caring based on earth literacy, it is important to enable learners to understand the environmental crisis on two interdependent levels, personal and societal. The countless personal decisions people make day by day determine whether their lifestyles are part of their world's ecoproblem or part of its solution. It is important to learn that many everyday actions can help in small ways to save a livable planet. Although personal actions by more and more people are invaluable, by themselves they are not enough. We also need to equip learners of all ages to challenge and work with others to change the dominant socioeconomic and political systems that are the root causes of planet-damaging individual and group behaviors. This dimension of ecoeducation includes highlighting the vital role of developing sound public policy, economic strategies, and legislation to reduce the prevalence of environmental illness.[7]

Earth literacy education must make economic justice a robust emphasis. Without this emphasis it may be critiqued as a middle-class concern and perhaps luxury. Justice-based ecoeducation must address the special problems and responsibilities of individuals, groups, and nations that are poor as well as those that are more affluent. The British rock musician Sting has done many benefit concerts for environmental and justice causes in numerous countries. He declares, "If I were a Brazilian without land or money or the means to feed my children, I would be burning the rain forest too."[8]

Al Gore highlights the justice-ecology connection insightfully:

> In today's world, the link between social justice and environmental degradation can be seen everywhere: the placement of toxic dumps in poor neighborhoods, the devastation of indigenous peoples and the extinction of their cultures when the rain forests are destroyed, a disproportionate level of lead and toxic air pollution in inner-city ghettos, the corruption of many government officials by people who seek to profit from the unsustainable exploitation of resources.[9]

As American playwright Paul Harrison states: "The poor tread lightest on the earth. The higher our income, the more resources we control and the more havoc we wreak."[10]

In ecoeducation it is important to mention examples of how ecoviolence against the earth is interrelated with other types of human violence in the global pandemic of violence. The dynamics of intrapersonal, interpersonal, intergroup, and ecological violence are intertwined, flowing from some of the same causes. Reducing violence against humans, on the one hand, and reducing violence against animals, plants, and the environment, on the other, are both essential for saving a healthful earth. Ecofeminist poet and writer Betty Roszak observes that the gender stereotypes that are deep in our society are at the roots of the "manhandling" of both women and nature.[11]

An encouraging sign of growing contextualizing (deprivatizing) of professional therapy and education is the multiplicity of professional social responsibility groups. Teachers, psychologists, psychotherapists, physicians, social workers, pastoral counselors, and many other professionals have formed such groups to focus on social context issues.[12] The needed contextualizing and politicizing of privatized counseling, therapy, medical healing, and education is continuing, as evidenced by these movements and by a wealth of recent writing in the journals of these fields. All these developments reflect efforts to join personal and social healing and growth.

Balancing the Creative and Destructive Sides of Nature and Life

Ecotherapy and ecoeducation must deal realistically with the darkness as well as the light, the violence as well as the serenity, the death as well as the life, found everywhere in nature. The healing energy and nurturance of growth in nature are found in the balance and interaction of these two sides of reality. To ignore the darkness is to sentimentalize nature and to miss the ways in which healing occurs in the darkness as well as in the light. People who are struggling in the shadows of personal crises need to discover that healing is available in their darkness as well as in the light of the interpersonal and natural world.

I remember various recovering alcoholics, at open meetings of Alcoholics Anonymous in its early years, saying, in effect:, "I'm thankful I'm an alcoholic because it has forced me to learn what life is really about." In the same vein I have been touched by clients saying things like, "I feel awful about what has happened in this terrible crisis. But life has taught me some valuable things I sure didn't want to learn." I think of a plant in our yard that seems like a botanical miracle—the night-blooming cereus, whose gorgeous, lilylike flowers open at night in the late summer and fall.

I know well, as you probably do, that human life, like the natural world of which we are an integral part, includes suffering and struggle and eventual death. But, as nature illustrates so dramatically, the pain is wondrously intermingled with joy, the suffering with celebration of the good gift of being alive. Wendell Berry, the earthy farmer-poet with deep roots in his rural Kentucky, acknowledges that the struggle for survival in nature is often violent. But, "it is not all fear and flight, pursuit and killing. This is part of it, certainly; and there is cold and hunger; there is the likelihood that death, when it comes, will be violent."

Berry goes on to illustrate the other side of nature's destructiveness and death with a delightful story:

> I sat one summer evening and watched a great blue heron make his descent from the top of the hill into the valley. He came down at a measured pace, stately as always, like a dignitary going down a stair. And then, at a point I judged to be midway over the river, without at all varying his wing beat he did

a backward turn in the air, a loop-the-loop. It could only have been a gesture of pure exuberance, of joy—speaking of his sense of the evening, the day's fulfillment, his descent homeward. He made just the one slow turn, and then flew on out of sight in the direction of the slew farther down in the bottom. The movement was incredibly beautiful, at once exultant and stately, a benediction on the evening and on the river and on me. It seemed so perfectly to confirm the presence of a free nonhuman joy in the world—a joy I feel a great need to believe in—that I had the skeptic's impulse to doubt that I had seen it. If I had, I thought, it would be a sign of the presence of something heavenly in earth. And then, one evening a year later I saw it again.[13]

Modifying Attitudes, Feelings, and Memories

Ecotherapy and ecoeducation seek to modify both conscious and unconscious attitudes, feelings, and memories that influence the dynamics of people's relationships with the natural world. This is accomplished by utilizing both rational, cognitive, left-brain methods and intuitive, holistic right-brain methods.[14] Ecotherapy often deals with painful childhood alienating-from-nature experiences that are no longer remembered. They have been repressed into closed rooms in the unconscious mind, called metaphorically the basement of consciousness. But, as Carl Jung and some other psychotherapeutic pioneers made clear, the unconscious mind also has rich hidden resources that need to be reclaimed in the process of healing and growth. Ecotherapy draws on these creative resources and problem-solving capacities of the deeper unconscious levels of the psyche.

Theodore Roszak's concept of the ecological unconscious is much like my understanding of the deeper mind. Starting with the premise that the ecological unconscious is the core of the mind, Roszak states: "The goal of ecopsychology is to awaken the inherent sense of environmental reciprocity that lies within the ecological unconscious. . . . The ecological unconscious is regenerated, as if it were a gift, in the newborn's enchanted sense of the world." Ecopsychology (and also ecotherapy) seeks to help adults recover this gift as a resource for developing what Roszak calls the "ecological ego."[15] I would add that a strong ecological ego is formed as the wilderness within that Freud called the id is integrated with the conscious mind's multiple capacities for relating intentionally with present realities.

People's wounded childhood feelings, attitudes, values, and experiences of nature and of their bodies sometimes are among the root causes of their overall sense of alienation. Both ecotherapy and ecoeducation therefore utilize methods designed to deepen body-mind-earth bonding by healing the inner child who is an enlivening part of our deeper selves. As in all holistic therapy and teaching, it is essential to balance and blend cognitive, analytic, logical, "scientific" left-brain approaches with a variety of right-brain approaches such as guided imaging, dreamwork, and projective drawing or painting. Right-brain methods provide major access paths to enabling people to recover both the pain and the positive resources of the unconscious mind.

Love, Hope, and Laughter for Transformation

Ecotherapy and ecoeducation recognize that love, hope, and laughter are key energies for empowering ecological healing and learning. Fear and guilt often are appropriate responses to the environmental crisis, but they should play a limited role in motivating basic ecological change The most effective source of motivation and energy for facilitating ecological change in oneself and others does not come from these painful emotions. Rather it is from one's love of the earth, from generating reality-based hope, and savoring the satisfactions of deepening one's relationship with the earth and being nurtured by it. These benefits provide positive reinforcement of one's ongoing earth-caring.

Wise love built on a foundation of reality-based hope and laughter is an energizing resource for change in both ecotherapy and ecoeducation. Such love has many appropriate objects, including love of ourselves, of our children, of the children of humankind and of other species, the biosphere, and for the spiritually aware, the Creator-Spirit of this incredibly wonder-full world. To express holistic love in today's world often requires courage. As many of the present and past ecopioneers demonstrate, only such courageous and passionate love can provide ongoing energy sufficient to empower the difficult changes in personal lifestyles and social policies that are required in our present ecocrisis.[16]

To love ourselves and others fully involves loving in widening circles of concern leading to becoming lovers of the planet. The widening circles of our loving concern, reaching out to the whole human family and the entire biosphere, can link us as de facto allies in our communities and also to peoples in other lands with very different backgrounds who also care deeply about Mother-Father Earth.

Earth-healing love, like all genuine love, is love-with-muscles—often called "tough love." This understanding resonates with Erich Fromm's definition. In his classic book *The Art of Loving*, he wrote that love is "active concern for the life and growth of that which we love."[17] This concept challenges the superficial, narcissistic understandings of love rife in our love-hungry society. American poet Angela Morgan, early in the twentieth century, highlighted this unsentimental understanding of healing love: "Love is not alone for pleasure, love is not alone for bliss, love is for the rousing of the nations, the healing of the world!"[18] Healing persons by healing the world is what ecotherapy is all about.

What is the task of all of us who love people and the planet? It is to give our passions arms and legs and a voice. This means expressing the love in down-to-earth environmental, justice-generating, and peacemaking work.

The tough challenge is to do this on as many levels of our lives as possible, beginning with small earth-friendly changes in our personal and family lifestyles. This should be followed by doing everything we can to reach out to our schools, congregations, and neighborhoods, as well as partnering with other groups to enhance the environmental health of our hometowns, nation, and world.

Why is hope so vital in ecotherapy? In a "Despair and Empowerment" workshop I once co-led, a deeply concerned graduate student voiced feelings about the nuclear crisis that are comparable to the feelings many people struggle with today in the earth's crisis. Exploding with frustration, having become acutely aware of the reality-based terror of the nuclear crisis, he cried out, "I feel so damn helpless! What can I do? What can any one person do about problems so gigantic and overwhelming?" This frustrated student seemed to be hovering on the brink of radical futurelessness, a paralyzing mind-set devoid of hope.

To become aware of the depth and complexity of the planet's violence and agony can cause earth-caring people to feel pulled down into a vortex of helplessness. It is tempting, though ultimately futile, to try to defend ourselves against the pain of feeling helpless by fantasies of denial. A prevalent denial fantasy that is temporarily comforting but ultimately dangerous is: "They [meaning the government, or religious leaders, or God] will prevent the unthinkable from happening to our precious planet home." Such fallacious hopes cut the nerve of committed, responsible action.

A therapist's or educator's task is to be an intentional awakener of reality-based hope that motivates action. Only such hope can enable people to respond constructively to reality-fed fear. This kind of hope, in sharp contrast to superficial optimism, includes a clear awareness of the threats to our only home, the earth, if collectively we stay on the present environmental trajectory. This hope is not dichotomized, in that it rejects the false dualisms that are deeply embedded in our culture's patterns of coping with the agony and the ecstasy of life. It recognizes that despair and hope, weakness and strength, darkness and light are complementary energies from the same paradoxical reality. Reality-based hope is not the opposite or the absence of appropriate fear. It is hope that integrates the energy of this fear in a fresh synthesis that brings new strength to the person. Like tough love, this is *tough hope*. It is hope with muscles. It is hope grounded in a transcending spiritual reality that is its ultimate wellspring.

Reality-based hope includes awareness of the tenacity of human resistance to making necessary changes to save the earth and ourselves from such a fate. It is hope grounded in some understanding of the steps required to overcome this resistance so that we can move toward helping to resolve the ecological crisis. It is also hope rooted in commitment—to put the ounces of our influence into collaborating with others to heal the living planet's wounds and enhance her health. It is hope that is grounded in awareness of nature's incredible regenerative capacities, tempered by the recognition that these capacities are already beginning to be overtaxed. It is hope that gains strength from at least a partial image of a better future toward which one is drawn to take small steps. And, very important, it is hope rooted in an awareness of the divine Spirit's commitment to life in all its fullness for the planet and all of its inhabitants.

At times, when I look into the complex depths of the planetary environmental crisis, my characteristic optimism about the future wavers. I feel like the young man in the workshop mentioned above. At such times I am tempted to reverse the title of Sigmund Freud's classic critique of religion, changing it to "the illusion of a future." But then the energizing energy of hope begins to trickle back within me, probably from sources in my spiritual life that can help pull a person out of the pit of despair.

Is it realistic to hope that profound ecological changes in the psychology and philosophy of human healing and growth will be accepted widely in established professional guilds with so many basic assumptions that ignore this dimension? Undoubtedly it will be an uphill struggle. But the remarkable greening of the thinking of growing numbers of us in healing and teaching is a hopeful sign that the climate in these fields is gradually changing.

Al Gore articulates a prophetic truth about the faith foundation of hope for a viable future:

> For civilization as a whole, the faith that is so essential to restore the balance now missing in our relationship to the earth is the faith that we do have a future. We can believe in that future and work to achieve it and preserve it, or we can walk blindly on, behaving as if one day there will be no children to inherit our legacy. The choice is ours; the earth is in the balance.[19]

Playfulness and humor are valuable but often neglected resources in many kinds of healing and growth work.[20] In ecological education and therapy they are of particular value. These bumper stickers illustrate what I have in mind:

HONK IF YOU DON'T LIKE NOISE POLLUTION!

STOP TREATING OUR SOIL LIKE DIRT!

Some may feel that the global ecology crisis is too serious to be the subject of lightheartedness. In my view, the more desperate a situation is, the more important it is for people to have frequent mini-vacations of laughing with others and laughing at the absurdities in the situation. An example of these absurdities was recently reported on world news from Beijing. In that increasingly polluted mega-city, young professionals can now unwind at the end of a hard day by stopping at an "oxygen bar." There they can inhale fresh air for around $6 an hour.

Tears and laughter are not opposites but complementary aspects of the ability to care. To laugh and to cry about the planet's promise and pain are appropriate responses from the same energy source. This source should be written as Source. It is the transcending and undergirding ecological spirituality that is the heart of the ecological circle.

Thomas, the wise old man in Brian Swimme's *The Universe Is a Green Dragon*, puts playfulness in its cosmic context. He declares that the universe insists that we become adventurously playful the way all life forms are. The healthy development of the earth depends on our species celebrating our unique capacities for "play, fantasy, the imagination, and free exploration of possibilities: these are the central powers of the human person."

Thomas then lifts up the transforming power of relating intimately with nature: "When you have been in the forest for a while, deepening your sensitivity to the forest's presence, the richness concentrated in you cannot be contained. You radiate forest wherever you go, whether or not you say a single word." Returning to playfulness, he adds: "Every song has tremendous value! Learn to sing, learn to see your life and work as a song by the universe. Dance! See your most ordinary activities as the dance of the galaxies and all living things."[21]

In *The Universe Story*, Brian Swimme and Thomas Berry (the real Thomas) spell out the unique role of humans in celebrating the joy of being:

> Within the context of celebration we find ourselves, the human component of this celebratory community. Our own role is to enable this entire community to reflect on and to celebrate itself and its deepest mystery in a special mode of conscious self-awareness. In the various stages of the human, from its earliest tribal form throughout the shaping of the more complex civilization, this celebratory experience is consistently associated with the sense of the sacred.[22]

Many of us have difficulty "keeping on keeping on" in the uphill struggle of work for social change, even though we believe deeply in the need for change. For this reason, ecojustice and ecological action burnout is widespread. Incorporating playfulness, laughter, and a spirit of adventure in education, therapy, and living increases stress-reducing protection against all types of burnout including ecology action burnout. And, most important, taking time to reconnect with the natural environment for regular refueling experiences is a key to avoiding this burnout.

People suffering from ecoalienation often have a diminished sense of healthy playfulness. Their "recreation" does not really re-create their mind-body-spirits. Alienation from the earthiness of childhood is linked with what transactional analysis describes as alienation from the natural inner child, a valuable presence within all of us throughout adulthood. Finding the lost inner child often helps one rediscover and befriend the lost earth connection, including the healthy wildness associated with certain stages of childhood. When this happens people's wellness can be enhanced by the flow of creativity, sensuality, playfulness, and sexual passion along with passion about many other things. Dull sex regains some of the fire of passionate, earthy playfulness.

Earth-Caring Includes Peacemaking

In ecoeducation and ecotherapy, holistic earth-caring must include peacemaking as well as ecology action. There is an obvious and urgent need for effective peacemaking in today's violent world. On both the levels of individual and social problems, peacemaking involves teaching skills for nonviolent conflict resolution, fair compromising, and bridge-building by increasing respect for differences. Most counselors and many educators are trained in these skills. Many of the skills can be applied to the process of understanding and preventing violence against nature.

A personal earth-caring lifestyle is a vital part of holistic peacemaking. Emphasizing the interdependence of ecological and peace issues is a promising way to reenergize peace movements, which are suffering from low levels of support currently resulting from very premature relaxation of peacemaking concerns with the ending of the cold war. Humankind certainly can rejoice that everyone on planet earth can breathe much easier because the insanity of the superpowers' nuclear confrontation has ended. But the threat of a nuclear disaster, either accidental or intentional, is still enormous. The scientific know-how to create "the bomb" has been acquired by a wide variety of nations, some with unstable leadership, and it is increasingly possible that terrorists will acquire this capacity.[23] Furthermore, the threat to the health of the biosphere from chemical and biological weapons of mass destruction is still widespread.

Quite apart from the nuclear threat, the continuing global violation of the earth by overpopulation, widespread species extinction, incredible earth-damage from massive military spending, rainforest devastation, mounting toxic wastes (including radioactive wastes), thinning of the protective ozone layer, pollution of the air, soil, rivers, lakes, and oceans—all these and many more earth atrocities are continuing. Together these are simply more gradual but nonetheless deadly forms of violent geocide.

Starting with Ourselves

In ecotherapy and ecoeducation, we who seek to encourage earth-caring in others need to begin by focusing on our own ecoalienation or ecobonding and on our own earth-hurting or earth-caring lifestyles. Some of our most healing and educational influences flow in subtle ways from our own attitudes, beliefs, and, most important, our behavior. To enhance effectiveness in ecoeducation and ecotherapy, therapists, teachers, healers, and parents need to begin by deepening their own relationship with the natural world, while examining their lifestyles to see if they express or contradict ecofriendly living. To the degree that these issues have been neglected, effectiveness in guiding others to experience ecohealing and ecobonding will be diminished. Furthermore, living a more earthcaring lifestyle is a precious gift we can give to our students, clients, patients, and families—and also to ourselves.

People's relationships with the natural environment always interrelate with both their human environment and their inner or intrapsychic environment. The principle of starting with ourselves, therefore, may involve dealing with alienation within ourselves or in our relationships. Clearing up emotional pollution in our internal world can release energy for bonding and mutual caring with the people close to us as well as with that little piece of the natural world in which we live. [Better self-care can enhance our ability to care for the earth.]

Six Transforming Perspectives for Ecotherapy and Ecoeducation

How people actually relate to the earth—the degree to which they live in earth-caring or earth-depleting ways—is influenced profoundly by their heart-level understanding of their relationship with the earth. Their human-earth perspectives influence deeply their self-definition, their consciousness, and their consciences regarding the earth. Therefore, fostering certain perspectives in ourselves and others is crucial to the overall effectiveness of the ecological dimension of counseling and education. In my experience, six interdependent ways of seeing our species' relation to the biosphere tend to generate environmentally constructive behavior.

Each of these interdependent perspectives is a variation on one common motif: *enhancing our species' sense of organic interrelationship and interdependence with the rest of the network of living things called the biosphere*. In the words of John Muir, inspirer and grandfather of the Sierra Club: "When we try to pick out something by itself, we find it hitched to everything else in the universe." To the degree that this fundamental principle is accepted by increasing numbers of individuals and groups, and is lived out in policies and practices, violence against humans and against nature will be reduced and the health of the biosphere helped to flower. As this happens, the earth's chances of survival as a life-nurturing place for future generations will be dramatically improved. Here is a description of the six transforming perspectives:

1. The View-from-the-Moon Perspective

As I write, I see above my desk a large photograph that many people (including myself) cherish. The photo was taken from the moon on July 20, 1969, showing an "earthrise." With the moon's barren, lifeless surface in the foreground, Spaceship Earth, our planet home, is a small gray-green ball 186,000 miles in the distance. The light of sunrise is creeping across its verdant surface. If you are more than thirty years of age, you (along with some 600 million avid viewers worldwide) undoubtedly watched such spectacular TV shots from the moon. The human family marveled at the incredible events—humans walking on the moon as we watched in our living rooms.

These pictures brought a dramatic new perspective to human consciousness worldwide. We saw the earth as one small, living globe with only a thin, vulnerable atmosphere of life-giving air. It is not coincidental that the environmental movement surged forward soon thereafter. The first Earth Day in North America, for example, was celebrated only a few months later. Twenty-five years after the first moon walk, 66 percent of Americans who were surveyed said that the moon landing was one of humankind's most significant achievements and adventures.

The late eminent physician-biologist Lewis Thomas gives this lyrical response to seeing the first earth photographs from the moon:

> Viewed from the distance of the moon, the astonishing thing about the earth, catching the breath, is that it is alive. The photographs show the dry, pounded surface of the moon in the foreground, dead as an old bone. Aloft, floating green beneath the moist, gleaming membrane of bright blue sky, is the rising earth, the only exuberant thing in this part of the cosmos. If you could look long enough, you would see the swirling of the great drifts of white clouds, covering and uncovering half-hidden masses of land. If you had been looking for a very long geologic time, you could have seen the continents themselves in motion drifting apart on their crystal plates, held afloat by the fire beneath. It has the organized, self-contained look of a live creature.[24]

The perspective from the moon can have a dramatic effect on how we humans see our troubled world. All the borders, boundaries, and barriers created by our species are invisible—except the Great Wall of China. On a study tour of the People's Republic of China, we learned that this wall was constructed as a defense, at tremendous cost of human life, between the fifth and third centuries B.C.E. As a defense it was a total failure. Perhaps this can be a healthy reminder that the walls we humans build and the boundary lines we have drawn—and waste countless young lives trying to defend or expand—disappear when the earth is viewed from a lunar perspective. It is also perspective-giving to be aware that all the buildings that reflect the power of left-brain achievements through the centuries also are not visible from a moon's-eye viewpoint.

This visual image confirms powerfully the impassioned declaration of scientists and peacemakers during much of the twentieth century that "It's one world or none!" In the long run, the only way to protect the well-being of individuals, families, groups, or nations on our tiny, living planet is to protect the well-being of this whole life-sustaining biosphere. Our Spaceship Earth is the smallest unit whose well-being can be protected and sustained. This is the bottom-line guideline for reinventing obsolete understandings of "national security" and "patriotism."

2. A Transgenerational Well-Being Perspective

If you are fifty-something or beyond, you may have chuckled at the bumper sticker "WE'RE SPENDING OUR CHILDREN'S INHERITANCE." In today's world, the following is an appropriate rewording: WE'RE SPENDING THE GLOBAL RESOURCE INHERITANCE OF A THOUSAND UNBORN GENERATIONS. Our species today is squandering our children's precious birthright. Consider the greedy ways we are consuming the earth's irreplaceable natural resources; the destruction of food-producing topsoil around the world; the dependence of our advertising-fueled economic systems on ever-increasing production, consumption, and obsolescence; the earth-destroying efforts of impoverished societies to follow the deadly example of the shopping-addicted, affluent societies. And certainly our generation is writing enormous checks on future generations' resources by accumulating astronomical national debts.

Environmentally concerned people frequently quote this wise Great Law of the Six Nations Iroquois Confederacy of Native Americans: "In our every deliberation, we must consider the impact of our decisions on the next seven generations."[25] But today, we must extend responsibility for the future far beyond seven generations. Consider the long-term environmental dangers of the tens of thousands of tons of radioactive wastes that have accumulated from cold war "nuclearism" and nuclear generators. Nuclear scientists have been warning us for years that these lethal wastes will be emitting potentially damaging radiation for at least half a million years, and that a major mishandling could threaten the gene pool on which all the earth's unborn generations depend. In light of such a reality, our personal and social decisions should take account of a much longer time-line. We must ask what the impact of today's decisions will be on the kind of world we are leaving for *all future generations*.

Astrophysicist Carl Sagan and his scientist partner Ann Druyan articulate well the vital need for transgenerational ethics:

> It is no longer enough to love, feed, shelter, clothe and educate a child—not when the future itself is in danger. . . We have been treating the environment as if there were no tomorrow—as if there will be no new generations to be sus-

> tained by the bounties of the earth. But they, and we, must drink the water and breathe the air. They and we are vulnerable to deadly waste, ultraviolet light and climatic disasters.[26]

I offer a personal reflection about a memorable day not long ago. I will describe it in gestalt therapy style by reliving it in the present tense:

> *I am climbing Kyexyong-San, a sacred mountain in South Korea, to celebrate my birthday with three Korean friends. The wildflowers are flashing their splendor as we hike through the meadow. We are walking through the woods on the mountain's flanks as a laughing brook serenades us and unfamiliar birdsongs float on the morning breeze. Now we're eating our lunches as we sit on the summit. The view is spectacular. Now, we are on our way down. We stop at a holy shrine and chat with the local native healer. We watch the women who are praying nearby. I talk with them and learn that they are praying to Sambul, the healing Earth Mother, for help with a variety of personal problems including sicknesses and business troubles. Now, it is afternoon. We are near the end of our hike and very tired. I ask my Korean friends what that large sign hanging over the trail says. They translate it:* "LET US LOVE NATURE AS WE LOVE OUR CHILDREN."

I reflected on this as we completed our mountain hike. It struck me that it is wise advice for the whole human family and that, in fact, *we cannot love our children fully unless we also learn how to love nature in ways that will leave them a healthy planet*. Implementing the Korean sign over the mountain trail is more important to humankind's well-being today than ever before.

3. The Whole-Biosphere Well-Being Perspective

One species can have optimal health only to the degree that the whole biosphere is made healthier. We are all on the same environmental spaceship whether we like it or not. The widespread fallacious belief that other species exist mainly or solely to serve the needs and desires of the human species is called anthropocentrism or species-ism. Expressing such species narcissism by our behavior actually contradicts the long-range well-being of our own species, because it reduces the life-supporting capacity of the biosphere. Furthermore, as the only species that threatens to destroy the biosphere, humans have the responsibility to do whatever is needed to make sure that the biosphere will flourish.

A brief item in the *Los Angeles Times* stimulated unexpected feelings of grief and anger in me. In a few matter-of-fact lines, it stated, "The last Palos Verdes blue butterfly is dead." The patch of locoweed that was the sole diet and last habitat had been bulldozed to create a baseball field. Puzzled by my wave of intense feelings, I stayed with them seeking their source. Slowly it dawned on me: They flowed from the awareness that my children and grandchildren, and all the other children and grandchildren of planet earth, will never see a Palos Verdes blue butterfly. That little species had just become extinct—forever!

I suspect that comparable feelings are shared, at least subconsciously, by millions of people around the planet, as more and more of us feel the personal grief that the ecological crisis is producing day by day. Pause for a moment and become aware of what it may mean to your children and grandchildren that at least ten species (according to scientific estimates) will became extinct while you are reading this chapter. Consider the estimate that at least seventy-two species will become extinct somewhere on the planet in the next twenty-four hours, and all future twenty-four-hour periods. *And extinction is forever.* It seems fitting to paraphrase a much-quoted line from John Donne: No living creature is an island. When a single thread is ripped from the wonderful web of living beings, the fabric of our individual lives is somehow torn and diminished.[27]

I invite you to reflect on Theodore Roszak's statement of this truth, in the context of mental health:

> In the century since psychology was first staked out as a province of medical science, we have learned a troubling lesson. The sanity that binds us one to another in society is not necessarily the sanity that binds us companionably to the creatures with whom we share the Earth. If we could assume the viewpoint of nonhuman nature, what passes for sane behavior in our social affairs might seem madness. The Earth's cry for rescue is our own cry for a scale and quality of life that will free each of us to become the complete person we were born to be.[28]

"Thomas," the wise old man in *The Universe Is a Green Dragon*, expresses the biosphere perspective when he calls for moving from the species egotism of anthropocentric values and choices (which are destroying the earth) to biocentric choices. This means struggling to adopt the viewpoint of the earth as a whole and letting the earth become our teacher. The hardest struggle is to surrender the dangerous belief that our species is the center of everything. Instead we must adopt a *biocentric* and *cosmocentric* orientation in which the earth's web of life and the universe are understood as the fundamental referents of everything. This transformation of perspectives is mandatory if a healthful earth is to be preserved. Thomas Berry, whose ecophilosophy inspired Swimme's *Green Dragon* tale, voices a similar theme: "Our best procedure might be to consider that we need not a human answer to an earth problem, but an earth answer to an earth problem. The earth will solve its problems, and possibly our own, if we will let the earth function in its own way. We need only listen to what the earth is telling us."[29] Reflecting on efforts to save an endangered species of bird, someone observed astutely that learning to do this may help us discover how to save another endangered species—humans.

4. The Whole-Human-Family Well-Being Perspective

This is a logical extension of the earlier perspectives. Individuals, ethnic groups, regions, nations, and a plethora of self-interest groups view life from a single perspective—what they see as their self-interest, values, and world views. From an ecological viewpoint, this shortsighted perspective does not even serve humans' *real* self-interests because these interests cannot flourish in the long run unless the needs of people and the well being of all lands are considered. Therefore, it is clear that exclusivistic, jingoistic patriotism is no longer healthy (if it ever was) for our interdependent world.

What is required on our shrinking planet is an inclusive understanding of the world that sees the economic and ecological well-being of one's own group in the context of a healthy community, nation, and world. Such inclusive caring about the world need not lessen our love for our own community, homeland, language, ethnic group, or religion. In fact, it can enrich this love by grounding it in the reality that to love one's own place and people *wisely* today is not possible unless one also cares about the planet and all its interdependent peoples.

This reality-grounded perspective is causing some people to redefine the meaning of patriotism. *Planetary survival patriotism* understands that what is good for one's own group, country, or natural environment, but bad for the biosphere and health of others, is bad for one's own group, country, or natural environment in the long run. This inclusive ethical perspective logically leads to doing whatever is needed to help rectify the gross economic injustices and the nonsustainable development practices in poor countries, and the unfair, nonsustainable resource-depleting consumption in rich countries.[30]

The ancient tribal lore of the Chumash people, Native Americans in my part of southern California, includes this wise saying: 'All the world is a canoe, and whether paddlers or passengers, we are all one people together in that vessel." This Chumash wisdom reflects centuries of expertise in constructing large dugout canoes and navigating them between the mainland and the islands that are now the Channel Islands National Park, some twenty-five miles off the coast. Paraphrased to reflect the modern situation, the statement would declare that *humankind is all in the same boat—Spaceship Earth*.

5. The Wise Woman/Wise Man Perspectives

Psychologist Lesley Irene Shore describes insightfully how she has integrated feminist and nature-oriented methods of practicing psychotherapy. She does this in her book *Tending Inner Gardens, The Healing Art of Feminist Psychotherapy*. In discussing the feminine model of the healing relationships, she declares:

> The wise woman tradition is an old tradition. Its roots travel back to at least 25,000 B.C., a time when we paid homage to the immanent Mother Earth goddess and saw ourselves intertwined with nature. . . . Healing was linked with

> the tasks and spirit of motherhood. This nurturing model relied on sensory observation and intuition for gathering information which was handed down from one generation to the next. And although wise women frequently administered herbal remedies, whole natural substances, they also typically remained with their patients during the healing process, touching them, and connecting with them in a variety of ways.[31]

This wise woman heritage is sorely needed today. If a healthy planet-home is to survive, the riches of the earth wisdom of 52 percent of human beings—women—must be prized and used as much as men's earth wisdom. The urgent need is to use more fully the best insights of both women and men in the ecological struggle. The dominance of male understandings and values in traditional psychotherapy (as well as local, national, and international leadership) is comparable to trying to drive an automobile on a crowded freeway with only half of its cylinders firing. Such a dangerous loss of power must be transformed if the unprecedented challenges of the eco-justice crisis are to be met constructively. Insights from ecofeminists, feminist therapists, and feminist theologians are greatly underused in the public sphere. Election of many more ecologically aware women to political positions can enrich the global ecological leadership pool dramatically. The recent voting records of women in the U.S. Congress shows that they are much more consistently pro-environment than their male colleagues.

It is also clear that men's earth-caring and earth-protecting wisdom must be prized. They often control decision-making power on ecological issues. This wisdom is repressed in many males, but earth-loving men like Pierre Teilhard de Chardin, Loren Eisley, John Muir, and Francis of Assisi provide evidence that it is recoverable by loving men.

In both the private and public spheres, ecologically liberated women can challenge men who suffer from male sexist programming that is both self-damaging and planet-damaging. Members of each gender must be the prime agents of their own self-liberation, but this is an important way that liberated women can lubricate the process of male self-liberation. I heartily affirm psychologist Charlotte Kasl's view that men need to look critically at their own internalized oppression caused by being programmed to always try to be tough, independent, domineering, in control, and, when frustrated, to be violent. To be programmed to devalue matters of the heart and the skills of loving violates their own psychological, spiritual, and physical well-being. It also is a major contributor to the need to control, dominate, and exploit the environment.[32] The wisdom of the wise man in men's deeper minds can help them find constructive alternatives.

It is an uphill struggle for most men to gain freedom from social programming that produces behaviors destructive to their health and that of the environment. The pains of male sexism are less apparent and the gains of self-liberation take longer to appear in most men's lives than is true of women. Only when the pains outweigh the gains of traditional sexist programming will either gender be motivated to change. Ecopsychological awareness may be a resource in tipping the scales toward change by men. My desire to change my programming as a man has been augmented by the discovery that my sexist male identity and the resulting behaviors are hurting not only my own health but the health of the planet that my children and grandchildren will inherit.

Playing the blame game between the genders is counterproductive. It militates against the kind of creative partnership that is vital for earth-caring. Both genders have contributed to the sexist patterns that have helped damage the earth. The patriarchal sex role schema, which benefits as well as oppresses both sexes, requires two to tango. It can be maintained only when both men and women play their interactive roles in continuing it. But as members of the gender that is socially defined as one up in that schema, men have special power and responsibilities to change its destructive features.

6. An Interfaith, Inclusive Ecological Spirituality Perspective[33]

Divisive religions with exclusivistic beliefs that only their understanding of faith has spiritual validity tend to block the cooperative global earth-saving and peacemaking efforts that will be required to reverse the ecological crisis. Ecological spirituality, as spelled out in the next chapter, is inherently interfaith and unifying, bridging the many beliefs and value chasms that exist in the world's religions.[34]

It is immensely hopeful for the planet that there is a widespread greening of the thinking and action of an increasing number of religious individuals and groups (including Christian evangelicals), as well as religiously motivated environmental groups in many faith traditions. Many religiously committed people are recognizing that a green consciousness and conscience, green theological beliefs and ethics, have deep roots in their faith traditions. They are becoming proactive in what they describe as saving God's creation. Many are aware that the power and passion, consciences, and commitment of religious individuals and groups are essential to facilitate the radical lifestyle transformations that will be required to save the biosphere.

Experiencing the One-Earth Perspective

This awareness exercise is designed to enable persons to bring alive in here-and-now experience one of the transforming perspectives described above. Experiential methodologies like this can be used in both teaching and clinical practice. Experiencing it yourself is the most expeditious way to discover if it may be a resource in your work.

Instructions: Relax your body-mind organism to increase your receptivity to this use of creative imagination. Continue reading until you come to this mark /, then close your eyes for as long as you need to do what has been suggested.

In your imagination form an action picture of yourself as an astronaut walking in your marvelous spacesuit down the ladder of the landing craft to the moon's surface. / Now, for the next few minutes, be the astronaut on the moon. Explore your surroundings, staying in touch with all your feelings. / Now look across the lunar horizon and see your planet home for the first time from this vantage point. Be aware of your feelings as you look at our small, gray-green, living planet in the vast, cold, dark immensity of space. See it as home—your only home and the only home of everyone you love, and the one home of the whole human family as well as all the living creatures who together breathe its air and share its resources and miraculous aliveness. / Keep glancing toward earth occasionally as you explore further around the landing craft. /

Now, climb the ladder back into the landing craft, lock the air hatch, and blast off for home. / You are now speeding through space away from the sterile lunar environment and toward the one living planet of the sun. As you look back and ahead, stay in touch with your feelings and thoughts. Try thinking about your personal problems from the perspective of what you have just experienced. / Now reflect on the huge problems of your world from this perspective, including the global environmental crisis threatening the planet's precious biosphere. Stay with the flow of your images and feelings for a few minutes, discovering where they take you. See if you can get in touch in a fresh way with the reality you have long known: that it truly is "one world or none." Be aware that this is the most holistic perspective from which to work for wellness. /

When you are ready, open your eyes and jot down what seems important to remember from your experience. Be particularly aware of anything you now are aware that you need to do as a result of what you have experienced. / Discuss your experience with a family member or friend and then plan whatever action is appropriate to implement what you have decided. /

As outlined above, these six perspectives can facilitate ecobonding and earth-caring in ecotherapy and ecoeducation. When these are accepted, they can nurture a gradual ecological transformation of consciousness.

Needed: A New Self-Understanding for Humankind

As insightful conceptualizations of the needed transformation, Jeremy Rifkin's writings are among the most stimulating. Of various innovative thinkers focusing on the global ecocrisis, he has generated one of the most challenging interdisciplinary visions of a viable future in the twenty-first century. The basic hypothesis set forth is that humankind is suffering from a failed attempt to solve the problems of our world because of a seriously outdated self-understanding or consciousness. In *Biosphere Politics—A New Consciousness for a New Age*,[35] Rifkin uses the biosphere as the basic framework for the earth-centered reconceptualizing of psychology and personality theory, culture, economics, all the sciences, and politics. He sees this reconceptualization as essential for a sustainable future.

Rifkin builds on British philosopher-lawyer Owen Barfield's outline of three stages of human consciousness.[36] In the first stage, through most of the human story, our species enjoyed direct bodily participation with the earth and humans' bodily nature. There was little sense of individual identity and independence either from nature or the group.

Beginning with the Neolithic revolution of agriculture and the domestication of animals, humans began to assert their control over nature and develop an embryonic sense of individual self-awareness. This new consciousness increased with the development of industrial production. It accelerated exponentially with the commercialization of nature, privatization of human social life, and the explosive development of science and technology. As Rifkin states, "The human race has evolved from a state of undifferentiated oneness with Mother Nature to a detached self-aware isolation from her. We have won, at least, the illusion of independence from nature, but we have lost a sense of relationship and kindred spirit with the earth."[37] Humankind is now entering the biospheric era in which we must reparticipate fully and cooperatively with the body of nature. Rifkin shows how this new consciousness is uniquely different from the undifferentiated oneness with Mother Earth in the early experience of humans and the life of every baby:

> The key difference is the ability to choose. . . . To re-participate with nature out of an act of love and free will rather than out of fear and dependency. . . . By making a choice to re-participate with the body of nature, the individual and the species is affirming life, accepting the inevitability of death, and celebrating the organismic unity that binds the parts to the whole. By freely choosing, each person retains his or her own unique identity while reveling in the oceanic oneness of the biosphere . . . the consciousness of the individual and the species reach out from a self-imposed exile to re-embrace the earth, secure in the fullness of their grounding in the biosphere. Today, it is ours to choose. . . . To heal the wounds we have inflicted on our planet and our souls. . . . To make the world secure. To secure ourselves and our being. To take the leap."[38]

Exploring This Chapter's Issues Further

For fuller information about these books, see the annotated bibliography at the end of this book.

Charles Birch and John B. Cobb, Jr., *The Liberation of Life*

Herman E. Daly and John B. Cobb, Jr., *For the Common Good*

Al Gore, *Earth in the Balance*
Jeremy Rifkin, *Biosphere Politics*
Theodore Roszak, Mary E. Gomes, and Allen D. Kanner, *Ecopsychology*
Lesley Irene Shore, *Tending Inner Gardens, the Healing Art of Feminist Psychotherapy*

Notes

1. "When the World Hurts, Who Responds?" in *Ecopsychology: Restoring the Earth, Healing the Mind*, ed. Theodore Roszak, Mary E. Gomes, and Allen D. Kanner (San Francisco: Sierra Club Books, 1995), 17.

2. *World as Lover, World as Self* (Berkeley, Calif.: Parallax Press, 1991), xii.

3. *The Voice of the Earth* (New York: Simon & Schuster, 1992), 321.

4. Quoted in an advertisement for *The Progressive*, no date.

5. Environmental racism refers to the fact that, because of white racism, people of color in the USA (particularly Hispanic, African American, and Native American) are impacted disproportionately by environmental pollution and by the location of the production, treatment, disposal, and storage sites of hazardous waste. They usually are among the least able either to influence where such sites are located, or to afford housing in a healthier setting.

6. Norman Cousins described this experiment in "The Winners in Healing" lecture at the Cambridge Forum, 19 March 1986. I shall always be grateful to him for his moral support during the founding of the Institute of Religion and Wholeness at Claremont, California.

7. A landmark book illuminating these social, political, and economic ecological issues is *For the Common Good*, (Boston: Beacon Press, 1989). This book is "must" reading for ecologically informed teachers coauthored by philosopher-theologian John B. Cobb, Jr. and economist Herman E. Daly. The subtitle, describes the book's main message: *Redirecting the Economy toward Community, the Environment, and a Sustainable Future.*

8. *International Herald Tribute*, Paris, 14 April 1989. (Sting's full name is Gordon Matthew Summer.)

9. *Earth in the Balance: Ecology and the Human Spirit* (New York: Penguin Books, 1992), 247.

10. *Columbia Encyclopedia of Quotations*, New York.: Columbia University Press, 1993, 282.

11. "The Spirit of the Goddess," in *Ecopsychology*, 288–300.

12. The most influential of these, the International Physicians for the Prevention of Nuclear War, was awarded the Nobel Peace Prize several years ago. A decade ago, several of us in the field of pastoral psychology and counseling formed what became the International Pastoral Care Network for Social Responsibility, an action network to promote peace, justice, and ecology communication. The group now has representatives in some fifty countries on all the continents (except Antarctica). This group is a contemporary flowering of the century-spanning prophetic heritage, within the Hebrew and Christian traditions, concerned with the interdependence of healing persons, healing social institutions, and healing God's creation.

13. *Recollected Essays* (Berkeley, Calif.: North Point Press, 1981), 110–12. I thank Mary Evelyn Tucker for highlighting this story in her *Education and Ecology* booklet

(Chambersburg, Pa.: Anima Books, 1993), where I first encountered it. Paul Shepherd makes a point similar to Wendell Berry's: "The zebra watches the lion, but it does not bolt except when it is chased. It does not fear the lion . . . in the sense of [human's] obsessive terror of the unknown. Zebra evolution does not lead to escape from lions, but to more defined affiliation in which the moment of truth is directed through certain individuals at the appropriate state of life. None of this is visible to non-totemic societies because they do not watch zebras—or, more correctly, it is apparent to any society to the degree that it watches wild zebras" (*Thinking Animals: Animals and the Development of Human Intelligence* [New York: Viking Press, 1978], 162).

The lodgepole pine offers an example from nature of the remarkable capacities with which many living beings are equipped to cope with adverse circumstances. It has two types of pine cones. One kind releases its seeds under normal environmental circumstances. The other opens to release its seeds only after a forest fire. Given sufficient rain, new seedlings take advantage of the open spaces and access to sunlight created by the fire. They begin to regenerate the forest. This process is paralleled by the capacity of many persons to use losses and adversity as an opportunity to grow.

14. The terms *left brain* and *right brain* are used to distinguish the logical, linguistic, quantitative, rational, lineal thinking functions centered in the left hemisphere of the brain (in right-handed people), and the intuitive, imaging, holistic functions of the right hemisphere.

15. *Voice of the Earth*, 320.

16. One weakness of peacemaking-ecological education and action since the dawn of the nuclear age is that much of it has been terror-oriented. As theologian Henri Nouwen once observed, many of the fine, courageous people working for peace-with-justice are "living in the house of fear." This is also true of some who are deeply involved in ecology action, which is a major reason why many people burn out in the often frustrating work of seeking to overcome resistance to the basic changes needed to heal an agonizing planet. To paraphrase Nouwen's apt advice, only as we earth-carers move from the house of fear and guilt to the house of love and hope, will we be able to "keep on keeping on."

17. New York: Harper and Row, 1956, .6.

18. William L. Stodger, *Flames of Faith* (New York: Abingdon Press, 1922), 16.

19. *Earth in the Balance*, 368. Gore then points to a critical fork in humankind's road: "Now that we have developed the capacity to affect the environment on a global scale, can we also be mature enough to care for the earth as a whole? Or are we still like adolescents with new powers who don't know their own strength and aren't capable of deferring instant gratifications? Are we instead on the verge of a new era of generativity in civilization, one in which we will focus on the future of all generations to come? The current debate about sustainable development is, after all, a debate about generativity." Ibid., 364–65.

20. For an in-depth discussion of the role of playfulness in healing and wellness, see "Using Laughter and Playfulness for Healing and Health," chapter 7 in Howard Clinebell, *Well Being: A Personal Plan for Exploring and Enriching the Seven Dimensions of Life* (San Francisco: Harper San Francisco, 1992).

21. Quotations from *The Universe Is a Green Dragon* (Santa Fe, N.M.: Bear & Co., 1984), 107, 119, 123, 145, 148, 150.

22. *The Universe Story: From the Primordial Flaring Forth to the Ecozoic Era, A Celebration of the Unfolding Cosmos* (San Francisco: Harper San Francisco, 1992), 264–65.

23. As I write this (December 1994), the World Watch Institute has just reported encouraging news: Peacemaking by governments has increased dramatically in the last five years, with much more peacekeeping by the United Nations, military base closures, and so on. But expenditures on armaments worldwide are still $770 billion—forty times the total peacemaking amount of $16 billion. I feel anger mixed with shame that my country is the world's leading distributor of deadly armaments to developing countries.

24. From Lewis Thomas, *The Lives of a Cell*, quoted by Kirkpatrick Sale in *Dwellers in the Land: The Bioregional Vision* (Philadelphia: New Society Publishers, 1991), 11.

25. Quoted by Senator Paul and Jean Simon in a recent Christmas greeting.

26. "Give Us Hope" (pamphlet), Boston Council for a Livable World Education Fund, 1988, 1, 3.

27. I remember that passenger pigeons were once the most abundant birds on the earth. Then the railroads and the telegraph enabled hunters to find their concentrated flocks and eventually wipe them out. The same extinction-bound process is occurring today with many species. For example, in recent years fish populations on both coasts of the United States have been radically depleted by huge nets, sonar, and fossil-fueled factory ships, threatening the future of some fish species that have been a major source of protein for people in many countries.

28. *Voice of the Earth*, 13–14.

29. *The Dream of the Earth* (San Francisco: Sierra Club Books, 1988), 35.

30. With only 20 percent of the world's population, rich nations produce more than 50 percent of all ozone-depleting gases. This is only one of many examples of the inequity and the danger to all countries emulating from affluent countries' lifestyles.

31. New York: Harrington Park Press, 1995, 50.

32. See Charlotte David Kasl's outstanding book on recovery from addictions, *Many Roads, One Journey: Moving Beyond the Twelve Steps* (New York: HarperCollins, 1992).

33. The term *spirituality* is used here to suggest that both religiously minded and secular-minded people have meaning-creating beliefs and values that can have either a healthy or an unhealthy influence on their lifestyles and on the biosphere.

34. One of the most destructive misuses of religion in the Western world today is the religiously sponsored antienvironmental groups that seek to block efforts to save a healthy earth. This misuse of spiritual passion and energy tends to aid and abet the individuals and groups in continuing the very earth-exploiting policies that produced the global ecojustice crisis.

35. New York: Crown Publishers, 1991.

36. Barfield's stages are similar to those of other thinkers, including Walter Christie (see chapter 2).

37. Rifkin, *Biosphere Politics*, p. 325.

38. Ibid., 325–26.

4

Spiritual, Ethical, Cosmological, and Meaning-of-Life Issues in Ecotherapy and Ecoeducation

Nature is the art of God.
—Sir Thomas Browne, *Religio Medici*, 1642

Humans and nature belong together, in their created glory, in their great tragedy, and in their salvation.
—Paul Tillich; on the memorial at his grave in New Harmony, Indiana

If ye dwell in the light which was before the earth was, with it ye will preserve the tender plants.
—George Fox, founder of the Society of Friends, 1624–1691[1]

On February 23, 1944, hiding from the Nazis with her family in a secret Amsterdam loft, fifteen-year-old Anne Frank wrote these wise words in her journal:

> The best remedy for those who are frightened, lonely or unhappy is to go outside, somewhere they can be alone, alone with the sky, nature and God. For then and only then can you feel that everything is as it should be and that God wants people to be happy amid nature's beauty and simplicity.
> As long as this exists, and that should be forever, I know that there will be solace for every sorrow, whatever the circumstance. I firmly believe that nature can bring comfort to all who suffer.

It is impossible for most of us to even begin to imagine how intense her desire to go outside must have been.[2]

This chapter explores ecological spirituality as the unifying center of the ecological circle and the power of the transforming perspective described earlier—a perspective of interreligious, inclusive ecological spirituality. It spells out why ecological spirituality and ethics are so crucial in earth-caring by individuals as well as public policy makers. The dynamic roles of spiritual, meaning, and ethical issues in causing as well as curing destructive alienation from the earth will be examined. However, the focus will be limited to only those spiritual and ethical issues directly relevant to doing effective ecotherapy and ecoeducation.

Four assumptions undergird this chapter. *First*, individuals and societies around the globe are suffering from a pandemic of spiritual malaise characterized by spiritual confusion, value vacuums, religious pathologies, and ecoalienation. *Second*, these spiritual and ethical pathologies are key factors among the complex, intertwined roots of our species' global addiction to violence. This includes the rape of women and children and of Mother-Father Nature, as well as the countless wars fueled by ethnic hatred. *Third*, only to the degree that ecologically healthier spiritualities and ethical priorities replace spiritual and ethical pathologies will healthier people create a healthier society and planet. Religiously inspired and motivated biophilia and ecoethics are the most dynamic and energized expressions of these much-needed human responses. *Fourth*, therefore humankind must grow spiritually and ethically by moving toward globally unifying, ecofriendly spiritualities and ethical commitments to the well-being of the whole human family and the whole biosphere. Such growth may be the human family's most urgent need but also the most difficult challenge in the ecojustice crisis.

The unprecedented UN Conference on Environment and Development in Rio de Janeiro in June 1992, was the largest assembly of world leaders ever to meet. Their goal was to devise international strategies and working agreements for reversing the destruction of the biosphere. Their relative success in this is a promising harbinger of hope. The initial session of this Earth Summit began with two minutes of silence for an endangered planet. This silent ritual probably was a symbol of a hopeful awareness growing among millions of people around the planet—people who understand life's meaning from many diverse religious perspectives, ethical philosophies, and cultures. This is the awareness that there are crucial dimensions of depth and height as well as breadth in the earth's crisis.

The pivotal representative from the United States was Al Gore, then a senator and subsequently vice president. In his enlightening book *Earth in the Balance*, he highlights the centrality of spiritual and ethical issues, describing the problem as "a spiritual crisis in modern civilization that seems to be based on an emptiness at its center and the absence of a larger spiritual purpose." He declares:

> Whether we believe that our dominion [over the earth] derives from God or from our own ambition, there is little doubt that the way we currently relate to the earth's natural systems is wildly inappropriate. But in order to change, we have to address some fundamental questions about our purpose in life, our capacity to direct the powerful inner forces that have created this crisis, and who we are. . . . These questions are not for the mind or the body, but the spirit.

Gore points to the dangers the earth faces from our species: "Once again we might dare to exercise godlike powers unaccompanied by godlike wisdom."[3]

The terms *religion* and *spirituality* in this book are used as rough synonyms and are defined generically from a perspective of the disciplines of psychology of religion and sociology of religion. They refer to whatever beliefs about ultimate reality and values people use to provide themselves with some sense of meaning and direction. These beliefs enable them to cope with the often-pressed awareness of the tragedy and death that are inherent in the lives of all humans as living-dying creatures. The term *ethics* refers to the functional values and priorities that actually guide people's day-to-day behavior. Such implicit inner guidelines often are radically different from the values people espouse in their belief systems.

People's heart-level religious beliefs and functional ethical commitments, in contrast to their head-level beliefs and values, influence profoundly how they relate to themselves, other people, their natural environment, and the divine Spirit. Religious and ethical dynamics are powerful influences in determining whether individuals and groups will change their behavior enough to save the biosphere in the years ahead. Spiritual and value dynamics often are keys to enhancing earth-caring behaviors as well as to openness to being nurtured by nature. Furthermore, experiences in nature can be pathways to spiritual awakening and healing. For these reasons, spiritual-religious issues and behavior-guiding ethical commitments are crucial variables in both ecotherapy and ecoeducation. Of course, in secular cultures like ours, religious issues are expressed in nonreligious language more often than not. The issues are present, at least implicitly, whenever people question the meaning of their lives, the brevity and vulnerability of life, the death or disability of their loved ones or themselves, or struggle with discovering what is really worth living for.

Al Gore uses the apt metaphor of a gyroscope in describing the crucial importance of faith:

> I have come to believe in the value of a kind of inner ecology that relies on the same principles of balance and holism that characterize a healthy environment. . . The key is indeed balance—balance between contemplation and action, individual concerns and commitment to the community, love for the natural world and love for our wondrous civilization. . . . If it is possible to steer

> one's own course—and I do believe it is—then I am convinced that the place to start is with faith, which for me is akin to a kind of spiritual gyroscope that spins in its own circumference in a stabilizing harmony with what is inside and what is out.[4]

Some Therapists' and Teachers' Views about Ecospirituality

The presence and prominence of spiritual and ethical issues in the ecological understandings of psychotherapists and teachers vary tremendously, of course. But, in spite of the sampling limitations of the survey done for this book,[5] the responses to one of the questions were striking: *Do you regard issues of meaning, value, or spirituality as important factors in determining people's lifestyles, including how they treat the environment?* Most respondents affirmed this view. Here are some of their responses:

- "I see these as at the core of people's deepest motivation from the smallest decision to the greatest."
- "Meaning, value, or spirituality are core issues which determine lifestyles. Of necessity they influence how one treats the environment, even if it isn't consciously recognized. It's in the fabric of one's life and cannot be separated."
- "I think Matthew Fox is correct in his belief that what we lack is an appropriate cosmology. It is this lack that allows us not to love but instead to rape the earth, which leads to raping other people and ultimately, ourselves."
- "People with more universalized value/meaning systems tend to be more sensitive to the environment. The more narcissistic clients "use" the environment and value it as something to serve them."
- "I believe that meaning, value, and behavior rise from our experience of the holy. How we understand and experience the holy has a great effect on lifestyles and how we relate to nonhuman nature."

Ecospiritual and Ecoethical Objectives

What are the concrete objectives to keep in mind when one is dealing with implicit or explicit spiritual and ethical issues in ecotherapy and ecoeducation? Whether (and how) any objectives are implemented depends, of course, on where individuals are on their spiritual and their earth journeys. But in general the three primary objectives are: (1) *facilitating healing of spiritual and ethical pathologies and emptiness that feed damaging treatment of nature*; and (2) *strengthening the transcending or spiritual dimension of nurturing bonding with nature*, and thus (3) *to help people discover, in their own way and time, the spiritual heart of ecologically whole, environmentally caring living*. This third objective often involves awakening religious passions and commitments to caring for God's creation where the individual lives and works. The following subobjectives are ways of implementing these primary objectives.

• It is important to help people understand how their earth values (and the underlying beliefs that determine these), influence and motivate them everyday in relating to the natural world, in constructive and-or destructive ways.

• We need to encourage people to move from ecodestructive beliefs, values, and behaviors that cause harmful alienation from nature, toward ecofriendly beliefs, values, and behaviors that make for nurturing bonding with nature. Earth-friendly beliefs, values, and behaviors can be facilitated simply by affirming them as they emerge. The self-reinforcing rewards of such changes may include reduction of *appropriate* ecological guilt, plus the longer-range benefits to mind-body-spirit wellness from healthier lifestyles. Firm but caring challenges of earth-alienating beliefs, values, and behaviors are important at the appropriate time. These challenges can be done by pointing out how these beliefs and behaviors are contrary to the well-being of the immediate environment on which their well-being depends.

• Another objective is to encourage people who find meaning in particular religious and ethical traditions to search for and hopefully discover earth-respecting beliefs and values in their own traditions.

Human Beings Are Spiritual and Valuing Creatures

To understand the role of religions in earth-caring, it is necessary to explore an observation by various cross-cultural scholars that humans are "incurably religious." This simply means that ours is a religion-creating species. All known cultures have generated some cosmological beliefs and religious practices. Secular counselors, therapists, teachers, or parents may find it illuminating to understand how the meanings people create by their spiritual beliefs influence their ethical values, and how ethical commitments have a powerful impact on how people think, perceive, feel, and behave in all areas of their lives, including their relationships with the earth.

We humans are spiritual and valuing creatures, whether or not we have any interest in organized religions. We live in our meanings and in our beliefs about what is ultimately important to ourselves and our world. By our beliefs, ethical commitments, and the religious practices that may express them, we create some sense of meaning and order in our turbulent lives. Maintaining at least a minimal sense of meaning and purpose enables us to stay relatively sane and whole, even in times of tragedy.

My understanding of human personality's spiritual roots has deep affinities with the thought of Carl G. Jung and Roberto Assagioli. In what became his most-quoted statement, Jung once wrote: "Among all my patients in the second half of life . . . every one of them fell ill because he had lost what the living religions of every age have given their followers, and none of them has been really healed who did not regain his religious outlook." [6]

Assagioli was an Italian psychiatrist who introduced psychoanalysis to his country and then moved beyond this to generate his own growth- and spiritually oriented approach to psychotherapy called psychosynthesis. Assagioli accepted many of Freud's brilliant insights about the unconscious mind, but he came to understand human personality and the goals of therapy in ways that go beyond Freud's tripartite model of id, ego, and superego. Assagioli believed that the "I" (self or ego) of everyday experience is not one's ultimate or true identity. Rather the core self is a reflection of what he called the transpersonal or spiritual Self, which is the powerful source of spiritual energy for growth.[8] It is noteworthy that Jung also distinguished the self or ego from the Self.

The Self, in my understanding, as in Assagioli's and Jung's, and more recently in therapist Thomas Moore's thought, refers to that transcending dimension of personality called soul in traditional religious language. This dimension is the major reason why humans are inherently religious. The goal of spiritually centered counseling, therapy, education, and parenting is to enable people to make their transpersonal, spiritual Self the unifying, enlivening center of all aspects of their lives. This Self is the channel by which the inspiration and empowerment of the divine Spirit become active in human lives. And, most important from an ecotherapeutic perspective, *I understand the transpersonal Self, intertwined with the deep ecological self as constituting the core of our very being.*

The most illuminating understanding of why humans are inherently religious, in my view, derives from the empirical reality that all persons have existential or meaning-of-life needs. These are deep hungers of the heart that can be satisfied adequately only by spiritual, religious, or philosophical systems of belief and practice. For this reason, all therapists and teachers, whatever their professional setting or personal religious orientation, deal regularly with clients or students who are struggling to satisfy often unrecognized spiritual needs. The spiritual goal of therapy and education is to help them do so in life-enhancing rather than life-diminishing ways.

The following are the basic spiritual or existential needs of all humans that are directly related to ecotherapy and ecoeducation:

- The most basic earth-related spiritual need is for *continuing interaction with the profoundly creative resources of the core of personality, the Self or Soul that is both transcending and earth-based.* Interaction between everyday consciousness and this deeper eco-Self is mediated by transcending experiences in nature, as well as by living symbols, earthy trust-renewing rituals and earth-rooted stories of meaning (myths). It is via this Self that transcending experiences in the natural world are mediated.
- All humans need *vital beliefs that give life meaning and renew basic trust in*

the midst of both the tragedies and triumphs of life. The fact that humans are the species who know they will die some day is a significant motivation for generating religious systems. Spiritual beliefs grounded in an earth-based cosmology that enables people to feel "at home" in the universe are trust-engendering for many people at a deep body-mind-spirit level.

• *A human survival need in a creative but violent world is for ethical values that provide guidelines for behavior that is both personally and socially responsible, including responsibility to end violence against the biosphere*. A global, multireligious ecological ethic clearly is a pressing priority today. An invaluable asset in facilitating change to such a spiritually energized ethic is the science-based cosmological paradigm often called the "common creation story." The pioneering thought of Pierre Teilhard de Chardin has been continued by scholars like Thomas Berry and Brian Swimme. *The Universe Story* by Swimme and Berry can inspire learners of any age, if they are open to its message, to enlarge their understanding of their human-earth interaction. It does so by illuminating the wondrous process of the continuing evolution on our planet and wider universe, a process by which our species eventually came into being. Using different terms, Swimme and Berry articulate the dimensions of what I call the ecological circle. A perceptive graduate student in theology wrote, after reading their book, "For me, this story provides a radical sense of connection to all that is. Seeing the emergence of the story was important for feeling more connected with nature. We should learn to celebrate this story."

• *All humans need healing, growth-stimulating, hope-energizing experiences that psychologist Abraham Maslow called peak experiences*. These are mystical moments of self-transcendence in art, beauty, relating in depth to other people, and to the divine Spirit. Many people find their most energizing and healing peak experiences by encountering the divine Spirit in the awesome beauty, wisdom, and wonder of nature.

• Another earth-related spiritual-ethical need is for *loving respect for all life*. This flows from the spiritual awareness of the mysterious gift of being part of the wonderful web of all living things, including the whole human family.

In discussing caring for the soul, Thomas Moore adds an insight that illuminates the spiritual basis of earth-bonding. Drawing on the ancient concept of the *world soul*, he declares: "Our soul, the mystery we glimpse when we look deeply into ourselves, is part of a larger soul, the soul of the world, *anima mundi*. The world soul affects [and is found in] each individual thing."[9] Moore uses the pregnant term "soul-ecology," reminding us that the Greek word *oikos*, from which *ecology* is derived, means home:

> Speaking from the point of view of soul, ecology is not earth science, it is home science; it has to do with cultivating a sense of home wherever we are, in whatever context. The things of the world [including nature] are part of our

> home environment, and so soulful ecology is rooted in the feeling that the world is our home and that our responsibility to it comes not from obligation or logic but from true affection.[10]

He observes: "If the world soul and our own souls are one, then as we neglect and abuse the things of the world, we are at the same time abusing ourselves."

In my view, the *soulful living* that reawakens the aliveness of the soul, for which Moore calls, must include loving interaction with the biosphere. This means being nurtured by nature and reciprocating by nurturing nature in one's everyday life. From an ecotherapeutic perspective, Moore's related point is on target:

> If we are to attempt to develop sound practice of ecology, we need to tend our own inner pollution at the same time, and if we are to attempt to clean up our personal lives through therapy or some other method, we will need at the same time to tend to the neuroses of the world and to the suffering of things.[11]

In my view, the key to the inner preparation for more earth-caring living often is finding healing of the spiritual and ethical wounds that are barriers to loving earth-bonding.

One of the international graduate students interviewed for this book was Il Koo Cho of Korea. He recalls that in his childhood, "most Korean villages had a very old symbolic tree. The people took care of such trees very carefully because they believed they had their own spirituality. They never cut or destroyed these. People performed religious services under the trees, bringing many kinds of food for these spirits. When somebody did not have children, they made earnest requests to these deities in trees or a river, or other nature things, because they believe all things in nature have their own spirituality." Although he no longer accepts the worship of nature deities, he still believes that nature has its own spirituality and that harmonizing with nature is important.

Those who are seeking fresh ways of connecting more integrally and caringly with the biosphere can learn from such a perspective, drawing on those aspects of animism that are valid. Like the nature mystics in Christianity and other religions, they can accept spiritually energized respect for nature and enjoy the transcending reality in nature, without accepting a belief in nature deities or seeking to manipulate them magically for their own desires.

There is widespread awareness that survival ethics today must include a global ecological ethical paradigm. During the years that the threat of nuclear holocaust was hanging over humankind like a global sword of Damocles, Erik H. Erikson did an illuminating exposition of the ethical implications of psychoanalysis for that crisis. His challenge is still crucial today in the ecocidal crisis: "We live at a time in which—with all the species-wide destruction possible—we can think for the first time of a species-wide identity, of a truly universal ethics, such as has been prepared in the world religions, in humanism, and by some philosophers." [12]

Ecopathogenic vs. Ecosalugenic Beliefs and Values

The assumption that all religion is good for people is as fallacious as the opposite assumption by Sigmund Freud that all religion is an expression of illusory infantile longings that blocks acceptance of reality. What Freud did not recognize is that religion can be either a source of healing and health, or of spiritual immaturity as well as personal and societal pathology. If ecotherapy and ecoeducation are to help reduce inner blocks to nurturing earth-bonding, a vital distinction must be made between health-fostering (salugenic) and sickness-engendering (pathogenic) religious beliefs and ethical practices. This distinction has a long history in the discipline of psychology of religion (in which the author is trained). It goes back to near the dawn of the twentieth century when psychologist-philosopher William James, father of this discipline, in his classic *Varieties of Religious Experience*, contrasted "the religion of the healthy minded" and "the religion of the sick soul." My use of this distinction contrasts religious expressions by individuals and groups that heal and enhance well-being for individuals, other groups, and the biosphere, with religious expressions that hurt and even destroy persons, other groups, and the biosphere. All around the world, as I write, rigid, fanatical, authoritarian systems of faith and values fuel and justify the pandemic of violence, including violence against the natural world. When this occurs the tremendous potential healing and earth-caring energies of healthy systems of believing and valuing are subverted, caricatured, and transformed into violence and destructiveness. I have developed general criteria for distinguishing pathogenic from salugenic religious expressions elsewhere.[13] Here I will focus only on those that militate against ecowholeness.

Pathogenic religious beliefs and ethical values often are among the root causes of behaviors directly damaging to the biosphere, and of the social, economic, and political injustices that are at the roots of the ecojustice crisis. The human family probably will not make the radical changes in values and lifestyles necessary to heal the biosphere, unless the passion for and commitment to what the World Council of Churches calls "love, justice and the integrity of creation" is widely mobilized among millions of people in all the world's major religions. The process of doing this involves understanding and reducing the epidemic of pathogenic beliefs and values that militate against earth-caring, justice-enhancement, and peacemaking. Until we humans make our religious and ethical systems healthier, saving a viable earth for future generations is not likely to happen.

Gerald O. Barney, head of the Millennium Institute of Arlington, Virginia, delivered an address to the 1993 Parliament of the World's Religions, challenging religious leaders from many faith traditions.[14] He overviewed the powerful evidence that the earth and its peoples are in profound trouble, and that if present trends continue, the earth will be increasingly unable to meet the basic needs of an expanding population with diminishing arable land and growing environmental problems. Holding that the basic problem is spiritual, Barney described the crucial fork in humankind's path into the future:

> Down one path, I see steep, slippery ground. I see desperately poor people losing their footing and slipping into a downward spiral of eating their seed grain and burning their shade trees. Then they migrate. Wealthier areas guard their borders to keep refugees out. Many die at the borders. Some manage to slip across to work illegally in dreadful conditions and to live in fear of hate and violence. Distrust destroys any sense of community among peoples and nations. Petroleum is exhausted. Pollution of all kinds worsen. Groups of heavily armed bandits begin roaming about, stealing, raping, burning, and killing.
>
> Down the other path I see a very different Earth. Sustainable, just and humane development for the whole Earth has become the principal goal of every nation and people. Peoples, cultures, and faiths are united in planet-wide efforts to understand the Earth and its peoples and to envision what Earth can become. Protection of Earth has become a top priority of every person. Human ignorance, poverty, and bigotry are recognized everywhere as primary threats to global security. Population growth slows and stops. Earth begins to heal, and gradually a sustainable economy emerges which meets even more than the basic needs of everyone.[15]

Barney summarized the essential changes necessary for our species to leave the path to death and move onto the path to life. Then he challenged leaders of the world's religions:

> Here we desperately need help from you, our religious leaders. We need you to help us acknowledge that our old dream has failed. We need you to help us dream a new dream—a dream that is not only true to ancient traditions, but also true to the revelations emerging from what we are learning about the Earth [from the sciences]. But I must tell you honestly that many people now wonder if any of our faith traditions have the wisdom we need for the future. Many people doubt that there is a "sustainable" faith tradition on Earth today, and faith such that if everyone adopted and followed it, we would be assured a sustainable, just, and humane future for Earth and her people. Many feel that our faith traditions have become a very central part of the human problem.[16]

Barney then cited evidence that some contemporary religious beliefs are contributing significantly to the crisis. He mentioned: the prevalence of wars (fifty when he spoke), most of them fed by interreligious hatreds; faith groups that continue to oppress women, blocking the need for women to make their full contributions in all cultures; extreme anthropocentric faith views that justify continuing the "devastating war with Earth"; the otherworldly focus of some faiths that makes protecting the earth unimportant; opposition to birth control at a time of runaway population growth; preoccupation with preserving ancient traditions, blocking both needed critique of these and creating a new ethics and earth-saving dream.

Barney summarized the factors blocking the essential spiritual, economic, political, and social changes required for our species to move from the path to death to the path to life. The largest barrier is the fact that the prevailing human dream has failed and has not yet been replaced by a sustainable, workable dream. The failed dream that still dominates hopes and public policy around the globe is the development model—the affluent, high-consumption, wasteful-lifestyles model prevalent in Western industrialized countries. The dream has failed because its underlying beliefs are false. The poor and women reject the beliefs that the dream is just, equitable, and all-inclusive. Beyond this, the whole earth rejects the belief that the planet exists for the use and abuse of as many humans as humans choose to conceive.

Our rudderless dilemma is described by Barney: "We are now a species without a vision. We cannot act together because we no longer know our goal. We can no longer guide our countries and our corporations because we no longer know what (sustainable) 'development' is. We no longer know how to distinguish 'progress' from 'failure.'"[17]

Fortunately, some religious leaders, struggling to keep their faith communities relevant to the ever-changing human situation, are recognizing the urgent need to help generate a new earth-saving guiding vision or ethic for society. In retrospect, history may show that this re-visioning was the most vital single contribution of religious people of all faith groups in the global ecojustice crisis.

Understanding Ecologically Destructive Religious and Ethical Beliefs

A closer look at the spiritual and ethical roots of the environmental crisis is in order. It is important to see that these interdependent ecodestructive spiritual beliefs are part of the contemporary global spiritual crisis that produces all manner of problems in individual lives and in societies. In doing ecotherapy, it is appropriate to deal with spiritual and ethical problems explicitly, when such problems are discovered to be creating ecoalienation for particular persons. Treating extreme expressions of these pathogenic issues may require the professional expertise of pastoral counselors trained in helping persons trapped in pathogenic spiritual and ethical beliefs. But it is well for therapists and teachers in general to be aware that this cluster of issues may be diminishing people's ecological and overall wellness. The following ecopathogenic problems overlap. In a sense, all are rooted in the first alienation.

Root of the Crisis: Human Alienation from the Earth

The fundamental psychospiritual root of the ecological crisis is alienation of humans from an awareness of our organic connectedness with the planet's marvelous network of living, inspirited things. This alienation fuels narcissistic feelings of superiority, from which comes the prideful belief in humans' God-given right to exploit the rest of nature including other species. A wide variety of destructive I-it behaviors, to use Martin Buber's language, flow from this ecospiritual alienation. Healing involves enhancing awareness of the spiritual meaning of our profound interdependence with the whole interdependent community of life. A variety of methods of spiritual healing and reconciliation are described later in this book.[18]

Dualistic Western cosmologies and what could be called detached-from-the-earth spirituality contrast sharply with the earth-rooted cosmologies and spirituality of native peoples around the planet. Judaism and Christianity, by seeking to destroy "pagan" animism and shamanism, weakened their ties with the earth and helped open the door to exploitation of nature by industrial society with indifference justified by humans' species arrogance.

Leslie Gray, an Oneida/Seminole clinical psychologist, trained with medicine people from various tribal backgrounds. She observes that humans have more than 40,000 years of shamanic experimentation concerning how to live in a healthy, healing relation with the earth. Whereas many models of sustainable indigenous societies exist, there are no models of sustainable industrial societies. Her point is well taken that, in generating models of a sustainable future, it would be tragic to waste this accumulated knowledge by which more than 300 million indigenous people still live today.[19]

Psychospiritual Idolatries

A plethora of psychospiritual idolatries are among the deep causes of ecoalienation. Idolatry in this context means making an object of ultimate devotion (de facto worship) of something that is not worthy of worship.[20] This takes many forms, including idolatry of oneself, idolatry of one's species (discussed above), ethnic group, or nation; and idolatry of one's in-group values and religious beliefs. Idolatry of one's nation and economic system is used to justify economic exploitation of both the natural resources and the poorly paid workers in impoverished nations.

Among numerous destructive in-group idolatries are exclusivistic, prideful, ghettoized religious beliefs and ethnocentric ethics that constrict the application of one's caring to one's ethnic, religious, or species boundaries.

The flourishing of rigid fundamentalism in world religions is an expression of in-group idolatry that blocks global ecospirituality. As a result, the basic meaning of the word *religion*—"to bond together"—is lost.

It seems clear that an inclusive global spirituality, unifying spiritual identity, and ethical commitment are needed to draw the human family together across our deep differences. This can provide the spiritual vision and energy to help us cooperate wholeheartedly in saving our precious imperiled planet home. This is why helping people to liberate themselves from constricting ethical and theological ghettos is an essential goal of ecotherapy and preventing such spiritual exclusivism is a primary goal of ecoeducation.

Creation theologian Matthew Fox articulates this spiritual challenge:

> How can humans deal with cosmic energy and their responsibility for it, without a cosmic spirituality? . . .The human chauvinism that so narrows our vision that . . . we spend a million dollars a minute on weapons of cosmic destruction must cease. It is religion's task to reintroduce a cosmic vision, a less arrogant and less humanly chauvinistic way of seeing our world. Today it is more and more evident that the time has come for humanity to let go of war, to admit that it has outgrown war, and to move beyond war as a way of settling differences. Just as humanity one hundred years ago outlawed slavery, so it is capable of outlawing war. [21]

Those who have an interest in the health of religious institutions have an opportunity to influence these socioreligious systems in ecosalugenic directions. To do this effectively, we must start by critiquing our personal understanding of our particular religion to make sure that we are contributing to ecospiritual sanity rather than to earth-damaging spirituality.

The Pathology of Materialism

Another spiritual-ethical pathology contributing to the planet's ecojustice crisis is the idolatry of things, money, and "success" defined in materialistic terms. The worshipers of things, sometimes called "thing-a-holics," to some degree includes most of us who are relatively affluent. This is the most widespread addiction in affluent societies, driven by the costly illusion that the road to real happiness is via consumption. Disillusionment results in less, not more, happiness. The process of acquiring, consuming, wasting, and throwing away things produces massive destruction of precious, limited natural resources. It also exacts a high toll of human health by consumption of expensive, unhealthy foods and the mega-production of industrial pollution.

Ecopsychologists Allen Kanner and Mary Gomes observe that the mega-consumption habits of our culture have "a vast fantasy life that now rivals dreams as paths to insights about the irrational depths and everyday life—the mesmerizing power of media and advertising industry stimulates these fantasies to grow into a vast collection of projected desires, fears and aspirations."

This is a part of the "narcissistic wounding of the American psyche" by the advertising industry.[22]

The addictive idolatry of things blinds consciences as captives seek vainly to allay their anxieties by overdosing on consumption. Ethical tunnel vision prevents many affluent *oppressed oppressors* from seeing how their overconsumption is paid for by impoverished masses in the two-thirds world, on whose backs they ride economically. It tends to blind consciences to the ethical obscenity of 20 percent of the world's population consuming 80 percent of its limited natural resources. In addictive consumerism, the affluent contribute disproportionately to environmental destruction—for example, by pouring immense amounts of pollution, greenhouse-effect and ozone-depleting gases into the pool of oxygen from which all living things must breathe.

Healing the idolatry of things requires dealing with the vacuums in spiritual meaning and value that people try vainly to fill by compulsive consumption. People who become energized by responding to the ecological crisis (or other societal needs) often find an integrating cause that constructively fills the spiritual emptiness from which most of us suffer to some degree in our spiritually impoverished society.

In economically oppressed megacities of the two-thirds world, the Fourth World of native peoples, and the urban slums of rich countries, the problem is not, of course, overconsumption. Rather it is the fact that economic squalor forces the poor to sell their souls to acquire enough food and shelter to even survive, while they long for the privileges of the rich. In such settings in a variety of countries I have been appalled by stark evidence of both economic injustice and ecological destruction everywhere—in stifling air, disease-infested drinking water, dead rivers, growing noise and toxic pollution, and the extreme toxicity suffered by the poorest, including countless street children begging, stealing or selling their bodies in prostitution.

Spiritual Alienation from the Feminine

Another spiritual problem that contributes to ecoalienation and earth-damaging behavior is male-dominated religion. Both women and men are spiritually impoverished by religious institutions dominated by male authority figures (past and present), guiding images, meaning-making sacred stories and hymns. Both genders are impoverished by not learning to value the spiritual wisdom of female spiritual pioneers in many cultures and religions. Their inspired and inspiring writings often have been suppressed by powerful patriarchal religious leaders. Feminist scholars have shown that the self-esteem and spiritual empowerment of many women is enhanced when they are encouraged to use female images of the divine such as "Spirit," "Goddess," and "Sophia" (wisdom), as alternatives to male images such as "Lord," "King," and "Father." In my experience, men's spiritual well-being can also be strengthened by more androgynous understandings of religion. Both genders can be spiritually enriched by the complementary spiritual and ethical insights of women as well as men, in history and living today. Reclaiming the so-called "feminine" side of divinity and of themselves can open some women and men to deeper experiences of God.

How does all this relate to ecoalienation and earth-damaging lifestyles? Religiously sanctioned sexism reinforces powerfully our society's endemic sexism. As discussed earlier, when women's self-esteem and economic and educational opportunities are constricted, they tend to contribute more babies to our overpopulated planet. They also tend not to use their potential earth-saving insights and influence to help protect their bioregional environment by efforts in the public sphere. This is a great loss to their communities because women's experiences of birthing, nursing, and nurturing seems to keep many more of them connected with the earth and earth-rooted spirituality.

Men who accept religiously reinforced sexism are less likely to engage in earth-caring behavior, because the sexism alienates them from their nurturing side and from valuing male nurturing of nature as well as of children. Healing the sexist-generated alienation between the genders in ecotherapy can facilitate earth-caring partnerships between women and men. Healing sexist spirituality can help to heal ecoalienation, a process that may be furthered by inviting people to savor the wisdom of ecofeminists[23] and creation theologians like Hildegard of Bingen.

Rejection of Wildness

A related psychospiritual pathology discussed earlier is the spiritually motivated rejection of wilderness, of so-called "wild" animals, and the potentially valuable wildness within us humans. Dualistic cosmologies (world views) and religions often equate being "religious" and "ethical" with rejecting what Jung called the shadow side of the psyche and being "supercivilized," meaning conforming rigidly to cultural norms. Such spiritual-ethical systems feed the growing spiritual impoverishment caused by alienation from both internal and external wildness.

A significant goal of ecotherapy and ecoeducation is encouraging people to develop spiritualities and ethics that affirm and celebrate the wildness in their shadow side, channeling this valuable energy in ethically and ecologically creative directions. Sad to say, those suffering from severe alienation from wildness miss the energizing of their spirits, as well as of their bodies, minds, and sexuality, that comes from bonding with nature's wilderness. Helping people in therapy and education develop a positive bond with wildness in themselves and in nature can enliven and liberate their self-encapsulated lives, including their shallow, boring spirituality. It also can diminish defensive, paranoid projection of inner wildness onto persons and groups whose ethical and spiritual understandings are radically different from one's own, making them seem spiritually weird or dangerous. In our econuclear age, when global collaboration is required to save the planet, this age-old enemy-izing is more dangerous than ever before in the long, tragic saga of human violence.

Ecotheologian Thomas Berry, speaking to a group of artists, identified significant spiritual and artistic losses resulting from alienation from wildness:

> I speak just now of the wild dimension of existence and the reverence and fear associated with the wild since precisely here is where life and existence and art itself begins. When Thoreau said, "In wildness is the preservation of the world," he made a statement of unsurpassed significance in the history of human thought. I know of no more comprehensive critique of civilization itself than this immense effort that has been made over these past ten thousand years to bring the natural world under human control. An effort that would even tame the inner wildness of the human itself, to reduce those vast creative possibilities of the human to trivial modes of expression. . . . Wildness, we may consider as the rootedness of the authentic spontaneities of all being. It is that wellspring of creativity whence comes the instinctive activities that enable all living things to obtain food, to find shelter, to perform their mating rituals, to sing and dance and fly though the air and swim through the depths of the sea. This is the same inner tendency that evokes the insights of the poet, the skill of the artist, the power of the shaman.[24]

Jungian analyst Lynda Wheelwright Schmidt (mentioned in chapter 2) spent much of her girlhood on the huge Hollister ranch near Santa Barbara, California. In an essay entitled "Wilderness and the Sacred" she remembers:

> I grew up without organized religion or mention of God, so I turned to wilderness and nature instead. . . . It seems to me, though, that had I had "God" and an organized way of relating to God, I still would have needed a physical manifestation to give me an active connection with the sacred. . . . The wilderness, both as an idea and as a concrete reality, may be necessary to help balance the vast abstraction of religious imagery. Wilderness is big enough for the job of incarnation. . . . Thus, the wilderness is a sacred place. It offers a quest that one can enter into body and soul . . . [and] the wilderness is perhaps the richest place for the body to participate in the quest. And without the physical experience of the sacred, we can lose our way.[25]

She then describes her liberating experiences of self-transcendence in nature:

> I have been coping for years with natural inner and outer forces beyond me. The consciousness I gained from these efforts finally became a normal part of me that found its outlet in efforts . . . to be myself. This usually happens when I am in a natural setting. Perhaps one could say that this "beyond ego" experi-

> ence was and is an experience of the archetype of origin, and that its inward-outward manifestation is a source of creativity. I feel sure that this is so, because I nearly always come to life at those times when the inner, self-regulating system connects with the outer reality—when I am in Tepitates with wilderness around me. This is the time when the outer, elemental forces stimulate the inner, and the inner greater-than-me enhances the extraordinary meaningfulness of the outer. It is an experience of completion.[26]

A variety of methods (described in chapters 8 and 9) can help people enrich their lives, including their spirituality, by meeting and befriending their own wildness. These include dreamwork and guided daydream techniques; reliving memories of a time when the person felt safe, if not at home, in nature and was aware of the presence of the divine Spirit there; pet therapy may help in some cases, if a cared-about animal's wild roots are not totally obscured by domestication. Positive experiences in petting sections of zoos may help adults as well as children.

Magical, Manipulative Religious Beliefs

Another religious-ethical pathology is reality-denying beliefs. These enable some people and religious groups to maintain magical rescue fantasies of being saved from the brink of the ecological abyss by God, or perhaps rescued by the power and omniscience of religious or political leaders onto whom godlike expectations are projected. In our spiritually impoverished, violence-terrorized world such dangerously naive, denial-spawning beliefs are proliferating. Expecting a deity or godlike leader to fix things has the dangerous effect of deadening appropriate ethical responsibility to do all that can be done to help prevent ecosuicide of the planet. Such magical beliefs often are clutched rigidly like a terrified person floundering in a rushing river grasping at a floating straw. Confronting such dangerous defenses is an essential objective of ecological therapy.[27] Such beliefs often are interrelated with privatized religious beliefs, and with dichotomized thinking that favors either-or thinking and distrusts both-and thinking.

How does earth-caring relate to those understandings of Christians who believe that the present life is only a preparation for getting to an eternal home in heaven? A focus on the future life becomes problematic only when it is used to justify not doing everything possible to live responsibly in the present. "God so loved the world . . . " is the first clause of a scripture verse often quoted by Christians (John 3:16). This can be understood as supporting loving God's whole creation as God does, both the natural and the human world in the present.

Linking Social Justice and Ecology

In spiritual hyper-individualism devoid of concern for issues beyond individual ethics and salvation, issues of institutional injustice and ecology are often regarded as nonreligious problems. These approaches understand the arena of encountering the divine as limited to the world within individuals. They fail to see that justice-based earth-caring is vital to healthy spirituality today. The justice taproot of commitment to both social salvation and ecological salvation is severed. Spiritually motivated ethical concern for God's earth may be dismissed as dangerous "nature worship."

Those whose faith is so constricted fail to hear the spiritual imperative that confronts all humans today, whatever their faith. They miss what many religious people understand as the clear call of the divine Spirit to join hands with other people of goodwill in all faiths, to help save God's wonderful gift—life on the earth—by justice, peace, freedom, and environmental work. Spirituality today cannot be whole if it encourages no-shows in the struggles to create a just, healthy society in a healthy environment so that all persons will have the fullest possible opportunity to develop their God-given potentialities. Holistic spirituality offers the challenge of becoming co-creators with the God who loved *the world*, not just the souls of individuals in it. What is missing in privitized half-faiths is the prophetic ecological dimension of religious responsibility. As Al Gore astutely observes: "The idea of social justice is inextricably linked in the Scriptures with ecology. In passage after passage, environment degradation and social justice go hand in hand." [28]

Expanding Ethical Horizons

Another religious-ethical ecoproblem is the lack of a dynamic ethical concern for saving a viable planet. Some religious individuals and groups defend old, venerated beliefs and ethical guidelines that are no longer adequate to respond to current ecological realities. Even if old ethical concerns are still relevant to today's world, as many are, the unprecedented ecological crisis requires expanding our ethical horizons. Brian Swimme and Thomas Berry highlight this point: "In morality we are expanding our moral sensitivity beyond suicide, homicide, and genocide to include biocide and geocide, evils that were not recognized in our civilization until recently."[29]

The lack of a sense of ecological responsibility is a deadly dimension of widespread gaps in awareness of social responsibility. William Sloane Coffin, in his 1993 message to U.S. churches, paints a powerful picture of environmental irresponsibility. He declares that "we live in our environment as in a hotel, leaving the mess for others to clean up."[30]

Spiritual Deadness

As a result of spiritual-ethical ecoproblems like those described above, many people suffer from spiritual deadness. This prevents them from experiencing and celebrating the incredible gift of being alive and aware as living beings, organically interacting with a dynamic, living planet. They miss what holistic, health-enhancing religions often give people—*a love affair with life*, including both all of the biosphere and the Source of life. Healing spiritual deadness involves opening oneself to the energy of the living presence of a loving God, available in every here-and-now moment, including experiences of nurturing bonding with God's continuing creation. Some religious people are discovering that regular spiritual renewal, by opening themselves to the ultimate Source of love for the earth, is essential for sustained earth-caring and peacemaking. People who are having a love affair with life and the Source of life, will not tolerate passively the continuing destruction of the natural environment.

Contributions of Psychologists of Religion and Pastoral Counselors to Healing Spiritual Ecopathology

Professionals in the interdependent disciplines of psychology of religion and pastoral counseling usually have special training that provides the insights and skills that may facilitate the healing of complex religious problems, including some ecospiritual issues. Psychology of religion uses psychological insights and tools to understand salugenic and pathogenic religious beliefs and behaviors. Pastoral counselors (also called pastoral psychotherapists and pastoral psychologists) are a growing specialty within the ministry. They are trained to integrate insights from four sources: psychosocial research, contemporary psychotherapies, psychology of religion and the healing wisdom of their own (and often many other) religious traditions. They function in a variety of settings, including pastoral counseling centers, congregations, mental health programs, theological seminaries training clergy, and hospitals and prisons, as chaplain supervisors of seminarians and clergy in clinical pastoral education programs.

At an Asian Conference on Pastoral Care and Counseling in Indonesia, I was asked to lecture on "Healing Persons by Healing the Earth." The responses of pastoral counseling specialists from many Asian countries to the ecotherapeutic model and the challenge I offered was striking: "Our primary task as pastoral caregivers of the earth is to help heal the multiple religious and ethical sicknesses that feed humankind's cruel, exploitative treatment of the planet."[31] This response by one young man from India was typical: "In the village where I grew up everyone grew their own food. We took for granted that unless we lived in an earth-caring way we probably would not eat nor have healthy lives."

Theological teacher Larry Kent Graham is one theorist in pastoral psychology and counseling who also has recognized care of nature as a component of human caring. In his development of a psychosocial, process-oriented model of pastoral care and counseling, he observes that the ministry of care takes place at the interface between the six components comprising everyone's world: person, family, society, culture, nature, and God. He writes:

> Human life, society, and culture cannot function without nature. . . . Nature provides the pervasive context for our greatest achievements as human beings, as well as the greatest threats. We are born as a result of the interaction of social and natural processes. The threat to our existence as represented most ultimately by death itself, is fundamentally an inevitable natural process, however affected by human social experience. . . . If nature has given rise to human beings and shaped their social and cultural life, so has human society and cultural life shaped and defined nature.

It is clear in Graham's thought about caregiving, as in ecotherapy, that we must "replace our current philosophies of domination of nature with ecosystemic philosophers of partnership and bondedness with nature."[32]

Well trained pastoral psychologists and counselors are equipped to think in scientific ways about religious and ethical issues and pathologies, including those involved in earth-alienation. Their clinical and graduate academic training is designed to equip them to understand in critical, psychosocial ways the dynamics of people's beliefs and values, and to use the special skills needed to facilitate healing of pathogenic religious and ethical beliefs, as well as enable growth toward more salugenic spiritual and ethical living.[33]

In the current environmental crisis, a new partnership between scientists and religionists is needed. Mutual learning by pastoral psychologists and counselors and those in parallel secular professions should become a busy two-way street. In fact, there is an urgent need for a dynamic partnership between scientists in all the people and earth disciplines, professionals in all the healing and teaching disciplines, and scholars of religion in all the major faith traditions. All academic and professional disciplines, and particularly women in those fields, have special contributions to make in resolving the ecology crisis. Ecology is the ultimate multidiscipline field. Therefore, unprecedented collaboration among professionals (as well as lay people interested in those fields) is needed to create and implement overall public strategies for saving a viable planet for future generations.[34]

A pioneering theoretician in creating an ecologically relevant theology is John B. Cobb, Jr. He uses the rich resources of Alfred North Whitehead's process understanding of reality in this crucial intellectual task. Cobb identifies the key barrier that must be overcome if pastoral counselors and other religious thinkers are to be agents of sustainable living:

> With regard to the Earth, the economic and dominant modern theological paradigms have been all too similar. Both have been anthropocentric. The church as yet hardly knows what it means to implement its teaching of caring

> for creation. . . . When it [caring for creation] becomes a basic part of our regular worship and educational programs, the tension between its implications and the normal life of the congregation, as well as the normal lives of individual members, will be more keenly felt. Our history shows that this does not guarantee that change will follow, but it does offer hope. . . . Since our society operates primarily out of the economic paradigm, the critique of that paradigm and the way it shapes our institutions, beginning with the church, is urgent. As the church articulates the Biblical paradigm more clearly, it must also make explicit what is wrong with the economic one. It is as true today as ever that we cannot serve both God and mammon. Mammon's hold must be challenged and broken if the church is to be truly God's servant. It must also be broken if the Earth is to be saved from continuing degradation.[35]

Spiritually-Based Healing in Nature

In my clinical practice and workshops, spiritual awakening and healing frequently results from encouraging encounters with loved places in nature. This can be done by recommending that people revisit these places, either in fact or in their imagination. The two cases that follow illustrate how one skilled pastoral psychotherapist used an innovative imaging approach to facilitate spiritual growth via nature, when he was chaplain and therapist in an in-patient chemical dependency treatment program. The program integrates the twelve step recovery program of Alcoholics Anonymous, personal counseling, and medical aspects of a holistic treatment modality. His roles on the therapeutic team includes offering patients counseling help with the spiritual aspect of their recovery program. He describes how he worked with two patients around spiritual blocks in their recovery struggles.

One male patient was a self-described "program failure," having returned repeatedly to addictive drinking after completing several rounds of treatment. The chaplain-therapist sat with the man as he talked about his despair and his lost spirituality. "I know that the higher Power thing is what I need, but I just can't get it," the patient said. Sensing that he was near the deep despair of believing that his recovery was impossible, the chaplain suggested that they pass over the idea of God for the time being while the man told about himself, including the times when he had felt most sane. He recounted stories of horror in Vietnam, from where he had returned spiritually and emotionally empty and addicted, to then experience a series of job and marriage failures. He said he felt emotionally dead.

The man mentioned the death of his father and was about to pass over this huge loss quickly, when he said he had cried just once. After his father's funeral, he was camping with his current girlfriend near a rushing creek with a tiny island of granite boulders. He told of being able to cry as he sat alone on one of the rocks, surrounded by rushing water. When asked how that felt, he said, "It was so good to just let go." The chaplain-therapist, aware of this man's unhealed grief, invited him to close his eyes and "sit on that rock again." For long moments the patient was still as he relived this memory. When he finished, he said that it was just like being there again and that he had found the peace he had seldom felt during his addiction. Chaplain: "It sounds like you've found your higher Power." Patient: "Is it that simple?" Then, answering himself, he said, "It *is* that simple. It's really here inside me. I can sit on that rock any time."

Another young male patient had entered the hospital twenty-eight days earlier, drunk and utterly discouraged. Now he was about to leave, a profoundly changed man having completed the treatment program. Early in his treatment, the chaplain recalls, "I had asked about his faith and he had talked wistfully about believing in God when he was young." A series of disappointments had left him feeling alienated and hostile toward the meaningful faith he had learned as a child. The chaplain recalls: "As we surveyed his life, the man lit up as he spoke of a small ranch he once owned. It had a tiny house set back from a rural road. His favorite time of the day was walking to the mailbox, under an arch of trees lining the road. I invited him to follow that joy, to imagine himself under those trees once again. As he did so, he reexperienced that joy again. I suggested that he invite God to take that walk with him, to see what might happen. In the warm emotional environment of a much-loved natural setting, the man found that he was able to experience God's presence quite easily. He discovered that God's presence actually enhanced the feelings of well-being in revisiting that setting." The man made a daily routine of his comforting "walk with God." He told the chaplain, "I was afraid of God. On those walks I told him how angry and frightened I had been. I found that God could handle all I had to say. . . That walk you sent me on was the best thing that happened to me here in the hospital."[36]

Finding Help in Nature with Death Anxiety

While walking with a friend in a blossoming springtime apple orchard, Freud once commented that only those who can grieve effectively can enjoy such beauty fully because they know that it will soon pass. Existential anxiety, distinguished from neurotic anxiety, stems from the human awareness that we, and all those we love, are living-dying creatures. Counselors, psychotherapists, and health professionals deal with existential anxiety regularly, whether or not they are aware of this. People are struck now and again with painful experiences that remind them of their mortality—serious illness; accidents; the death of a loved friend, a relative, a dear pet, a dream; a birthday that reminds them of the swiftly shrinking future; or the countless everyday griefs of smaller losses, disappointments, failures, and esteem depletions. It is then that the volume gets turned up on the unwelcome background music that plays constantly and that is ordinarily ignored. This background music is the awareness that we are all mortal creatures. The existential anxiety triggered by this persistent but usually denied awareness is one of the subconscious motivations that make religions so need-satisfying and ubiquitous in all cultures.

Olivia Goldsmith, in her novel *The First Wives Club*, expresses the existential anxiety of one of her characters whose mother has just died: She now had "no one standing guard against her own mortality. There was no one now between her and eternity."[37] This awareness is what theologian Paul Tillich called humanity's "heritage of finitude." Philosopher Søren Kierkegaard, father of existentialism, once observed wisely that the anxiety about our mortality may be transformed into a "school"—an occasion for learning and creating. But only a vital religious faith or deeply meaningful philosophy of life can enable this to occur. During existential crises and despairing times, experiencing the divine Spirit of all life in nature, as one bonds with the living world, may awaken hope and transform this anxiety into a school. It is noteworthy that in his powerful poem "Canticle to the Sun," nature mystic Francis of Assisi (1181–1226), celebrates not only the brightness of the moon and stars, the divine light in Sister Water and Brother Fire, and the royal Mother Earth with all her bounty, but also Sister Death. As Matthew Fox observes, "Death for Francis is not to be feared but celebrated as part of the cosmic process."[38] Francis saw death, like all the little deaths of his life such as embracing and kissing the leper, which he considered his greatest act, as an experience of death and rebirth in the cosmic plan.

In some aware people, existential anxiety is triggered when they confront the profound (though usually ignored) mystery of time—the experience of the momentary present connecting the forever-past and the unknown future. This may be a major subconscious reason why many older people, whose future seems to shrink more and more rapidly, find nurturing comfort as well as pleasure in gardening and other hands-on work with living plants and the soil. Some grieving or depressed people discover healing-in-nature spontaneously. If not, it may be helpful to suggest that they spend extra time in their favorite natural setting. Walking and talking with such persons in such a place sometimes facilitates their grief work.

Pioneer naturalist and Sierra Club founder John Muir spoke insightfully of the learning opportunities in nature for children:

> On no subject are our ideas more warped and pitiable than on death. Let children walk with nature, let them see the beautiful blendings and communions of death and life, their joyous inseparable unity, as taught in the woods and meadows, plains and mountains and streams of our blessed star, and they will learn that death is stingless indeed, and as beautiful as life, and that the grave has no victory, for it never fights. All is divine harmony.[39]

The same experiences may bring healing to death-fearing adults.

During my many years of interest in healing grief (my own and that of others), numerous experiences have confirmed that being nurtured by nature often helps grief sufferers handle agonizing losses by generating earth-rooted spiritual healing. When people connect with the life-death-rebirth cycles of nature, they often discover that they are viewing death from a different perspective. Though still profoundly saddened, they are no longer devastated or immobilized.

Spiritual-Ethical Resources for Earth-Caring in Jewish and Christian Creation Theologies

Numerous resources for earth-caring can be found in the sacred writings of the Jewish and Christian traditions. The intertwined biblical view of human and earth well-being is articulated clearly. We humans do not own the planet but only have it in trust from God—"The earth is the LORD's and all that is in it, the world and those who live in it" (Ps. 24:1, NRSV). Respectful stewardship toward all creation is celebrated in the ringing refrain repeated after each epoch of the first creation story—"God saw that it was good" (Gen. 1:25). And at the end of the process: "And God saw everything that he had made and indeed, it was very good" (Gen. 1:31). The psalmist declares: "O LORD, how manifold are your works! In wisdom you have made them all; the earth is full of your creatures" (Ps. 104:24). In discussing dietary laws, the apostle Paul's first letter to the young Timothy states: "Everything created by God is good, and nothing is to be rejected, provided it is received with thanksgiving" (1 Tim. 4:4).

The second creation story (in Genesis 2) symbolizes the inherent earthiness of humans by picturing us as created out of the dust of the earth, into which we are recycled at death. There is a semantic linkage of the Hebrew words for humanity (*adam*), and soil (*adamah*). The universe is depicted there as a beautiful garden where humankind's responsibility is "to till it and keep it" (Gen. 2:15). This implies working to make the garden productive with watchful, protective care. Apparently this is what "have dominion over the earth" really means. Unfortunately two creation themes—"Be fruitful and multiply" (Gen. 1:28) and "have dominion over all the earth" (Gen. 1:26)—have been misunderstood or applied apart from the very different historical context of the story, to justify the overpopulation that today is producing ecological disasters on our polluted planet, crowded with "wall-to-wall people," and hurting or destroying other creatures with no ethical restraints.

Both Jews and Christians think of themselves as people of the covenant (or sacred agreement) with God. All the biblical covenant stories include care of every other living creature along with humans. In the rainbow covenant of the Noah story, humans are charged with insuring the future of all living creatures by taking a productive pair aboard the ark. At the conclusion, the covenant, restated seven times, includes both humans and all creation (Gen. 9:1-17). The covenant that Moses is said to have negotiated with God on Mount Sinai was the same. It included care for the land, regulations for sanitation in the camp, and feeding wild animals. The "new covenant" for the future (Hos. 2:18 and Isa. 11:6-9) includes responsibility for all creatures, and the establishment of shalom, meaning wholeness and health for God's whole creation. Mainline Christians believe that Jesus is the fulfillment of this new covenant that Paul describes as setting all creation free from its bondage to decay (Rom. 8:21).[40]

Jesus grew up immersed in this Hebrew ecoprophetic world view. He lived in nourishing intimacy with the earth. This is reflected in his frequent use of teaching images from nature, like flowers of the field, birds of the air, weeds in a wheat field, and a farmer planting seeds. He used such images to express truths about God's coming "shalom" community, which in the Jewish world view that he shared included both the human and ecological communities. He did his teaching outdoors, for the most part, and on several occasions, his healing work involved the symbolic use of water and earth.

A passionate concern about the ecological brokenness produced by oppressive institutions (the ecojustice theme) is an ancient refrain in the prophetic heritage of the Hebrew Scriptures. An example is the prophet Isaiah's century-spanning awareness of the ecological devastation resulting from spiritual alienation and sin: "The earth dries up and withers. . . the heavens languish together with the earth. The earth lies polluted under its inhabitants; for they have transgressed laws . . . broken the everlasting covenant" (Isa. 24:4-5). Speaking for God, Jeremiah laments: "I brought you into a beautiful land to eat its fruits and its good things. But when you entered you defiled my land, and made my heritage an abomination" (Jer. 2:7). The prophet Hosea held that all creation suffers because of humans' sins against each other, including "bloodshed following bloodshed": "Therefore the land mourns, and all who live in it languish; together with the wild animals and the birds of the air, even the fish of the sea are perishing" (Hos. 4:3). Throughout the Bible, individual well-being and salvation are seen as occurring in the context of the well-being of the community of faith and of creation. Thus justice and ecological issues are inextricably intertwined.

Spiritual Healing in Nature Hymns

Nature images and music have long communicated nurturing spiritual energy for many Christians. From the thirteenth century, this poem-song by Francis of Assisi communicates the way in which nature mystics experience nature's celebration of the divine Spirit:

All creatures of our God and King, lift up your voices and with us sing,
O praise ye! Alleluia!
O brother sun with golden beam, O sister moon with silver gleam!
O praise ye! O praise ye! Alleluia! Alleluia!
O brother wind, air, clouds, and rain, by which all creatures ye sustain,
O praise ye! Alleluia!
Thou rising morn, in praise rejoice, ye light of evening, find a voice!
O praise ye! O praise ye! Alleluia! Alleluia!
O sister water, flowing clear, make music for thy Lord to hear, Alleluia! Alleluia!
O brother fire who lights the night, providing warmth, enhancing sight,
O praise ye! O praise ye! Alleluia! Alleluia!
Dear mother earth, who day by day, unfoldest blessing on our way, Alleluia! Alleluia!
The flowers and fruits that in thee grow, let them God's glory show!
O praise ye! O praise ye! Alleluia! Alleluia![41]

A more recent hymn (1864) has been sung and loved by countless Christians:

For the beauty of the earth, for the glory of the skies,
for the love which from our birth, over and around us lies;
Lord of all, to thee we raise, this our hymn of grateful praise.
For the beauty of each hour, of the day and of the night,
hill and vale, and tree and flower, sun and moon and stars of light;
Lord of all, to thee we raise, this our hymn of grateful praise.[42]

Theology's Contribution to Healing

Clergy and others who find that theological reflection illuminates their religious experiences and beliefs are fortunate today in the rich variety of theologies that shed light on the spiritual and ethical roots of the ecojustice crisis. The major theological approaches that I find most relevant are creation theologies, the ecofeminist theologies (see books by Sallie McFague and Rosemary Radford Ruether), process theologies (see the writings of John B. Cobb, Jr.,[43] David Griffin, and Jay McDaniel), and liberation theologies that recognize that liberating society from injustice-bred oppression is essential for individual human liberation and for saving the environment.

For ecotherapy's spiritual growth work, it is fortunate that creation theologies are flowering. For those of us who regard modern sciences as a source of continuing insights for rethinking our cosmologies and theologies, creation theologians are valuable sources. Brian Swimme and Thomas Berry express a motif of creation theology: "Religion begins to appreciate that the primary sacred community is the universe itself. . . . To preserve the natural world as the primary revelation of the divine must be the basic concern of Religion."[44]

Creation mystics like Thomas Berry and those who lift up the ancient image of the cosmic Christ, like Matthew Fox, offer invaluable resources for getting in touch with spirituality that is grounded in the earth. I find ecological illumination in the writings of Hildegard of Bingen, Meister Eckhart, Julian of Norwich, Francis of Assisi, and (currently) Matthew Fox. Hildegard's green creation mysticism speaks deeply to my ecospiritual hungers with her teaching that every creature is an expression of God's radiance, that we humans "have a natural longing for other creatures and feel a glow of love for them. We often seek out nature in a spirit of delight."[45] Matthew Fox shows how the great creation mystics were filled with "cosmic wisdom," which they sometimes called the "Cosmic Christ." This ecospiritual wisdom is shared by many native spiritualities and is described with various other terms in the world's major religions. This ancient wisdom honors the earth, the sacredness of life, the human-earth bond, sexuality and women. Fox holds that returning to this Cosmic Christ wisdom is essential for healing the spiritual causes of violence against the earth and its people, particularly against women and youth. He holds that recovering the Cosmic Christ can help spark the dawning of a global renaissance of peace and justice.[46]

Earth-Caring Resources in Other Religious Traditions

Other major religious heritages of the human family also offer rich resources for people in many cultures who are searching for spiritual grounding of the global environmental ethic that is urgently needed today. Fortunately, Mary Evelyn Tucker and John A. Grim have edited an in-depth study of the contributions of various religions to such an ethic.[47] Al Gore's *Earth in the Balance* provides a succinct summary of ecological resources in different religions. Here are some highlights:

Sikhism, a monotheistic offshoot of Hinduism in northern India dating from 1500 C.E., emphasizes the spiritual lessons humans can learn from nature. Guru Nanak, the Sikh founder, wrote: "Air is the Vital Force, Water the Progenitor, the Vast Earth the Mother of All: Day and Night are nurses, fondling all creation in their lap."[48]

Islam is a major world religion often misunderstood in the West. In its sacred book, the Qur'an, the voice of Allah states that the earth is the sacred creation of Allah:

> Have We not made the earth as a cradle and the mountains pegs? And created you in pairs, and We appointed your sleep for rest; and We appointed night for a garment, and We appointed day for a livelihood. And We . . . appointed a blazing lamp and have sent down out of the rain clouds water cascading that We may bring forth thereby grain and plants, and gardens luxuriant.

The Prophet Muhammed declared, "Whoever plants a tree and diligently looks after it until it matures and bears fruit is rewarded."

As one who loves rivers, lakes, and oceans, I resonated to Al Gore's observation that the sacred quality of water is a shared motif in many religions:

> Christians are baptized in water, as a sign of purification. Qur'an declares that "we have created everything from water." In the Lotus "Sutra," Buddha is presented metaphorically as a "rain cloud," covering, permeating, fertilizing, and enriching "all parched living beings, to free them from misery to attain the joy of peace, joy in the present world and joy of Nirvana . . . everywhere impartially without distinctions of persons." The sacredness of water receives perhaps the greatest emphasis in Hinduism. According to its teachings, the "water of life" is believed to bring to humankind the life force itself."[49]

The Eastern religions and world views have numerous nondualistic, ecological, and animistic themes. The nonviolent ones—Hinduism, Jainism, and Buddhism—discourage violence against nature. The Dalai Lama has said: "If peace does not become a reality in the world and the destruction of the environment continues as it does today, there is no doubt that future generations will inherit a dead world." Taoism, the most ecologically aware Eastern perspective, reflects a creation cosmology. In *The Way of Life* by Lao Tzu, Taoism's founder provides guidance for relating to the earth as well as to people in harmonious, noninvasive, nonmanipulating ways.

Anand Veeraraj, a Hindu interviewed in preparation for this book, pointed out that "Hinduism is a nature religion. Many entities in nature—animals, trees, plants, birds—these are personified and woven into folk stories. Images and idols are personifications of these nature entities."

> A modern Hindu environmentalist, Dr. Karan Singh, regularly cites the ancient Hindu dictum: "The earth is our mother, and we are all her children." And in the Atharvaveda, the prayer for peace emphasizes the links between humankind and all creation: "Supreme Lord, let there be peace in the sky and in the atmosphere, peace in the plant world and in the forests; let the cosmic powers be peaceful; let Brahma be peaceful; let there be undiluted and fulfilling peace everywhere."

During my teaching and learning experiences in a variety of African, Eastern, South Pacific, South American, and Asian cultures, I have been struck repeatedly with how organic bonding with and respect for nature are basic in their cultural world views and religious life. As in the Western religious traditions, however, adherents often do not follow in practice the earth-caring beliefs and ethics of their faith systems.

Earth-Caring Resources in Ecofeminist Theology

A central and vital message of ecofeminists, including the ecofeminist theologians, is to challenge the dualism and hierarchies that are deeply entrenched in Western patriarchal thinking and replace them with an awareness that all things are related and interdependent. Hierarchical dualism splits apart the sacred and the secular, mind and the earth, men and women. The first item in each of these dualisms is more elevated and valued, and the second is subservient and less valued. This dichotomizing leads to social and religious domination by males and violence against both women and nature. Historically, the major Western religions often have reinforced this hierarchical dualism by theological justifications.

Chung Hyun Kyung, a Korean Christian theologian and author of *Struggle to Be the Sun Again*, has developed an Asian women's theology that incorporates strengths from the rich, often-ignored spiritual heritage of women in Asia. Recovering this heritage often empowers women and also results in efforts by them to save the planet. She gives evidence that Asian women believe that God is creator of all the splendor and endless variety of our beautiful universe. They experience this creating God in their own creating—giving birth, gardening, preparing food from the earth, and communicating.

> When Asian women touch their own creativity and create their own healing, they touch the life source—God. . . . Women get out of their oppression and create alternative structures which are life-giving for them, they meet their God through the process of liberation. . . . In their faith, Asian women know they are invited as God's partners to the covenant of "justice peace and the integrity of creation."[50]

In *The Body of God: An Ecological Theology*, feminist theologian Sallie McFague offers rich spiritual and ethical resources that can be used in ecotherapy and ecoeducation. She artfully balances the traditional Christian emphasis on God's transcendence with not only an immanental model, but also with an understanding of God's transcendence in immanental ways. McFague does this by viewing the world organically as our meeting place with God and seeing humans as inspirited bodies deeply interrelated with all other bodies (meaning everything else in the universe), but with unique responsibilities to be partners with God in helping all life prosper. Hers is an "earth-up" rather than a "sky-down" way of understanding God's relation to us and all the biosphere.

Another prominent ecofeminist theologian, Rosemary Radford Ruether, uses Gaia, the Greek earth goddess, as a metaphor in creating an ecofeminist theological critique of the ecological inadequacy of Western patriarchal religious traditions to understand our relation to our living planet. She declares: "A healing relation with each other and to the earth then calls for a new consciousness, a new symbolic culture and spirituality."[51] This will help transform our minds and thus facilitate healing relations between women and men, humans and the earth, the divine and the earth, and thus humans and the divine.

Another wellspring of ecofeminist spirituality resources is found in Part

Three of *Healing the Wounds: The Promise of Ecofeminism*, edited by Judith Plant. Entitled "She Is Alive in You: Ecofeminist Spirituality," this section includes chapters on "A Gandhian Perspective in the Himalayan Foothills" and "Sacred Land, Sacred Sex." The annotated bibliography of this volume is a helpful guide to other ecofeminist books. Another fruitful source of ecofeminist insights on spirituality is *Reweaving the World, The Emergence of Ecofeminism*, edited by Irene Diamond and Gloria Feman Orenstein. It includes chapters on reconnecting ethics and politics, deep ecology and ecofeminism, bioregionalism and ecofeminism, and one on "Goddess in the Metropolis: Reflections on the Sacred in an Urban Setting." I find feminist process theologian Catherine Keller's piece on "Women Against Wasting the World" particularly illuminating.

Earth-Caring Resources in Native American Spirituality

Viewpoints paralleling those of some ecofeminists are found in the spirituality of many native peoples. Paula Gunn Allen, a Laguna Pueblo-Sioux Indian scholar, identifies a variety of spiritual ecoresources from a female Native American perspective. She draws on her own learnings as a little girl:

> When I was small, my mother often told me that animals, insects, and plants are to be treated with the kind of respect one customarily accords to high-status adults. "Life is a circle, and everything has its place in it," she would say. That's how I met the sacred hoop, which has been an integral part of my life.[52]

Native American world views are of a nondualistic, circular, and dynamic universe in which all things are related and parts of one family. Songs, legends, and ceremonies remind Indians that all creatures (including humans) are interrelated parts of one whole, living universe. This view is the spiritual wellspring of all Native American healing.

> In American Indian thought, God is known as the All Spirit, and other beings are also spirit—more spirit than body, more spirit than intellect, more spirit than mind. The natural state of existence is whole. Thus healing chants and ceremonies emphasize restoration of wholeness, for disease is a condition of division and separation from the harmony of the whole. Beauty is wholeness. Health is wholeness. Goodness is wholeness. . . . The circle of being is not physical, but it is dynamic and alive. It is what lives and moves and knows, and all life forms we recognize—animals, plants, rocks, wind—partake of this greater life.[53]

Ed McGaa Eagle Man, an Oglala Sioux lawyer with expertise in Native American spirituality and rituals, also emphasizes earth-grounded spirituality as what the human family needs for healing. In his *Mother Earth Spirituality: Native American Paths to Healing Ourselves and Our World*, he expresses his appreciation "to all the strong dominant grandmothers of all tribes who would not let the Mother Earth Spirit die and kept whispering it into our ears when we were little ones." He asks the question that apparently had been directed to him: Why is he willing to share the spiritual heritages of native people? His response:

> We do not have any choice. It is one world that we live in. If the Native Americans keep all their spirituality within their own community, the old wisdom that has performed so well will not be allowed to work its environmental medicine on the world where it is desperately needed . . . all us two-leggeds will have to work together to get ourselves out of it. A spiritual fire that promotes a communal commitment to a worldwide environmental undertaking is needed. Native or primal ways will fuel that fire and give it power.

He quotes with approval these words of Frank Fools Crow, an Oglala holy man and ceremonial chief of the Teton Sioux tribe, regarding Native American healing rituals: "These ceremonials do not belong to Indians alone. They can be done by all who have the right attitude . . . and who are honest and sincere about their beliefs in Wakan Tanka (Great Spirit)."[54]

Native American leader Vine Deloria has written:

> The Indian is confronted with a bountiful earth in which all things are experienced as having a role to fill. . . . Primarily in the world with which he is confronted is the presence of . . . the manifestation of life energies, the whole lifeflow of a creation. Recognition that the human being holds an important place in such a creation is tempered by the thought that he is dependent on everything in creation for his existence.[55]

Returning to Saint Francis's creation-loving, singing spirit, I savor heart-warming memories of a 1994 visit to his city in Italy. It became clear to me there that his compassion for poor, oppressed humans in his violent times was rooted in his profound spiritual bonding with nature. Being in the woods, the mountain-side grottos and olive orchards in Assisi put me in touch again with the spirits of Francis and Clare, two ecological saints. It is well to remind ourselves, now and again, what is really at stake for each of us in greening our attitudes, values, and spirituality. A poem by two contemporary lovers of St. Francis' celebrative spirit (entitled simply "Earth") does just this:

> Who is not moved by the Beauty of the Earth?
> Newborn each day,
> Dew-drenched each night,
> New yet old as time.
> The Earth
> Who feeds and governs us—
> We love you!
> Praise be for Mother Earth![56]

If your sense of wonder at the universe's grandeur needs refreshing or your perspective on personal and world problems needs enlarging, try this experiment. On a moonless night, away from the pollution of city haze and lights, look up to the night sky. Find Polaris, the North Star, around which the stars and constellations seem to rotate in the Northern Hemisphere. Then say to yourself: *That star is actually a huge sun 5,200 times brighter than our sun. Through the millenia, its light has guided sailors over the trackless seas. The light that I see tonight has been coming through space, traveling at 186,000 miles every second of every day for the last 316 years—so fast it could have circled our little planet seven times in one second. The light I see left the North Star only about forty years after the pilgrim refugees from England landed at Plymouth. Yet Polaris is a nearby star in our galaxy's immediate neighborhood.*

Now, if it's a crystal clear night, find the Milky Way and say to yourself: *What looks like a path of hazy star dust across the sky is actually a side view of our gigantic galaxy—our little solar system's home. It takes light, traveling at 186,000 miles per second, 130,000 years to go from one side to the other of this gigantic swirling star city of 100 billion suns.* Then let yourself reflect on a recent discovery of the Hubble space telescope, which enables astronomers to estimate that there are 50 billion galaxies in the whole universe. Light from the farthest of these has been flashing through space for some ten billion years before being photographed by the Hubble.

Now, for awareness of one down-to-earth wonder, say to yourself: *Each of the human brains that created that telescope—and my brain that seeks to grasp its mind-boggling findings—has some 10 billion nerve cells. Collectively, in our species, these make all human science and culture possible.* Such a cosmic perspective may help make our lives and all life on our green, growing spaceship, wonderously precious, perhaps even joyous. As the Bible's ancient Book of Job declares poetically, "The morning stars sang together" (Job 38:7).

Exploring this Chapter's Issues Further

For fuller information about these books, see the annotated bibliography at the end of this book.

Carol J. Adams, ed., *Ecofeminism and the Sacred*

Thomas Berry and Thomas Clarke, *Befriending the Earth*

Charles Birch and John B. Cobb, Jr., *The Liberation of Life*

Howard Clinebell, "Facilitating Spiritual Wholeness: The Heart of Pastoral Care and Counseling," chap. 5 in *Basic Types of Pastoral Care and Counseling*; "Enriching and Enjoying Your Spiritual Life—Wellspring of Love, Well Being, and Joy," chap. 2 in *Well Being*

Charles Cummings, *Eco-Spirituality: Toward a Reverent Life*

Al Gore, *Earth in the Balance*

Catharina J. M. Halkes, *New Creation: Christian Feminism and the Renewal of the Earth*

Jay B. McDaniel, *Earth, Sky, Gods and Mortals: Developing an Ecological Spirituality*

Seam McDonagh, *The Greening of the Church*

Sallie McFague, *The Body of God: An Ecological Theology*

Thomas Moore, *Care of the Soul*

Judith Plant, ed., *Healing the Wounds: The Promise of Ecofeminism*, Part 3, "She Is Alive in You: Ecofeminist Spirituality"

Lewis G. Regenstein, *Replenish the Earth*

Elizabeth Roberts and Elias Amidon, eds., *Earth Prayers from Around the World*

Holmes Rolston, *Environmental Ethics: Duties to and Values of the Natural World*

Rosemary Radford Ruether, *Gaia and God: An Ecofeminist Theology of Earth Healing*

H. Paul Santmire, *The Travail of Nature*

Brian Swimme and Thomas Berry, *The Universe Story*

Mary Evelyn Tucker and John A. Grim, eds., *Worldviews and Ecology*

Notes

1. Linda Filippi highlights this quotation from her Quaker religious tradition.

2. Anne Frank, *The Diary of a Young Girl*, ed. Otto H. Frank and Mirjam Pressler, translated by Susan Massotty (New York: Doubleday, 1995), 196.

3. *Earth in the Balance: Ecology and the Human Spirit* (New York: Penguin Books, 1993), 267, 238–39, 240. It is noteworthy that Gore studied for a time at the Divinity School of Vanderbilt University, a scholarly graduate school, before studying law. His sophistication in dealing with spiritual and ethical issues in ecology may well reflect his learnings there.

4. Ibid., 367–68.

5. As noted in chapter 1, the importance given to spiritual and ethical issues undoubtedly was greater among this survey's respondents than among therapists and teachers in general, because of the high proportions of pastoral psychotherapists within the sample.

6. *Modern Man in Search of a Soul*. New York: Harcourt Brace, 1933, 229. Ruth Topping, a Chicago social worker, wrote to Jung in 1959 asking that he explain what he meant by "a religious outlook." Jung responded from Zurich, but his letter was not made public until it was published in the *New York Times*, 19 November 1993; republished in the *Santa Barbara News Press*, 2 January 1994, H2. He wrote:

> When you study the mental history of the world, you see that people since times immemorial had a general teaching or doctrine about the wholeness of the world. The teaching had always a "philosophical" and ethical aspect.
>
> In our civilization this spiritual background has gone astray. . . . Thus one of the most instinctual activities of our mind has lost its object. As these views deal with the world as a whole, they create a loss of wholeness of the individual, so much so that for instance a primitive tribe loses its vitality when it is deprived of its specific religious outlook. People are no more rooted in their world and lose their orientation. They just drift. This is very much our condition, too. The need for a meaning of their lives remains unanswered, because the rational, biological goals are unable to express the irrational

wholeness of human life. Thus life loses its meaning. That is the problem of the "religious outlook" in a nutshell. C. G. Jung"

8. "Spiritual" in this context includes religious experiences but also the whole range of ethical, aesthetic, and uniquely human values.

9. *Care of the Soul: A Guide for Cultivating Depth and Sacredness in Everyday Life* (New York: HarperCollins, 1992), 270. Soul, in Moore's view, is not a thing but rather a dimension of experiencing ourselves and life. It has to do with value, depth, beauty, and relatedness. It is noteworthy that Theodore Roszak also discusses *anima mundi*, the great mother soul of the earth, in *The Voice of the Earth* (New York: Simon & Schuster, 1992), chap. 5. It is noteworthy that the word "soul" is female in most if not all cultures.

10. Moore, *Care of the Soul*, 271.

11. Ibid., 273.

12. *Insight and Responsibility* (New York: W. W. Norton & Co., 1964), 242.

13. See "Diagnosing and Treating Spiritual Problems," in *Basic Types of Pastoral Care and Counseling* (Nashville: Abingdon Press, 1984), 116–24.

14. Barney directed the *Global 2000 Report* for President Jimmy Carter, the only report to any national government on the economic, demographic, resource, and environmental future of the world. It sold 1.5 million copies in eight languages. The full report of his challenge to religious leaders appears in *Global 2000 Revisited: What Shall We Do?* (Arlington, Va.: Millennium Institute, 1993), distributed by Public Interest Publications, P.O. Box 229, Arlington, VA 22210.

15. Plenary address, ibid., 24–25.

16. Ibid, 24–25.

17. Ibid., 24–25, 31.

18. For an in-depth presentation of a variety of methods for treating spiritual and ethical pathologies, see two of my earlier books, *Basic Types of Pastoral Care and Counseling*, chap. 5, "Facilitating Spiritual Wholeness"; and chap. 6, "Counseling on Ethical, Value, and Meaning Issues"; and *Well Being*: chap. 2, "Enriching and Enjoying Your Spiritual Life—Wellspring of Love, Well Being, and Joy."

19. "Shamanic Counseling and Ecopsychology," in *Ecopsychology: Restoring the Earth, Healing the Mind*, edited by Theodore Roszak, Mary E. Gomes, and Allen D. Kanner (San Francisco: Sierra Club Books, 1995), 182.

20. Psychologist Chellis Glendinning provides evidence that substance abuse has characteristics comparable to the addiction in urban industrial society to technology and its products. She labels this "techno-addiction." See "Technology, Trauma, and the Wild," in *Ecopsychology*, 41ff.

21. Fox, *Original Blessing: A Primer in Creation Spirituality*. (Santa Fe: Bear & Co., 1983), 13.

22. "The All-Consuming Self," in *Ecopsychology*, 78, 88.

23. See books by Carol Christ, Rosemary Reuther, Sallie McFague.

24. Unpublished paper, "The Wild and the Sacred," presented to an artists' group in New York City. October 1993, 2, 3.

25. *The Long Shore*, 191.

26. Ibid.

27. In discussing methods of spiritual counseling elsewhere, I suggested: "Helping people transform their wholeness diminishing beliefs is seldom easy. All of us tend to resist tenaciously because immature, authority-centered beliefs give an illusion of security that comes from avoiding the risk of living out the freedom of an open, grow-

ing faith. Facilitating spiritual self-transformation usually requires all the counseling skills one can muster. In this process, it may be necessary to deal with the pathogenic, growth-inhibiting beliefs directly, to question [gently] their validity, and suggest and model by one's attitudes and behavior, more growthful beliefs" (*Basic Types of Pastoral Care and Counseling*), 120.

28. *Earth in the Balance*, 247.

29. *The Universe Story: From the Primordial Flaring Forth to the Ecozoic Era, A Celebration of the Unfolding Cosmos* (San Francisco: Harper San Francisco, 1992), 257.

30. *A Passion for the Possible* (Louisville: Westminster/John Knox Press, 1993), 10.

31. See Mesach Krisetya, ed., *Pastoral Care and Counselling in a Pluralistic Society* (Bali, Indonesia: Fifth Asia Conference on Pastoral Care and Counseling, 1993).

32. *Care of Persons, Care of Worlds: A Psychosystems Approach to Pastoral Care and Counseling* (Nashville: Abingdon, 1992), 61–62, 44. I would add to Graham's point that the philosophies of domination of nature often associate women with the earth, thus linking domination of nature with domination of women

33. Some professionals in secular teaching and healing disciplines are becoming aware of the special competencies of pastoral psychologists and counselors, and are drawing on their expertise when they encounter complex or enigmatic spiritual and ethical issues troubling clients or students.

34. A large international group of scientists generated this Appeal on the Environment and sent it to leaders of all the world's major religions:

> As scientists, many of us have had profound experiences of awe and reverence before the universe. We understand that what is regarded as sacred is more likely to be treated with care and respect. Our planetary home should be so regarded. Efforts to safeguard and cherish the environment need to be infused with a vision of the sacred. At the same time, a much wider and deeper understanding of science and technology is needed. If we do not understand the problem, it is unlikely we will be able to fix it. Thus there is a vital role for both religion and science. We know that the well-being of our planetary environment is already a source of profound concern in your councils and congregations. We hope this Appeal will encourage a spirit of common cause and joint action to help preserve the Earth.

35. John B. Cobb, Jr., "The Church and Sustainable Living," *A New Communion, Justice, Ecology, Spirit, Community* 10, (February 1995): 4, 17.

36. I am grateful to Robin Crawford, Ph.D., for sharing these accounts of his spiritual therapy work.

37. New York: Bantam Books, 1992, 427.

38. Matthew Fox, *The Coming of the Cosmic Christ: The Healing of the Mother Earth and the Birth of a Global Renaissance* (San Francisco: Harper and Row, 1988), 114.

39. Peter McWilliams and John-Roger, *You Can't Afford the Luxury of a Negative Thought* (Los Angeles: Prelude Press, 1988), 118.

40. For further information about biblical attitudes toward the earth, see James A. Nash, *Loving Nature: Ecological Integrity and Christian Responsibility* (Nashville: Abingdon Press, 1991), and a monogram entitled "The Bible and the Earth," in *God's Earth, Our Home*, a packet for congregational study and action on the environment, economic justice, and hunger, edited by Shantilal P. Bhagal and published by the National Council of Churches of Christ in the USA in 1994.

41. *The United Methodist Hymnal* (Nashville: The United Methodist Publishing House, 1989), 62.

42. *Ibid.*, 92.

43. An major foundation-laying volume for developing an ecological theology and approach to life is *The Liberation of Life: From the Cell to the Community* (Cambridge: Cambridge University Press, 1981). This important book is co-authored by Australian biologist Charles Birch and American process theologian John B. Cobb, Jr. It integrates the expertise of the authors, who spell out a unified, ecological model for understanding the interrelations of the wide-ranging issues related to the global environmental crisis. They describe a valuable biological and theological model for a just and sustainable world.

44. *The Universe Story*, 243, 257.

45. *Hildegard of Bingen's Book of Divine Works, with Letters and Songs*, edited by Matthew Fox (Sante Fe, N.M.: Bear & Co., 1987), 128, 129.

46. See Fox, *The Coming of the Cosmic Christ*. Richard Voss reports that the Episcopal Cathedral of Saint John the Divine in Manhattan (sometimes called the "Green Cathedral") now has a side altar dedicated to the earth, on which several aquariums represent the tide of life. One is filled with algae and bubbling water.

47. See *Worldviews and Ecology* (Lewisburg, Pa.: Bucknell University Press, 1993). Scholars with expertise in these religions discuss their beliefs about ecological questions: Native American, Judaism, Christian, Hindu, Buddhist, Jainism, Taoism, Confucianism. Also included is an illuminating chapter by David Ray Griffin on the most ecologically relevant contemporary philosophy, the world view of Alfred North Whitehead.

48. Gore, *Earth in the Balance*, 261.

49. Ibid, 261.

50. Maryknoll, N.Y.: Orbis Books, 1993, 49. It is heartwarming to me that Chung Hyun Kyung was a graduate student at the School of Theology at Claremont, California, while I was teaching there.

51. See *Gaia and God: An Ecofeminist Theology of Earth Healing* (San Francisco: Harper San Francisco, 1992), 4. Her use of Gaia builds on the work of biologists James Lovelock and Lynn Margulis, who use the concept to refer to their thesis that the entire earth is an organism functioning as a living system.

52. Paula Gunn Allen, *The Sacred Hoop: Recovering the Feminine in American Indian Traditions* (Boston: Beacon Press, 1986), 1.

53. Ibid., 29, 60–61. Allen shows how the powerful images of women in Native American cultures are in sharp contrast to Western patriarchal images of women. "Pre-Conquest American Indian women valued their role as vitalizers. . . . Through their own bodies they could bring vital beings into the world. . . . The power of mothering was not reduced to simply biological creativity. It was understood as "the power to create and transform, a spiritual power, the power of Creation Thinking." The mother-ritual based Indian cultures are, according to Allen, "also cultures that value peacefulness, harmony, cooperation, health, and general prosperity, they are systems of thought and practice that would bear deeper study in our troubled, conflict-ridden time."

54. Ibid, p. 28, 29.

55. Vine Deloria, *God is Red*, 102.

56. Francis Raymond Line and Helen E. Line, *Man with a Song: Some Major and Minor Notes in the Life of Saint Francis of Assisi* (Garden City, N.Y.: Image Books, 1982), 130.

5

Resources for Ecotherapy and Ecoeducation from Other Therapies and Cultures

When my office was [moved]. . . closer to nature, I listened to clients in new ways. And between sessions I gardened. . . . I gradually learned about a growing and healing process that was more encompassing and powerful than any of the psychotherapeutic techniques I'd previously struggled to emulate. The garden was my teacher.

—Lesley Irene Shore, *Tending Inner Gardens, The Healing Art of Feminist Psychotherapy*[1]

Ecotherapy is only one but an often missing dimension of any truly whole-person approach to healing. It utilizes insights and methods from a variety of contemporary psychotherapies, adding its ecological perspective and methods to these. This chapter examines resources from other therapies, insights, and methods that have potential usefulness in enabling people to move toward the objectives of both ecotherapy and ecoeducation. The chapter also presents perspectives on earth-respecting attitudes and values from a variety of nonindustrialized cultures. These are viewed as ecotherapeutic resources from which those of us in industrialized, urbanized cultures may well learn in our own efforts to heal both persons and the earth.

Using Ecotherapy with Other Therapies

It is clear that methodologies for facilitating ecological healing and bonding must be integrated with approaches designed to enable constructive changes in other facets of peoples' lives. Here are two clinical illustrations of how therapists can do this in responding to the multiple needs of their clients.

An ecologically aware marriage, family, and child therapist tells of counseling a woman in her late thirties suffering from chronic fatigue syndrome.

It soon became clear that the woman was suffering from multiple-level burnout—from the exhausting struggles to keep up with the demands of a "controlling career in which she had been overly compliant," the load of caring for her two-year-old daughter, and her self-care needs that were often neglected. Fortunately this therapist could understand many of these pressures from inside her own experience as a woman. She also was aware of the crucial need to help such clients increase their self-care and sense of esteem and inner power.

In the process of her therapy, the client decided that she needed to be more assertive with her boss. She discovered that she must be assertive enough to arrange for changing the location of her office. It had been on an upper floor of a high-rise building with windows that did not open and new carpeting to which she learned she was highly allergic. The therapist encouraged her to implement this change and to try a variety of self-help methods designed to reduce both external and internal sources of her stress. The client found that a particularly helpful de-stressing approach was to rearrange her schedule so that she could spend regular time outdoors playing with her daughter and, in the process, "connect" with her favorite tree in their yard. Like many people, this woman found that being outdoors while enjoying something with someone she loved, brought surprisingly stress-reducing and renewing energies.[2]

A case drawn from my own clinical experience also illustrates using a variety of therapeutic methodologies, in addition to ecotherapeutic approaches, to deal with different facets of a client's complex problems in living. A male client in his late twenties reported that he had spent his childhood in a crowded, polluted urban ghetto. He had struggled long and successfully to create a new lifestyle for himself in the suburbs. In one session the man described a recent vacation spent camping in a remote wilderness area of the Sierra Nevada mountains. To be in such a setting was a first for him. He wondered why being in this "desolate place" during a mountain thunderstorm had filled him with feelings of panic. He was even more mystified by strange feelings of fear after the storm cleared, when he stood outdoors alone under the diamond-studded night sky.

As we explored these feelings, it became apparent that the little boy inside him was responding with a child's terror, in an unfamiliar natural setting, to the awesome grandeur of nature's violence, and grandeur. Systems theory provided the conceptual map for helping him explore how he had internalized the attitudes of his urban family and culture that regarded wildernesses as dangerous, desolate places to be feared and avoided if possible.

The client was invited to relive in memory his recent scary experience in the Sierras. The healing process involved helping him reprogram his frightened inner boy, using the method described in chapter 8, designed to help heal traumatic childhood memories. Then he was invited to reexperience in his imagination the recent Sierra thunderstorm. But this time he was encouraged to "re-parent" his inner child by being the strong, caring, competent adult he now is, protecting, reassuring, and comforting the little boy within him. Later, the man reported that he had taken another vacation in the mountains and was pleased to discover that his fear had diminished dramatically during his experiences there, and in its place he felt some exhilaration.

The young man had been reared in a family he described as having "heavy religion, lots of don'ts in it." He had absorbed religious beliefs and attitudes that created unbridgeable chasms between spirituality and nature, between the human and divine, and between his "bad" body and his always-trying-to-be-good soul. These dualistic religious beliefs were linked with moralistic ethical beliefs that spawned irrational guilt feelings related to his body's pleasurable impulses, especially his sexuality. As therapy progressed, it became clear that the young man's childhood religious indoctrination was producing many of the self-rejecting feelings that made his life relatively joyless. I gently invited him to consider more unifying ways of understanding the religious and ethical dichotomies that his inner child still accepted uncritically. At my suggestion, he began to read a book describing nondualistic, embodied, creation-oriented religious attitudes.[3] As his responses to the book were discussed in therapy, it became clear that this "bibliotherapy" was having a salutary effect, helping him bridge the false dichotomies of his religious upbringing. This spiritual and ethical change helped free him for better self-care. It enabled him to enjoy his sensuality and sexuality unencumbered by irrational guilt feelings. When he phoned after a wilderness experience several months later, he said his "new religious life" had helped him feel more at home in nature.

Ecotherapy Resources from Other Therapies

The ecotherapy model of healing and growth utilizes, builds on, and complements many of the theories and methods of the seven earlier clusters of psychotherapies discussed below. It affirms the significant contributions each of these approaches can make to overall therapeutic and educational work. What is needed today is a more holistic model that integrates relevant insights and methods from these and adds the missing ecological orientation and methodologies.

Each of the major clusters of contemporary psychotherapies offers valuable insights and methods that can be useful in ecotherapy and ecoeducation. I have written in depth elsewhere about the resources for growth, healing, and spirituality in some twenty-five different approaches to psychotherapy, including those mentioned below. Space constraints make it impossible to describe the methodologies in detail here. Only a brief overview will be given. For readers who are interested in knowing more about these therapies, notes will point to relevant chapters in my earlier book, *Contemporary Growth Therapies*. At the end of each chapter of that book is an annotated bibliography of key books about the particular therapies discussed therein.

Throughout the twentieth century, personality theories and the psychotherapeutic and educational theories derived from them have continued to evolve. Profound changes have occurred in both the guiding images and major methods as new approaches proliferated. When overviewed historically, the most striking feature of this development is how the parameters of healing and growth work have been broadened and made more inclusive.

A water image comes to mind as I reflect on the changes in psychotherapies and their underlying personality theories during the twentieth century. I visualize a stone dropped into a pool of water. A series of concentric circles spreads out from the point of impact. The stone represents the powerful influence of the theories of Sigmund Freud, the psychological genius who founded the psychoanalytic movement, and the other depth psychologies and therapies in the early years of the modern psychotherapy era.

As one moves out from this impact circle, each of the other concentric circles reflects a more holistic understanding of human brokenness and wholeness. Each represents a needed broadening of personality theory and methods of healing and growth. Furthermore, each of the diverse clusters represented by the circles continues to serve as a useful (though partial) cognitive map for guiding current therapy and education. Each contributes something to our understanding of how to make healing and growth facilitation more effective and whole, including the facilitation of the ecological dimension of therapy and education.[4] The ecological model continues this increasingly holistic movement, complementing the inner circles. This categorization of therapies, like all such schema, has major limitations.[5] But, in spite of these, the schema is offered with the expectation that it may be a useful framework for understanding the development of therapeutic cognitive maps leading up to the emergence of ecotherapy.

Ecotherapy Methods from Traditional Insight-Oriented, Analytic Theories

The modern psychotherapeutic movement began around the beginning of the twentieth century with an in-depth focus on seeking to repair the problems of deeply troubled persons. At its dawning, psychotherapy was profoundly shaped by the remarkable insights of Freud and his early disciples regarding the deeper reaches of the mind. Psychoanalytic therapies (also called psychodynamic therapies) focus on changing intrapsychic dynamics. Methods were developed aimed at making unconscious psychic forces conscious, leading to liberating self-understanding and the resolution of repressed early-life conflicts that block psychological maturation.

These therapies include, in addition to the Freudian approach that continued to evolve throughout his life, those of Freud's disciples. Many of these early followers eventually critiqued and modified his theories radically. Among these are the therapies of Alfred Adler, Otto Rank, Carl Jung, neo-Freudians such as Erich Fromm, Karen Horney, the ego analysts and object relations therapists.[6] The basic psychoanalytic model still is at the heart of a variety of contemporary therapies.

Because Freud and several of his close disciples were physicians, the dominant early mode of understanding psychotherapy was the traditional medical orientation, emphasizing diagnosis and treatment of pathologies. But, because their therapeutic goal was to resolve growth blocks and facilitate psychosexual maturation, these therapies also included a psychological growth orientation. Rank, Adler, and Jung moved into even more growth-oriented approaches that tend to be in greater harmony with ecotherapeutic insights and methods. Rank's emphasis on the birth trauma, for example, puts his thought in touch with a natural, maternal process that is both very bodily and very earthy.

Jung developed a controversial concept that seems to me to shed light on the ecological body-mind core of human beings. This is the idea that the deeper or collective unconscious is a psychic repository holding remnants of the long evolution of our species. The theory is that the content of our deep mind parallels the history of evolution written in our bodies. The collective unconscious, according to some Jungian analysts, holds ancient psychic residuals from the eons our species were hunter-gatherers in the wilderness; deeper still, the collective unconscious is a repository of our animal ancestors in the prehuman world. This view may shed useful light on our earthy body-mind organisms in which the evolutionary story is deeply imprinted. One can build a case supporting Theodore Roszak's view that Jung's idea of the collective unconscious may prove to be the most serviceable concept available in modern psychology for the creation of an ecopsychology in that it points to the ecological self at the core of human personality.[7]

Insight-oriented analytic therapies and models of personality (psychodynamic theories), though generally not ecologically oriented, provide invaluable illumination concerning how ecological alienation and bonding occur in children's early development. Ecoalienation probably takes place as children experience nature-related trauma that arouses intense anxiety feelings that become repressed. Conversely, ecobonding probably results when pleasurable, earth-friendly experiences of closeness with nature occur. Ecoalienation and ecobonding also happen as children automatically internalize, as in the case illustration above, alienating or earth-friendly attitudes and feelings from parents, from other significant adults such as teachers and clergy, and from their subculture. Another important resource from psychodynamic theories is an understanding of the dynamic interrelationships of ecoalienation and ecobonding with interpersonal alienation and bonding, each of which tends to reinforce the other.

Various psychodynamic methods can facilitate bringing back into awareness and resolving unconscious ecoalienation or reclaiming ecobonding memories. In particular, dream analysis and active imagination (or guided daydream) techniques are valuable methods. Therapeutic approaches based on object relations theory focus on the early-life interpersonal environment as the key to changing internalized attitudes and feelings. These methods lend themselves, in longer-term therapy, to use in enabling clients or patients to experience healing of these early roots and move toward reconnecting with their natural as well as their human environments.[8]

A classic illustration (which may be apocryphal) of the use of an ecologically oriented therapeutic methodology is from the lore attributed to Sigmund Freud. According to the story, as I recall it, he once was consulted by the internationally renowned orchestra conductor Bruno Walter, who came for help when suffering from a paralyzed conducting arm. Freud, who had a background in neurological research, probably ascertained that the paralysis was caused by stress and emotional pressures rather than neurophysiological problems. Instead of recommending long-term psychoanalytic therapy, Freud prescribed what I would describe as an "ecotherapeutic vacation." He recommended that Walter go to the beautiful Mediterranean island of Sicily and spend considerable time relaxing on the beach. The conductor followed Freud's advice and apparently experienced the healing energies of nature—the sun, the wind, the waves, the rain showers, the birds, and the quieting rhythms of the sea. As predicted, his hysterical paralysis was healed as he experienced the deep relaxation of attuning with nature. Apparently the paralysis did not return in his ongoing life as a conductor.

Several of the innovative psychodynamically oriented therapists of the twentieth century seemed to have an intuitive awareness of the earth-groundedness of human personality. At least they made some therapeutic use of nature images and the healing potentialities in nature. Milton H. Erickson and Carl Whitaker were both raised on farms and both use earthy farming images in their practice of therapy. Erickson, who was raised on a dairy farm in Michigan, practiced and taught in Phoenix, Arizona, in his later life.

On the first day of one of his famous teaching seminars for therapists, he instructed the participants to climb to the summit of Squaw Peak near Phoenix, a little before sunrise the next day. He added, "I believe that patients and students should do things. They learn better, remember better. Besides, the climb is worth it." He also asked the seminar participants to visit the magnificent botanical gardens in Phoenix and to make sure they saw two things—a boojum tree and the creeping devil cactus. He commented enigmatically, "The Boojum Tree will give you a problem. . . . You will know intellectually that it is a tree, but you won't want to believe it. . . . And you will have a great deal of respect for Creeping Devils."

Later in that seminar, he described a revealing case example of a psychiatrist and his wife who had come to Phoenix from Pennsylvania for marital therapy. The couple had been in psychoanalysis with the same analyst, the husband for thirteen years, the wife for six years. After learning about their unhappy lives, strangulating relationship, and interminable therapy, Erickson recommended that the husband take three hours to climb nearby Squaw Peak and that the wife spend the same amount of time at the botanical gardens. When they reported to Erickson the next day, the man said that climbing Squaw Peak was the most wonderful thing he had ever done. In contrast, the woman reported that the botanical gardens walk was the most boring three hours of her life. She hated every minute of it. Erickson then reversed their homework instructions. The contrast in the husband's positive and wife's negative reports was equally dramatic. The third day they were told to pick their own tasks. The husband returned to the botanical gardens and luxuriated in it again. The wife climbed the mountain again, turning the air blue cursing Erickson and the mountain all the way.

After this, Erickson told the couple that their therapy was completed. They flew home utterly baffled. Later the husband phoned Erickson to say he had fired his analyst and started putting his practice in shape for the first time in thirty years. The wife had decided that she was "tired of climbing the mountain of marital distress day after day," so she filed for divorce. Both were much happier. As simplistic as it seems, their encounters with nature and their contrasting responses apparently had triggered awareness that they needed to escape from their toxic, mutually damaging marital environment.[9]

Some Jungian analysts apparently use their theories and methods to help overcome alienation from nature and contribute to individuation, the maturation goal of that therapy. Jungian therapist Lynda Wheelwright Schmidt declares:

> Entering the wilderness and its microcosms—gardens and parks—gives us an opportunity to connect with that instinct (the earth-groundedness of personality) and rests our fragile psyches from the exhaustion of trying to stay intact in the civilized world, which is so alien for many of us. It is necessary to know a wild area that feels right. For some, a park is just right; for others, it may be a garden. My grandmother's garden was such a place for her.[10]

Ecotherapeutic Methods from Growth-Oriented Therapies

The diagnostic repair model, with its major roots in medicine and psychoanalysis, gradually began to be challenged and broadened by increasingly growth-oriented approaches. Most of these therapies make the healing of psychopathology the penultimate objective of therapy, but define the ultimate objective as moving beyond repair to maximize growth toward wholeness. Their robust emphasis on growth makes them especially compatible with ecotherapeutic theory and practice.

The therapeutic philosophy of Carl Rogers is thoroughly growth-oriented, having been developed in a child-guidance clinic. Unlike psychoanalytic theories, his does not focus on unconscious dynamics. Its primary objective is to facilitate more life-affirming, potentializing inner attitudes, feelings, and self-concepts. In his later thought, Rogers became more relationship-oriented and so could also qualify for that cluster of therapies. So far as I can discover, his evolving growth theories and therapy did not come to include the ecological dimension.

More recent nontraditional therapies such as transactional analysis, gestalt therapy, and psychosynthesis are strongly growth-oriented and invaluable resources for the relearning phase of ecotherapy and for ecoeducation. Some of their methods can enable people to bridge their inner chasm of ecological alienation and enhance their ecobonding.[11]

Psychosynthesis, a therapy created by Roberto Assagioli, begins with analysis and goes on to synthesis. It provides methodologies for meeting and befriending what it calls one's subpersonalities. These semiautonomous parts of the personality are comparable to, but much more numerous than, the Parent, Adult, and Child of transactional analysis (TA). In my view, they often includes the ecoalienated inner child—as in the case of the young man terrified by being in a mountain storm. Psychosynthesis seeks to enable people to integrate their whole personalities around their spiritual core, which Assagioli called their higher Self. This approach is useful in healing the spiritual roots of ecoalienation and fostering the self-transcending, spiritual experiences that can be the unifying center of the ecological circle.

In the practice of eclectic, growth-oriented psychotherapy and teaching, including the ecological dimensions of these, I find several methods from TA, gestalt, and psychosynthesis particularly useful. For example, the nonanalytical, reliving, gestalt approach to dreams, described in chapter 8, is a significant methodological contribution to ecotherapy. This same approach can also be adapted to work with guided daydreams or imaging techniques, both guided and spontaneous. Gestalt therapy's methods for heightening awareness of this moment in time—rather than living in the past or future—can help enhance people's empathetic experiencing of the natural world in the here-and-now moment.[12]

Ecotherapy Methods from Cognitive-Behavioral and Learning Theory Therapies

The stone-in-the-pond image must be modified here because this cluster of therapeutic approaches was not derived from the thought of Freud or other insight-oriented psychologists. It came from a very different stone impacting the waters of psychological therapeutic theories. This was the thought of behaviorists, beginning with Ivan Pavlov and John B. Watson, and their successors, the learning theorists, especially B. F. Skinner and his followers. Like the influence of psychodynamic theories, this very different impact has produced expanding circles of influence in both therapy and education. Learning theories have dominated academic psychology in many settings and have generated a variety of useful therapies as well as educational approaches.

The behavioral and cognitive-behavioral therapies reject key psychodynamic theories, including the unconscious and insight. They all share two basic assumptions. First, problems brought to therapists are understood as simply unconstructive behavior resulting from faulty learning. Second, the objective of therapy is not insight but changing dysfunctional behavior to more constructive and adaptive behavior, using the basic principles of learning, unlearning, and relearning.

The cognitive-behavioral theorists and therapists have moved far beyond Skinner's reprogramming-by-external-reinforcement (rewards and punishments) model. They have done this by recognizing that cognitive functioning (including perceptions, thoughts, and beliefs), adds an element of potential self-determination to the human equation. Within this cluster are Aaron Beck's cognitive therapy, Albert Ellis's rational emotive therapy, and David Knox's behavioral marital therapy, as well as William Glasser's reality therapy and several approaches to crisis intervention and sex therapy. They include cognitive methodologies designed to enhance constructive behavior via reeducational methods, rather than aiming at changes in inner feelings via insight. (Feelings change as behavior changes.)

The cognitive-behavioral therapies can also be called action therapies because they utilize direct, action-oriented methods to enable clients to learn new behavior. By illuminating the basic principles by which all learning, unlearning, and relearning occur, including earth-related feelings, beliefs, and behaviors, they can facilitate understanding and reprogramming such alienations. In particular, the methods for using positive and negative reinforcements to extinguish maladaptive behaviors, attitudes, and concepts, and to reprogram more constructive ones in their place, can enable people to unlearn biophobic patterns and substitute more biophilic learnings. For example, a client's phobia of house-rattling thunder may have resulted from one or more traumatic experiences in early life that are still imprinted in the deeper mind. These methods often are the treatments of choice for enabling people to reprogram the ecoalienating religious beliefs, prejudices, and values they learned from parents, other caregivers, and authority figures in early life or later.[13] The emphases of ecotherapy-ecoeducation on learning-by-doing and the growth-enabling effects of earth-caring actions are in harmony with the behavioral thrust of these therapies.

Ecotherapy Methods from Relational and Social Systems Therapies

Therapies in this cluster understand personal dysfunctions and dysfunctional relationships as two sides of the same dynamic process. Sickness and health in relationships tend to spawn personal pathology or wellness in the individuals involved, and vice versa. Relational healing methods are designed to make close relationships such as marriages, families, and other intimate social systems, healthier. Broadening of the intrapsychic, individualistic focus of psychoanalysis led a growing number of theoretician-therapists to develop the conjoint model of marriage therapy.[14] In this approach, the marital dyad is the "client" and the couple are often seen together for therapy focusing on improving their interaction. This relational circle continued to expand, becoming conjoint family therapy, the goal of which is to increase the health of the whole family systems, usually by seeing them together. As the health of the whole family system is enhanced, everyone in it is freer to function in healthier ways. In the educational field, the systemic-relational growth approach is the most productive model for family life education, as well as marriage and family enrichment.[15]

How do relational therapies apply to ecotherapy and ecoeducation? Because the ecological model of personality focuses on the dynamic interdependence of natural and human systems, the social systems model provides invaluable resources for both ecotherapy and ecoeducation. We humans are an inherently relational species. The health of our interpersonal and natural environments is intertwined throughout our lives. The quality of our relationship with the natural environment is profoundly influenced by feelings, attitudes, and behavior patterns learned in early experiences with significant adults and reinforced by current relationships. As noted earlier, the brokenness and toxicity of the planet is in us and our children. Our alienation from the natural environment also lessens the aliveness of our intimate relationships as well as the wider social systems that impact our lives constantly. The quality of our current relationship with nature can be enhanced by sharing enriching experiences in nature with the significant people in our lives, and vice versa.

For all these reasons, the relationship-oriented therapies are useful approaches for diminishing ecoalienation and strengthening ecobonding. In marriage counseling, conjoint family therapy, and family enrichment work, ecodiagnostic questions should be one aspect of the assessment process and, depending on the findings, an ecological focus a part of the healing and growth work.

Human sexuality illustrates the interrelationship between body-alienation or bonding and earth-alienation or bonding. Enjoying the rich pleasure potentials of sexuality involves all dimensions of our organisms—body, mind, spirit, and relationships—each of which is influenced by the quality of relating to the earth. The earthy grounding of sexual orgasms suggests that a therapeutic dimension could well be added to both sex therapy and sexual enhancement education. The relaxation of ego boundaries required for experiencing deep merging of two persons during sexual intercourse and orgasm seems to be facilitated by feeling positive grounding with the energies and rhythms of nature.

The sensual love story that is the Song of Songs in Hebrew Scriptures describes a couple who probably are luxuriating in making love outdoors in the blossoming springtime. Couples who feel at home in nature often make the delightful discovery that lovemaking in a beautiful, secluded natural setting can awaken sensuality and sexual passions. Unfortunately most people in urban society seldom have access to outdoor places devoid of mosquitoes, poison oak, police officers, and a crowd of other people. Finding such a setting often requires a major effort. Fortunately, there are other options. For example, celebrating nature's and God's joyous gift of sexuality can be enhanced by making love in a room with lush living plants or a picture window or skylight with a view of the ever-changing sky or tree limbs.

Ecotherapeutic methods can enrich marriage and family therapy. I sometimes recommend that conflicted couples and families, as their communication gradually improves, plan and share vigorous outdoor activities. As was true with the couple seen by Milton Erickson, the natural setting seems to give clients fresh perspectives on the dysfunctional areas of their family interaction. These new perspectives can have relationship-renewing effects, as the following case illustrates.

A dual-career couple came to an ecofeminist marriage and family therapist for help with their dysfunctional marriage. As their relationship improved, they reported that they were searching for some type of shared activity they might find mutually satisfying. Knowing that they lived near a large lake, the therapist asked if they had considered taking brisk walks together in the evenings, on the beach or along the trails of a nearby woods, perhaps being aware of their surrounding as they walked. They liked the idea. Doing this proved to be health-giving in several ways. It provided daily aerobic tune-ups, a chance to talk as they walked, and a shared connectedness with each other in nature. In a later session, they reported that they were surprised to find themselves holding hands while they walked.

Ecotherapy Methods from Social Context or Radical Therapies

The multifaceted radical or social context therapies (also described as liberation therapies) are a logical extension of the relational model. Feminist psychologists, theologians, and ecofeminists are foremost contributors to the ongoing development of the radical therapy model. Liberation psychologies and theologies—including Latin American, Asian, African American, womanist, and feminist, and those of native peoples—can provide valuable resources for understanding and correcting pathogenic beliefs and value systems that produce violence against people and nature. Environmental racism and classism are examples of issues that radical ecotherapists seek to address.

These varied approaches are derived from the awareness (spelled out in chapter 3) that society's injustices produce individual and family brokenness on a wholesale scale while therapists and teachers are working overtime to heal and prevent these on a retail basis. Ecotherapeutic theory draws substantially on the radical therapy perspective. The central concepts of radical therapies are essential aspects of ecotherapeutic theory. But, with the exception of the ecofeminist therapies, ecotherapy goes beyond the other radical therapies by focusing on environmental context issues in addition to issues of societal context.

Radical therapies recognize that individual solutions alone are never enough to solve social systems or environmental problems. Therefore, both ecoeducation and ecotherapy use consciousness-raising methods designed to empower individuals and families to work together to change, by means of sociopolitical action, the unhealthy laws, institutional policies, and systemic practices that breed not only individual and family problems, but also environmental problems.[16] The ecotherapeutic objective is not to encourage people to adjust to pathological institutions or toxic human or natural environments. Rather it is to develop the motivation and empowerment required to help them change the wider causes of the social and environmental problems suffered by themselves and countless others.

Empowering consciousness-raising means enabling wounded people to do three things: (1) experience healing that causes them to discover their inner worth and strength; (2) learn through consciousness-raising, by other victims or by a teacher or therapist, to identify the social and environmental causes that contribute to their distress; (3) use whatever empowerment they have gained, to work with others to help heal the underlying social, economic, political, institutional, and ecological causes of their own and others' brokenness.[17]

Lesley Irene Shore's garden-grounded feminist therapy is radical in another way that reflects the basic meaning of the word *radical*—going to the roots. Her earthy roots obviously influence her professionally functioning profoundly. The personal impact of her linking her natural and the human environments is reflected in the dedication of her book, *Tending Inner Gardens:* "To the people who planted my garden, protected it, nurtured it, and ensured it would thrive: Hanna Fischmann Shore and Felix Francis Shore." Her dedication awakened my awareness of my gratitude to my parents, who did the same for my inner garden, and to my family, who help keep it growing.

Ecotherapy Methods from Spiritually Oriented Theories

Because spiritual and value distortions are among the root causes of the destruction of the biosphere, methods of correcting these, drawn from the spiritually oriented therapies are essential tools in both ecotherapy and ecoeducation. Furthermore, professionals like myself who believe that people's overall well-being is enhanced when their spiritual lives are enriched can use ecotherapeutic methods to facilitate this.

This cluster includes any therapy that gives high priority to dealing with meaning-of-life issues and guiding values—which is what is meant here by "spiritual issues." These therapies understand spiritual issues as crucial in facilitating all healing and growth. The Jungian, existentialist, and psychosynthesis approaches, together with my growth counseling and therapy approach, all fit in this category as well as the growth therapies category. The spiritually oriented therapies also include pastoral counseling and psychotherapy and several of the Eastern approaches to enhancing consciousness, often called Eastern religions. These approaches to wholeness have many parallels with the growth-oriented Western psychotherapies. They are perspectives from which numerous therapists in the West, including this author, draw insights and methods that balance and complement Western therapies.

Ecotherapy adds to other spiritually oriented therapies an emphasis on inclusive, earth-grounded spirituality as one hallmark of spiritual health, especially in our time of global ecojustice crisis. Methods of dealing with spiritual and value issues in ecotherapy and ecoeducation were discussed in chapter 4. To illustrate such an approach, an ecofeminist pastoral psychotherapist with roots in the Inner Light spirituality of the Society of Friends describes her uses of nature this way:

> I have encouraged clients to spend more time in places of beauty, openness and meaning as they move more deeply into their healing processes. For example, during a particularly stressful time one man with whom I have worked would sit quietly outside in the evening air. In time he calmed, opening to the night sky in ways that allowed him to feel less alone and more connected with himself and God. In another situation I worked with a woman whose garden grew and thrived as she dealt with early trauma in her life. The garden became both a "place" of healing and a reflection of her own deep healing. I often encourage clients to walk in open places, play while outside, set up a special place—an *altar area outside where they can go for quiet, safety and prayer and/or be with that which is not human in ways that feel nourishing and deepening*. Often these ways will arise spontaneously when given openness and attention.[18]

Ecotherapy Resources from Body Therapies

The emergence of the diverse circle of body therapies widened the objectives of counseling, therapy, and education to include healing body alienation and enhancing bodily well-being by direct focus on the body. Various factors opened the eyes of some psychotherapists and many educators to see that ignoring bodies while seeking psychological wholeness can be a costly myopia of professional focus that unwittingly may cause psychotherapy to deepen the chasm between people's minds and bodies. The error of ignoring bodies and the profundity of body-mind interactions was explored in depth initially by Wilhelm Reich. He was one of Freud's early disciples who developed direct body interventions to open up what he described as the muscular armor that embodies the psychic conflicts and sexual repression. Subsequent advances in psychophysiological (psychosomatic) medicine and in physiological psychology and psychiatry have provided abundant evidence of the inextricable interdependence of bodies and minds.

These therapies include such varied theories and methodologies as Alexander Lowen's bioenergetic therapy; Ida Rolf's structural integration (called Rolfing); autogenic training for deep, full-body relaxation; movement therapies including dance therapy; and Eastern consciousness and body disciplines such as t'ai-chi, aikido, and hatha-yoga. Gestalt therapy often includes emphasis on body awareness and empowerment.[19] Feminist therapies almost always include therapeutic interventions to enhance body awareness and body-based self-esteem. The development of body therapies has helped many of us who were trained in traditional psychotherapy to become more cognizant of this vital dimension of whole-person healing. Unfortunately body therapies have not yet received the emphasis they deserve in the understanding and practice of many less-than-holistic therapists.

The embodiment model has major relevance to ecotherapy and ecoeducation theory and methodology. Because our bodies are the most obvious, ever-present, and inescapable part of the natural world in our lives, leaving out this therapeutic thrust is a significant omission. It is impossible to survive if one is not nurtured by nature every moment of our lives. Our body's health is impacted continuously by the health or sickness of our environment. Therefore, our bodies are paths par excellence to awareness of the earthy, ecological core of our being. Alienation from our bodies—meaning the lack of integrated body-mind-spirit experiencing—usually is linked dynamically with alienation from the biosphere. Overcoming body alienation also helps people overcome ecoalienation, and vice versa.

By helping people overcome body alienation, body therapies can facilitate recovery and enjoyment of their inherent human bonding with the earth. These therapies may be paths to deeper connecting of our embodied selves with the energizing body of nature. They can help people's bodies awaken to the aliveness of the rest of the natural world. Body alienation causes people to neglect physical self-care and misuse their bodies as if they were things to be exploited rather than vital aspects of themselves to be cared for lovingly. Body therapies aim at awakening sensuous awareness in people's whole bodies. In this way they aid recovery of body-mind-spirit aliveness with its spontaneity, sexual enjoyment, playfulness, and creativity, all of which have been diminished by overintellectualizing and lopsided hyperrationality.

When people come alive in their bodies, they cease to feel like disembodied spirits with disenchanted bodies, separated from the earth. Recovery of the body's inherent playfulness releases the capacity for enhanced mind-play (creativity) and creativity-stimulating, playful interaction with the natural environment. Ego identity is weak and vulnerable when divorced from body feelings and the earth. When the sense of self is reconnected with one's earth-grounded body, people tend to reclaim their forsaken body and with it their connection with Mother-Father earth.

It may seem surprising to some that body therapies can also help heal spiritual alienations for people who discover that *disembodied* spirituality is *disempowered* spirituality. Bioenergetic therapist Alexander Lowen observes, "As long as the [body-alienated] ego dominates the individual he cannot have the oceanic or transcendental experiences that make life meaningful."[20] Non-incarnated spirituality, disconnected from bodies and nature, often contributes to depriving people of a sense of the profoundly spiritual connection of their whole being with the earth.

People who get reconnected with their bodies frequently experience simultaneous reconnection with the natural world, and vice versa. Their experience of themselves becomes grounded more firmly in the earth. Connecting with their previously rejected bodies tends to deepen their connection with the physical world outside their bodies, and with transcending spiritual reality in it. For women and the poor whose intimate connection with the earth has been used as a put-down, integrating earthiness with their minds and spirituality may help liberate their self-identity.

A variety of breathing, body-posturing, and movement methods from body therapies can facilitate this body-earth grounding.[21] The exercise that follows is an example of this.

Body Energizing and Grounding Exercise[22]

To experience a body therapy exercise, try this simple grounding approach now, stopping at each / to do what has been suggested.

Begin by relaxing your whole body-mind-spirit organism. Do this by tensing all your muscles for five seconds or so, and then relaxing them fully, as you continue to breathe deeply. Do this repeatedly, pausing briefly between each cycle, until your whole body-mind-spirit organism is deeply relaxed, like a limp rubber band, but keenly alert, like the flexible elasticity of such a band. / Now, deepen your relaxation by concentrating your awareness only on your breathing. / As you breathe in, let yourself experience the reality that each inhalation you are receiving is a life-giving gift from the biosphere—a literal gift of life-renewing oxygen from the plant communities around the planet that generate it. / Let yourself experience your connection with the earth and with all other oxygen-breathing creatures in your community, your country, your world. / Now let yourself experience the reality that each exhalation is a life-giving gift of carbon dioxide to the oxygen-producing plants. They all must receive this gift to live, blossom, reproduce, and, in the process, produce oxygen for us animals. Be aware of the miracle of interacting with all living things as you drink from the life-sustaining global pool of oxygen and help to replenish the life-sustaining global pool of carbon dioxide. Continue this and experience your life-sustaining interdependence with all other living things. /

Now, as you experience being deeply relaxed but very energized, become aware of the dependable pull of gravity holding your body securely in its place on our spinning planet. / To feel more firmly grounded, put both feet solidly on the floor and straighten your spine to free the flow of life energy throughout your body as you continue relaxing, energizing breathing. / Experience your buttocks supported by the chair and your feet by the floor. / Your chair being supported by the floor. / The floor supported by the entire building. / The building supported by its solid footings and foundation. / The foundation supported by the solid earth beneath it. / Let your weight sink toward the earth and feel supported by it. / Now, as you are firmly grounded, feel and enjoy the flow of earth energy up through the foundation, the building, the floor, through your feet and chair, permeating and sustaining your whole being. / Enjoy feeling empowered by the energy from the earth merging with the spiritual energy flowing from the creative Spirit who is under, and above, and around you and within your whole body-mind-spirit organism. /

If you found this body grounding-energizing exercise helpful, repeat it regularly, particularly when you are feeling uncentered ("beside yourself"), depressed, or "up in the air." Try it outdoors in your favorite natural setting. Discover how being immersed in the natural world and literally sitting on the ground may change the experience for you.

Ecotherapeutic Resources from a Variety of Cultures

In our increasingly multicultural world, perspectives on the healing, teaching and growth-enhancing arts are urgently needed. Certainly, we in the people professions need to learn from each other across the many ethnic, class, racial, national, language, and cultural boundaries that separate humankind. One of the joys of my career through the years has been the privilege of teaching and leading workshops in some thirty different countries and cultures. More than half of these were developing countries in Africa, South America, Asia, Eastern Europe, and the South Pacific. These horizon-stretching experiences have provided many humbling examples of a teacher learning far more than he taught.

One encouraging cross-cultural discovery is that my holistic, earth-grounded, social systemic approaches to counseling and education seem to fit well the experiences of many professionals in nontechnological cultures. Most people in those cultures take for granted that their own well-being is interwoven inextricably with the well-being of the natural environment. Repeatedly I have discovered that many of the earth-based theories underlying ecotherapy and ecoeducation seem in harmony with indigenous views in non-industrialized societies.

Because environmental problems transcend all human-made boundaries, effective responses must involve networking among people all around this small, precious planet. Ecological well-being or destructiveness in one country impacts the lives of millions of people in other countries. Only a global environmental ethic will suffice if all children are to inherit a healthy earth. In light of this, the research for this book explored the earth-related experiences, beliefs, attitudes, and values of young adults from a variety of countries. Sixteen persons, mainly graduate students preparing for counseling, teaching, and ministry professions, were interviewed in depth about their ecological experience. In addition, among the people who responded to the research survey for this book were eight professionals from countries other than the USA. These twenty-four people reflect life in eighteen different countries and cultures. Many were from these non-Western, developing countries: Korea, Tonga, Nepal, Kenya, Mexico, Zaire, India, Jamaica, Japan, Venezuela, Costa Rica, Hong Kong, Fiji, Philippines, Tanzania, and Bolivia. One woman is an African American who reflects that earth-connected subculture; another woman lives in a Western, developed country (Italy) and is included because of her clear ecological awareness and caring. Undoubtedly she is an example of countless professionals in developed Western countries who are tuned to ecological issues and committed to ecocaring. The religious backgrounds of the twenty-four persons included widely diverse Christian denominations, Hinduism, Buddhism, shamanism, and several other indigenous religions. Only one respondent said, "I don't believe in God."

The responses by those whose homelands are nonindustrialized countries make it clear to me that we in the industrialized West can learn much from other cultures about healthy ecological values and lifestyles. Their responses confirm an impression I had repeatedly when I taught in such countries: that ecoalienation is very rare. Most people there seem to be more intimately bonded with the natural world than are most Westerners. They know that they must live in a respectful, cooperative way with nature if they are to prosper or perhaps even survive. The poor and oppressed in all cultures, of course, frequently struggle to survive, whether they live in rural or urban settings.

The respondents revealed a remarkably high degree of ecological knowledgability, concern, and caring. This probably was due in part to the fact that many of them were young adults and graduate students. Furthermore, nearly all those raised in developing countries expressed beliefs and attitudes with a high degree of compatibility with the major ecotherapy themes. I was pleased to find that, what I describe as the three dimensions of the "ecological circle" were implicit in many of their responses. To illustrate this, here are the ten interrelated themes that emerged most frequently in responses by these twenty-four persons. I have summarized the major themes that were stated in different ways by the interviewees.

• *I grew up very close to nature and my early experiences with nature had a strong influence in shaping who I am today.* This was the one theme articulated by all respondents. Mpyana Nyengele of Zaire remembers the cycle of their two seasons, wet and dry. "As I grew up, I remember being involved in this cycle. When the longed-for first rain came, after the dry season, I automatically thought of planting seeds and cutting grass on the farm. This also meant the coming abundance of food—peanuts, corn, beans, sweet potatoes, green vegetables, pumpkins. My mother took care of the farm, as my father also had to teach and pastor a church. We children all worked with her when not in school. By the way, although we worked very hard, we enjoyed it very much because of the sense of community and closeness to each other we enjoyed. While working, we sang songs and told stories."

Musician Anil Edward Veeraraj remembers this from growing up in India: "In my early teens, I used to take hikes with my friends in the country. The serenity and calmness I felt could not be topped. Those were the best inspiration I received that caused me to write my best musical works."

• *All things are interconnected. Therefore the well being of any part always involves the interdependent network of life.* This, the fundamental ecological premise, stands in sharp contrast to the dominant Western valuing of hyper-individualism, autonomy, nature-human dualism, and anthropocentrism—the view that human desires and needs are the only significant values. Kenyan Peter Mwiti Rukungah wrote his Ph.D. dissertation on an ecological model of pastoral counseling from the perspective of his Bantu people.[23] He explained that the proper frame of reference for counseling and therapy is the ecological community. Here is his description of how nature molded his personhood: "The land is a life force that has shaped my personality and my understanding of God and the universe. . . . Like most Africans, I understand change and healing and well-being in terms of interconnectedness. Effective pastoral counseling with the Masai of Kenya aims at assisting the person to open the channels of blocked relationships with God, people, Mother Earth, and with other beings in the cosmos."

Zairian Mpyana Nyengele remembers: "There is a saying in Zaire that 'The palm tree shelters him that is afar off.' So, it is encouraged not to cut a tree without an acceptable reason." Respect for trees is derived from the wider awareness of human's dependence on all nature.

Conrad Shayo, a Tanzanian teacher, was steeped in the emphasis on interdependence and communal responsibility found in what anthropologists call the "we culture" of his homeland. He contrast this with the individualistic "I culture" of America: "As I grew up, other people came first, I came second. The 'me' I had to learn when I came to America, where putting others first at times leads to your being taken advantage of."

• *The earth is the mother of life to be respected as a mother, the provider of everything.* This view was expressed in different words by numerous respondents. Nephil M. Maskay from Nepal declared: "The earth is the mother of all living things." Conrad Shayo from Tanzania remembers: "In my childhood, I saw nature as mother and provider of our life. I was taught that you must enrich the soil by adding organic manure and plant two trees for each one you uprooted. I remember an old Tanzanian adage, 'You don't eat the planting seeds.' " Anand Veeraraj (from India) observed wisely: "One does not understand nature or its processes unless one is close to nature, living in harmony with it."

A Kenyan proverb, recalled by one respondent, emphasizes the importance of stewardship of all animals and avoiding the wanton human exploitation of them: "Do not dispose of the monkey's tail before it is dead."

• *To care for the earth or to harm it, is to care for or harm oneself and one's family.* Anand Veeraraj states the theme insightfully in describing what humans should learn from nature: "There is so much of sharing and recycling that goes on in nature. Every entity, after having lived for itself, turns itself over to be used by others. . . . This lesson from nature we humans ought to learn and practice in our communal life. Unfortunately, humans are the only species in the whole wide world who block being recycled. We draw more resources from nature than we return to it. This is one of the fundamental causes of environmental crisis."

Elias Joel Mbaabu of Kenya remembers learning the theme from his parents: "My mother desires not only interacting with nature but struggling with it, enjoying it and letting it remain intact for the benefit of future generations. My father prefers a just maintenance of nature as the only profound philosophical foundation for natural existence. Hence, a just and good action toward nature is a just and good action toward the self, humanity, and creation."

Ana Maria Dominguez of Mexico wrote, "Destroying nature is destroying what we are as human beings—earth and spirit. Nature symbolizes our beginnings. If it is destroyed, we cannot return to our origins."

• *One's connectedness with nature is deeply rooted in the soil of the particular place where one was raised.*

Jamaican Authisa Spencer Miller points to a ritual practice there, occurring after an infant's birth: "The experience of burying the umbilical cord ties you to the land of your birth. One's sense of belonging is tied to being a part of an identifiable people with a particular land."

Peter Mwiti Rukungah reflects this theme when he comments on his birthplace (Kenya): "Having been born and raised in an agricultural setting, the things we raised on the farm were our existence. We raised our food and clothing and were educated with the resources from the farm. I was very much aware of my connection with Mother Earth as my existential life force that has not only given me means to exist, but also shaped my personality. This is true of many Kenyan people who have a deep respect of the land where 85 percent of them live and earn their livelihood."

• *Nature and spirituality cannot be separated, for spirituality is earth-grounded and sensual. Spirituality is related to everything in everyday life. It is both expressed and enhanced by its interrelationship with nature.* Aldhea Spencer Miller (Jamaica) says: "I have come to think of the interconnectedness of all things as a harmonizing flow where the heart of God is to be found."

Elias Joel Mbaabu (Kenya): "God is the one who graciously offers me *Usalama* [the East African Swahili word meaning wellness within the creation], increasing through teaching and counseling the *Usalama* in people's lives and in nature. Life in the world could be greener, happier, newer and more fulfilling if we lived in an intimate, caring relationship with God and creation." Doing pastoral counseling for *Usalama* would be to have the people of Chogoria share in communication about their faith, hope, feelings, views, attitudes, and values focused on God, humanity, and nature. The format would be divided into six phases: (1) experiences with nature; (2) nature's influence in their life; (3) analogies, metaphors, parables, and stories about humans and nature; (4) human violence against nature; (5) restoration of nature; (6) a state of wellness with God, community and nature."

Roman Catholic Sister Gloria Ruiz's bonding with the earth apparently has increased as a result of her work in Far Eastern countries for many years (most recently in Hong Kong). She writes: "After I left the States, nature has always been close at hand. In the last few years, trees and patches of foliage are my daily companions as I walk to work. I notice things I never saw before each day. A spider web with dew on it, sunlight outlining each thread, is awesome. It speaks to me of God's presence; so much life is actively in process. The cosmogenesis of all that I encounter each day helps me to live in the mystery of God's love, which is to live without answers, especially to the pain and suffering that is shared daily with me in my hospital work."

She continues: "Only recently have I come to understand that the destruction of our forests and victimization of people and communities is the ongoing crucifixion. Nature shares this message with me. Nature is also teaching me to understand and use psychic healing methods in pastoral care work with cancer patients. The results are sometimes dramatic with the debilitating pain seeming to decrease."

Several respondents in earth-grounded cultures mentioned how indigenous beliefs and spiritual practices have been harmonized by them with their Western religions. Psychologist Shakunthala David of India writes: "Even though coming from a Christian background and growing up in a large city, the Sun was a 'God' who gave energy, the water, especially the Ganga, the source of life. Putting my feet and sitting in the river were the best things."

From his Hindu background, Anil Edward Veeraraj remembers: "They used to worship nature. On a superficial level, it seems ridiculous, but I realized how much importance they placed on nature, unlike other religions."

• *My experiences of nature make me aware of both the beneficent, nurturing side and the dangerous, destructiveness side of natural processes.* Musician Anil Veeraraj grew up in a Tamil region of India, a beautiful rural area with tropical forests and seven lakes. He recalls: "As a boy I played with my peers on the open fields, in the woods, swam in the lakes, slept under the open sky looking up at the starry canopy. However, the village we lived in had lots of cobras, scorpions and poisonous insects. So for me, nature contained both good and evil. We also had seasonal epidemics like cholera and smallpox, as well as starvation and death when the monsoon rains failed. However, I was not aware of the distinction between humans and nature as it is today, nor was I aware of environmental crisis till I moved to an urban center to get higher education and a job."

Mpyana Nyengele (Zaire) says: "What I cherish today is the awareness of the importance of the rain. After a windy and dry season, the first rain brings refreshment and also life to the plants that have lost their leaves. The glow in the sky changes and the smell of the soil is so wonderful. This smell reminds me of the good times with my family as a boy, and it refreshes me in a very unusual way. When we planted seeds, we expected the rain to continue as usual. But sometimes there was a long drought, which meant that there would be no hope for the crops to grow."

Jamaican Aldhea Spencer Miller told of an experience that illustrates how painful ecological experiences can give beneficial learnings: "Being caught in a wave and tossed ashore like so much flotsam, helped me learn my limitations." Another Kenyan proverb sheds light on potential valuable learning from nature about life's dark experiences: "When the moon is not full, the stars shine more brightly."

• *The destruction of nature associated with the global environmental crisis causes me deep personal pain.* African American Gloria Johnson declares: "Violence against nature feels like a violation of my core, my person." In a similar vein, Chu Seop Song of Korea says: "When I see people throwing trash in the mountains, I feel like a bug is biting me in my back."

Tongan Ahid Kalatini expresses his pain about the plight of his country and other Pacific island nations: "As a boy my experiences with nature were marvelous—fishing in the ocean and growing food—papaya, guava, bananas. The ocean still feeds my people. But developed countries like USA made atomic bombs. They took the atomic wastes and dumped them in the Pacific Ocean where my people find their life."

Elias Joel Mbaabu of Kenya was raised at Chogoria, near the foot of Mount

Kenya, where he remembers beautiful forests, valleys, rivers and streams, waterfalls, pools, fruit, flowers, and birds. "I fell in love with life at Chogoria, with nature and with its compelling beauty." With sadness, he reports that now most of the forest has been cut to make room to grow coffee, tea, and dairy products, to satisfy the colonial export market. Modern agriculture, with chemical fertilizers and imported animals, has destroyed much of the rich topsoil. "Many birds and insects are either killed or pushed out of Chogoria, violence against nature has increased soil erosion, drying up water pools and perennial springs, frequent drought and famines and a continuing process of desertificaton. Today, most of Chogoria's green trees, colorful flowers, and natural beauty is gone. I miss the green paradise of forests, animals, and plants. I miss the music of the birds and insects. I miss Chogoria's natural beauty." Then he expresses this heartfelt longing: "I would love to celebrate memories of a healthy Chogoria, within a healthy nation and in a just world."

• *Economic and political injustices often are at the roots of environmental problems as well as problems of human exploitation, with which environmental problems are intertwined. Therefore, political and economic eyes are needed to understand and deal with these problems.* Several respondents from developing countries expressed this justice theme: Sister Gloria Ruiz describes the Chinese belief that "nature is a living, breathing organism. . . . Mother Nature's voice is being lost in the frantic rush to build Hong Kong and for profits."

Tongan S. Toa Finau, now teaching in Fiji, reports: "My students and I are very concerned about and protest the testing of nuclear weapons which continues by the French in the Pacific." Filipino clinical pastoral educator and hospital chaplain Narciso Dumalagan is active in efforts to make his country nuclear free and to persuade the U.S. Congress to clean up the environmental disasters left at their now abandoned military bases. He remembers: "I grew up on a farm where the plants were green, the air clean, and you could drink the water in the river without fear of contamination. Now I live in Manila, which is very polluted, and no green trees exist. I have witnessed the autopsy of a man living in the city. He was not a smoker, but his lungs were black from the city's air pollution."

Adriana Cavina heads a ministry of social justice for her denomination in Italy. She remembers: "I spent long summer vacations with my grandmother in her village in the hills of southern Tuscany. It was bliss: to be free, to enjoy playing and walking outdoors, to come to know plants, animals, and to live the simple lifestyle of country people." When asked if she has a close relation with nature now, she responds sadly: "It is one of my longings. Unfortunately, I live in a very crowded city (Rome). My relationship with nature is limited to a few pots of flowers on my balcony and the trees I can see out of my window."

• *Nature influences the way I do teaching and counseling:* When asked how her relationship with nature influences her human relationships, teacher Ana Maria Dominguez (Mexico) responded: "I am not alone My person, my wholeness depends on my harmony with others, nature and the world. I'm aware that we are not single individuals, but part of a global system which has to exist in harmony and balance. I am teaching and trying to do this."

In her ministry, Adriana Cavina (Italy) responded: "My early relation with nature taught me what real freedom is and gave me a strong sense of independence, which have both helped me a lot in my personal and professional life [which includes much work on peace and justice issues]. I believe in holistic health, and this includes a harmonious relationship with the environment we live in, and with the global unity that includes all living beings and our planet. Encouraging clients and students to get involved in environmental, peace, and justice work helps them pull down the barriers of a limited, inflated ego, and gives them a better, more proportioned view of their own lives and problems."

S. Toa Finau recalls: "Growing up in Tonga, I knew that we were living out of the land and sea, planting coconut and other fruit trees; working to prevent poisoning of fish and pollution of the ocean and beaches. In my teaching and counseling I use many analogies and metaphors, images to encourage growth and reconciliation."

Fijian teacher-author Paula Niukula writes: "I grew up in a village beside a river near the sea. I have warm, happy memories of exploring the forest, swimming in the river, fishing with my mother and working with my father in his garden. Now, I feel very much at home in the open, on the ground, with trees and flowers around. I enjoy counseling and teaching outdoors."

Psychologist Shakunthala David of India recalls: "In my childhood I learned the therapeutic value of plants and how to use different herbs and leaves to cure illnesses. I still grow such plants in my apartment and use them for common ailments. I am in the process of discovering the scientific value of certain rituals related to nature to reinforce my belief in nature. My strength in counseling comes from touching nature and rejuvenating the self. In my counseling, I sometimes suggest to clients that they try things related to nature, e.g., offering flowers and doing gardening. Some clients needing these experiences come from states where communal problems and human rights violations cause heavy stress."

Anil Veeraraj, speaking from his Hindu roots in India, articulated the theme of playfulness: "Teaching and counseling first and foremost have to be fun, playful, and full of enjoyment, for both student and teacher, the counselor and the counselee. Ancient Greeks and the gurus of the orient knew the art of teaching through play. . . . Animals employ this technique of learning and growth through playful exercise. Industrial cultures lost the art of play and enjoyment. . . . New methods of counseling and teaching have to be devised to achieve this basic goal of enjoyment. When this is achieved, healing and growth will begin to occur spontaneously."

I trust it is clear from these illustrations that many of the concepts and attitudes about nature, among persons from nonindustrialized cultures, are in harmony with key themes in ecotherapy. They also are consistent with the approaches to ecoeducation that are most likely to enable children, youth, and adults to learn biophilia, ecological literacy, and earth-caring.

To close this brief examination of multicultural resources, here is an interfaith prayer of healing from the United Nations Environmental Sabbath Program. This pledge could well be a shared commitment of people in all cultures and religions who are concerned because the earth is in painful jeopardy.

We join with the earth and with each other
To bring new life to the land
To restore the waters
To refresh the air

We join with the earth and with each other
To renew the forests
To care for the plants
To protect the creatures

We join with the earth and with each other
To celebrate the seas
To rejoice in the sunlight
To sing the song of the stars

We join with the earth and with each other
To recreate the human community
To promote justice and peace
To remember our children

We join together as many and diverse expressions
of one loving mystery: for the healing of the earth
and the renewal of all life. [24]

In a similar spirit, here is a native Hawaiian prayer for the earth:

May the earth continue to live
May the heavens above continue to live
May the rains continue to dampen the land
May the wet forests continue to grow
Then the flowers shall bloom
And we people shall live again. [25]

Exploring This Chapter's Issues Further

For fuller information on these books, see the annotated bibliography at the end of this book.

Howard Clinebell, *Contemporary Growth Therapies*
Mary Evelyn Tucker and John A. Grim, eds., *Worldviews and Ecology*
John Welwood, ed., *The Meeting of the Ways, Exploration in East/West Psychology*

Notes

1. New York: Harrington Park Press, 1995, 15. Lesley Shore is an outstanding example of a feminist psychotherapist who is pioneering in what I call ecotherapy.

2. Personal communication from Huntley Lewis, MFCC.

3. The book was Matthew Fox's *Original Blessing: A Primer of Creation Spirituality*.

4. The most effective therapists and teachers today use eclectic theory and methodologies drawn from a variety of sources.

5. A few caveats about the limitations of this schema. First, there obviously are numerous ways to group contemporary psychotherapies in meaningful categories. Second, no categorization is fully adequate because they all inevitably oversimplify and hide the complexities of these interacting, overlapping clusters. Third, each category includes a diverse variety of therapies linked only by certain common characteristics. And finally, numerous therapies could be placed in more than one category.

6. For further information about Freud, see chap. 1 in Howard Clinebell, *Contemporary Growth Therapies: Resources for Actualizing Human Wholeness* (Nashville: Abingdon Press, 1981). Chapter 2 discusses Adler and Rank. Chapter 3 overviews Fromm, Horney, and Harry Stack Sullivan. Chapter 4 examines Jung, Carl Rogers, and the existentialists.

7. Theodore Roszak, *The Voice of the Earth* (New York: Simon & Schuster, 1992), 301–2.

8. As was pointed out in chap. 2, the object relations expression of psychodynamic theories is especially relevant to ecotherapy because it offers a cognitive map for exploring the interpersonal process by which the internalization during early relations with Mother-Father Earth was incorporated into our identity at unconscious levels. Deep ecological attitudes, like other deep attitudes, are more often caught from adults in early life than taught by formal instruction. Formal teaching however, can reinforce or help to correct the spontaneous internalizing of bonding or alienated experiences with the natural world. For earth bondedness to occur, teachers, clergy, and parents must be earth-friendly in their own deeper ecological feelings. The reversing of early earth-alienating learnings from persons in families, schools, or churches tends to take place in relationships with earth-affirming teachers, therapists, family members, and close friends. Thus the object relations model, the learning-theory model, and the relational model each throw complementary light on understanding the process by which ecohealing and positive ecolearning can occur.

9. Jeffrey K. Zeig, ed., *A Teaching Seminar with Milton H. Erickson* (New York: Brunner/Mazel, 1980), 145–47.

10. *The Long Shore*, 199, 201. Another Jungian analyst whose theory is ecopsy-

chologically oriented is Stephen Aizenstat. See his "Jungian Psychology and the World Unconscious," in *Ecopsychology: Restoring the Earth, Healing the Mind*, edited by Theodore Roszak, Mary E. Gomes, and Allen D. Kanner, 90–100.

11. For more information about methods of transactional analysis, gestalt therapy, and psychosynthesis, see Clinebell, *Contemporary Growth Therapies*, chapters 6, 7, and 11.

12. See ibid., 183. A gestalt psychologist who is active in the bioregional movement is William Cahalan. See "Environmental Groundedness in Gestalt Therapy," in *Ecopsychology*, 216–23.

13. A description of behavior and cognitive-behavioral methods is found in chapter 5 of *Contemporary Growth Therapies*. Using a cognitive approach (rational emotive therapy) to correct pathogenic beliefs is described in Howard Clinebell, *Basic Types of Pastoral Care and Counseling*, 120.

14. On a personal note, during my neo-Freudian training in psychotherapy in the late 1940s, the interpersonal orientation of Harry Stack Sullivan and the sociocultural orientation of Erich Fromm (both of whom were associated with the White Institute of Psychiatry where I trained) nudged me toward an increasingly relational orientation.

15. For more information about family systems therapies, see chapter 9 in *Contemporary Growth Therapies*.

16. For more information about radical therapies, see chapter 10 in *Contemporary Growth Therapies*.

17. Consciousness-raising that produces empowerment occurs as teachers, therapists, or small-group members share experiences of victimization that awaken in others the awareness of the social roots of their individual pain and problems.

18. Personal communication from Linda Fillipi, Ph.D.

19. For an overview of body therapies, see chapter 8 in *Contemporary Growth Therapies*.

20. Alexander Lowen, *The Betrayal of the Body* (London: Collier Macmillan, Ltd., 1967), 259.

21. The psychospiritual potential of breathing is suggested by the fact that many meditative disciplines use focusing on one's breathing as a way of focusing and quieting the noisy stream of consciousness and gaining access to the profound serenity that is discoverable in our deeper minds. It is also relevant to note that the same word is used for breath and spirit in a variety of languages, including Greek, Hebrew, and Sanskrit.

22. The exercise that inspired only the second of this one is in Anne Kent Rush, *Getting Clear, Body Work for Women* (New York: Random House, Inc., 1975), 49–50.

23. Rukungah's dissertation, "The Cosmocentric Model of Pastoral Psychotherapy: A Contextualized Holistic Model from a Bantu African Worldview, A Perspective for Post-Modern Pastoral Psychotherapy" (School of Theology at Claremont, Calif., 1994), describes an alternative paradigm to the anthropocentric, dualistic, individualistic understanding that has dominated Western science, Greek philosophy, Western religions and healing models, including psychotherapy. The proposed alternative paradigm is derived from the Bantu world view. Rukungah tells of his personal awakening to the down-to-earth implications of the interconnectedness of all things. At age ten, while looking after the family's grazing flock, he cut a tree to make a toy cart for himself from the wood. When his father discovered what he had done, he was

indignant about his son's "brutal activity" toward the tree. Rukungah recalls: "I sensed in him a pain which I was not aware of, as he talked to me. . . . For the first time I realized how deeply he valued and respected trees. From his perception, it seemed to me that the cutting of that tree was not only a violence against nature, but also violence against him. This was an experience that opened my eyes to honor, love and respect for the natural environment of which I am an intimate and intrinsic part." (Personal conversation with the author.)

24. *Only One Earth*. New York: United Nations Environmental Programme, 1991, no page number.

25. David Malo, *Hawaiian Antiquities*. Honolulu, Hi: Bishop Museum Press, 1983. Used by permission.

Part Two

Methods of Ecotherapy and Ecoeducation

6

Greening and Cleaning the Settings of Counseling, Therapy, Education, and Living

SMOG IS AN APHRODISIAC
NEVER TRUST AIR YOU CAN'T SEE!

—Bumper stickers seen on smog-stifled days in Los Angeles

That which is not good for the bee-hive cannot be good for the bees.

—Marcus Aurelius, Stoic philosopher and Roman emperor A.D. 161-180 (*Meditations*, Book 8)

Environmental factors are among the causes of many, perhaps even most, illnesses. Diane Porter of the National Institute for Occupational Safety and Health declares: "Years of research have taught us that environmental hazards in offices, hospitals and even classrooms can make people sick." Elissa Feldman of the Environmental Protection Agency's Indoor-Air Division adds: "The air in many office buildings can be more seriously polluted than the outdoor air in even the largest, most industrialized cities in the country." It is clear that the environmental toxins that diminish our wellness often are in the homes, offices, schools, and communities where we spend most of our lives. Therefore, greening and cleaning the places where we live and work is the most immediate contribution we can make to the ecological well-being of ourselves, our families, and those we serve in our professional lives.[1]

It is obviously counterproductive to do counseling, psychotherapy, education, or medical healing in settings where the physical environment militates against healing and growth. Yet, the environmental impact of the rooms, buildings, and neighborhoods where such activities often take place clearly works against facilitating healing and maximized learning. Counseling and therapy rooms, classrooms, and homes infected with toxins, lacking outside light and adequate fresh air, clearly are ecohazardous to people's health. Being in such settings impacts the body-mind-spirit organisms of healers and their clients, teachers, and learners, whether or not anyone is aware of this deleterious influence. Fortunately, detoxifying and "naturizing" homes and work places is one of the simplest and often the least expensive ways of increasing ecowellness in people's lives.

An oncology ward hospital chaplain shared a vivid illustration of the ecotherapeutic possibilities in one medical setting. She had been present at a conference where I had lectured on grief about the earth and using ecotherapeutic methods in working with the dying and their families. After returning home, I received an affirming letter from her. She told of her work with patients in a high-rise, super-modern hospital with nonopening windows. One seventy-eight-year-old male attorney suffering from Parkinson's disease was described by the hospital staff as "tight, stuffy, cranky, nontalkative, and uncooperative with his treatment." The chaplain reported that when she took him outside in a wheelchair, "As might be expected, the sun, wind, flowers, and life of nature changed his attitude ninety-nine percent." She added, with a touch of humor, "when he was shut up inside in a 'controlled climate' environment, he couldn't help being stuffy. Ha!"

The chaplain described several patients who were ranchers who had worked outdoors all their lives until they were hospitalized with cancer. She wrote: "They grieve and pine for fresh air, earth smells, the feel of wind and rain. As I fight to get them outside, doctors tell me, 'We must protect them from the environment—viruses, germs, etc.' Many of them die after weeks in the big city hospital with no connection to what was a major part of their being, the earth!"[2] As I read her letter, I thought: How misguided to keep them "safe" by cutting them off from contact with earth in an artificial environment that for them was deadly. And how fortunate they were that this insightful chaplain is aware of the healing energies of contact with nature. Later I visited the chaplain at her hospital. She told of one rancher who was longing so intensely for fresh air that she found him out of bed trying to pry open a narrow louver at the bottom of the sealed window with a fork. I was not surprised that she frequently brings living plants and nature motifs into her devotional services for patients and staff.[3]

Understanding Ecological Illnesses[4]

It is important to encourage people to increase their knowledge of what medical professionals now label ecological illnesses or environmental diseases. These include allergic reactions and biochemical intolerance to a wide range of foods, pollens, food additives, tobacco smoke, pesticides, cosmetics, and other products flowing from the expanding cornucopia of modern chemistry and industrial pollution. Some patients, described as having "allergic toxemia" or "allergic tension-fatigue syndrome," live and/or work in "sick" buildings. They suffer from symptoms such as continuing fatigue, wooziness, hyperirritability, hopelessness and depression, confusion, growing anger, flulike aches, disruption of intimate relationships, and, in some cases, problems of thinking and memory.

The onset of environmental diseases often occurs when persons are heavily exposed to toxic chemicals. This chemical overload may come from massive exposure, such as the pesticide poisoning suffered by migrant farm workers who pick vegetables and fruits in California and Texas (for the minimum wage or less). Often the biochemical crisis is derived from a gradual buildup of toxins in people's bodies from pollutants they unwittingly consume in water, food, and air, and the toxins absorbed from chemically treated building materials in new structures. Common sources of toxic airborne chemicals in offices include office furniture made from pressed wood (which emits formaldehyde), certain photocopiers that emit ozone, air conditioning systems that harbor mold and bacteria, and fumes from pesticides, cleaning solvents, glues, and deodorizers.

Drinking tap water in developed countries is generally considered safe, at least with respect to contracting major illnesses such as cholera. But the U.S. Environmental Protection Agency (EPA) estimates that up to half of the water supplies in the United States do not measure up to its minimal standards. The agency considers exposure to drinking-water pollution one of the four top-priority health problems in this country. In a recent year numerous people got flulike symptoms from the public water supply of several large American urban areas (among them Milwaukee and the District of Columbia). This illness was caused by protozoa found in animal and human feces. Numerous community water systems, mainly in rural areas, are contaminated by pesticides, agricultural runoff, carcinogenic chemicals, and heavy metals. Bottled water sales have skyrocketed in the last decade, driven by public fears. Unfortunately, studies reveal that some of this expensive water may be more contaminated than tap water. So it is prudent to have your water (tap or bottled) tested by an EPA-approved company.

The etiology of environmental illnesses is not yet fully understood. Genetic vulnerability may play a significant role. So may protracted severe emotional stress and grief that overwhelm the ability of the immune system to cope with the heavy load of chemical stressors. This enormous cluster of medical problems was first described by a physician in 1951. But many health professionals are not yet aware of the nature of environmental illnesses or prepared to treat them. Consequently, the symptoms of environmental diseases are often misdiagnosed and treated by psychotherapists and physicians as exclusively psychological problems, or the sufferers are dismissed as hypochondriacs.

Effective diagnosis and treatment of suspected environmental illnesses can best be done by physicians who are trained in clinical ecology. Because such specialists are in short supply, treatment can be difficult to find, long-term, and expensive. The avoidance of toxic substances often requires such radical transformations of lifestyles as to render the success of treatment problematic. The changes required may involve alterations in diet, clothing, home furnishings, social life and recreation, as well as changing one's residence and workplace.

No one really knows how big the problem of environmental illness is. Physicians in the American Academy of Environmental Illness estimate that as many as 10 to 15 percent of the population in this country suffer from environmental illnesses or are vulnerable to developing them.[5] Environmental diseases among otherwise normal people have increased dramatically since World War II—a period of astronomical increases in multichemical exposure. Thousands of people now belong to support groups for sufferers of environmental illnesses.

Consider this case illustration of an effectively treated patient reported by a hospital chaplain:

> Ethel is a woman in her sixties. A widow, she spends much time caring for her brother. Her life was "a series of one physical calamity after another," she said. She showed a long list of medicines, many of which were tranquilizers, that had been prescribed by a number of physicians. Finally when she was about to give up she learned of the ecology unit. After treatment there she resumed a more or less normal life simply by avoiding chemicals and rotating her food in a restricted diet. Her biggest complaint was that she can no longer attend her church because "Sunday morning is mass pollution of perfume." She expresses a longing to go to church, and wishes the members would be more considerate of people like her for "they won't really believe that perfume can make someone sick." [6]

Making Buildings Environmentally Healthy

It is appropriate for all of us to question how safe the buildings are where we live and work, play and worship. The "sick building syndrome" has received increasing media attention in recent years. The World Health Organization estimates that up to 30 percent of all recently remodeled and new buildings may be sick buildings, some of whose occupants suffer from health problems caused by indoor pollution. Furthermore, the National Academy of Sciences holds that between 15 and 20 percent of the American population have symptoms of chemical sensitivity that make them unusually vulnerable to sick buildings.[7] Such buildings often are "built tight" on the laudable motivation of conserving energy. In such nearly sealed environments, toxins and disease-causing organisms may be distributed via the air conditioning systems and generated by synthetic carpeting, disinfecting cleaning solutions, and office equipment such as fax and copying machines. Workers in sick buildings report a high incidence of symptoms such as headaches and fatigue.

In many cases, the problems could be cured by adequate fresh air. Unfortunately, in many cities the outside air is so polluted that opening the windows does not help. To be really healthy places for humans, many such buildings need four times as much fresh air as they are getting via their high-tech ventilating systems. This situation is difficult if not impossible for most individuals to remedy, even if they are aware of the cause of their distress. In most such cases it takes planned action at the top of the organization, or organized pressure on management from the grass roots or from a union, to correct the problem.

Fortunately, in recent years, an increasing number of ecologically aware architects have been designing hospitals, medical offices, clinics, and office buildings that are greener and livelier with more homelike atmospheres. The buildings also are no longer boring and aesthetically sterile with institutionalized colors and decor. Living plants have been added to many treatment settings and offices. As we will see below, plants have the potential to improve air quality as well as the mental and physical health of occupants.

Making Schools Environmentally Healthy

It is appropriate for all teachers and parents to be alarmed about the widespread health hazards in many school buildings. Children are at greater risk from environmental toxins than are adults because "they have fewer detoxifying enzymes, breathe more air for their body weight, absorb toxic chemicals more readily, and have cells that divide more rapidly, making them more prone to cancer."[8] Unfortunately, there are no regulatory standards for acceptable levels of indoor contaminants for schools.

The media has carried numerous exposés of the problem. In an elementary school in Indiana, 50 out of 170 children were suffering from recurrent bronchitis and extreme fatigue. They were coughing up "black stuff." Complaints by two mothers were written off as "female hysteria" by school authorities, a typical sexist response. But when 200 parents found a lawyer in a local church to represent them, the allegations that the school was toxic were taken seriously. The source of the problem was the insulation. Fortunately, it was removed. Unfortunately, this was done with the children present. The workmen had masks to protect them from the toxic dust—the children did not![9]

Toxic substances frequently found in schools include pesticides (often applied when students are present), asbestos, lead, radon, food additives, and general indoor air pollution. Tests were made for radon (an odorless radioactive gas arising from certain rock formations) in some three thousand classrooms in sixteen states. The tests revealed that the EPA's safety level for radioactivity was exceeded in nearly one out of five classrooms. (No one knows if there really is any safe level of radon.)

Recent studies have revealed that much lower blood levels of lead than previously believed can pose severe health hazards for children.[10] Many experts have now concluded that there is no such thing as a really safe level of lead in young children. Elevated levels of lead in the blood can interfere with the body's ability to make red blood cells and can cause irreversible harm to still-developing and therefore more vulnerable nervous systems. Children affected by lead exposure often have reduced attention spans, more behavioral problems and learning difficulties, impaired hearing, and reduced intellectual abilities. Scientists now regard lead poisoning as the number-one public health crisis for children of *all* socioeconomic levels. Levels are much higher among poor, inner-city children. The major sources of dangerous lead levels in air, soil, drinking water, and food include incineration of municipal garbage (which contains large amounts of lead), exterior paint (lead is not permitted in interior paints), some lead-glazed ceramics and china, water in old plumbing systems (mainly in houses built before 1930), and house dust.

Numerous school systems have instituted programs to reduce all types of toxins. For example, Maryland's State Department of Education, through its school facility office, has an impressive indoor air program. It offers testing equipment and an indoor air management program, together with workshops for administrators and teachers to train them in identifying such problems.[11]

Making Our Cities Environmentally Healthier

The ecological concern of therapists and teachers should not stop at the walls of their homes, schools, and workplaces. Such a limited concern militates against the health of the wider environment thereby reducing the wellness of the environments closest to people. The interest of professionals and parents in enhancing the ecowellness of their neighborhoods and communities is a clear example of enlightened self-interest.

Some city planners and environmental architects point out that the way most modern cities are designed ignores or downplays all humane values, including wellness. The term "predatory city" is sometimes used to describe such cities. The problem obviously includes urban slums but is much wider. The paucity of parks, flowers, trees, and green corridors in many cities makes them less healthy for everyone. Other noxious features include endless rows of boxlike apartments and condos without green courtyards, and urban rivers used as dumps or made into lifeless concrete ditches.

Consider noise pollution and other wellness-diminishing effects of living and working, driving and playing in a congested city. If you are in such a city, pause and let yourself become aware of the constant din from honking taxis, screeching brakes, ear-splitting motorcycles, cruising teenagers blasting their own ears (and the ears of everyone within 100 feet) with hard rock music, all mixed with a toxic cocktail of exhaust fumes and industrial pollutants that produce stifling smog.

At a concert in San Francisco by the heavy-metal rock group Magadeth, 13,000 pairs of earplugs were distributed to the fans. It seems that this group is supporting the program of HEAR, an effort devoted to preventing the permanent hearing loss from which many adolescents suffer by the time they reach adulthood.[12] (I wonder why the rock group did not simply turn down the volume twenty decibels. This undoubtedly reflects the multiple-generation gap between myself and rock-loving teens.)

Without our being aware, constant loud noise gradually ages our ears, robbing us of hearing acuity permanently by damaging the inner ear. The U.S. Public Health Service estimates that 10 percent of persons of all ages have significant hearing loss, defined as sufficient to interfere with their ability to understand ordinary conversation. One study of adolescents found that just over one in five suffer from significant hearing loss in both ears, a much higher rate than a few decades ago. Half of us over sixty-five have significant loss from the cumulative impact of lifetimes of noise pollution added to what may be the effects of aging.

Loud noise also stimulates the body's fight-or-flight stress responses, elevating blood pressure and heart rates. If this is chronic, it can be a factor in causing fatigue, irritability, and diminished ability to concentrate. The U.S. Occupational Safety and Health Administration has set 90 decibels for eight hours as "allowable." A rock concert produces 105–110 decibels; a jet taking off or disco dance music averages 120 decibels. Perhaps all this makes you long for more gentle quietness in your living space. It obviously behooves us professionals to reduce noise pollution to a minimum in our work setting.

Another toxic feature of many cities is the absence of community meeting places like town plazas or old-fashioned town squares with trees and flowers. City planner Morneen Kamiki Bratt has done illuminating research on ways a green plaza can make a city more life-enhancing for people.[13] She points out that plazalike public spaces existed in most premodern villages around the world, making them "nurturing villages." People were nurtured by gathering there to talk; eat and drink; hold public meetings, concerts, and dances; celebrate major religious and secular holidays; as well as hold community rituals of all kinds. Bratt believes that widespread efforts are needed today to bring the renewal of nature (including wild nature) into the settings where people live and work. In our modern Western world the vast majority of people live and work in cities. Rather than defining as "civilized" cities that seek to largely exclude nature, we need to understand that to be truly humanizing and health-nurturing, cities must incorporate nature into their fabric.

A hopeful sign is that a growing number of cities are doing just this. They are discovering that creating contemporary equivalents of green village squares can make the whole urban atmosphere healthier and more people-nurturing. Experience has shown that such public gathering places also become valuable economic assets that enhance property values in adjacent areas as well as tourism. A dramatic example of this is the Riverwalk in San Antonio, Texas, where a smelly, polluted stream was transformed to become the focal point of a dramatic urban redevelopment project. What has evolved is a beautiful Venice-like area where riverside shops, restaurants, and a public library line the banks; flowers, sculpture, and other art festoon the area; and small boats carry tourists along the stream. Wandering musicians, jugglers, and mimes enliven the atmosphere, while friends chat over their favorite drinks and enjoy people-watching. Urban pedestrian trails connect Riverwalk with adjacent commercial and residential areas of San Antonio.[14]

Bratt points out that green, nurturing city plazas are good neighbors for colleges and other schools, post offices, parks, book stores, restaurants, museums, art galleries, libraries, offices, shops, theaters, and bakeries. She gives examples of how green, people-nurturing plazas often stimulate the creation of other green public spaces, wildlife corridors, mini-parks, and playgrounds throughout larger cities.

Among other encouraging efforts to enliven cities are the emerging urban reforestation education-action programs such as The Tree People in the Los Angeles basin. Countless children and adult volunteers have worked as urban foresters in planting tens of thousands of trees. Inner-city gardening programs for children and adults, such as the one pioneered by Rachel Bagby in the Bronx (New York City), are also helping to green the places where millions of people live.

New buildings and communities need to include green spaces and to be in harmony with the ecosystem, the climate, and the historical heritage of the region. A friend who is a builder-developer shared his experience in creating an earth-friendly community in the Monterey, California, area.[15] He employed a biologist to do a flora and fauna survey of the area before it was developed. The layout was designed to preserve as many as possible of the native plants and animals by the use of green corridors and parklike areas. As an enlightened business person, he was pleased to discover that concern for the environment paid off financially, making the homes more attractive to potential buyers. It is encouraging that some developers of less expensive homes also are discovering that earth-friendly planning is a profit maker.[16]

Therapists, educators, health professionals, and parents have a high stake in doing everything possible to make their neighborhoods and wider community settings greener, cleaner, and more life-giving. As Bratt makes clear, any community or building can be made greener, healthier, and more people-nurturing if the people involved have the vision and understanding to make even small changes.[17] Because professionals have visibility and respect in most communities, their support of struggles to make buildings, neighborhoods, and cities healthier tends to carry considerable weight. Making ecologically nurturing changes will also benefit those who seek their professional services. Simply being in a healthier environment will increase in subtle ways the benefits of their therapy, healing, or learning. Town planner and architect Carl Anthony emphasizes the ecological role of cities as moral barometers of our society. He holds that without justice, solutions to problems of wilderness, endangered species, open spaces, and natural beauty in cities will be impossible.[18]

Plants as Living Air Cleaners

Plant lovers have long intuited that an abundance of living plants improves the intimate world where they spend their days and years. Recent evidence confirms this and suggests that choosing certain plants to share your space will remove more pollutants and make the air healthier to breath.[19] The three most common air pollutants are formaldehyde, benzene, and trichloroethylene (TCE). Formaldehyde, found in everyday items like plywood, foam insulation, room deodorizers, facial tissues, paper towels, and permanent-press fabrics, can irritate the nose, eyes, and throat. Prolonged exposure may cause serious medical problems. Benzene is emitted from some inks, paints, plastics, detergents, and pharmaceuticals. Inhaling high levels can cause eye and skin irritation, headaches, nausea, irregular heartbeat, and even kidney and liver damage. TCE is used in dry cleaning and in some paints, inks, varnishes, and adhesives. The National Cancer Institute identifies TCE as a carcinogen with a negative effect on the liver

In 1989, scientists at the National Aeronautics and Space Administration, searching for ways to clean the air in space stations, made a surprising discovery. They found that certain popular tropical and flowering houseplants can reduce the concentration of these three toxins in the air by up to 90 percent. Air-cleansing plants include philodendron, aloe vera, spathiphyllum, English ivy, spider plant, African daisies, chrysanthemums, azaleas, dieffenbachia, and poinsettias. Most of these plants are widely available and require minimal care to thrive. Some do well even in low light. Spider plant, one of the hardiest and most common houseplants, seems to be the most effective.

No precise numbers have been established as to how may plants are required to purify the air in an office or home. B. C. Wolverton, a NASA research scientist, offers these guidelines: If your "space" has an eight-foot ceiling, it is well to have two or three plants (8- or 10-inch pot size) for every 100 square feet of floor area. If you use larger plants (for example, in 14-inch pots), you can use only half the number. The more plants the cleaner the air—and the larger the plants, the better. So if it is healthful to do everything possible to eliminate indoor air pollutants, it is also prudent to surround ourselves with flourishing plants.

On a personal level, I feel fortunate indeed that the place I call home has such an abundance of flowers and plants flourishing inside and in the surrounding yard. This peaceful, energizing green environment exists thanks to my long-time partner's passion for plants, her hard work, and her two (at least) green thumbs. Other ecohealing parts of our personal environment include a giant live oak that spans the width of our lot behind the house, and the ever-changing vistas of the Pacific Ocean in the distance. In all this I rejoice and am very grateful, particularly when I remember the issues highlighted in the following section.

Social Causes of Environmental Problems

Teachers, parents, and healers who are concerned about the roots of ecological problems in the wider social context should be knowledgeable about environmental racism and ecopoverty classism. Environmental illnesses proliferate among ethnic minorities and the poor in both developed countries like the United States and in the developing world. Social and economic injustice issues are inextricably intertwined with environmental racism. They are also linked with the systemic economic injustice that produces the culture of poverty in rich countries and the chasm between these nations (mainly in the Northern Hemisphere) and the impoverished nations (many in the Southern Hemisphere). Poor families are forced by their economic circumstances to live in the most polluted areas of cities. To illustrate, it is estimated that half of slum-dwelling African American children in cities suffer from lead poisoning. This is compared with only about 7 percent of middle-class white children. Robert D. Bullard, African American sociology professor at the University of California at Riverside, is spearheading the growing grass-roots movement of urban environmentalism. He has authored two revealing books on racism and environmental problems. When he researched the waste dumping records of Houston, Texas, in 1979, he discovered that all five of the city's garbage dumps and six of its eight garbage incinerators were located in predominantly African American communities. This pattern, he learned, obtains throughout America. He declares, "Communities that are the most polluted have the highest poverty rates, poor health care, and high unemployment rates." Then he adds, "Kids playing in a park that's contaminated with lead is an environmental problem as serious as any you can find, though not as exotic and attention-grabbing as other important issues like saving the whales and spotted owls."[20]

Bullard observes that comparable dumping patterns exist worldwide. Some transnational industries in affluent countries move their production to poor countries with nonexistent or nonenforced environmental standards, as well as an abundance of workers whose poverty forces them to work for pitifully low wages with zero fringe benefits. Some cities and companies ship their pollution to poor, developing countries such as Haiti or China. American companies are required by law to ship their manufacturing waste produced in poor countries back here, along with finished products. But instead many dump it into sewers, abandoned lots, and rivers and streams.

The worst industrial accident in history occurred in 1984 in Bhopal, India, where a chemical leak at an American firm (Union Carbide) killed several thousand people, caused many times that number to suffer serious health problems including blindness, and produced an estimated 200,000 semipermanent refugees. The Boston-based National Environmental Law Center says that America is at risk for a Bhopal-type tragedy. More than ten billion tons of extremely hazardous chemicals are stored in factories in this country, according to its report, and accidents still happen here.[21]

Making Your Living and Working Places Greener and Cleaner

Here are some practical steps you can take to enhance the healthfulness of your work and home environments:

1. *Start by carefully checking your workplace and home for toxins, pollutants, and other health hazards. Then take whatever action is needed to remove these.* In addition to having your drinking water tested for lead and other toxic substances, it is also wise to check for asbestos in the insulation and for various carcinogenic chemicals that may be present in paints, building materials, and cleaning solutions. Make sure there is an abundant supply of clean, fresh air, free of tobacco smoke and other pollutants. If the building has sealed windows, devise ways to open them or in other ways to increase the flow of fresh air. Make sure the water and the food consumed there are as healthful and as free from chemical toxins as possible. Use only washable drinking glasses or mugs or cardboard cups (never plastic) for beverages. Check the cleaning and office supplies. Remember that numerous office supplies such as cleansers and copying materials may be hazardous. What is the impact on people's wellness of the lighting, colors, textures, and visual images used in home and office decor? The yellow pages in your phone directory probably list companies providing environmental testing and remedial services.[22]

2. *Naturize the places where you live and work with an abundance of living plants and, whenever feasible, natural light from the sun, plus open windows. If possible, create easy assess to outdoor areas where plants and birds flourish.* If artificial illumination is essential, use energy-efficient halogen or fluorescent bulbs. Make sure there is plenty of clean, fresh air. Eliminate or drastically reduce noise pollution. Views of nature can promote healing and serenity. If

seeing the sky or an outdoor natural setting is not possible, photos or paintings of natural beauty can help. It is noteworthy that physician Bernie Siegel has observed that hospital patients heal more rapidly after surgery if they can see the sky through a window from their beds. Logo-therapist Victor Frankl recalls that during his solitary imprisonment in a Nazi death camp, he found solace and strength when he was able to glimpse from his cell the limb of a tree which blossomed in the spring. I find anxiety reduction in the view from my dentist's chair—not my favorite place. Just outside the window, there is an ivy-covered wall enclosing a small space with a well-patronized bird feeder and an abundance of plants.

Greening your indoor space with living plants is among the most enlivening things you can do to enhance the physical atmosphere and mental climate of your most-frequented environments.

3. *Arrange to do some of your work—counseling, therapy, teaching— and some of your relaxing and reading while sitting or walking outdoors*. Try brief brisk walk-breaks outdoors, mid-morning and mid-afternoon. You'll find that these provide more energizing mini-vacations than the problematic caffeine-and-sugar breaks. Such brief ambulatory outdoor escapes are particularly valuable if you work in a high-stress job or in a sealed, "climate controlled" building. During professional conferences in hotels with windowless meeting rooms, I have found that even brief outdoor walks increase energy and mind-body feelings of well-being significantly.

At the pastoral counseling and growth center where I was clinical co-director for many years, we discovered that some resistant, angry, or frightened adolescent clients opened up to therapeutic communication when they walked and talked with their counselors amid the native California plants in the nearby botanical garden.[23] For some clients, just being away from a formal office setting was probably as important as being outdoors. The occasions when I move classes outdoors or do peripatetic outdoor counseling provide convincing evidence that healing- and wellness-stimulating energies are available there that are scarce or absent in many indoor settings. Teaching under a tree on the campus, as I occasionally did in Claremont, was a welcome release for both learners and teacher. Student feedback revealed that the advantages more than offset the occasional problems of noise pollution from the nearby avenue or from an airplane overhead. (The same could not be said for the time the automatic lawn sprinkler system gave all of us uninvited showers.) Unfortunately, in the Los Angeles region, as in many other metropolitan areas, it is important to escape into the outdoors only on days when there is no smog alert.

4. *Include a question or two about environmental pollution in the initial assessment or diagnosis, especially when clients, patients, or students live in obviously toxic environments or describe what may be ecological distress caused by pollution.*

Then, depending on their responses, include this dimension in your unfolding teaching or treatment plans. Refer those you suspect or know may be suffering from environmental illnesses to a physician trained in ecological diagnosis and treatment.

It is impossible for most of us to avoid some noxious chemicals in our everyday life. In the days before electronic meters, canaries were used in coal mines to warn of dangerous gases, to which they are more sensitive than humans. In similar ways, persons who suffer from environmental illnesses are the "canaries in the coal mine" of our polluted world. This image (from physician Gary Obert, an environmental illness specialist in Illinois) should remind us that an underlying cause of these health problems is the prolific use of countless ecologically untested chemicals in our consumer-driven manufacturing-marketing-throwaway society.

5. *Urge clients, patients, students, and parents to do everything they can to clean and green their own most-occupied settings as well as their communities*. Help them understand why doing this may be an essential way of improving the self-care and health-care of themselves and their families. Communicate the message that ecology must begin at home. Making our professional settings cleaner and greener will encourage persons who perceive us as knowledgeable, caring authority figures, to clean and green their own work and home settings. To increase this emulation effect, prepare a brief handout describing why and how you have made your office, classroom, or treatment center a healthier place. Action ideas from this chapter can be among items listed in such a handout.

6. *Check for environmental health hazards in the neighborhood, and the wider community where you live. Remember that the sickness or the health of these places impacts yourself, your family, and the people you serve*. The angry, self-serving attacks of anti-environmentalists should not deter you from creative ecoaction to benefit all the people and other living beings in your community.

7. *Use a carrot-and-stick approach to motivating yourself and others to do essential ecological cleanups*. The positive rewards of the vitality and pleasure of working and playing in healthier, greener places (the carrot) are at least as important as the negative challenges of the reality-based ecological health problems (the stick). Remember that stories concerning what others have done to clean and green their places can inspire reality-based planning and action.

8. *If the sheer number of things that need doing is overwhelming, start with one important cleaning-and-greening action*. Lots of people discover that as that single action becomes routine and produces some benefits, it becomes easier to add other environmentally healthful actions.

Exploring This Chapter's Issues Further

For fuller information see the annotated bibliography at the end of this book.

Robert D. Bullard, *Confronting Environmental Racism: Voices from the Grass Roots*, and *Dumping in Dixie: Race, Class, and Environmental Quality*

Environmental Education and Health Services, 3202 W. Anderson Lane, Austin, TX 78757 (Tel. 512-288-2369), can provide information about innovative programs to make schools healthier environments.

National Center for Environmental Health Strategies, 1100 Rural Avenue, Voorhees, NJ 08043 (Tel. 609-429-5358). Resource center for persons with environmentally and occupationally related illnesses.

Superintendent of Documents, United States Printing Office, Washington, DC 20402. Offers a variety of pamphlets on environmental health, including *Lead in School Drinking Water*.

Notes

1. These two quotes are from "Is Your Job Making You Sick?" *Glamour*, January/February 1993, 72. This article represents an encouraging trend: the increased coverage of environmental issues in the popular media. I quote such media elsewhere in this book.

2. I am indebted to Chaplain Carol Meridith, H. C. H. Wesley Medical Center, Wichita, Kansas, for sharing her moving experiences.

3. Reflecting on the chaplain's experiences, I recall hospitals I have visited in India and Africa. They lacked much of Western hospitals' high-tech equipment. But they all had windows that not only could be opened but were opened wide most of the time, except during monsoons. At most medical facilities, patients could see trees, flowering plants, and animals, as well as numerous people, through the windows. Perhaps Western hospitals will gradually discover that "clean" does not have to mean super-sterile, except for patients who are especially vulnerable to infections. The benefits of less rigid rules may enhance morale and the will to live, as well as activate patients' immune systems, in ways that counterbalance the legitimate need to prevent infections.

4. I am indebted to Earle P. Barron of Erwin, Tennessee, for much of the information in this section. See his paper "The Pastoral Care of Environmental Disease Patients," *Journal of Pastoral Care* 38, no. 1 (1984): 44–51.

5. *Sierra*, a periodical of the Sierra Club, January–February 1993, 23.

6. Barron, "Pastoral Care of Environmental Disease Patients," 47.

7. Laurie Goering, "Understanding Environmental Illness," *Santa Barbara News Press*, 7 August 1991, D-1.

8. Nancy Sokol Green, "America's Toxic Schools," *E Magazine* 3, no. 6 (November/December 1992): 31.

9. Ibid., 31–32.

10. Michael Castleman, "Lead Again," *Sierra*, July–August 1992, 25. In 1970, scientists thought that the toxic threshold was 60 micrograms of lead per deciliter of blood. By 1992, it had been lowered to 10 micrograms, and this so-called safe level is still falling.

11. See publications offered by the School Facility Office, Maryland State Department of Education, 200 W. Baltimore St., Baltimore, MD 21201.

12. *Santa Barbara News Press*, 3 December 1992, A-2.

13. See Bratt's master's thesis in urban and regional planning "An Analysis of the Ecofeminist Question: Does the Minoan Central Courtyard or Plaza, Serve as a Model for the Postmodern Plaza and Provide a Mirror into the Possible Nurturing City?" (Pomona: California State Polytechnic University, 1993). Bratt uses an ecofeminist analysis to explore the question of how the ancient courtyard areas of the Bronze Age Minoan cities on Crete can serve as models of the use of plazas to help modern cities become more people-nurturing.

14. Other big-city examples are Seattle, Tacoma, and Spokane, Washington. A healthy, economically productive city metamorphosis occurred in Spokane a few years ago when its downtown river was cleaned up and the whole depressed area transformed into a lovely city park. An example of the plaza principle in a smaller city is Claremont, California, where my family and I lived for three decades. Claremont Village is an attractive area with outdoor eating, drinking, and shopping places where people enjoy gathering near some of Claremont's many beautiful tree-lined streets. (Unfortunately, some of the outdoor eating places are so near streets that people breathe the tailpipe toxins from numerous passing cars.)

15. The developer is Mike Fletcher.

16. The EPA has discovered that it is financially cost-effective to invest in improving indoor air quality and eliminating other pollutants in businesses, because cleaning up the air reduces the enormous costs of lost productivity from major pollution-caused illness. The same is undoubtedly true of making schools environmentally safe and, in the long run, of cleaning up the air, water, and soil of whole communities and regions. More important, in terms of the human values that are of top priority for healers and teachers, it is very cost-effective humanly to reduce the unnecessary price that people pay when they live and work in ecologically unhealthy communities.

17. Bratt, "Analysis of the Ecofeminist Question."

18. Carl Anthony, "Ecopsychology and the Destruction of Whiteness," in *Ecopsychology: Restoring the Earth, Healing the Mind*, edited by Theodore Roszak, Mary E. Gomes and Allen D. Kanner, 263.

19. By improving the quality of the air, plants actually make it possible to "breathe a little easier." The presence of plants as well as animals may touch our psyches at a deep levels, awakening ancient, genetically based unconscious memories of our species' eons in wilderness areas.

20. "Not in My Backyard," *Modern Maturity*, October–November 1993, 22.

21. Reported in *USA Today*, 2 December 1994, 3-A. See also Alfred de Grazia, *Cloud Over Bhopal: Causes, Consequences, and Constructive Solutions*, Palmoale, CA: Kalos Foundation, 1985.

22. Some toxins are not easily detected. While I was writing this chapter, we had our home tested for radon, knowing that in some sections of the foothills around us this odorless radioactive gas is emitted from rincon shale. We were shocked to learn that high levels were present in our house. Fortunately remedial work by a radon contractor corrected the problem. The EPA estimates that, even though the effects are long-term, radon causes more than 15,000 deaths from lung cancer each year, second only to cigarettes.

23. A psychologist known to Richard Voss aptly calls these "walk and talk" as compared with "sit and talk" sessions.

7

Assessment, Timing, and Process in Ecotherapy and Ecoeducation

Love all of God's Creation, the whole and every piece of sand in it. Love every leaf, every ray of God's light. Love the animals, love the plants, love everything. If you love everything, you will perceive the divine mystery in things. . . . you will begin to comprehend it ceaselessly more and more every day. And you will at last come to love the whole world with an abiding universal love.

—Fyodor Dostoyevsky, *The Brothers Karamazov* [1]

This is the first of three chapters describing methodologies of ecotherapy and ecoeducation. These methods are drawn from those the author has developed in his therapeutic practice, teaching, and workshops, and methods described by therapists interviewed for this book. This particular chapter will focus on assessment and the sequence of steps frequently involved in ecotherapy.

The Importance and Objectives of Ecoassessment

Because people's well-being is impacted continually by their relationship with the natural world, it is essential to include an ecological component in the overall assessment of their healing and learning needs. Theodore Roszak comments on a major gap in psychiatry's diagnostic schema: "There is not a single recognized disease of the psyche that connects madness to the nonhuman world in which our environmental responsibility is grounded."[2] This serious omission reflects a widespread view of sickness and health that ignores the influence of people's relationship with the earth. Environmental medicine has been focused almost exclusively on ecological illnesses resulting from toxicity. It is not surprising, in light of this, that no scientific measurement tools have been developed for ascertaining the degree of alienation from or healthy bonding with the earth. Furthermore, focusing on the ecological problems of individuals without also assessing the pathogenic contexts that contributed to their problems is seriously deficient diagnostically. Such hyperindividualistic diagnosis does not provide guidance for interventions to treat their problems holistically.

The ecodiagnostic methods described in these three methodology chapters are rough measurement tools designed to gain diagnostic impressions for use as clues in planning ecotherapeutic and ecoeducational approaches. It is clear that scientific assessment tools are needed to measure ecoalienation and healthy ecobonding. As ecotherapeutic methodologies evolve, it is to be hoped that more precise assessment instruments will become available for use in healing and wholeness work.

In the meantime, whole-person diagnosis in therapy and assessment in education should discover as many clues as possible concerning the quality of clients', patients', and students' relationships with the natural world. The overall objective of the diagnostic methods described below is to gain clinical impressions of people's degrees of nurturing connectedness or depriving alienation from the natural world. This assessment process involves attention to two interrelated dynamics—one external, the other within persons: What is the quality of people's relationships with the natural environment, and what are their internalized images, attitudes, and feelings about the natural world, all of which influence profoundly how they relate to nature? It is important to gain some awareness of feelings of ecological alienation, angst, grief, and despair, on the one hand, and feelings of positive bonding with the biosphere, on the other. It is from their sense of bonding that comfortable "at home" and "I belong" feelings in the natural world flow, along with the attitude that nature is to be cared for respectfully as the sustaining source of all living beings. The external and internal ecological dimensions of people's lives reinforce each other in reciprocal interaction.

Ecological diagnosis aims at shedding maximum light on how well people are living out the three dimensions of the ecological circle. To review, these include: (1) facilitating the healing of ecoalienation and strengthening the bonding with the natural world that enables people to be nurtured by nature; (2) facilitating awareness of the self-transcending spiritual reality in nature; and (3) facilitating earth-caring action that continues the process of healing and bonding. As people experience more fully the three dimensions of the ecological circle, their overall self-care for wellness tends to be strengthened.

Ecodiagnosis obviously is only one relatively small part of overall therapeutic and educational assessment. It is important to gain some understanding of how the findings of ecological diagnosis relate to the findings of the psychological, social, cultural, physiological, and spiritual dimensions of diagnosis. This understanding will illuminate how people's relations with the natural world influences the learning or growth needs for which they seek therapy or education.

The Timing of Focusing on Ecological Issues

It is important to emphasize that ecotherapy and ecoeducation are much more than an explicit focus on ecological issues when appropriate. They reflect an orientation of therapists and teachers that influences much that they do in practicing their arts. Clinical psychologist Sarah A. Conn expresses this orientation insightfully. She holds that disconnectedness from the earth expresses itself in clients as a "materialistic disorder" characterized by addictive-compulsive consumption. Regarding her clients, she states: "I also look at what they describe as personal, inner pain not only as an expression of their unique personal history or circumstances, but also as an expression of the Earth's pain." Using an apt sailing metaphor, Conn says, "I try to tack back and forth between the personal level and the level of larger cultural, political, and more often now, ecological systems." To raise their consciousness of these wider systems, she directs their attention to the social and ecological contexts of their anxieties and distress. She invites emotional responsiveness to the world by asking about their reactions to the natural world and validating their responses as sources of valuable energy.[3]

Whether to focus explicitly on ecological issues, and if so, when, are key decisions in both therapy and teaching. The nature and timing of these interventions depend on the setting and process of which these are a part. Ecotherapeutic issues may be of relatively large or small relevance in particular situations. When they are relevant, appropriate timing of ecological interventions varies more widely in settings of counseling and therapy than in educational contexts. In some psychotherapy and medical problems, such issues may be of little or no direct relevance to the pressing issues that bring people for help. But, it is noteworthy that as professionals' awareness of the prevalence of ecological issues in people's well-being increases, they often are surprised at how frequently these issues play some significant role.[4]

Is it ever appropriate to focus even minimally on ecological issues with people in painful personal or family crises or dysfunction? It may be, but only after the people have been helped to mobilize their coping resources and gain enough relief from crisis pressures to allow them to attend to other issues. If their responses to one or two initial ecological questions indicate that high levels of pollution may be present in their living or working spaces, it is clearly important to focus on this dimension earlier. More detailed exploration of ecological issues ordinarily should be done only when therapists have clinical evidence that such issues may be relevant to people's problems in living. In any case, it is well for crisis counselors and therapists to keep in mind the real possibility that ecological issues may be hidden behind presenting problems. Asking an ecodiagnostic question or two is an expeditious way to explore hidden toxicity or ecoalienation problems.

In contrast to crisis counseling situations, teachers should consider introducing relevant ecological assessment earlier in many learning contexts. Where environmental issues are one focus of a class or workshop, tools such as the checkup and the ecology storytelling methods described below can well be used near the outset. In classes and workshops in which ecological issues have some relevance but are not stated in the course description, the teacher should point out, as the class objectives are being discussed, how these issues are related to the course's objectives.

The following checkup has proved to be a valuable tool, particularly in classes and workshops. I have not used it in the ecotherapeutic dimension of counseling and therapy, except to keep some of the items in my mind as if this dimension of people's problems is explored. The findings have both diagnostic and therapeutic usefulness. As a do-it-yourself tool for persons seeking to enhance their ecological self-care, as well as in ecoeducation,[6] the findings can provide a quick, subjective assessment of the relative health of persons' relationship to the earth. It also provides an extensive "shopping list" of concrete options that can be implemented to enhance self-care and earthcare by strengthening one or more of the three dimensions of the ecological circle. The checkup suggests a variety of ways persons can help heal that small part of our wounded planet where they live, by actions that help make their community's ecosystem healthier for themselves and others. The findings obviously are merely suggestive, rather than scientifically diagnostic, because they reflect the subjective self-perception and self-reporting of the person taking the test. But, in spite of this limitation, the checkup is a useful tool in ecological education and in the educational dimension of counseling and therapy. It often is not feasible to use the entire checkup in therapy, but individual statements from it can be rephrased as questions and used to explore ecological issues in therapy.

An Ecological Wellness Checkup

Instructions:[7] In front of each item write one of three initials:

E: I am doing Excellently in this area.

OK: I am doing OK but there is definitely room for improvement.

NS: My life and lifestyle definitely Need Strengthening in this area.

Ignore those items that do not seem relevant to your situation.

____ I love the natural world and feel a deep connection with the wonderful network of living beings of which I am a tiny but significant part.

____ I find healing energy and sometimes joy in getting close to plants, animals, and beautiful places in nature. I can get high when I am in a lovely garden or other natural beauty spot.

____ Being in forests or mountains, or by unspoiled rivers, lakes, oceans, or in wilderness places, brings me refreshing renewal.

____ Being in wild nature or near wild animals does not arouse inappropriate anxieties in me. Rather it stimulates my inner connection with my own creative inner wildness.

____ I know how to open myself to be nurtured by nature, even when I am in a city—in a park, garden, or near a tree or growing plant or a blooming window box.

____ I feel stirrings of pain in my body-mind-spirit organism when I'm in a polluted place dominated by environmental ugliness.

____ I know how to protect myself from the violence of nature—for example, tornadoes, floods, or earthquakes—without being paralyzed by fear.

____ I know how to "ground" my body and my grief and pain, when these occur, in awareness of the dependable supportiveness of the earth.

____ I like to share the enjoyment of nature with the people I love. Sometimes we experience our love being deepened by such sharing.

____ I am aware that my own and my family's levels of wellness, at any given time, are inextricably interwoven with the level of wellness of our place in the natural world, as well as the wellness of the society around us.

____ I regularly examine my lifestyle and the values that guide it, and make changes to enable these to express more fully my loving respect for the health of the environment. I often make such changes even when they require sacrifices.

____ I know that there are no individualistic solutions to the societal causes of the ecological crisis. So, in addition to my personal ecological changes, I work for political and economic justice. I choose to be politically involved in efforts to help heal the planet—for example, by voting only for candidates who favor strong local, national, and international programs of justice and earth-caring.

____ I have a solid support group of friends, family, and others who share my passion for loving the earth by caring for it individually and collectively.

____ My spiritual life is enriched with aliveness both when I open myself to be nurtured by nature and when I engage in action to help save a healthy earth for all living creatures. I often experience the lift of spiritual awe when I become aware of the mystery and wonder of nature and of the eon-spanning process of continuing creation called evolution.

____ My sense of partnership with God in working for a world of wholeness is a source of hope, challenge, and serenity in what is sometimes frustrating earth-caring work.

____ I practice ways of enhancing the love of life that is at my spiritual core, making this love the primary energy source for planet-loving work, rather than guilt or fear.

____ I am seeking the continuing growth of my spirituality and value commitments to make them more bridge-building with those in different faiths, nations, races, languages, and cultures. I cherish the awareness that they are sisters or brothers in one species, the global human family, and much-needed potential partners in saving the biosphere.

____ People-caring and earth-caring are interdependent and mutually reinforcing processes for me, beginning in my home and community, but reaching out to the larger national and international levels. I experience a sense of caring, compassionate relatedness with both people and nature when they are suffering.

____ I know that the dream of saving a viable planet can only be realized by mobilizing broad international, intercultural, and interreligious collaboration.

____ My ways of expressing my love for and loyalty to my own country do not contradict my more inclusive love for and loyalty to the well-being of the biosphere and of the whole human family. I know that my nation's long-term wellness can be protected and enhanced only if the wellness of other nations also is protected and enhanced.

____ I try to stay current and knowledgeable about the complex, rapidly changing global ecojustice crisis and the creative ways being developed to help resolve it.

____ I am aware that earth-caring and peacemaking are two interdependent sides of the same earth-healing process.

____ I am actively involved in working in and/or supporting financially some local, national, and international groups committed to healing and protecting the natural environment.

____ When I observe or see photographs of violence against women, children, nature, or animals, I experience intense feelings such as anger, grief, guilt, and commitment to help prevent such violence. And I use these feelings to motivate me to engage in constructive geojustice and peacemaking action.

____ I am aware that among the social causes of escalating damage to the whole planet's environment are the widening chasm between rich and poor nations, the population explosion, most pronounced in poor countries, lifestyles of unfair consumption in affluent countries, and the tragic resource waste of the planetwide arms race.

____ I recognize that violence against Mother-Father nature and against persons socially defined as lesser, weaker, or "other"—for example, women, children, minorities—is rooted in some of the same psychosocial causes, injustices, and inequities of power, prestige, and property.

____ The nurture I receive from nature is healing and valuable in itself; it also provides energy for preventing burnout and sustaining earth-caring and peacemaking when the going ahead is difficult.

____ Whenever I experience feelings of despair, denial, and powerlessness concerning the enormous, complicated ecojustice problems, I use methods for transforming these numbing feelings so as to recover the hope and energy required to do effective earth-caring and justice-making.

____ I often use my sense of humor and laugh with my earth-caring and peacemaking partners as a pressure release valve and an energy-renewing method for ecoaction.

____ I am practicing parenting and friending for peace, justice, and ecological wellness in my family, extended family, and other close relationships.

____ I am finding ways to practice earth-caring, peace-nurturing, and justice-making in my work, my social life, and my faith community—ways that enable me to think and act both locally and globally.

____ I am seeking a hopeful image of a transformed future of eco-wellness for the planet. I sense that this vision lures me toward such a future and helps energize and guide my ecojustice action.

____ I resist taking long checkups like this one, particularly if they threaten to increase my guilt for not doing more for a cause in which I believe fervently. (If so, welcome to the club! If you avoided taking the checkup, let me encourage you now to quickly scan the list and note the items that seem especially important to you.)

____ Add any additional checkup items here:

Instructions for using your findings: To gain the most from what the checkup has enabled you to discover, follow these steps. Remember that the slash mark (/) means to pause while you do what has been recommended.

1. Be aware of how you felt while taking the checkup and reflect on what you have learned from the experience. /

2. Scan the initials you placed in front of the checkup items and get an overall impression concerning the well-being of your lifestyle as it relates to the health of the environment and society. / Discuss what you learned with a friend, family member, therapist, teacher, or group member, depending on the context in which you are using the checkup. /

3. Affirm yourself for items that you honestly scored E. / Write two lists of items you scored either OK or NS (Needs Strengthening). These are areas in which you and the earth can benefit most from your constructive changes. / Now select not more than three items that seem particularly important to you and to your "place" on the earth. / Beside these items, jot down your thoughts concerning what you need to do to enhance your wholeness in these areas. This is your "to do" list. /

4. Write out a Self-Earth Care Plan responding to the items you have selected. / In your imagination, picture yourself implementing your plan effectively, in spite of resistances within yourself, in other people, and in institutional structures. / Share your plan with a friend, family member, teacher, therapist, group, or class to receive critical but friendly feedback and suggestions for improving it. /

5. Implement one part of your plan, keeping notes about what you learn in the process about caring better for yourself by caring for the earth. If your efforts fail, waste no time with self-punishing guilt feelings and postmortems. Instead go back and redesign the plan to make it more workable. / Be sure to reward yourself for each step you take toward implementing your plan. /

6. After implementing one part of your plan, repeat these steps with another item on your "to do" lists.

The Healing, Growth-Enhancing Process of Ecotherapy and Ecoeducation

This section presents an overview of the steps that frequently are involved in doing ecotherapy and ecoeducation. The process is flexible, with movement back and forth on the path, and many individual variations. This sequence of steps often is an effective path for enabling people to move from denial, disempowerment, despair, and paralysis about the world and the ecojustice crisis, toward hope, empowerment, and motivation for earth-caring action. These steps represent a drastic modification of a process developed first by Joanna Rogers Macy.[8] I have made substantial changes and additions to her process to make the following directly relevant to the ecological crisis.

It should be emphasized that the full process is most useful with people who need to move from ecoalienation to ecobonding. The individual parts of the process, however, can be used productively for self-help as well as for facilitating growth when working with others. For example, encouraging people simply to spend additional time being nurtured by nature can be health enhancing. The same is true of telling one's own earth story or implementing decisions to live a more earth-friendly lifestyle.

Although the series of steps in the overall process can be used most readily in educational settings, elements of it can be selectively adapted for use in counseling and therapy. Following are the major steps of this process.

Step 1: An invaluable way to initiate the process of ecohealing is to invite people to tell their ecological story. Doing this also serves to throw diagnostic light on ecoalienation and ecobonding. If there is time in a particular teaching or helping process, the individual's story may well begin with earliest memories of nature and how nature has influenced the person's life at each subsequent stage including the present. By inviting people to recall both disturbing memories and nurturing ones, the autobiographical roots and development of their attitudes, feelings, identity, and ways of relating to social and natural environments may become available for use in therapy and education.

Ask them how they became aware of the environmental crises. Also inquire about both their negative and positive experiences with nature. The latter may become healing and renewing resources for them.

To prime the storytelling pump, ask people to write between sessions about their relationship to nature and the outdoors. Suggest that they be aware of their feelings as they write and then that they reflect on what they can learn from this recall. In some cases, this exercise produces breakthroughs in awareness of how profoundly their lives have been shaped by early positive and negative experiences with the earth. Whether or not people have written their personal ecological history beforehand, it is important to invite them to tell the therapist or the class group what they regard as the highlights of their story. If time permits, provide opportunities to read portions of their stories aloud. Actually saying the words aloud with others listening sometimes adds to the therapeutic impact of their storying.

A simple way to prime the pump of storytelling, is to ask a few questions such as these: Where were you raised (on a farm, in a village or town, or in a city, large or small)? What were your most memorable childhood experiences with nature? Did you help raise a garden, or animals, as pets or for food? Can you remember frightening or fun experiences in nature, alone or with your family, as you were growing up? Did you learn to enjoy being in the outdoor world, or to fear and avoid it? How did all these early experiences influence your present feelings and relationship with nature? What do you recall about your parents', grandparents', older siblings', or teachers' atttitudes and feelings about relating to the natural world—the soil, plants, and animals (wild and domesticated), the wilderness? How would you describe your relationship with nature now? What are your feelings when you are in a beautiful garden or park? In a violent storm? By a river, a lake, or ocean?

If oral or written storytelling awaken memories of what seem like shaping encounters with nature—frightening or satisfying—it is often helpful to guide people in reliving these formative experiences briefly in the here-and-now. Simply invite them to close their eyes and reexperience the nature triumph or trauma, including all the feelings that still cling to these significant events. Invite them to describe what they are experiencing in the present tense. This reliving often surfaces suppressed feelings. It may bring into awareness ways in which their family's, congregation's, or culture's earth-related feelings, attitudes, and practices shaped their own early and continuing feelings, attitudes, and practices. In discussing the reliving experience, it is important to inquire about how their earth-orientation was shaped by the earth attitudes and behaviors of the adults with whom they were most deeply bonded in early childhood.

In ecoeducation, an interactive way to help people access information about their earlier relations with nature is to ask them to tell their earth stories in small groups. For example, in triads members B and C may say how they imagine member A would respond to questions such as those listed above. This is done around the circle until all members have had turns. Then individuals share their own answers. This approach is effective in enabling group members to understand and engage with each other's experiences.[9]

Step 2: Help clients, students, and family members become more fully aware of and express both their painful and positive feelings about their natural environment. Intense and often conflicted feelings about the earth crisis are widespread among people of all ages today.[10] In both therapeutic and educational work, it is important to ask about ecological anxiety, grief, guilt, shame, anger, despair, and powerlessness in the face of the planet's huge, complex problems. Only occasionally are such feelings among the presenting problems that bring earth-caring people to therapy. More frequently they are among the feelings that motivate people to enter ecological learning experiences.

Asking questions that invite the participant to open up, such as the following, may bring subconscious feelings to the surface: How do you feel about the environmental situation? Do the feelings that brought you to counseling (or this class) have any connection with the world situation or to environmental problems? One or two such queries by a teacher or counselor often elicit awareness of the way stressors in the social and environmental context are among the factors involved in the person's problems-in-living. Unless such questions are asked, most people who come for counseling or therapy assume that it is not appropriate to discuss such issues in that setting. The teacher's or therapist's knowledge that such painful feelings may be present, often at a subconscious level, can encourage people to express their pain in the knowledge that they will be heard. A variety of methods for enabling people to get in touch with and express their earth-oriented feelings are presented in the three methodology chapters. For example, taking the ecological wellness checkup in a class usually increases students earth-awareness significantly. The consciousness-raising methods and the storytelling approaches described in this chapter usually awaken feelings about the earth.

What are the most prevalent feelings reported by youth and adults when they think about the impact of the global biosphere crisis on their own lives? According to the therapists and teachers surveyed for this book, the most common feelings are variations on one theme: "I can't make any real difference because the problems are so gigantic and complex." Increased awareness of the magnitude of global environmental problems often feeds feelings of powerlessness, helplessness, hopelessness, denial, and despair. When people feel overwhelmed by such feelings, they often reassure themselves by internal self-talk messages such as "It can't be as bad as some scientists think it is." Or rescue fantasies like, "If it is that bad, someone will rescue the planet from its impending fate." The "someone" frequently refers to God or to national or religious leaders whom they foolishly endow with godlike omniscience and omnipotence. Tremendous psychological energies are wasted by such magical psychological defenses, and a valuable sense of personal responsibility to do whatever *is* possible is lost.[11] The therapeutic challenge is to use methods that enable people trapped in such paralyzing feelings to recover and rechannel these energies for use in earth-caring and people-caring.

It is very important to validate people's painful reality-based feelings as appropriate and also to affirm the inner strength required to face rather than simply ignore the present ecological reality. This validation of feelings can be done by saying, in effect, "You are not weird to feel intense pain about the damage caused by environmental destruction even though a lot of folks seem not to be aware or to care. On the contrary, facing these feelings is an essential step toward the hope that is necessary for doing all you can to care for the earth." Of course, feelings appropriate to the environmental crisis usually are intertwined with people's idiosyncratic feelings from their personal problems. It is important to encourage people to distinguish the sources of their feelings because those from personal problems require different approaches to healing than those from the ecojustice crisis.

Step 3: Encourage people to strengthen their sense of organic connectedness with the natural world by intentionally opening themselves to be nurtured by nature more often and in more depth. As they do this, they may well discover their lost earthy Self and also experience the self-transcending spiritual dimension in nature. Opening themselves to experience awareness of their profound connections with the biosphere usually increases people's awareness of the earth's suffering. But the experience of belonging to and being nurtured by the earth community, and of being energized by the spiritual reality in nature, however this is understood, can provide much-needed power and hope for using the pain to motivate earth protecting and nurturing action.

An essential goal of this step is to help people rediscover their inherent sense of connectedness with the earth, which many have lost. Recovering this profound earth-community connection will enable them to experience nature not as an inert thing, but as an interdependent living organism that is in them and of which they are a vital part. A vital reward of this bonding process is earth-grounded love. They discover that loving their own bodies, as well as loving their children and grandchildren, family, and friends, must involve loving the natural world. Getting reconnected in this way lets them become what all of us should be—lovers of the earth.

It is an important step toward healing when people become aware that a major source of their pain in living is their caring about the suffering earth and their own violated connection with it. As they get in touch with their lost earthiness, they often feel violated in their body-mind-spirit organisms by the violations of nature. It is a valuable though painful experience on an organic level to feel the violation that impacts all living things on planet earth every hour of every day. Because the earth community is in us and we are in the biosphere, it is appropriate to feel personally violated in ourselves when the living things that are dear to us are violated.

People increase their earth bonding by a variety of experiences, including deepening their awareness of the wonders of the natural world, learning to enjoy experiencing their spirituality being enriched by nature, and investing themselves in helping to care for nature more lovingly. Earth bonding also may be strengthened by deepening awareness of how we are a part of the marvelous, eon-spanning evolutionary process by which the entire biosphere, including ourselves, came into being and continues to change and grow. [12]

To gain a taste of this approach, invite a person with whom you feel comfortable to join you in experiencing the following segment of an exercise from Macy's *Despair and Personal Power* (pp. 101–5). This awareness exercise, which some people initially find embarrassing but also meaningful, involves taking turns cradling and examining each other's hands.

Lift gently your partner's arm and hand. . . . Cradle it, feel the weight of it . . . flexing the elbow and wrist, note how the joints are hinged to permit a variety of movement. . . . Look, look as if you had never seen it before, as if you were a visitor from another world. . . . Turning the palm and fingers, note the extraordinary intricacy of their structure. . . . What you hold is an object unique in our cosmos: it is a human hand of planet earth. . . . In the primordial seas where once we swam, that hand was a fin—as it was again in its mother's womb. . . . Feel the energy and intelligence in that hand—that fruit of a long evolutionary journey, of efforts to swim, to push, to climb, to grasp. . . . Note the opposable thumb, how clever and adept it is . . . good for grasping a tool, a pen. . . . Open your awareness to the journey it has made in this lifetime . . . how it opened like a flower when it emerged from the birth canal . . . how it reached out to explore and to do. . . . That hand learned to hold a spoon . . . to tie shoelaces . . . to throw a ball . . . to write its name . . . to give pleasure . . . to wipe tears. . . . There is nothing like it in all the universe

Now continue this cradling exercise by switching roles with your partner. / After this, share what each of you experienced in each role. I hope that the exercise generates feelings of appreciative wonder for the human body and the evolutionary process of which it is one amazing result.

Another way to deepen earth bonding is to spend a significant segment of time alone in silence in a natural setting. This is akin to the ancient "vision quest" practice of Native Americans that involves fasting for several days alone in the wilderness, during which time they may get a fresh vision or sense of their lives. I have experimented with very abbreviated versions of this self-discovery experience, with workshop participants and a class. These begin with a brief discussion of purposes of the experience and suggestions concerning getting the most from it. People then go off by themselves and find a place to be silent and close to nature while they focus their awareness in a meditative spirit, on the many things going on in themselves and in nature around them. They return after an agreed-upon period of time and share their learnings in small groups. Some people find that such experiences help them get in touch with one or more of the three dimensions of the ecological circle. Others simply slow down and let go of some of the self-stressing of being continually "on the go," striving to "get somewhere," and are nurtured by the aliveness of nature.

Step 4: As their energy and motivation are generated by ecobonding, encourage people to reciprocate by earth-caring actions, thus expressing grateful awareness for the earth's bountiful gifts to their earthy Self. As noted earlier, active outreach to help heal the earth is an essential part of mutually nurturing interaction with the earth. For example, becoming involved in environmental reading, discussion, strategizing, and action can facilitate the continuing healing of inner earth alienation and enhance earth bonding while one is helping to heal the earth.

An in-depth look at this essential step is in order. Clinical experience demonstrates that becoming involved in some type of healing-the-planet action may contribute significantly to the overall well-being of students and clients, children and families. Many teachers, parents, grandparents, and others are passionately concerned about the health or sickness of the world we adults are leaving our children. A hope-eliciting expression of such concern is to become proactive by working with others in earth-caring causes. It is more satisfying than many people suspect to thus make a gift toward healing the earth for the coming generations. Furthermore, firsthand involvement in earth-caring is an effective way to reinforce people's concern and raise their ecological awareness to a more informed, motivated level. Perhaps you remember the prince in Antoine de Saint-Exupéry's charming story *The Little Prince*. He could not decide whether he cared for his little planet, because he loved it so much, or loved it so much because he was so invested in caring for it. Comparable feelings often are expressed by people giving time and energy to earth-caring. Psychologically, loving anything—a person, a cause, or a thing—and actively caring for it are mutually reinforcing dynamics. Each enhances the other.

Facilitating ecological outreach can be done in a variety of ways. One approach is to suggest small action steps as a part of "at-home" work between classes, counseling, or grief healing sessions. As people become more aware of and concerned about ecological issues, it usually makes sense to them when a teacher or therapist recommends such action. They find it fulfilling to try their hands at a close-to-home earth-caring need, particularly if the action options were generated by them from their knowledge of the community's needs. Suggestions by the facilitator can prime the pump with a list of local possibilities from which to choose. In practice, asking two questions after a discussion of ecological issues may suffice. One is simply, "What are some of the things you are already doing about local environmental problems? The second is, "Are there other things you want to do to help lessen environmental problems and enhance the health of your neighborhood and community?" It is often helpful to affirm people's positive responses with a statement like one of these: "Folks often discover that doing something, however small, to help improve their environment is a meaningful way to express their concern about these issues." Or, "However much or little our actions accomplish immediately for the environment, it probably will enhance our well-being to join hands with like-minded people in some down-to-earth action."

A pastoral psychotherapist told of counseling with three depressed, suicidal young adults. Despair about the world situation was one factor in their painful depression, intermingled with personal and interpersonal conflicts. She reported that the most helpful therapeutic intervention she made with these clients was to recommend that they get involved in some type of study-and-action peacemaking group. Similar therapeutic interventions with several depressed individuals and families who came for help to our pastoral counseling and growth center produced comparable hope-awakening results.

Responding to the survey done for this book, one psychotherapist reported that she encountered deep despair in several of her counselees during the Persian Gulf War of 1991. Their depression was in response to the widespread, media-promoted attitude of glorifying that war as totally "heroic" on the allies' part, when they were aware that it had both heroic and disastrous aspects. (Examples of the downside included the killing of countless innocent children by our so-called smart bombs and the spending of a billion dollars a day to liberate a rich, authoritarian government that had a dismal record of undemocratic oppression of women and others.) The therapist reported that the best therapy was to encourage such clients "to take their distress to heart, not to minimize their pain, but to work actively with others to search for more constructive solutions to the problems that caused this international fiasco."[13]

To summarize, it often is healing to challenge students (and occasionally clients), gently but firmly, to consider becoming actively involved in doing something to help create a more sane, ecologically sustainable society. Many people are hungering for hope about the world's crisis, and action is a way of empowering hope by giving it arms, legs, and a voice. Fortunately some counselors, therapists, teachers, and parents are enabling the awakening of reality-based hope by encouraging such outreach action. In doing this, they are creating a much-needed bridge of mutually strengthening interaction between personal and societal healing.

What about the firm resistance to getting involved that many people express by denial and inaction? Frequently such resistance flows from a severe *hope-vacuum* expressed by one client this way: "I can't make any real difference because environmental problems are such a big, complicated mess!" Although ecological action can awaken hope, it requires a minimum degree of hope just to get moving. For some, sufficient hope to energize action comes from hearing a true story of courageous, inspiring ecoaction. Like many others, I find the story of Wangari Maathai and the women and children of Kenya to be such an account. Here, in her own words, is what happened:

"I became exposed to many of the problems women are facing—problems of firewood, malnutrition, lack of food and adequate water, unemployment, soil erosion." She knew that all these problems were rooted in ecological devastation such as the massive deforestation and the resulting topsoil depletion in her country. Fortunately, when she became aware of this huge problem, she avoided saying to herself, "What can one person do?" or "Someone ought to do something!" Instead she began by planting seven trees in her own backyard on June 5, 1977. She knew that women were paying the highest price for her nation's ecological disaster, "so we went to the women and talked about planting trees and overcoming, for example, such problems as lack of firewood and building and fencing materials, stopping soil erosion, protecting water systems. . . . The women agreed." Maathai went to schools and involved children in digging holes, walking to tree nurseries to collect trees, planting them, and taking care of them. It was the children who took the message home and gradually got their mothers and groups of women interested. Rural women collected tree seeds in the folds of their skirts. Thus Maathai began what came to be called the Green Belt Movement in Kenya. In the first sixteen years more than three thousand schools with over one million children have become involved. More than ten million trees have been planted by children and women throughout Kenya. Fifteen hundred nurseries, almost all of them operated by women, are producing trees. Maathai describes the women in her grass-roots ecology movement as "foresters without diplomas." As she puts it, "The weight of the environmental crisis the rural women have been carrying on their backs is being lessened, one seed at a time."[14] The topsoil is being saved, firewood is grown, and food from the trees is feeding hungry families.

There are many other inspiring stories of innovative and courageous ecological action. A teacher or therapist may discover that telling a story that communicates energy and power is a highly effective way to help liberate people from ecoaction paralysis. It may motivate people to begin by planting one tree in their own yard or community open space.

Step 5: Encourage people to develop a self-care fitness plan that includes a robust earth caring dimension. An important objective of all wellness-oriented counseling, therapy, and education is to encourage people to develop and commit themselves to implementing personalized fitness self-care plans. By including earth-caring actions, they can help make the natural environment in which they live and which lives in them, day-by-day healthier. Enlarging their self-care plan by making it a self-earth care plan, they obviously practice *enlightened* self-interest. I find that the following approach often encourages clients or students to include ecological components in their fitness plans:

• In a class or counseling or therapy session, as the objectives of the experience are being discussed and the healing or learning agreement ("contract") is being formulated, I mention the potential benefits of developing a self-care wellness plan that may include some outreach. This seed-thought is planted early.

• At an appropriate point in the class or therapy, people are encouraged to jot down notes to themselves about actions aimed at implementing any concerns that have emerged in their minds. These concerns may include ecological issues, and suggestions for ecoaction can be made. It is well to encourage them to pick an action option that really appeals to them, and begin doing something, however small, in that direction between sessions.[15]

It often encourages people to move ahead in implementing their self-earth care plans when counselors, teachers, or parents affirm the small successes they achieve in earth-caring action. With some people, it is also helpful to put what they are doing in a larger context by reminding them that their actions may be small but significant contributions to a healthier, saner future for themselves, their children, and for Mother-Father Earth. If they are religiously oriented, reminding them that they are helping save God's continuing creation may be meaningful.

Books for exploring in more depth the issues raised in chapters 7, 8, and 9 are listed at the end of chapter 9.

Notes

1. *The Brothers Karamazov.* (New York: Grosset and Dunlap, 1957), 354-55.

2. *The Voice of the Earth* (New York: Simon & Schuster, 1992), 15. His comment refers to the American Psychiatric Association's *Diagnostic and Statistical Manual III* (DSM III). The gap still exists in DSM IV.

3. "When the Earth Hurts, Who Responds?" in *Ecopsychology: Restoring the Earth, Healing the Mind*, edited by Theodore Roszak, Mary E. Gomes, and Allen D. Kanner, 156ff. Conn asks new clients to fill out a questionnaire that includes questions about their feelings concerning society and the earth. She also teaches a course at Harvard Medical School entitled "Ecopsychology: Toward New Models of Mental Health and Psychotherapy." Peacemaking and feminism were important paths that led her to ecopsychology, as they were for several ecotherapists including myself.

4. In any case, the counselor-therapist's first task is to hear people out concerning their "presenting problems". This is the place to begin the helping process, but not to end it. It is important to encourage them gradually to move beyond these issues to examine wider aspects of their life situation. Often they are not consciously aware of these initially, or have not developed sufficient trust to discuss them openly. Such hidden issues may include ecoalienation as one part of their overall feelings of disconnectedness with themselves, others, and God.

5. This checkup is adapted from the one in Howard Clinebell, *Well Being: A Personal Plan for Exploring and Enriching the Seven Dimensions of Life* (San Francisco: Harper San Francisco, 1992), 182–84.

6. If you are using this checkup in a class, therapy session, or family conference, here are some guidelines for maximizing its productiveness:

• Ask persons to self-administer the checkup and then follow the first four steps of the post inventory instructions between sessions.

• Offer them an opportunity to share their findings in written form and/or orally in the next session, inviting and receiving affirming but also challenging feedback from you or others, particularly on their self-earth caring plan.

• Suggest a brief ritual (without labeling it that) by which they commit themselves to implementing their plan.

• In subsequent meetings, ask how their implementation is going and offer suggestions for increasing their success.

7. Pre- and post-checkup instructions are directed to an individual using the checkup. Therapists, teachers, and parents can adapt these to use with clients, students, or family members.

8. Macy's valuable book, *Despair and Personal Power in the Nuclear Age*, guides readers through her original process. A useful part of that book for counselors and teachers is a brief section on doing despair and empowerment work in counseling and therapy contexts. It includes a discussion on "Obstacles to Counseling about Concerns for the World" (see pp. 60–66). This approach was created at the height of the cold war in the context of the reality-based nuclear angst spawned by the superpowers' insane policy of "Mutual Assured Destruction." In our contemporary world, where the threat of environmental geocide is more in people's consciousness, it is fortunate that many of Macy's experiential exercises are earth-centered. So, the basic despair and empowerment process can be readily adapted to both the continuing nuclear crisis and the other huge dimensions of the earth's pain in the ecojustice crisis. A more recent book coauthored by Macy, focuses more directly on helping people open themselves to experience deep ecology. It is entitled *Thinking Like a Mountain: Toward a Council of All Beings* (Santa Cruz, Calif.: New Society Publishers, 1988). Coauthored by John Seed, Pat Fleming, Arne Naess, it uses healing methods similar to some in Macy's earlier work and adds other ecologically oriented exercises. See also Joanna Macy's "Working Through Environmental Despair," in *Ecopsychology*, 240ff.

9. Richard Voss suggested this method, saying that it is amazing how much participants will intuit about each other.

10. A survey of specialists in pastoral psychotherapy taken several years ago at the height of the cold war, by Jim Farris and myself, revealed that anxieties about the nuclear crisis had been raised by 69 percent of their adult clients and 37 percent of the adolescents seen by them during the previous year. Anxiety about the nuclear threat has diminished sharply (as well as prematurely) with the melting of the cold war. But the responses of teachers and therapists in the survey done for this book

show that anxieties about the environmental crisis have increased markedly during this period and are present in many clients and students. Suppressed feelings about the earth can steal valuable creative energy, making it unavailable for much better things.

11. The anti-environmental movements are expressions of a variety of dynamics including greed and dangerous defenses against facing and changing the planet-damaging behavior patterns of our society.

12. One technique for deepening awareness of our place in the evolutionary process is described in *Thinking Like a Mountain*. The meditation called "Evolutionary Remembering," by John Seed and Pat Fleming, is a mind-stretching way of getting some feel of the 13,500-million-year continuing miracle of which we humans are a remarkable contemporary expression. One is invited to identify with each stage of the continuing creation that is the evolution of organic life (see pp. 45–51). Other exercises include a ritual "Honoring Endangered Species"; "Identifying with Another Life-Form"; "The Council for All Beings." The aim of all these exercises is to help people expand their narrow ego-centered sense of self to their larger, more ancient true ecological Self.

13. Her report reminds me of someone's apt description of the post–cold war global chaos as the "new world disorder."

14. Reported by Terry Tempest Williams in "The Wild Card, A Woman's Call to Community, A Poet's Vision of Environmentalism Reinvented," in *Wilderness*, published by the Wilderness Society, Summer 1993, pp. 28–29.

15. Also mention that any self-care plans they make will be more likely to be implemented if they: plan on small things they really want to do; find a friend who wants to join in the action; and include a realistic time line of when they will begin implementing their plan and by when they expect to achieve certain objectives. It is also good to encourage them to plan rewards they will give themselves at each step toward a particular objective. Of course, there is an automatic reward for many people in the satisfactions of contributing to a better future for themselves and their children by their actions. Another idea to suggest is that they keep a day-by-day record of the ups and downs of implementing their self-earth caring plan. Such a record can guide needed mid-course corrections to keep their actions headed toward what they want to achieve.

8

Basic Methods of Ecotherapy and Ecoeducation: Grounded Healing and Growth Work

Min Taku Oyasin.

—Lakota Native American saying meaning "All things are related."

You do not have to be good.
You do not have to walk on your knees
for a hundred miles through the desert, repenting.
You only have to let the soft animal of your body
love what it loves.
Tell me about despair, yours, and I will tell you mine.
Meanwhile the world goes on.
Meanwhile the sun and the clear pebbles of rain
are moving across the landscapes,
over the prairies and the deep trees,
the mountains and the rivers.
Meanwhile the wild geese, high in the clean blue air,
are heading home again.
Whoever you are, no matter how lonely,
the world offers itself to your imagination,
calls to you like the wild geese, harsh and exciting—
over and over announcing your place
in the family of things.

—"Wild Geese" by Mary Oliver[1]

This chapter and the next one present a variety of psychotherapeutic methods. Most of these techniques can be used in ecoeducation as well as ecotherapy. A few of the methods are designed to be used by trained counselors and psychotherapists. But the vast majority of them are relatively uncomplicated. These can be used by teachers, clergy, and parents. Many people may shy away from a few of the approaches—for example, the sweat lodge, because they seem strange or awkward. As you read, you can identify methods that appeal to you and seem relevant to use in your work or family.

Ecological Consciousness Raising (CR)

In our earth-alienated culture, many people are so out of touch with the natural world that they are unaware of both the deep pain and healing energies of the biosphere. Some are focused so intensely on "getting there"—on reaching their immediate objectives—that they fail to smell the flowers along the way. This unawareness often is reinforced by the ego defense of denial of the seriousness of environmental issues.

Visual psychologist Laura Sewall, who teaches ecopsychology and women's studies at Prescott College in Arizona, points out that our five sensory capacities are our avenues of connecting with the world. She holds that the heart of the ecocrisis is the deadening of these senses:

> The Earth calls continually. She calls us with beauty, sometimes truly breathtaking, sometimes heart wrenching, and always provocative and visceral. . . . If we are receptive to the ways in which the landscape speaks to us, or the ways in which perception serves as a channel for communion, we may reawaken and preserve a sense of human integrity within the family of all relations.[2]

What is an effective response to people who suffer from ecological alienation causing a lack of awareness of the natural world? Utilizing consciousness-raising (CR) techniques, when the timing is right, often brings to the surface subconscious feelings about nature. Methods discussed in chapter 7 can help raise ecological consciousness and caring; for example—taking the ecological wellness checkup, telling one's ecological story, or a teacher or therapist pointing out the interrelations of social and ecological context factors with one's personal issues. Consciousness frequently is raised when people encounter respected teachers, therapists, grandparents, or parents who model ecological literacy, passionate concern, and active involvement in earth-caring.

Another CR method is enabling symptoms of humankind's earth-damaging beliefs and behaviors. In educational contexts a simple experience like stepping outside on a smoggy day may be sufficient. Consciousness may be raised by viewing a videotape of a TV nature documentary on a topic such as the global devastation of forests, extinction of species, the environmental impact of the population explosion, or one of the many planet-saving programs. Showing a selected segment of such a videotape during a class or loaning a tape for homework viewing in therapy, after ecotherapy issues have arisen, may lift earth awareness. Visual images often have a greater impact than printed messages for people of all ages, especially young people.[3]

Bibliotherapy can be an effective CR tool with those who enjoy reading. This involves encouraging people to read newspaper or journal articles, pamphlets, or relevant sections from novels, nonfiction, or poetry volumes about the ecojustice crisis. The annotated bibliography at the end of this book includes many bibliotherapeutic resources. Whatever the stimulus to increased awareness, it is important to reinforce the experience by providing opportunity for feedback and open dialogue during the next class or therapy session. Let us look now at two approaches that often facilitate both increased environmental awareness and motivation to do more earth-caring.

Using Ecological Stories

Sharing inspiring stories of courageous earth-loving people like Wangari Maathai or Rachael Carson is an effective teaching method. It may also be a tool for therapy with people of all ages.[4] Carefully chosen stories illuminate the special issues in which particular clients, students, or children are interested. True stories of individuals or families who learned something from a frightening, destructive, inspiring, or enlivening experience with the natural world may be effective. Stories can be drawn from one's own or other people's nature experiences, from novels, films, nonfiction writings, from multicultural folklore and myths, or simply from one's own fertile imagination. Composite stories that are reality-based but drawn from more than one person's experience can illuminate particular issues.[5]

Storytelling by teachers, therapists, and parents can begin with phrases such as, "I once knew a person [or family] who responded this way to the environmental crisis . . . "; or "I wonder if your experience is anything like a person I read about in the newspaper, who. . . "; or "I recall a statement by a character in this film (or novel), who declared. . ."

In addition to human stories, tales of plants, animals, or other natural beings can communicate metaphorically a truth regarding humans and the earth. Such tales are used to transmit cultural traditions, lore, and wisdom to children in many cultures. Adults also remember and learn from such stories. To get in touch with the power of nature metaphor stories, consider this dervish "Tale of the Sand" from ancient Persia. In an innovative book about right-brain methods of marriage and family counseling, pastoral psychotherapist Douglas A. Anderson uses this tale effectively to communicate insights about family dynamics. [6]

> A stream, from its source in far-off mountains, passed through every kind and description of countryside, at last reached the sands of the desert. Just as it had crossed every other barrier, the stream tried to cross this one, but it found that as fast as it ran into the sand, its waters disappeared. It was convinced, however, that its destiny was to cross this desert, and yet there was no way. Now a hidden voice, coming from the desert itself, whispered, "The Wind crosses the desert, and so can the stream."

The stream objected that it was dashing itself against the sand and only getting absorbed, that the Wind could fly, and this was why it could cross the desert.

"By hurtling in your own accustomed way you cannot get across. You will either disappear or become a marsh. You must allow the Wind to carry you over, to your destination." But how could this happen? "By allowing yourself to be absorbed in the Wind." The idea was not acceptable to the stream. After all, it had never been absorbed before. It did not want to lose its individuality. And, once having lost it, how was one to know that it could ever be regained?

"The wind," said the sand, "performs this function. It takes up water, carries it over the desert, and then lets it fall again. Falling as rain, the water again becomes a river."

"How can I know that this is true?"

"It is so, and if you do not believe it, you cannot become more than a quagmire, and even that could take many, many years; and it certainly is not the same as a stream."

"But can I not remain the same stream that I am today?"

"You cannot in either case remain so," the whisper said. "Your essential part is carried away and forms a stream again. You are called what you are today because you do not know which part of you is the essential one."

When he heard this, certain echoes began to arise in the thoughts of the stream. Dimly he remembered a state in which he—or some part of him, was it?—had been held in the arms of the Wind. He also remembered—or did he—that this was the real thing, not necessarily the obvious thing, to do.

And the stream raised his vapor into the welcoming arms of the Wind, which gently and easily bore it upward and along, letting it fall softly as soon as it reached the roof of the mountain many, many miles away. And because he had his doubts, the stream was able to remember and record more strongly in his mind the details of the experience. He reflected, "Yes, now I have learned my true identity."

The stream was learning. But the sands whispered, "We know because we see it happen day after day and because wet sands, extend from the riverside to the mountain."

And that is why it is said that the way in which the Stream of Life is to continue on its journey is written in the Sands.

Anderson observes that this story illuminates metaphorically these family principles: Families are moving, developing entities that are always changing on the ups and downs of the family life cycle. Like a stream, families often get blocked when they come to a "desert," when old ways of coping are no longer effective. The unique identity of a family is preserved only if they become open to learning new ways of letting go in order to traverse obstacles. Anderson states: "The transforming agent in the above story is the Wind, the dominant biblical image for the life-giving Spirit of God, Author of all change and growth."

Stories of animals, plants, or natural forces such as the wind or sun are part of the cultural riches of many Native American nations. A delightful example is the story of the jumping mouse by Hyemeyohsts Storm.[7] The wise

Native American storyteller shares the story with some children of the tribe, weaving a teaching tale about a courageous, adventuresome mouse named Jumping Mouse. The mouse has many learning-rich encounters, including conversations with other Mouse People, and later, as he ventures out, with his other brothers including Raccoon, Frog, Buffalo, and Wolf. Each of the animals is an embodiment of desirable characteristics and insights in Native American wisdom. The conversations occur as Jumping Mouse risks going far beyond the relative safety of the usual, isolated Mouse Place, to Medicine River and the Sacred Mountain. His transformations are interwoven with his periodic sacrificing of something precious in order to help other animals. Jumping Mouse eventually is transformed into an animal representing a being who marvelously transcends mousehood.

A Pastoral Psychotherapist's Earth-Story

To get in touch with some ways earth-bonding may influence professionals' work with people, consider Tom Summers, a teaching chaplain and pastoral psychotherapist. His teaching is with seminarians, clergy, and mental health professionals who are enhancing their caring and counseling skills by practicing these arts under careful supervision. His healing work is with the mentally ill. Tom is one of an innovative breed of clergypersons who bring together enlivening spirituality, psychotherapeutic skills, and commitment to use their training to deal with social context issues including helping to save a viable planet. His ecotherapeutic understanding of personality is reflected by these thoughts about letting ourselves be nurtured by nature:

> We have a tremendous inner capacity to take in the beauty of a forest, the awesome and quiet quality of a hazy moon shining on ocean water, or the peacefulness of a river, so that consequently these memory events "live" in us. They can become sustaining features and threads in our ministry of being with others, for the earth supports us, empowers us with its beauty, strength and color. [8]

As I write, out of the corner of my eye I see a colored photo of the Edisto River that Tom sent to me. It is a small, tree-lined river in South Carolina that he calls his spiritual home. He describes how "self treasures of the earth" came into his life from this river:

> Many of my summer childhood and adolescent days were spent either in the depth of those cool, dark waters, exploring its rocky bottoms; or else dashing through the muddy swampland, relishing in the delight of ever-present mud fights with peer friends, and especially "enemies." I learned to swim in that river at the age of three, by first experiencing in its swift currents the opportunity to hold fast to my father's neck as he swam back and forth across the river, allowing me to experience trust and support in him and the water. Then there

> was the cold winter in my adolescence when four of my teenage friends and myself adventurously vowed to one another that, regardless of the river's cold waters, we bravely would jump into those freezing waters each month. This daring feat, I might add, was always accompanied by a quick leap back on the river bank to sit by a log fire! [9]

Tom cherishes his memories of enjoyable summer picnics by the river and floating peacefully down it on an inner tube. But his earth treasures from his river are not all light and carefree. He remembers one of his first encounters with life's tragedies when he witnessed a drowning in the river. And in recent years, he has been passionately involved in seeking to protect his river, organizing protest demonstrations aimed at saving the river from the deadly poisons of radioactive wastes from the Savannah River nuclear weapons plant that is located in its watershed. This plant produces the tritium triggers for America's thousands of nuclear warheads.

How has all this influenced who Tom is as a person?

> My experiencing of the river has stayed with me very vividly as an earthy thread and is a deep part of my history. I now find in my adult living when I am in need of solitude or when I might feel especially perplexed over deep questions, I often desire to be near water, like at the ocean, or some mountain lake, or some river. My adult needs in these respects are rooted deeply into the nourishment and the healing comfort found in the inner earthy thread, entwined about my early Edisto River involvement. . . . You and I live in the earth, and our history may include enriching moments of taking into ourselves the passion, wonder, and beauty of the earth about us. [10]

Has Tom's bonding with the earth influenced his work with deeply troubled people? He articulates the conviction that is at the heart of this book when he declares:

> The earth has the potential of nourishing us, more deeply integrating us, and preparing us to "be" with others. For it is from our earthy places and the receiving of care and nurture for ourselves that we can go forth further into the world to be more present with others. . . . In my attempts to offer ministry with others, the living background of the Edisto River feels profoundly available to me at times. There are moments when I hear some frightened person share such concerns as wondering whether there are any supports to trust in amongst dreadful floundering in the midstream of life or whether survival will result by risking some icy plunge into a new and shaky venture of living. The thread with the river certainly gives me no answers, but its deep presence allows me, hopefully, to risk being more available and sustained as I experience the searchings and the questions of others in the human mystery of inner and outer voyaging in life. [11]

Tom's remembrances of his river awaken fond memories of my boyhood experiences on another small river, the Sangamon in central Illinois. The rowing, floating, and camping trips with my dad down its meandering course from Springfield past its confluence with the much larger Illinois River are cherished experiences of male bonding in nature. And my joyful and sometimes scary adventures along it with a high school buddy have a valued Tom Sawyer–Huck Finn feeling associated with them.

Encouraging Increased Nurture by Nature

Experiencing the healing energies of nature more often and in greater depth is an essential objective of the ecotherapeutic process. If professionals do nothing more than encourage people to increase time spent relating intimately with nature, they probably will contribute to their total well-being. Teachers and therapists who recommend this often are surprised by the psychological, physical, and spiritual benefits that accrue. This section and the next one (on stress reduction) describe earth-intimacy methods that can be used, whatever else counselors, therapists, or educators do in the ecological area.

A clear example of how many socially and psychologically wounded people spontaneously turn to nature for healing is the childhood of novelist-poet Alice Walker, best known for her powerful novel *The Color Purple* (for which she received the Pulitzer Prize). Walker was born in a family of impoverished African American sharecroppers in Eatonton, Georgia. Her mother worked as a maid and was paid seventeen dollars a week. Her father's earnings were never more than three hundred dollars a year. At age nine her duty was to cook the family's meals and clean the house, as well as work in the fields.

Now in her late forties, she remembers, "Looking back, I can see that it was an intolerable situation for everyone. But then I was suicidal. I thought of jumping off the roof of the barn." Fortunately she was able to find temporary serenity and healing by relating intimately with nature. She remembers, "We had terrible housing in the middle of nowhere. But it was smack in the middle of the countryside. I bonded with the earth, with the spirit of nature. I walked miles and miles through the forest and beside the streams. I felt accepted by nature. I saw myself as an earthling and that gave me peace."[12] About her own earthy identity, Alice Walker declared, "In search of my mother's garden I found my own." The central theme of this chapter and, in fact, the whole book is implicit in Walker's account of the healing, shaping power of nature during her wounding childhood.

In his book *Plants as Therapy*, Robert Steffen highlights some keys to understanding the healing power of close contact with plants:

> Gardening is good therapy for young and old. The earth has great healing power. It is the plant, of course, which makes it all possible. Realizing we could not exist on this planet without the plant is significant. Learning how and why

> this is true can occupy much of a lifetime and be only a beginning. Plants are miraculous creations. They hold so many secrets that they present a challenge and a hope for many people who are disturbed, frustrated, and concerned about the future. Knowing and understanding plants can give them hope and reassurance that with death there follows life and the great cycles of the seasons are part of the even greater rhythms of the universe that are not dependent on mortal man's manipulations. [13]

There are numerous ways to incorporate experiences of nature in therapy and education. I frequently ask students and clients where and how often they spend time in natural settings. Some folks respond to this simple eco-diagnostic query by saying that they never or seldom do this. We then may discuss how increasing the frequency and length of such experiences can be a healthy, satisfying way to enhance their self-care.

Several case vignettes illustrate how various therapists integrate ecotherapeutic approaches in their healing work. A male client suffering from a severe middle-age crisis came for therapy with me because he was in the early stages of both professional and marital burnout, each of which was exacerbating the other. After we had thoroughly explored his concerns and were discussing what he could do to address them, I asked him how much time he spent each week in natural surroundings. He reported that he worked long hours, but on weekends he occasionally played a fast round of golf or worked as briefly as possible in his yard. Otherwise he spent most of his time in a climate-controlled office building, an air-conditioned auto, or inside his home. In the days following this discussion, he discovered that spending as little as ten minutes a day sitting quietly or walking around his yard, while he experienced the plants and birds, gave him a much-needed lift. Especially renewing was walking twenty to thirty minutes with his wife in a beautiful nearby park. This proved to give some help to both his personal and his marital burnout.

At a Nashville counseling center where I was invited to discuss ecotherapy, two staff members told of the healing impact of simply taking clients outside. A male pastoral counselor described a twelve-year-old boy who was very depressed and resistant to talking when seated in the counselor's office. Then the two of them went outside and walked among the trees while they talked. The lad began pouring out his feelings, a healing flow that continued in subsequent sessions both indoors and out. A female pastoral counselor described the turning point in therapy with a painfully depressed woman. It occurred when they moved a session outside on a beautiful spring day. In the next session, the client reported that she had planted tulip bulbs in a pot, expecting that they would bloom about the time of her birthday.

A clinical social worker from whom I have learned much about ecofeminist methods (and life) has a deep personal sense of her own connectedness with the living earth. She enjoys spending time almost every day working in her garden and with her abundant houseplants. She uses numerous ecotherapeutic approaches in her individual and marriage therapy. During a severe drought in her region of southern California, she discovered that some clients had intense feelings in response to the plight of the thirsting earth. Some of her clients found that this was a powerful symbol of their own enormous sense of emotional thirst. This issue then could be explored therapeutically. As the seasons change or when cycles of the moon are obvious, this therapist sometimes mentions such changes to clients in passing. She has discovered that this increases awareness of the renewing cycles of nature, which creates opportunities to reflect therapeutically on the clients' using the cycles and transitions in their own lives, as well as the cycles of nature.

A women client came to this therapist for help with severe post-traumatic stress disorder. She was choking on bottled-up rage from childhood sexual abuse and from recent memories of almost daily sexist put-downs. The therapist suggested that she go for walks along an ocean beach, in early mornings when she probably could be alone. She discovered that she could scream out her rage and frustrations on the beach. The crashing waves helped to release her rage and sense of devastating betrayal by those she had loved and trusted. This proved to be more than a powerfully healing emotional catharsis. Being alone immersed in the powerful energies of the ocean beach facilitated a kind of cleansing psychological exorcism.

This therapist also used ecotherapy effectively in vocational counseling with a young adult male client. He was employed outdoors "temporarily" in creative landscaping work, but was engaged in long-term, frustrating preparation to qualify for a desk job. In response to his hoping to increase the satisfactions of his present job, the therapist suggested that he try to become more aware of what he was experiencing while working outdoors. He gradually began to sense the differences of his experiences on foggy, sunny, and rainy days. This sharpened awareness of his natural work surroundings put him in touch with the strong affinity for the outdoors he had developed while growing up on a farm. All this became relevant as he wrestled with the pros and cons of his plan to find an indoor job rather than continue his outdoor work.

A pastoral psychotherapist in the Southwest tells of a client who was suffering from the winter onset of the depression known as SAD (seasonal affective disorder), which caused her to feel "closed inside." When spring came, she still felt depressed and stayed inside sleeping during the day. As she explored her depression, she got in touch with the conflicts she felt around the influence of two key people in her life. One was her Native American maternal grandmother, who was very sensitive to human communication with nature and the elements of earth, rain, wind, and fire. The other was her rigid German father, who was oriented to ignoring, even abusing, the environment in his technological- and industrial-minded centering on his work life. The therapist reported that part of the client's therapy was to "encourage her to go outside and reacquaint herself with the Indian, natural side of herself through outdoor walks and other activities in daylight in parks, etc."[14] This case illustrates the way conflicts in cross-cultural attitudes toward nature, when internalized in a person's identity, can contribute to problems in living.

Ecotherapeutic Methods for Reducing Stress

I sometimes ask stressed-out students and clients questions such as these: "Do you have any place to experience real silence in a natural setting? Do you ever long for such a place?" Then, depending on their response, I may say, "It probably will lessen your stress load if you find such a place and plan your life so as to spend a little time there every day. If you can't go there on a particular day, try imagining being there during your quiet or meditation time. This can be a welcome mini-vacation from your pressures."

Clinical evidence suggests that chronic noise pollution and deprivation of quietness may contribute to diminished wellness. Studies of stress levels in overcrowded, polluted, noisy cities suggest that chronic noise, with no respites of relative quiet, may contribute to our society's epidemic of chronic stress overload. On the other hand, consider the healing impact of certain sounds as demonstrated in the expanding field of music therapy. Professionals in that field are discovering the healing benefits of experiencing music that either quiets and inspires or stimulates aliveness in people suffering from emotionally caused energy depletion. Since it is clear that having regular times of serenity in nature tends to enhance the body-mind-spirit well-being of many people, it is important to encourage people to give themselves the gift of slowing down long enough to become really still in a natural setting. It is important to become quiet enough to open oneself to the rhythms of nature and the nonverbal messages of trees and clouds, of rocks and animals.

It is sad to contemplate how we humans seem to adjust to all kinds of stresses, including places with high noise pollution. Years ago, my family and I lived in a congested section of New York City. I remember visiting friends whose apartment was only a few yards from the tracks of a heavily used elevated train line. When the "el" rumbled by, their living room trembled with a mini-earthquake. Conversation stopped because it was impossible to hear each other. When I asked how they coped with the frequent intrusion of noise, family members looked surprised by the question. They shrugged their shoulders, saying they hardly noticed when the trains went by.

More and more people around the planet live in noisy, polluted, congested cities. Increasing percentages of the human family have few if any opportunities to be in places where they can experience either stillness or the therapeutic energies of nature. People obviously are able to exist in congested cities, often because they are poor and have no options. But we humans belong to a species whose ancestors lived for eons in wilderness, rural, or tribal settings where quiet times in nature were abundant. Probably because of these genetic roots in relatively quiet natural surroundings, chronic noise with no times of relative quiet can diminish the quality of life significantly. A sensitive young adult I know observes that she gets "irritable, uneasy, and off centered" when she lacks regular times in places of quiet.

It is no coincidence that people of special spiritual and psychological depth, through the centuries, in a variety of faith traditions, have given themselves the gift of silent retreats in beautiful places of natural serenity. There they report enjoying the "still small voice" of the divine Spirit, and also the many voices and music of nature, more clearly.

Regrettably, many people are addicted to the stimulation and stress of constant loud music—which many of us experience as raucous noise—and nonstop television, much of which is intellectual garbage.[15] Those so addicted seem to be *silence-phobic*, experiencing elevated anxiety if deprived of their "fix" of nonstop TV chatter, blaring music, or their Walkman.

In light of all this, it may be helpful to offer stressed-out clients or students encouragement to give themselves more quieting times—times of listening and hopefully of hearing. Such occasions can be enriched significantly if people find a natural setting where they feel safe and at home. Often it is not feasible for people to go to an uncrowded place in a garden, park, or wild area outdoors. Fortunately, it is possible for many people to find the special quieting energies of nature by relating with a single tree, a flower, or a growing indoor plant, perhaps with the auditory energy of soothing music. Our daughter, for example, reports that she has found it relaxing to meditate between two large indoor plants (including a live Christmas tree) after her high-pressure work.

This is similar to a practice reported to be followed by some Native Americans. They choose a particular tree or other plant to spend time with every day, becoming tuned to its energies and rhythms. Some report that they receive helpful messages—gifts from their particular plant-companion. Are the "messages" they receive literally from plants or from heightened psychophysiological awareness of the organic relatedness with plants (and the rest of the biosphere) that they, along with all other humans, may have in their genes? Who knows? In any case, people who get in tune with nature often report that they learn simple lessons from its ways. For example, one person observes: "Why should we all dress after the same fashion? The frost never paints my windows alike twice."[16]

A marriage, family, and child counselor in southern California describes some creative ways she draws on nature both personally and in doing therapy. She spends time in her flower garden regularly and keeps fresh flowers and homemade potpourri in her office. Like numerous therapists in the survey reported earlier, she often uses metaphors, images, and stories from nature, as well as sharing her own experiences with nature, whenever these are therapeutically relevant. Here is her report: "Frequently I reinforce a client's positive experiences in vacation settings, where they are interacting with nature. For those with a visualization capacity, I use creative imagination to help them return to that place for respites." She encourages clients to bring to their sessions natural material such as earth, rocks, leaves, or other things with earthy sensate qualities such as smell, sounds, or moisture. Or she may simply ask clients what refreshes them in nature.

This therapist reports that only a small percentage of her clients who "graduate" from therapy have no quiet place inside themselves to which they can retreat. Few still receive no refreshment from nature. Many learn to utilize a combination of nature, music, art, and body movements to achieve a sense of inner serenity and renewal. Of those who discover a pathway to this peace during therapy, approximately three-fourths do so by connecting more intimately with nature.

The following case vignette illustrates how this therapist worked with a woman client who came to therapy suffering from multiple problems, including family dysfunction and other relationship problems, work impairment, rheumatoid arthritis, agitated stress, and depression. Over several previous years of work with other therapists, the woman had experienced only minimal improvement in a variety of traditional therapies. These included hospitalization, family and one-on-one therapy, and a twelve-step recovery program.

Gradually, as the client worked with this therapist, she experienced marked improvement in many areas of her troubled life. The therapist encouraged her to begin intensive organic gardening as both "recreation and a way to reconnect with a life-giving source." Despite pain in her knee joints, the client dug roots from a patch of earth infested by thorns and discovered that "gaining control over the weeds gave her a sense of possibility in her own recovery and renewal." She reported, "I grow herbs there, squashing them in my hands, smelling them and eating them as I work. My patch attracts the butterflies, and this is a life-affirming thing, close to nature in a literal sense." The client, who had grown up in the mountains and by the ocean, reported that now she was able to "be a part of them again." This reunion had, she said, enabled her to have a kind of spiritual rebirth.[17]

Using Imaging in Ecotherapy and Ecoeducation

Because several guided imaging exercises are presented in this book as illustrations of a productive methodology, further discussion of this approach is in order. Right-brain imaging (both guided and free-flowing), followed by left-brain reflection on the experience, often helps facilitate whole-brain healing and growth. Therefore, many therapists and teachers (including the author) have a wide repertoire of guided images available and also create new ones spontaneously in the flow of their work.

In ecological healing and growth work, the decision to use imaging and, if so, which ones, should be made in the light of individuals' differing therapeutic or educational issues. If a therapist or educator is working with people who have religious interests or are searching for more healing connections with the divine Spirit, a guided imaging exercise like the following may be helpful: "Recall an experience in a natural setting where you were in tune with the loving Creator, causing you to feel more alive and together, and deeply at peace. / In your imagination visualize that place as vividly as you can, and then return to that place and time. / Let yourself relive that healing, enlivening experience of God's deep, caring acceptance. Simply let go and relax in the loving arms of God's Spirit, taking as long as you choose. / To remember this experience and perhaps return to it when things get rough, make a quick sketch of the place in your journal. / "

The following is one of many earth imaging exercises drawn by feminist psychotherapist Diane Mariechild from ancient spiritual-based earth traditions of women.[18] In my experience, many of her exercises can be adapted readily for use by men. Mariechild calls this one simply "Seasons." It uses ancient goddess symbolism to help women deepen awareness of their profound connecting with the earth's cycling, re-creating energies and rhythms. It is worded for women and earth-aware men. Men undoubtedly will experience it differently, but it can easily be adapted for their use. As you try this imaging, you may wish to substitute your own preferred word(s) for deity, if "goddess" lacks any positive meaning for you. Using goddess symbolism, however, often stretches the horizons of women's and men's spirituality significantly.

"Relax, deepen and protect yourself. As a woman, you are in touch with the cycles of nature, intimately connected with the ebb and flow of the tides, the waxing and waning of the moon. Knowing of earth and matter, you will move deeper and deeper within. Move so deeply within that you will experience all of the rhythms and cycles of nature as they occur on the planet earth. You will experience the coming and passing of the seasons: the harmony and direction, the spontaneity and consistence, the continually rearranging balance. And you will feel them in ways that are at once universal and personal.

"Now it is spring and Persephone's return, the earth swells with the joyful reunion. There is an awakening, a new awareness, the joyous expectancy of new life. Pause about three minutes.

"Now the summer, the ripening, the coming harvest, the season of the corn mother, the sultry sensuousness of the summer. Pause about three minutes.

"Now the fall, the harvest. The days are growing shorter and cooler. Pause about three minutes.

"Now the winter, the goddess dons her wintry robes. The decay and dying, the cold and dying. Knowing your own dying. Pause about three minutes.

"Once again spring, the rebirth, the joy and the realization that life continues. Feel its continual ebb and flow within, knowing that you are maiden, mother, crone. Pause about three minutes.

"Now let your energy gently ebb and flow, moving you up and back to your waking reality. And you are wide awake and filled with energy. Open your eyes and stretch your body."

If you tried this exercise, take time to reflect on whatever you experienced, perhaps make a few notes to yourself of key learnings. Or make a rough crayon drawing to express your feelings about the experience. Do not plan the drawing; do it quickly, letting your fingers move wherever they want with colors that seem most expressive.

Once people are introduced to guided imaging by teachers or therapists, many begin to guide their own images intentionally. They sense which images are the most potentially productive for them as they deal with particular issues. The teacher or therapist should encourage this by offering to provide any guidance they request and by inquiring what they learned as they used a particular meditation between classes or sessions.

Using Dreamwork in Ecotherapy and Ecoeducation

Like waking images, nature images in dreams during sleep can be pathways for exploring and unlocking important personal issues, including earth traumas from early life. Such images may be carriers of emotional residues and unresolved conflicts from ecological traumas in childhood or later life. They may be symbolic vehicles carrying any manner of unresolved emotional and interpersonal conflicts, desires, hopes, and unresolved griefs. Images such as rivers, trees, mountains, storms, often are symbolic ways by which our unconscious minds seek to communicate deeper conflicts about our lives, work, or human relationships.

I frequently encounter nature images in continuing work with my own dreams, a practice learned during my training analysis. In my pre-ecology years of doing therapy, I assumed that dreams in general including nature dreams, were symbolic expressions of repressed conflicts or forbidden impulses in the unconscious mind. But it gradually became clear that some nature dreams are expressions of alienating trauma with nature, or pleasurable reliving of positive nature experiences, or longings from one's ecological core for regrounding of one's mind-body-spirit organism during times of feeling "up in the air." They may carry liberating messages about early experiences with the natural world—positive or negative—that need to be remembered and integrated with one's adult life experiences of nature. Therefore dreams, along with free associations and guided daydreams or images, may be a valuable path to recover repressed feelings concerning the natural world and to get reacquainted with the lost ecological self. As Freud once observed, "A dream that's not understood is like a letter that remains unopened."

To illustrate how dreams often have multiple levels of messages including those from the ecological core of our mind-body-spirit organisms, I share a turning-point dream of mine. Some years ago I had a frightening, baffling river dream while struggling to complete a book. I was resisting finishing the book, in spite of the fact that its message seemed important to communicate. With the help of my long-time partner (also a therapist), I used the gestalt therapy approach to the dreamwork (which I also use with clients).[19]

Here is my dream as I relived and retold it in the present tense: "I'm on a lonely journey in the winter and I'm not sure where I'm going. I'm off the road and wandering alone without shoes in a cold swampy area on the west side of the Mississippi River. I think I'm trying to find my way to my boyhood home across the river. I'm freezing and afraid I'll step in a deep hole of water in the swamp. Now I see the Mississippi through the bare willow trees on the bank that are partially under water. I know I have to get across but it's a terrifying flood. And the damn river keeps rising and may submerge the bank where I'm standing if I just stay here. I feel awful—utterly helpless and frustrated. There's no way to get across to the other side. If I try to swim it, I know I'll drown in the swirling current—if I'm crazy enough to try." At this point in my nighttime dream, I had awakened feeling terribly cold, defeated, and immobilized.

My partner asked if I wanted to continue the dream in fantasy. Reluctantly, I nodded my "yes." I first dialogued with the mighty river. When I spoke for it, I knew that it was a powerful symbol of ambivalence and fear that was self-blocking my finishing the book. But as I dialogued with it, the river told me what I needed to do to get across. I continued imaging the dream, saying, "I'm making a rough Tom Sawyer–Huck Finn type raft from logs that are lodged in the trees on the river bank. I'm feeling panic because I have no paddle or oars. But I must get across, so I'm lying on my stomach on the rickety raft as I push off into the river. I'm having an awful time struggling with both hands trying to 'swim' the raft across the rushing river. It's exhausting and I'm being swept far down river, but I'm making slow progress now." Eventually I got to the Illinois side, grabbed a protruding tree limb and fell on the bank face down in the mud, utterly exhausted. As soon as I was able to walk, I surprised myself by heading, not toward my boyhood home, but toward the city where I knew the publisher was. I had a contract with his company and I had felt considerable conflict and anger about his resistance to publishing what I knew I needed to write. My ambivalence about finishing the book was gradually illuminated, and I became aware of what I needed to do to complete the book (which I did six months or so later).

When I spoke for the river during my dialogue with it, it told me that it also carried messages about the residue of three terrifying early-life experiences with water. One was having my head repeatedly pushed under the surface of a swimming pool when I was very young, by an inept, sadistic swimming instructor who was frustrated by my fear of the water. Another was absorbing some of my nonswimming mother's terror of drowning, when she expressed it unintentionally and dramatically during a frightening experience on a lake when I was still very young. The third unfinished memory involved residual feelings from the teenage trauma during a river float trip with a friend (described earlier) in which I came close to drowning. As I worked with my dream, it became clear that I needed to do additional therapeutic work aimed at healing these still-haunting memories.

Methods of Resolving Ecoalienation Memories and Reliving Ecobonding Experiences

The process of overcoming ecoalienation often involves confronting and experiencing the healing of painful nature-alienating memories from earlier life, like those I relived by working with my river dream. Healing these ghost memories that still haunt one's relationship with the earth is frequently a crucial aspect of ecohealing.[20] When this healing of forgotten earth trauma occurs, people often recover repressed positive bonding experiences with nature. These probably are held out of awareness because they are intertwined with repressed traumatic memories. Thus, one of the potential benefits of walking the inner path from alienation to healing is that linked positive earth memories and feelings may come back into awareness—memories of satisfying, nurturing bonding with nature. These can become resources that enrich one's current life and relationship with the earth.

A terrorizing experience took place in my life when I was about five. My parents took me to view the utter devastation of a village on the central Illinois plains not far from our hometown. It had been wiped out by a "twister"—a violent tornado—the day before. Everything—houses and other buildings and their contents, automobiles, farm animals, and implements—was strewn about as though destroyed by a huge bomb. This painful experience was reinforced on later occasions during my childhood when warning came over the radio that a funnel-shaped cloud had been spotted in the area. People were told to seek shelter in their basements or in their storm cellars, if they were fortunate enough to have one (we did not). It is small wonder that I had at least one nightmare in which a twister was about to devastate us and our home.

I am grateful that healing of early storm traumas has occurred in later years. The evidence of this is that I have been able on several recent occasions to feel a lift of exhilaration (mixed with some reality-based fear) when I experience the awesome power of a storm. I remember being surprised and delighted by "getting a lift" from the challenge of struggling to navigate a twenty-eight-foot sailboat across twenty-five miles of open ocean between the California mainland and Catalina Island when a fierce storm came up with little warning. The wind was wild and the waves towering. When we finally made it to port on the island, the small-craft advisory flag still seemed about to blow away and our friends there were amazed that novice sailors had come across in the storm.

The most productive methodology for facilitating healing of painful childhood memories involves using imaging techniques like those outlined above. This process begins with reliving the experience in imagination and reexperiencing the full range of still-present feelings, followed by using an approach aimed at extinguishing the ghost feelings. This is done by reliving the experience again in one's imagination with both the positive resources of one's adult personality and a more positive outcome. Doing the following imaging exercise is an effective way of learning how to use such an approach in your own teaching and practice.

An Exercise to Heal Ecotraumatic Memories

Instructions: If this exercise is done in a group, designate one person to read the instructions, pausing when this sign / appears, long enough for everyone (including the reader) to do what has been suggested. If the exercise is done alone, the reader simply reads and closes her or his eyes when the sign appears for as long as needed to follow the instructions.

In your imagination, picture your present home vividly in as much detail as possible. / When this image is in clear focus, see yourself standing outside that building. / Be there now. *Walk around slowly, becoming aware of how the trees, other plants, birds, and perhaps other animals around your home (or the lack of these) affect the ambiance of that place for you. / Now, go inside your home and wander through it, becoming aware of how plants and perhaps pets, as well as the attitudes toward these living things of other people present there, impact the feelings you have about this shared living space. /*

Go back in time now and picture in your mind the first home where you lived during early childhood. (If you lived in more than one house, choose the home that seems most important to you.) See yourself approaching that home, looking as you and the house did then. / Be there now in the yard of your childhood home. What do you feel, being back? / Walk slowly around the yard, paying particular attention to the plants, birds, and animals there, or the lack of these. If there is a garden, walk among the flowers and vegetables, staying in touch with your feelings. If your childhood home is on a farm, walk among the farm animals and through the fields. / Now go inside your childhood home and wander through it, being aware of your feelings and thoughts about the place and the people who live there with you. Pay special heed to your feelings about any plants and pets that share that environment with you and your family. / Bring your parents and any other family members into the room with you, one at a time. As each enters be aware of that person's real feelings and attitudes about yourself, other family members, the world of nature outdoors, and the living things inside. If you are not consciously aware of their feelings, question them and listen to their answers. / Recall the most pleasurable times of relating to nature in your childhood and then enjoy reliving these, experiencing the joyous feelings that are still there. / Now, recall and relive the most painful or frightening experiences related to nature in your childhood. / Do anything else you would like to do while you are in your childhood home. /

When you are ready, leave that home and come back in your imagination to your present home. Discover if your feelings about the people, plants, and pets in your present home mirror similar feelings in your childhood home. / Reflect for a while on how your painful and joyful experiences with nature as a child may influence your current feelings and relationship with nature. Have they caused you to feel basically warm and friendly toward the natural world; or distant, detached, and anxious; or perhaps both ways at different times? / Return to the present time and place, and jot down what you want to remember from your experiences on this memory trip. / Share whatever you choose about your trip with a friend or family member, or perhaps a teacher or therapist.

The second part of this exercise focuses on healing ecoalienation and reclaiming ecobonding:

Go back to your childhood home and be yourself at that age. Then relive the most disturbing experience with nature you had during those years, including the people, sounds, physical pain, and—most important—the feelings you had when it first happened. / Now, see your adult self at your present age entering your childhood home and being alone with your child self. / Use all your mature know-how and caring skills to help your inner child cope more effectively with the frightening experience. Give support, comfort, and guidance, and whatever else is needed to make your child feel better about what is happening. Do or say whatever you sense that little person needs to help heal the old wound from that traumatic crisis. Listen as your inner child expresses his or her fears and other feelings, and respond in comforting, reassuring ways. As your inner child is feeling better about the natural world, you will know that the alienating feelings are in the process of being healed. /

Now, relive the most pleasurable experience with nature in your childhood, perhaps inviting your adult self to enjoy it with you. / Take time to enjoy playing a game in a beautiful natural setting with your adult self. Both your child and your adult self can enjoy doing something fun together there. You may discover that you both feel more at home there in nature as you play together. / Now, before you part, give each other a gift of something in that beautiful natural place, something to remember the experience you have shared. See if you and your child both laugh as you give and receive the gifts. / Say goodbye now with a hug, after you have decid-

ed when you will meet again to enjoy nature and perhaps another outdoor game together. / Return to the present place and time and reflect on whatever you learned during the experience that you want to remember or do something about. /

When you have finished this memory trip, it may be useful to sketch a rough picture of what you experienced, using the colors of crayons that best express your feelings. Do not plan what to draw or worry about how it looks. Sketching experiences like this may facilitate recall of feelings and learnings. / Take a few minutes to record in your self-care-earth-care journal "things to remember" from your memory trip. A few words or short sentences will suffice. In particular, note what you learned about the imprinted feelings, attitudes, images, and behaviors that may have remained with you through subsequent years. / Commend yourself for recovering both negative and positive memories, and taking steps to heal the painful ones. / Debrief the whole experience by sharing your feelings and learnings from this exercise with someone you trust. / If this exercise did not "work" for you, try it again later. It often takes several practice sessions to learn how to use such guided imagery effectively.

Doing an exercise like this one may help people get in touch with the early-life roots of their intimate or detached feelings about nature. It often brings some healing of ecoalienation while also making available positive experiences with nature that may be forgotten in people's memory banks. If people become aware of several strikingly traumatic or satisfying nature experiences that occurred at any life stage, it makes therapeutic sense to use the memory trip exercise several times, focusing on these one by one.

Healing Grief for Planet Earth

People often think of grief only in relation to losses from death and perhaps divorce. They do not realize that among their painful feelings concerning escalating environmental losses, genuine grief feelings often are present. Calling this pain what it really is—grief—can help identify amorphous feelings of discomfort, loss, depression, and sadness about the environmental crisis, and facilitate movement toward using that energy for earth-caring action.

Among earth-loving people, grief is an often-hidden component of the reality-based feelings triggered by the day-by-day evidence of our deteriorating environment. Sensitive people often feel such ecological losses very personally. Such grief feelings for our earth home, mingled with ecological anger and guilt, are appropriate for all of us today. In addition to environmental losses close at hand, these feelings are triggered by:

- the loss of green, clean, healthy places near where people live
- the continuing losses of the richness of biodiversity
- the crowding and loss of green space caused by exploding populations

• the extinction of plants with as-yet-undiscovered pharmaceutical healing powers

• and the painful loss of dreams—for a healthy world for oneself and one's children and grandchildren.

Many earth-literate people also feel considerable anticipatory grief about expected future losses in nature. Anticipatory grief is especially prevalent among children and adolescents as they become aware of the increasingly threatened future of the natural world in which they probably will live. Furthermore, most children and teens have not lived long enough in a deteriorating world to develop the denial defenses of adults. The hopeful side of all this is that many children and young people respond more eagerly to ecological education than adults.

Providing opportunities for people to do their "ecogrief work" is important in today's ecological crisis. Because the causes of ecological grief are ongoing, complete healing of this grief clearly is neither feasible nor desirable. What is important is healing those dimensions of grief that produce denial and action-paralysis. The ecogrief group leader's task is to facilitate interpersonal trust and group bonding so that members will feel free to express their feelings candidly, be heard empathically and without judgment, and be motivated to use the energy of their grief, guilt, and anger for earth-caring actions.

Here is what happened in one structured ecological grief group in which the author was a participant. The group was led by pastoral psychotherapist Robin Crawford as a training workshop during a professional conference on wellness and spirituality.[21] The eight-person group, called simply "Grief Group for Mother Earth," met for two intensive hours. After the facilitator explained the purpose of the group, we all introduced ourselves. He then asked us to reflect silently on the powerful poem "Wild Geese," which appears at the beginning of this chapter. After we had done this, group members shared intense feelings of ecological mourning awakened in them by the poem.

Then the leader told a little of his earth-story and invited all of us to do the same. Next the facilitator suggested that each person describe a recent experience in which he or she felt some pain for the earth. Heavy feelings of grief, fear, anger, and frustration flowed freely. Then we were invited to share a recent earth-encounter in which we experienced some measure of personal transformation. This put us in touch with positive, earth-bonding memories. The final sharing was in response to an invitation to tell what we planned to do to help protect and preserve the living earth. We then joined hands as the leader led us in a brief ritual of commitment and blessing. Based on my own experience and the comments of others after the group closed, it was clear that this brief, carefully planned grief group had produced some healing as well as renewed commitment to earth-caring.

Ecological grief healing can be facilitated in many formats and settings. It can be done in a structured group like the one just described, or a free-flowing, unstructured group in which people share spontaneously after initial seed-planting comments by the leader. Ecogrief healing can occur as one component of many study groups and classes; in individual and group counseling sessions; in marriage, parent, family, and singlehood enrichment events; in all types of general grief-healing experiences; and certainly in all events focusing on ecology issues. Grief issues often emerge spontaneously when people tell their ecological story.

To open a door to discussion of ecological grief, simply invite participants to share their feelings about painful problems in their communities and world. Such an invitation often brings needed attention to the interrelated violence against women and children, other ethnic, racial, and religious groups, and the poor, as well as against the earth.[22]

When our children were young, we took several winter camping trips to Death Valley, one of the most un-dead places in California in terms of its desert flora and fauna. I remember seeing the tiny desert pup fish in the little pools of water there. The ranger told us that these amazing fish had survived by adapting through the eons from the time when a huge fresh water lake filled the valley until now when only these tiny pools that are saltier than the ocean remain. Perhaps there's a lesson from nature about the capacity of all life, including humans, to cope with incredible losses and adversities. A friend of mine who has coped courageously with cancer and several other major health problems says that, like the old Timex watch ad, he "Takes a lickin' and keeps on tickin.' " We can use our grief about the earth to motivate most daring efforts to keep nature from being pushed beyond its capacity to adapt.

Encouraging People to Befriend Wildness—Inner and Outer

A variety of methodologies exist by which people can befriend their own wildness and the wildness of nature (discussed in chapter 2). How people relate to their inner wildness influences deeply their connection with or alienation from wildness in the external world, as well as whether they express this side of themselves in creative or destructive ways. Two interrelated therapeutic goals relate to people's responses to wildness. One is to help them discover, befriend, and affirm their inner wildness as a source of creativity and channel it into constructive behaviors. The other is to enhance a sense of connected affinity with nature's wildness that lets people enjoy but also respect and protect themselves from nature's awesome power, and also strive to preserve real wildernesses for future generations.

John Muir, pioneer nature lover, self-taught ecologist, and founder of the Sierra Club, tells in one of his books about climbing to the top of a limber pine tree during a violent High Sierra storm, shouting with joy as the wind whipped it and him back and forth. Muir's behavior certainly indicates a joyous, celebrative, nonfearful relationship with the wild side of nature—even though it clearly is not a smart thing to do in a lightning storm.

Diagnostic hunches about people's wildness can be gained in a variety of ways. The most direct is to trust your intuitive impressions about how repressed or integrated their wildness is in them. (Up-tight, hyper-controlled (and controlling) people often are over-invested in repressing their wildness.) Another way is simply to ask them how they feel when they are in wilderness places or near "wild" animals. It may be productive to ask how they feel about and relate to nature's wild, powerful, often violent side. Do they feel stimulated and challenged, or do these experiences trigger avoidance anxiety and the desire to withdraw? Diagnostic queries such as the following may produce useful clues: What are your feelings when you are in a fierce storm? How do you feel when you are alone in a wilderness area or near a wild animal who is not a threat to you at the time? Do you ever feel the urge to "run with the wolves" or other wild, free animals?

If teachers and therapists have naturized their places of work, with plants and perhaps pets present, people's comments about these and their ways of relating to them can be revealing. People who ignore the plants or animals or seem to sit as far as possible from them—or, in contrast, who touch a plant caringly or choose to sit close to it—may be communicating nonverbal clues about their degree of ecoalienation or ecobonding. As with all clinical hunches, it is crucial to check out with those involved the accuracy of the observations of human-nature interaction.

A potentially productive variation of this approach, which also can be used therapeutically, is to invite students or clients to take a walk in a garden, park or nature preserve where there are abundant plants and birds and perhaps other animals, making mental notes of changes in people's feeling tone associated with being in direct contact with the natural world. But, as suggested above, living things need not be outdoors to offer meaningful relationships Nor do the living things need to be beautiful. How people relate to a picture of a gorgeous sunset or a photo of a toxic dump, to a single flourishing flower or a terminally ill office plant, can be useful to explore with them at an appropriate time.

It often is healing and growth-enabling to help people discover, befriend, direct, and enjoy, rather than reject and repress, this high-energy, potentially creative "wild" side of their deeper body-mind-spirit organism. For those whose wildness is deeply repressed, often accompanied with wildness-denying religious sanctions, the discovery of inner wildness usually requires time and can cause anxiety. The discovery occurs as a gradual awakening in the process of work with a therapist or teacher whose own inner wildness has been befriended and valued. But when wildness is experienced as a lively, potentially exciting, and creativity-stimulating dimension of people's experience, the benefits often include surges of passionate love of life—their own life and that of the biospheric community.

Exploring This Chapter's Issues Further

Books for exploring the issues raised in this chapter in more depth are listed at the end of chapter 9.

Notes

1. Mary Oliver, *Dream Work* (Boston: Atlantic Monthly Press, 1986), 14. Emphasis added. I am indebted to Robin Crawford for acquainting me with this poem in the ecogrief group described later in this chapter.

2. "The Skill of Ecological Perception," in *Ecopsychology: Restoring the Earth, Healing the Mind*, ed. Theodore Roszak, Mary E. Gomes, and Allen D. Kanner (San Francisco: Sierra Club Books, 1995), 214.

3. For example, effective CR resources may include videotapes from the National Geographic television series on environmental issues, the CNN weekly series "Earth Matters," or a movie like *The Gods Must Be Crazy* or *A River Runs through It*.

4. Storytelling is a healing method that innovative therapists like Milton H. Erickson often use with superb artistry. It also is in harmony with the rich tradition of storytellers in folks' cultures.

5. If human stories are used in education, counseling, or psychotherapy, it is important to let clients know they are from nonclinical sources. This prevents arousing the anxieties of people who imagine that what they are sharing will be used in ways that violate confidentiality. Because some persons in educational settings may be considering seeking counseling or psychotherapy, the principle of confidentiality should apply there as well.

6. Idries Shah, *Tales of the Dervishes* (New York: E. P. Dutton, 1970), 23–24. Quoted in Anderson's A New Approach to Family Pastoral Care (Philadelphia: Fortress Press, 1980), 1–2.

7. See his Seven Arrows (New York: Ballantine Books, 1972), 68–85. On this theme, clinical social worker Richard W. Voss points out that families may find it meaningful after they have had an enjoyable outing to a zoological garden, if each person is invited to make up a story about a favorite animal, fish, or bird. I would add that, from the perspective of the rights of other species, the debriefing of such family experiences should include some reflection by the parents on the ethical ambiguity of keeping wild animals in such a setting, even if every effort has been made to create habitats designed with their health and well-being in mind.

8. Summers, "Implications of The Inner and Outer Growth Journey for the Mental Health Clergyperson." *Association of Mental Health Chaplains Forum* 30, no.1 (October 1977): 24–25.

9. Ibid.

10. Ibid.

11. Ibid, 25.

12. Quoted by Marian Christy, "Possessing Secrets of Joy," *Santa Barbara News Press*, 26 July 1992, G3.

13. Quoted in "Forever Growing, Horticulture as Therapy," Philadelphia Friends Hospital, n.d., 5.

14. This pastoral therapist asked to remain anonymous.

15. Why is it that so many people who have the luxury of being able to choose to be in a quiet place leave the TV or the radio blaring even when they are alone? Why is it that so many youth today injure their hearing by listening more or less constantly to high-volume rock music? There is a variety of reasons for auditory addictions. People who live alone and are also lonely (conditions that are not necessarily synonymous) may use electronic transmitted human voices or escape into sitcoms to assuage their loneliness. For youth, loud rhythmic music seems to serve as a tribal rite of belonging, and as a way of separating from their parents' generation, who often hate the volume and lyrics of their music. In some cases, external sound is used to drown out the painful messages from a conflicted inner world. These distressing inner voices often include the self-accusatory messages between different parts of the personality—for example, between what transactional analysis calls the inner Parent and the inner Child. In severely disturbed people, the painful inner sounds may also include delusional voices.

16. Lydia Maria Child, *Believing in Ourselves, The Wisdom of Women* (Kansas City: Ariel Books, Andrews and McMeel, 1992), 25.

17. Personal communication from Huntley Lewis, MFCC.

18. Diane Mariechild, *Mother Wit, A Feminist Guide to Psychic Development* (Trumansburg, N.Y.: The Crossings Press, 1981), 14.

19. Rather than analyzing a dream by taking it apart and seeking to understand each symbol by free association, "gestalting" dreams involves describing what is happening in the present tense and thus reliving the dream in the here-and-now. One also dialogues with all the things and people in the dream, one by one, speaking both to and for each part. When one reaches the point when anxiety of repressed meanings threaten to emerge and the dream censor in the unconscious mind cause one to awaken, the person relating the dream then continues to describe its conclusion past the waking point. This often illuminates crucial unfinished messages from the dream—messages that otherwise stay repressed in the unconscious mind.

20. Making peace with painful past experiences and memories that adversely influence present life and relationships is, of course, an objective of many psychotherapeutic and reeducational approaches.

21. I am indebted to Robin Crawford, a former Ph.D. student of mine, for providing my first experience of this type of grief group.

22. It is noteworthy that the grief-healing energy of connecting with nature is explored by hospital chaplain Phyllis Windle in "The Ecology of Grief," in *Ecopsychology*, 136–45. Also Joanna Macy describes the Council of All Beings as "a collective mourning ritual" that enables participants to get in touch with and work through their deeply repressed feelings about the ecological disaster. See her "Working Through Environmental Despair," in *Ecopsychology*, 240–59.

9

Methods of Ecotherapy and Ecoeducation: Grounded Healing and Growth Work (Continued)

WHEN THE GOING GETS TOUGH, THE TOUGH GO GARDENING

—Sign at the entrance of a small organic garden

O World, I cannot hold thee close enough!
Thy winds, thy wide gray skies!
Thy mists that roll and rise!
Thy woods, this autumn day, that ache and sag
And all but cry with color!. . . .
World, World, I cannot get thee close enough!

—Edna St. Vincent Millay[1]

Masculine and feminine are balanced in nature . . . psychotherapy relies on nature's growing and healing processes. Nature's rhythms, cycles, and balancing processes are our healing resources.

—Lesley Irene Shore, *Tending Inner Gardens: The Healing Art of Feminist Psychotherapy* [2]

This chapter explores additional ecotherapeutic methodologies, some of which also are useful in the assessment or diagnostic phase of the process. Like methods described in chapters 7 and 8, many of this chapter's approaches can be used in both ecotherapy and preventive ecoeducation.

Projective Methods in Ecotherapy and Ecoeducation

Projective methods, like storytelling and related techniques described below, are useful in understanding and helping to heal many aspects of people's inner lives, including their earth alienation. These tools can be used diagnostically to confirm movement away from ecoalienation and toward ecobonding. In addition they may be used to enable people to develop personal growth goals for themselves. It is important to explain to the participant how the exercise may serve the objectives of his or her therapeutic or educational experience.

Tell a Story about a Picture

This is a projective method modeled on the Thematic Apperception Test and a variety of similar diagnostic tools used by psychologists. If you are a therapist or teacher focusing on ecological issues, prepare for using this approach by collecting a variety of pictures in color (photographs or prints from periodicals) of nature in her different moods and conditions—beautiful, ugly, serene, nurturing, wild, violent, polluted. Your collection might include pictures of a quiet lake reflecting the beauty of a sunset; a volcano erupting; a swirling tornado; a scene from a devastating earthquake, fire, or flood; waves dancing in the sunlight; a dashing mountain stream; wild birds, perhaps a hawk with prey in its talons; wild animals, including dangerous species; a road-killed deer; a mountain wilderness; a polluted city; a treeless urban slum; a huge garbage dump; a nuclear power plant; a nuclear explosion; a beautiful flower garden; a baby animal or bird with its mother; a spring flower blooming next to a snowdrift.

Invite students or clients to choose a few pictures from the collection (or bring in their own favorite nature pictures) and tell a story about each one. According to projective theory, in creating stories about any ambiguous stimulus (such as pictures with no captions), people must utilize their imagination, fantasies, and memories. They thus project into their stories subconscious feelings, memories, and images. It is important to be aware of people's body language and the feeling tone of their voices as they tell their stories. If time is limited, it is acceptable to have people take pictures home and write their stories. Spontaneous stories communicated orally without reflection, however, may be more revealing of deeper feelings and memories such as frightening experiences in nature.

To check out a clinical hunch using this method, select pictures with images related to the particular issues with which individuals are struggling, and invite them to tell stories about them. In a class or workshop, it is productive to invite written or oral storytelling by students in response to pictures related to particular topics being discussed.

The Draw-a-Picture Method

A closely related projective tool is the draw-a-picture technique. To prepare for using this, assemble a pad of blank paper and a box of crayons or multicolored marking pens. Offer these instructions: "This is simply a way of getting in touch with some of your thoughts, feelings, and attitudes about the natural world that may be influencing your sense of well-being (or lack of it). This is not an art lesson, and no one is going to judge what you draw in terms of its artistic merit. Just let your fingers move quickly however they wish without planning—scribbling, doodling, or drawing. Use whichever colors your fingers are attracted to in order to express what you are feeling as you draw, in the next five minutes or so." Ecotherapeutic variations of the draw-a-picture technique include drawing a tree; feelings about a natural disaster; one's favorite place in nature; a violent storm; a wilderness place; an endangered species. It is important to note whether and how people include themselves in their pictures.

To stimulate ecological storytelling, it is helpful to invite people to quickly sketch their earliest memories of nature or the home where they were raised, including the plants, birds, and animals that were around it.[3] The draw-a-picture method can be used in conjunction with a variety of imaging techniques—for example, before or after visiting their meaningful places in nature. Here is an example of how projective drawing can complement a guided imaging exercise, in this case one that is useful for stress-reduction and renewal. (Remember that the / marks mean to pause in your reading and close your eyes while you do what has been suggested.):

Using your imagination, give yourself a mental mini-vacation by going to one of your favorite places in nature. / Be there now. / Lay aside whatever load of cares you have been carrying. Just release these as you exhale. (You can always pick up your load later, if you choose.) / Relax deeply and let yourself feel the vibrant serenity of this place as all your senses are touched by its aliveness. / As you are enjoying this beautiful place, remember that it is yours. You have it as part of the earth treasure in your mind, so that you can return whenever you need to be quieted, renewed, and relieved of stress. / Now complete the experience in your own way to give it a sense of closure. / Return in your thoughts to your present place and everyday life, bringing with you the feelings you found in nature. / Now, reflect on this experience and make a crayon sketch or two—rough drawings of your images and feelings when you went to the beautiful place in nature. / Now, you may wish to share this experience with someone and show them your drawings. /

Photographs in Ecotherapy and Ecoeducation

Because of the image power of pictures, still or video cameras can be valuable ecoeducation and ecotherapy tools. Like drawing, they provide a visual record to which people may return repeatedly for further learning. A student or client can place photos of places with positive or negative associations where they will be in view frequently, perhaps stimulating further processing of issues connected with them. It can be ecologically nurturing to carry pocket-size copies of pictures with special earth-friendly associations in one's wallet or handbag. Sharing photos of important people and places in one's life with a teacher, therapist, or class often adds special meaning both to the pictures and to those relationships.

These alternative instructions suggest other possibilities for uses of photos in therapy:

• "I suggest that you bring a photograph of your childhood home, as well as one of your childhood family to the next session. This may be useful in getting at the unfinished feelings connected with those people and that place."

• "I recommend that you take a picture of. . . . " This could be the person's childhood home; the place where he or she had a terrifying experience of nature's fury as a child or more recently; or the place where the person had a spiritual inspiration in nature.

• "I suggest that you take some close-up shots of little things in nature that catch your eye—for example, tiny wildflowers or patterns in a rock—things it is easy to overlook even though they are often around us. Doing this sometimes increases awareness of less conspicuous but interesting natural things at our fingertips."

Poetry and Music in Ecotherapy and Ecoeducation

Carefully chosen poems and songs can be resources in ecoeducation and ecotherapy. Powerful verbal images expressed in either poetry or prose often touch people at a deep-feeling level. Perhaps the poignant lines by Edna St. Vincent Millay that opened this chapter let you feel this power. Poetic images such as these can awaken awareness in many people of their longings for the energizing lift of intimate interaction with nature.

Some people in our frenetic world resonate with these images by William Wordsworth in one of his most familiar sonnets:

> The world is too much with us; late and soon,
> Getting and spending, we lay waste our powers;
> Little we see in nature that is ours.
> We have given our hearts away, a sordid boon!
> This sea that bares her bosom to the moon;
> The winds that will be howling at all hours,
> And are up-gathering now like sleeping flowers;
> For this, for everything, we are out of tune.[4]

These images of English poet-playwright John Drinkwater carry an ecological confrontation:

> When you defile the pleasant streams
> And the wild bird's abiding place,
> You massacre a million dreams
> And cast your spittle in God's face.[5]

How can poetry be used in ecotherapy and ecoeducation? In addition to suggesting that clients or students read and reflect on nature poetry that touches them, it often is growth-enabling to invite them to express their feelings about nature by penning a poem themselves. To do so can utilize feeling-laden images aroused by a photograph, a guided imaging session, a memory trip, or an actual visit to a place of natural beauty or disaster.[6] Some people will be pleasantly surprised by what they write and will want to have others hear or read their poems.

Ecologically concerned rock and folk singers have produced songs that combine the lyrics' verbal images with the right-brain impact of the music. Such songs can communicate energized earth-caring messages to some clients and students. Nature hymns, as was mentioned in chapter 4, can do the same for many religiously oriented persons. John Denver's passion for environmental issues, especially related to the Rockies, has been expressed feelingly in songs like "Rocky Mountain High" and "The Eagle in the Sky." Woody Guthrie's "This Land Is Your Land" has become a kind of unofficial anthem for the American environmental movement. Bruce Springsteen and other popular singers have recorded versions of this song. The Eagles, a well-known rock group, have popularized several environmental songs, including "The Last Resort." Sting has performed at least two songs about rainforests and given benefit concerts to help preserve them.[7] Holly Near describes her commitment to being more than simply a talented singer: "It's really important to me to be a good activist and a good thinker, a good musician, a good singer, and a good entertainer. You can't do it all, but I have walked those delicate lines as best I know how." [8]

Healing Rituals in Ecotherapy and Ecoeducation

Until more people around the planet recover respectful, caring attitudes toward all life, it is doubtful that our species will change our earth-destroying lifestyles in more than superficial ways. Earth rituals can be one means of strengthening attitudes of care for all living things. We humans are a ritual-creating species.[9] All cultures have living folk rituals that are group ways of acting out their meaning-creating history, beliefs and values, fears, hopes and dreams. These rituals are often earth-centered and form the central forces in these cultures' religious life. Some therapists and teachers are as surprised by the potential therapeutic power of relevant rituals as are their clients or students.

Earth rituals also are a part of many spiritually based healing traditions. The centuries-spanning Chinese traditional medicine (including acupuncture) focuses on five elements (water, food, fire, earth, and metal) and on five seasons of the year (winter, spring, summer, late summer, and autumn). The ancient healing heritage of women, as many women are discovering, is an abundant source of usable earth healing rituals.[10]

The healing rituals of native peoples, including Native Americans, are almost always earth-centered. For many such peoples these rituals still reflect ancient earth-bonding. To illustrate, the contemporary Pueblo Indians of the American Southwest continue to believe that their society's well-being depends in part on keeping in right relationship with Mother Earth. Feminist scholar, Morneen Kamiki Bratt, describes their annual ritual:

> A summer corn dance and feast enacted in the central and interior plazas of the village construct the collective self and ensure that the individual and collective selves are in right relationship with the shared spiritual beliefs of the community. Everyone is invited to partake in the corn festival including people from neighboring pueblos and tourists, because the collective crowd demonstrates success and future abundance.[11]

Environmentalist author Dolores LaChapelle observes that

> some people have the illusion that simply by taking humans out into the wilderness they will see the beauty and realize the importance of saving the environment. I think this is nonsense. If we want to build a sustainable culture it is not enough to "go back to the land." . . . If we are to truly reconnect with the land, we need to change our perception and approach more than we need to change our location. . . If we're going to rediscover a viable relationship with nature, it will not be by more ideas but through experiences where you know you are a part of nature, with no questions asked. When you've had these experiences, you know what you want and you can't be pushed around.

She has found that following "the wisdom of other cultures," particularly the sacred earth rituals of native peoples, often facilitates such transformation:

> Ritual is a focused way we can both experience and express respect. And most important of all, during rituals we have the experience of neither opposing nature nor trying to be in communion with it. Instead, we have the experience of finding ourselves within nature, and that is the key to sustainable culture.[12]

A client of mine introduced me to a book by Jungian analyst and pastoral psychotherapist John A. Sanford, a novel called *Song of the Meadowlark.*[13] It is the tragic story of Teeto Hoonod, a Native American involved in the Nez Perce Indian war. In the book's introduction, Sanford quotes an intriguing idea of Carl G. Jung: "The foreign country somehow gets under the skin of those born in it. . . . That would mean that the Spirit of the Indian gets at the American from within and without." My sweat lodge experience (see below) put me in touch with what I sense to be the truth in Jung's speculation.

The following tribute is from Matthew Fox:

> I can honestly say that I have learned more from Native Americans about prayer directly and indirectly (through dreams) in the past fifteen years than I have from any other source save for the feminist movement which in many ways is so similar. . . . The wisdom of Native American spirituality is a rich source for all who seek a more holistic, creation-centered, compassionate and powerful spirituality.

Fox holds that this spirituality includes at-oneness with humans' bodies, animality, and naturehood, and a "ritual that works." This means ritual that holds power for transformation by bringing us "down to places of depth and silence and darkness and healing," rituals that demand deep, bodily responses and celebrate a cosmology that "knows we *are* nature and are in nature and from nature."[14]

Ed McGaa Eagle Man, an Oglala Sioux scholar-lawyer with expertise in Native American spiritual rituals, has authored a remarkable book entitled *Mother Earth Spirituality: Native American Paths to Healing Ourselves and Our World.*[15] He illuminates the meaning of the sweat lodge ritual in an essay on "bringing forth your own Mother Earth wisdom." His insights awakened memories of a moving spiritual experience I had a few years ago. Several Lakota Sioux students in a crisis counseling class I was teaching at the Native American theological school in Tempe, Arizona, invited me to join them in their sweat lodge one evening. In what proved to be a deeply moving experience, my students became my teachers.

Our group of about a dozen men and women sat on the ground packed in a tight circle around the dark circumference of the tiny, domed sweat lodge. White-hot rocks from the roaring fire outside were placed in a shallow indentation in the center of the circle, uncomfortably near our feet from my perspective. Water blessed with Lakota prayers (graciously translated into English for the non-Indians), was poured on the rocks several times during the ceremony. It produced steam that made the intense heat and humidity of saunas I had experienced in Finland seem almost cool by comparison. The intense environment and anxiety narrowed my focus as cleansing sweat poured from every part of my body. The ritual's stages celebrated the four directions with their accompanying totem animals and chants. The earth on which we sat, the air, the rocks, and the fire, along with the cup of water and the ceremonial pipe on which we puffed periodically as it was passed around, all deepened my feelings of cleansing and healing connectedness with the sacred earth. Each ritual step was chanted by the ceremonial leaders in their Lakota language, accompanied with rhythmic drumming.

All this was strange and threatening at times, but the overall impact was profoundly cleansing and energizing of my body, mind, and spirit. I got in touch with my own need to ask for and receive forgiveness for my unwitting contributions, as a white man, to the destruction of both my native brothers and sisters and our shared earth. When it was over, I was aware that this experience had been an incredibly spiritual earth ritual for me. I suspect that it may have been even more so for the Indians present. Therapists and teachers can encourage learners to experience some of this energy by developing earth rituals that involve sitting on the ground and intense interaction with universal symbols of earth's creative and re-creative power (such as wind, water, and fire).

Some people spontaneously create earth rituals during workshops and psychotherapy that include an ecotherapeutic emphasis. A rabbi client created such a ritual as a part of his celebration of Yom Kippur, the Jewish high holiday of atonement. He and a dear friend, also a rabbi, created a powerful cleansing ritual by joining hands and walking ceremonially a short way into the ocean as they expressed appropriate Yom Kippur prayers in Hebrew.

A feminist clinical social worker reports that the earth-grounded rituals she uses in her practice sometimes have remarkable healing power for women as well as for some men. She encourages clients who are open to this to develop their own ritual ways of accessing "healing energy from Mother Earth," as she puts it. Among the natural objects that lend themselves to such therapeutic rituals are growing plants, roots, eggs, seeds, nuts, fruit, flowers, herbs, earth, shells, leaves, fire (candles), stones, feathers, and water (still or running).

According to this therapist, earth rituals have proved to be helpful to some women who have been traumatized by the profound violation of childhood incest or by rape in their adolescence or adult lives. One client, a victim of sexual violence during girlhood, smashed an egg the therapist had provided, letting herself "be messy" as she symbolically demolished her life-constricting old way of being. Later the therapist asked if she could "be the life" in another egg in her hand, experiencing herself growing in new, healthier ways. She was able to do this symbolically as a ritual movement toward doing it in everyday life.

For some, ritual contact with flowing water—a stream or waves or simply water poured from a pitcher—helps cleanse feelings of being dirty or defiled. As trust grows in the therapeutic relation, quiet times in a natural setting may help facilitate gradual recall of repressed memories of sexual violence. In working with women who have survived rape, nature rituals also may help heal the traumatized feelings of husbands or other men in the lives of these women.

Rituals may be useful for people struggling to enliven their spiritual health by increasing awareness of the mystery and the cornucopia that is beneficence of the natural world. As ecotheologian Thomas Berry states:

> For we will recover our sense of the sacred only if we appreciate the universe beyond ourselves and our role in the universe. As Saint Thomas says, "The entire universe participates in and manifests the divine more than any single being." . . . This is why our religious rituals are coordinated with the great liturgy of the universe itself.[16]

The Use of Trees in Therapy

Most people in preindustrial societies were in direct touch with forests and were thereby kept aware of the profound dependence of our species on trees. Today, as forests around the world are being decimated, the recovery of some measure of ancient veneration of trees may help our generation save these indispensable givers of oxygen, fuel, and building materials that we so easily take for granted. Tree festivals, involving the planting of young trees and celebrating magnificent mature trees, could encourage the recovery of tree-valuing attitudes.

Zen master, peacemaker, and poet Thich Nhat Hanh from Vietnam was nominated for the Nobel Peace Prize by Martin Luther King, Jr. In his book *Touching Peace, Practicing the Art of Mindful Living*, he describes this nurturing tree ritual:

> Ten years ago, I planted three beautiful Himalayan cedars outside my hermitage [in France], and now, whenever I walk by one of them, I bow, touch its bark with my cheek, and hug it. As I breathe in and out mindfully, I look at its branches and beautiful leaves. I receive a lot of peace and sustenance from hugging trees. Touching a tree gives both you and the tree great pleasure. Trees are beautiful, refreshing, and old. When you want to hug a tree, it will never refuse. You can rely on trees. I have even taught my students the practice of tree-hugging. . . . In the same way that we touch trees, we can touch ourselves and others, with compassion.[17]

As someone has said, "A person who plants a tree, plants hope." Since first reading these words, I have taken time now and again to give a caring hug to the huge trunk of the magnificent live oak whose branches shelter the entire width of the yard outside our home. (I still find myself checking first to see that no one is watching who might decide that this neighbor has really "lost it.") As I hug that tree, I often feel the gift of the tree's century-spanning strength and the aliveness of its continuing growth. I visualize its giant network of roots spreading in all directions, nourishing it and anchoring it when the fierce Santa Ana winds blow down the mountains. All this renews my own sense of being grounded in the earth. Perhaps, as Thich Nhat Hanh believes, the gifts between the oak and myself are reciprocal. Why not take a few minutes now to try an experiment: Give your favorite tree the gift of a caring hug. Who knows, you might discover that you and the tree both like this connecting.

On the subject of hugging trees, the social justice dimension of this practice is relevant. Remember the courageous women in northern India (see chapter 2) who are risking their lives by hugging the surviving trees in their area? This is their desperate effort to prevent the predatory chain saws of international lumbering corporations from destroying the remainder of the precious forests on which their families' well-being and livelihood depend.

Horticultural Therapy

When I reflect on my boyhood, my taste buds tingle as I remember the delicious red raspberries and vine-ripened tomatoes harvested each summer from our lush garden. My parents worked in their large vegetable, fruit, and flower garden early in the morning almost every day, though "never on Sunday." As I think of their passion for gardening, an insight dawns. For them, this was much more than a way of growing needed food during the Great Depression. Regular gardening was therapy that undoubtedly helped them cope with family tragedies and economic woes. Today, gardening might be called their horticultural therapy.

Gardening is a healthful hobby and stress-reducer for many people in our hectic, harassed world. Its psychological, physical and spiritual benefits can go far beyond the beautiful flowers and the healthy vegetables and fruits grown with no toxic chemicals in the soil or pesticides on the leaves. Healing-by-gardening has long been used in a wide variety of treatment settings with people who are mentally or physically ill. In the mental health field, therapeutic gardening is one of the oldest activity therapies. The healing impact of gardens was recognized as long ago as the glory days of ancient Egypt, when certain physicians prescribed walking in gardens as part of the treatment for some disturbed patients. In 1806 a mental hospital in Spain reported the benefits gardening offered patients. In 1812, physician Benjamin Rush, the father of American psychiatry (and signer of the Declaration of Independence), reported that gardening had a curative effect on the mentally ill. Five years later, America's first private, nonprofit psychiatric facility opened, a Society of Friends mental hospital near Philadelphia. It has pioneered in the use of horticultural therapy. From the beginning, its treatments included planting trees and vegetables in what was described as "rehabilitation through contact with nature and people." The current hospital brochure describes gardening therapy well: "When we are gardening, we are active participants in the process of nature, not passive observers. We can connect with the earth by putting our hands in the soil."[18]

Since around 1940 horticultural therapy has flowered in many places and is used with thousands of patients, including geriatric patients, alcoholics and drug addicts, and injured and disabled persons. Delynn Jones, a high school senior in Harrodsburg, Kentucky, suffered massive, debilitating head injuries in an auto accident. He credits horticultural therapy for his recovery. With the help of a professor at the University of Kentucky, Jones later developed a wheelchair-accessible garden of rows of tomatoes, corn, and bush beans.

Horticulture therapy became a distinct clinical profession in the 1950s when Michigan State University set up the first master's degree program in this field. Kansas State University established the first American undergraduate program in horticulture therapy in cooperation with the Menninger Foundation. Today more than a dozen colleges and universities have horticultural therapy courses or programs, and the American Horticultural Therapy Association is a flourishing professional guild.

Horticultural therapists at the Philadelphia Friends Hospital have discovered that patients often are helped in four areas as they care for plants. (1) *Cognitive improvements* result from awareness of the outside world and of oneself in it; learning new skills and language; acquiring decision-making and problem-solving skills and learning to work more independently and follow more complex directions. Patients also benefit by learning an enjoyable hobby that they can continue after their hospitalization. (2) *Psychological improvements* include feeling more productive and useful; increased self-esteem and self-confidence as projects are planned and completed; a sense of enhanced responsibility (because plants depend on the patients); release of anger and aggressive impulses constructively (through vigorous weeding, hoeing, and pruning); more peaceful inner feelings and more openness to talking through their feelings and problems. By this creativity and self-expression, patients may regain or strengthen their sense of self. (3) *Social improvements* come by working in a small group toward common goals, by learning communicating and compromising skills, and by talking about a neutral topic: plants. (4) *Physical improvements* are gained by exercise in a safe work environment with lots of fresh air. Patients' bodies improve as their muscles are retrained and coordination in both gross and fine motor activities is improved.

An assistant to the San Francisco sheriff, Catherine Sneed Marcum, became convinced that gardening could motivate prisoners to reexamine their lives and redirect their energies into something more productive and rewarding. Now, prisoners at the San Francisco Jail Number 7 tend twelve acres of land reclaimed from a prison junkyard. Their truck garden produces 120 tons of vegetables and fruits a year. Most of the prisoners are from the inner city and had never been near a garden or farm before. Many are school dropouts serving six-to-eight-month sentences for drug-related crimes. Many had been hardened to usual prison therapies by repeated incarcerations.

In a TV interview, Charlayne Hunter-Gault asked Marcum what in this program touches the young men in life-changing ways. She responded:

> I think that they are able to see themselves in another way, and I don't think anything else in their life could have done this. . . . I'm a gardener now, myself, and I wasn't when I started the program—I think watching something grow and being able to create something that good and valued helps people feel like they're good and valued. Because of their drug addiction, because of their criminal history, a lot of them have lost the sense that they can do something good. . . . When they come out to the jail program, most of them can't look you in the eye. They hold their heads down. You can barely hear them when they talk, but then gradually, as they're in the program for some time, they start

> walking up straight, and they're very excited about the spinach and the potatoes. . . . What I try to do with the spinach and the potatoes and the herbs and the flowers is really try to show them that just as we gave life to this stuff, this stuff that we need as people, you can give life to a new life. You can begin to grow a new path for yourself and use gardening to do that. . . . I've seen so many turn from being hardened crack dealers to someone who cares whether or not the roses have gotten just enough fish emulsion, or whether or not we're going to make sure they're pruned at the same time, and I really feel like gardening teaches you to care.[19]

The prison program has been so effective that it has been expanded to another garden for former inmates and a tree corps in which former inmates plant and care for trees throughout San Francisco.

What is the crucial factor in this program? It seems clear that it is the young prisoners' relation with Marcum, a person who believes in their ability to grow personally as they learn from growing plants. When asked what in her own background caused her to be interested in this approach, Marcum mentioned growing up in a family of fourteen children, and her own difficult life experiences. "My mother, I think, taught us that we had to do something to change the world. And I think the combination of that and having children made me feel like I have to do something to make the world a better place."[20]

Some home gardeners have discovered the satisfactions and multiple benefits of "biodiverse gardening" in which they plant a wide variety of native plants close together as they grow in the wild. In this way they create a mini-wilderness in their yard, a habitat that attracts more native creatures that do not feel at home in yards with manicured lawns and conventional gardens.[21] Raising many native plants may also help to save some of the countless species that are near extinction because of the narrow range of popular and imported plants in most of the 35 million "proper" gardens in the USA.[22]

Wilderness Therapy

Taking people into wilderness areas has become an increasingly valued method in both therapy and education. Psychologists Allen Kanner and Mary Gomes point out that similar therapeutic effects often accrue when people engage in intense urban ecological restoration projects:

> It is common . . . to report dramatic breakthroughs that shake individuals to their core. When the natural world reawakens in every fiber of our being the primal knowledge of connection and graces us with a few moments of sheer awe, it can shatter the hubris and isolation so necessary to narcissistic defenses. Once this has happened, ongoing contacts with nature can keep these insights alive and provide the motivation necessary for continued change. It is these experiences that will ultimately fill the empty self and heal the existential loneliness so endemic to our times.[23]

Comparable healing is reported by other teachers and therapists. Psychologist Steven Harper, who coordinates wilderness programs at Esalen Institute in California, describes how almost all participants experience an enlivening of the five senses in the wilderness, although the real work begins when they return.[24] Robert Greenway surveyed students who have taken his course on ecopsychology and wilderness at Sonoma State University in California, which includes three days or more of "alone time" in nature. Ninety percent of respondents described "an increased sense of aliveness, well-being, and energy." [25]

Clinical social worker and pastoral counselor Richard W. Voss leads week-long therapeutic canoe trips in Canada's Temagami wilderness for the purpose of offering participants ways of "finding balance by reconnecting and recommitting to self, neighbor, and earth." His invitation includes these words: "Imagine drinking pure, unpolluted, crystal-clear lake water during your early morning swim. Imagine watching the northern lights, dancing across the wide-open northern sky. Imagine stars in the millions. Imagine the lonely howl of a wolf, and the lilting call of the loon echoing across the lakes." Voss holds that the trips stretches people's boundaries as they travel into their own inner wilderness. "They reflect on their lives using the wilderness experiences with nature and each other as metaphors, instead of the classical analytical use of metaphors of the psyche."[26]

For the eight men on the trip, most of whom had been socialized to dominate nature rather than experience it, a week in the wilds often gives them the gift of opened eyes. The experience of "primitive camping and canoeing far away from electricity (except the lightning kind) and telephones is illuminating," Voss says. "I have found that men are inherently spiritual—I am amazed by the depth of the thirst or longing body-soul-spirit integrity that so many men have. The wilderness experience provides the necessary emotional, spiritual, and physical 'container' for the connecting process to occur. . . . I am also struck by how the wilderness and the use of traditional ceremonies helps (white middle-class) men get beyond their heads, not only in touch with their hearts, but with their bodies."

Voss shares this personal experience: "Upon completing a traditional (Lakota) hanbleceya (or Crying for a Vision) ceremony this summer, I learned how exposure to the elements actually enabled and allowed my body to accommodate, change, respond to the natural world around me in ways I never thought possible. We spend so much time and energy avoiding this kind of intimate contact with the natural elements. So the wilderness actually provided a totally new experience of body-self for me." He says that he has discovered similar transformations when he has offered "Therapeutic Rock Climbing" for fathers and adolescent males suffering from attention deficit disorder.

During one wilderness trip, Voss was moved to write a short poem titled "Earthdance." It reflects his intense earth-caring, a passion undoubtedly enhanced by the wilderness:

My bare feet
unbearably light
dance
to the pulse of the wounded earth.

Voss's moving account invites me to relive several intense earth-bonding experiences in my life—times that affirm the view that when we find wilderness we may also find a missing part of ourselves. One was the unforgettable day of climbing Longs Peak, a 14,000-foot mountain in the front range of the Rockies. I remember the scary ascent followed by cowering on a ledge near the summit during a violent afternoon thunderstorm. As I remember that day, I reexperience the unexpected gifts—an incredible blend of fear, excitement, challenge, and achievement, mingled with the joy of intimate oneness with that marvelous mountain. Psychologist Abraham Maslow's phrase "peak experience" acquired fresh and literal meanings that day for me.

A prized though very different event in my memory bank also brought the gift of body-mind-spirit enlivening. It involved a blending of interpersonal male bonding with nature bonding during a weekend in a rustic mountain camp with seventy men, all participants in a workshop on men's identity issues. It was co-led by poet-guru Robert Bly and a skilled teacher of hand drumming, a releasing art that was used with gusto by the whole group during several ritual celebrations. Bly read, with drum background, a grief-full poem about his troubled relationship with his alcoholic father, who had recently died. This awakened unfinished grief in me for my father, along with tender feelings of loving connectedness with my sons and daughter.

The healing and renewing potentialities of time spent in wilderness areas has been demonstrated with persons suffering from a wide variety of physical and psychological problems. Extensive empirical data has confirmed the self-esteem–building impact of the Outward Bound Program on emotionally disturbed adolescents.[27] A twelve-step recovery program in Arizona finds it helpful to take alcohol- and drug-dependent teenagers rock climbing or into the desert on horses and dirt bikes. A therapist in that program reports that these wilderness experiences, and their debriefing with staff members, often serve the youth as therapeutic metaphors of key issues in their lives and their recovery. A "Family Survival Project" takes Alzheimer patients camping in the Santa Cruz Mountains of central California. There some patients respond to experiences such as sitting around a campfire singing familiar songs like "Home on the Range." This setting also provides respites for family members exhausted by nonstop caretaking. A pastoral psychotherapist in Virginia has used therapeutic day-hikes with male clergy suffering from burnout. Pretest and posttest findings revealed significant reductions in the anxiety and stress levels of his experimental group, compared with nonhikers.[28]

The Role of Animals in Ecotherapy and Ecoeducation

Theodore Roszak shares his awareness of nonhuman animals: "Whenever I turn to an environmental issue, I find myself intensely aware that other, nonhuman eyes are upon me: Our companion creatures, looking on, hoping that their bewildering human cousins will see the error of their ways."[29] Encouraging animal bonding can be a useful intervention in ecotherapy and ecoeducation. Confirming what pet lovers have always known, scientific and clinical evidence is growing that shows that many humans find not only pleasure but also stress reduction and healing in relating caringly with other animals. The enormous popularity and love of pets attests to the fact that many people find them satisfying companions. A recent survey revealed that Americans own 60 million cats, 52 million dogs, 12 million birds, 5 million horses, and countless other animals such as fish, rabbits, and snakes.

During one of his recent canoe trips, therapist Richard Voss highlighted the resonance he felt between the inner and outer wilderness, with this poem entitled "Two Wolves":

> In the silent night
> I could hear the distant lonely howl of a wolf.
> Something howls within my loneliness, and the wolf within
> listens attentively to the wolf without.
> Something is remembered and restored
> within my earthen-body-soul-being.[30]

The healing energy many people experience in bonding with animals is illustrated by a marriage and family counselor's case report of one client, a thirty-year-old woman. The woman lived in a toxic interpersonal environment in which she felt she *must* respond to the demands of others. With embarrassed shame she told of her "very odd behavior" consisting simply of going to a stable to pet horses. She said that these contacts enabled her "to regroup and restore her sense of calm and worth." With the counselor's accepting response, the woman reflected on the meaning of horse petting for her. She was able to accept and enjoy her need to touch horses and ride them through the wild canyons of southern California. Gradually, in her therapy, she came to see this connecting with nature as providing the spark that ignited her freedom and power to respond or not respond to others' demands on her.[31]

Pets have served a therapeutic role for years in nursing homes, hospital wards, and geriatric programs. In a Missouri nursing home, an "Edenization" program using gardening and pet care, use of prescription drugs was cut by 50 percent. Pets attend to humans unconditionally, regardless of how the latter feel or look. A veterinarian, who is my nephew, tells of a large, docile stray cat who was brought to his hospital. The cat was treated and then given to a local nursing home when the staff asked for a mascot for their Alzheimer's ward. The cat makes his daily rounds of the patients' rooms, a therapy that the management says is very helpful to the residents. A stray dog, brought to the veterinary clinic after being hit by a car, had to have a leg amputated. After his full recovery and adjustment to his "tripod" stature, he was adopted by a medical professional who uses him occasionally as a visitor-therapist in the pediatric ward when young children face loss of a limb.[32]

A UCLA study by Judith M. Stiegel, professor of public health, discovered that people who own pets visit their physicians less frequently than those in their age group without pets.[33] Earlier studies showed that people with pets are both happier and healthier than those without pets. A personal example of this is our daughter's discovery that taking care of her two much-loved cats for several years boosted her ability to take care of human relationships as well as handle sometimes frustrating jobs. Irene Deitch, a psychology professor at the City University of New York, comments on the therapeutic role of pets: "This kind of animal and human bonding has been shown to increase cognitive skills, motor skills and social skills because it enhances self-esteem, a sense of competence, a sense of mastery. It is also a way of meeting a challenge." She reports the surprising finding that even watching fish in a tank has been demonstrated to lower blood pressure in persons with hypertension.[34]

Using pets to enhance ecological bonding in therapy and education need not be limited to people who have positive affinity with animals or consider themselves "naturally good with animals." Such persons, according to an expert in training animals, are usually those with lots of early experience with animals—those who were raised around pets or farm animals, were active in 4-H, rodeos, and so on. These persons often caught their comfortable, caring attitudes from animal-loving parents who kept pets or took their children to petting zoos and pony rides.[35] Persons who wish to overcome fear of animals may be helped to affirm their organic bond with animals by experiencing a gradually closer relationship to gentle, lovable animals. Painful animal phobias often respond in therapy to such systematic desensitization techniques developed by behavioral therapists.

The International Dolphin Watch in Britain has learned that swimming with dolphins can be therapeutic for persons suffering from mental and emotional illnesses, including depression and autism. The Dolphin Research Center in the Florida Keys uses swimming with dolphins to help children with disabilities strengthen their self-esteem while improving their motor skills.

Animal friendships certainly can diminish human loneliness. Pets sometimes seem to be placed in the position of a child, particularly with single or widowed people or childless couples. Psychiatrist Harry Stack Sullivan, pioneer in interpersonal personality development theory, was still practicing at the William Allison White Institute of Psychiatry in New York City when I trained there in the late 1940s. One of Sullivan's therapeutic eccentricities (which trainees were not encouraged to emulate) was keeping several dogs in his consulting room while seeing patients. Apparently Sullivan experienced high anxiety in close human relationships and felt more comfortable when his beloved dogs were also present. Pets give closeness and affection with minimal emotional demands on human recipients. For this reason, lonely people who have difficulty maintaining satisfying, emotionally intimate bonding with humans, may find their loneliness and sense of isolation assuaged by bonding with pets.

On a very personal level, I can identify with this. In my childhood I was painfully shy. High anxiety and low self-esteem caused "Junior" to spend much time alone, having few relationships with peers. My lonely self-isolation was mitigated by Bob, a loving Chesapeake water spaniel who joined our family circle as a puppy when I was about four and a half. Heartwarming memories return as I write. I remember the little ball of brown fur waddling across the carpet by the Christmas tree early in the morning after we opened other presents. Puppy Bob seemed to be propelled by his furiously wagging tail. With characteristic caring, my parents had arranged for this lively family gift a few months after the tragic death of my younger sister Ruth on her first birthday. Bob continued as a loved family member and a healing presence for me throughout my childhood and well into my adolescence.

Beyond the warm mutual loving Bob and I gave each other, I also received another priceless gift. It was the partial healing of my fear of dogs derived from a traumatic experience with a huge police dog a year or so before Bob and I became friends. Many years later, while I was away in college, the sad news came. Bob's life had to be ended because it had become a painful burden to him. The loss of Bob taught me that the death of a loved pet can bring profound and poignant grief, comparable to the death of a human member of a family. It is noteworthy that some veterinarians feel grief when an animal dies, which enables them to respond more empathically to pet owners' grief.

Jungian analyst Lynda Wheelwright Schmidt gives a moving account of her loneliness when she lived as a girl with her grandparents on their huge ranch with no other children nearby. She longed for a pet companion, but her grandmother did not allow pets in the house. Lynda's solution was to put a box of sand on the roof outside the window of her room. In the box she placed a horned toad. She developed a personal relationship with this lizard when she discovered that it would respond when touched.

This analyst's description of positive bonding with animals continues:

"Even now, when I am in open country or forest, I am attuned to animals—birds less so, although as I get older I am discovering my connection with them, too." She now believes that household pets, like humans, benefit from the renewal of being close to nature: "My dog used to bloom in the wilderness; she would change from her behavior in the city stance of being tall and on her toes, ready to cope with cars, people, and other dogs, to something less domestic. In the wilderness she would move differently, closer to the ground, her tail held low and not waving as it did in the city." In the forest the dog led with her nose and was clearly alert to the surroundings:

> Her heritage as a wild animal could work for her here. . . . Her eyes would change; they would become yellower and unrelated to me. Watching her in the wilderness, I found an increasing recognition of how the wilderness affected me. . . . When we came out of the forest, she would be tranquil and content to lie around and sleep. Whatever anxieties and disturbances she had been carrying, as evidenced by restlessness, scratching, or barking, had been washed away by her time in nature. She too had experienced its healing effect.[36]

Animal therapy has been used effectively in the mental health field. The Delta Society, founded in 1977, is an international nonprofit center for education, research, and community service on the interaction of humans, animals, and the environment, as well as animal-assisted therapy. The society publishes a popular journal entitled *People-Animal-Environment* as well as a scientific journal, *Anthrozous*. It also trains veterinarians, human health professionals, teachers, students, and the general public in human-animal bonding.[37]

The transforming power that often results from relating to animals is illustrated by the radical metamorphosis of convicted murderer Robert Strand. Having spent most of his adult life in prison, he was bitter, withdrawn, and difficult to handle. Then one day he discovered a baby sparrow that had fallen from its nest in the courtyard at Alcatraz. He took it to his cell and tried to nurse it back to health. He asked the guard for an orange crate to build a cage. The enlightened guard slipped it into his cell. In what may have been the first time in his life, Strand muttered, "Thank you." He helped other injured birds and read voraciously everything he could obtain about birds. Over time Strand became known as "The Bird Man of Alcatraz," a respected authority on this subject.[33]

The most popular staff member of a church in Atlanta is Buddy, a gray and white tabby cat, who is listed with the staff in Sunday bulletins as "Church Cat, Buddy T" (the T is for *the* cat). Buddy attends worship services regularly, is official greeter, and fosters relationships among visitors and church members. One time he befriended a mute child who had been severely traumatized by abuse. The child's relationship with Buddy became a path out of his isolation. To everyone's amazement, he began to speak, first to Buddy and then to people.[39]

In discussing crucial differences between a society living in ecological harmony and one living in ecological hubris, bioregional pioneer Kirkpatrick Sale describes what could be seen as an objective of both ecoeducation and ecotherapy. He declares that such a society must have "an ecological consciousness that reminds us that we are only one animal among many, and one species among millions, that continually seeks the balance between the human animal and human arrangement, on the one hand, and the natural world and natural arrangements, on the other."[40]

Enhanced playfulness is an attribute of animals that we humans could well emulate more fully. The young of all animal species, including humans, use play as the major way of learning important living and survival skills. Furthermore, the higher up the scale of animal intelligence a species is, the more it continues playing in adult life. Commenting on animals' playfulness, someone has suggested that perhaps we humans "need to learn to play less desperately and learn more playfully." "Party animals" might be well advised to consider playing more playfully and less compulsively.

Psychiatrist Stuart L. Brown reports on the playful behavior of both young and adult wild animals from the Arctic to Africa. He discovered how remarkably playful animals are, including creatures as diverse as polar bears, elephants, wolves, zebras, leopards, dolphins, cranes, and chimpanzees. He concludes that "exciting studies of the brain, evolution, and ethnology or animal behavior, suggest that play may be as important to life—for us and other animals—as sleeping and dreaming. Play is key to an individual's development and to its social relationships and status."[41]

Brown's study of the absence of play in humans was triggered by the 1966 killing of 13 people and wounding of 31 others by a Texas University student. When he explored the childhood of this Eagle Scout and former altar boy, Brown discovered that he had been abused and controlled so tightly by his father as to make play almost impossible. Brown studied 26 convicted murderers in Texas and found that 90 percent of these men had had largely playless childhoods or played only in destructive ways like bullying and cruelty to animals. All this made him realize "what a powerful, positive force play is."

Finding Spiritual Value in Nature

Spiritual pioneers often find depth renewal in nature. Take Jesus, for example: "But now more than ever the word about Jesus spread abroad; many crowds would gather to hear him and to be cured of their diseases. But he would withdraw to deserted places and pray." (Luke 5:15-16, NRSV). Many ecologically aware people today also find spiritual revitalization in nature. I have been surprised repeatedly by how many students and clients find a spiritual lift when they take time to let themselves be nurtured more deeply and often by nature. It is well to encourage people who are searching for spiritual revitalization to seek it in the beauty and aliveness of nature wherever they live.

Pioneer naturalist, essayist, and poet John Burroughs describes going to the bottom of the Grand Canyon in 1912:

> There is always satisfaction in going to the bottom of things. Then we wanted to get on more intimate terms with the great abyss, to wrestle with it, if need be, and to feel its power, as well as to behold it. It is not best always to dwell upon the rim of things or to look down upon them from afar. The summits are good, but the valleys have their charm, also. . . . It is always worth while to sit or kneel at the feet of grandeur, to look up into the placid faces of the earth gods and feel their power.[42]

Burroughs's reflections awakened in me vivid memories of an unforgettable experience in the Grand Canyon. In 1991, my long-time partner and I spent nearly two weeks on the Colorado River, white-water rafting on an inflatable oar raft. We traversed 183 miles through the Creator's handiwork.[43] Words cannot communicate the power of being intimately bonded to the river and canyon. Excitement was often mingled with a touch of fear as we shot countless rapids—some actually small waterfalls. Burroughs's words come to mind: "We had plucked the flower of safety from the nettle of danger, always an exhilarating enterprise." There also were numerous peaceful times of rowing on placid stretches and camping within a few feet of the river each night. There was a time-shrinking visit to the ancient Native American village ruins along the river, and hikes up several of the beautiful smaller side canyons adorned with fern grottos and waterfalls. Hilarious spice was added by floating on inner tubes down the section of the Little Colorado River just before it enters the Colorado. (Our rears mercifully were partially protected from the rocks in the rapids by our life jackets.) Quiet times around evening campfires often involved learning about the amazing geological story and the historical and botanical highlights of what we had seen that day. We learned about the flora and fauna of the canyon ecosystem and the threats to these from extreme fluctuating water levels dictated by how much water was released to generate hydroelectricity for Los Angeles and Phoenix day by day. Seeing the splendid star-studded sky from our sleeping bags and, at dawn, hearing the haunting melody of canyon wrens were among many inspiring moments. This experience of the divine Spirit's awesome handiwork was an enlivening and spiritually stretching time. When people develop eyes to see and ears to hear both the celebration and the pain of nature, spiritual awakenings are likewise available in commonplace natural settings close to home.

The most loved Christian saint, Francis, expressed a spiritually motivated caring about other animals in these words:

> Praise Thee, O Lord, for Our Mother Earth
> She who sustains us that we might
> Be led to love all creatures great and small
> As they show thy grace.[44]

Exploring the Issues of Chapters 7, 8, and 9 Further

For fuller information about these books, see the annotated bibliography at the end of this book

Douglas A. Anderson, *A New Approach to Family Pastoral Care*

Thomas Craddis, *The Bird Man of Alcatraz*

William S. Ellis, "The Gift of Gardening," *National Geographic*, May 1992, pp. 52–81

Diane Mariechild, *Mother Wit: A Feminist Guide to Psychic Development*

Carol Olwell, *Gardening from the Heart: Why Gardeners Garden*

Theodore Roszak, et al., eds., *Ecopsychology*, Part Two: "Ecopsychology in Practice"

John Seed, et al., *Thinking Like a Mountain*

Lesley Irene Shore, *Tending Inner Gardens*

Sara Stein, *Noah's Garden: Restoring the Ecology of Our Back Yard*

Barbara G. Walker, *Women's Rituals, A Source Book*

Terry Tempest Williams, "The Wild Card, A Woman's Call to Community, A Poet's Vision of Environmentalism Reinvented," in *Wilderness*, Summer 1993.

Notes

1. Quoted by William L. Stidger, *Flames of Faith* (New York: Abingdon Press, 1922), 126.

2. New York: Harrington Park Press, 1995, xi. This is a key book by a feminist ecotherapist.

3. Highly controlled, "rational" people often have difficulty in putting their rational left brain into neutral long enough to avoid careful planning of what they will say or draw. It may free the right-brain functions involved in spontaneous projection to suggest: "Get in touch with your feelings about this picture and just doodle quickly whatever your fingers want to do." Or it may help to suggest using their nondominant hand to express quickly whatever they are feeling.

4. "Miscellaneous Sonnet XXXIII," *The Poetical Works of Wordsworth* (New York: Thomas Y. Crowell & Co., n.d. [a very old edition]), "World" here, in contrast to its meaning in Millay's poem, apparently refers to the urban, commercial, industrialized arena.

5. This poem is entitled "The Defilers."

6. Suggestions to "write a poem" often arouse anxieties such as "I can't write poetry," or "It won't be very good." To reduce such blocking feelings, it is important to say:

"This is not an exercise in literary composition but simply a means of personal expression for you. Whatever you choose to write, in whatever form, will serve this purpose. Don't worry whether it rhymes or sounds like a poem. Nobody is going to evaluate it. Just let it flow and see what happens."

7. I am indebted to my musically knowledgeable son Don Clinebell for this information.

8. Quoted in a flyer from *The Progressive*, a journal of social concerns, inviting subscriptions. No date was given.

9. It is inaccurate to associate rituals only with highly liturgical religions or with some off-the-wall "California" belief system.

10. A wealth of earth rituals is available in Barbara G. Walker's *Women's Rituals: A Source Book* (San Francisco: Harper and Row, 1991). Particularly relevant to ecotherapy are her "Honoring the Earth Ritual" (p. 122ff.); "Changing the World Ritual" (p. 88); and her discussion of "Using Plants in Rituals" (p. 37ff.).

11. Moreen Kamiki Bratt, "An Analysis of the Ecofeminist Question: Does the Minoan Central Courtyard or Plaza, Serve as a Model for the Postmodern Plaza and Provide a Mirror into the Possible Nurturing City?" (Pomona: California State Polytechnic University, 1993), master's thesis.

12. "Sing to the River, Tell Stories to the Wind: An Interview with Dolores LaChapelle," by Jonathan White, *Shaman's Drum*, Summer 1994, 39. It is noteworthy that LaChapelle makes abundant use of sacred nature rituals of native peoples in the training experiences she leads. These include seasonal rituals, food rituals, dancing, chanting, drumming, rattles, singing, fasting, and passing around the "talking staff." She observes: "The physiological effect of these 'ways' is complex, but essentially what happens is they help to supersaturate the left brain so that the right brain is able to fully function. Bonding develops out of this tuning, and bonding is the real basis of all society, both human and nonhuman" (ibid.).

13. New York: Harper and Row, 1987. The quote is from the introduction.

14. "Native Teachings, Spirituality with Power," *Creation*, January–February 1993, 10–12.

15. San Francisco: Harper San Francisco, 1990.

16. Thomas Berry, "The Wild and the Sacred," unpublished paper delivered to an artists group in New York City, October 1993, 1.

17. Berkeley, Calif.: Parallax Press, 1992, 3–4.

18. Much of this information about horticultural therapy is from the brochure of the Philadelphia Friends Hospital. Today, on its one-hundred-acre campus in northeast Philadelphia, it has three registered horticultural therapists who are available to work with patients, ranging from adolescents to the elderly. About a third of all patients participate. For those who are unable to get to the greenhouse, a horticultural therapy room offers plant carts and materials for plant crafts. Patients care for numerous gardens, including a cut flower bed, an herb garden, an everlasting garden, a peony bed, a rose garden, and a shade garden.

19. *MacNeil/ Lehrer News Hour*, 16 December 1973, WNET, New York. See also Catherine Sneed Marcum, "Gardening Behind Bars," in Carol Otwell, *Gardening from the Heart: Why Gardeners Garden* (Antelope Island Press, 1990), 155ff.

20. In 1949, Percy and Ella Heron began what became a remarkable use of urban gardening. In a tough neighborhood in the Queens borough of New York City where they lived, they got permission from the city government to clean up a garbage-covered empty lot and convert it into a community garden. They planted a variety of flowers

and vegetables with the help of neighbors, including numerous children. In the process they taught about the soil and water and caring for growing things. New York City now has a Department Green Thumb that sponsors some 700 community gardens throughout the city. (See "Earth Day, 25 Years," *National Geographic*, April 1995, 138.)

21. A Maryland psychiatrist, for example, has planted shrubs, fruit trees, and "thousands" of plants from bloodroot to sea lavender into his half-acre yard. (He comments that some of his neighbors may think "it looks like Dr. Seuss has gone mad.") Instead of adhering to constricting, anti-ecological practices of formal landscaping, biodiverse gardeners remove lawns and "imported" plants that require lots of water, chemical fertilizers, grooming, and mowing. They replace these with a wide variety of plants indigenous to their region. The native plants produce flowers, seeds, berries, and shelter that attract native wildlife ranging from birds, butterflies, and small animals to earthworms and the fungi and bacteria that produce nutritious soil for the plants. One biodiverse gardener reports that his forty-two-inch lily and fish pond occasionally attracts more biodiversity than he bargained for. Recently, a heron dropped in and helped himself to some of the goldfish for lunch. (We have several raccoons who occasionally enjoy a midnight snack of some of the fish we keep in our tiny pond.)

22. Between 1800 and 1950, 90 native plant species became extinct. It is estimated that by the end of the twentieth century, another 475 species will have been lost forever. Sara Stein's *Noah's Garden: Restoring the Ecology of Our Back Yard* is a "how to" guide to biodiverse gardening, based on her own experience (New York: Houghton Mifflin, 1993).

23. "The All-Consuming Self," in *Ecopsychology: Restoring the Earth, Healing the Mind* (San Francisco: Sierra Club Books, 1995), 91.

24. "The Way of Wilderness," in *Ecopsychology*, 183ff.

25. "The Wilderness Effect and Ecopsychology," ibid., 128.

26. All the quotations about Richard Voss's program, including the poems, are from personal communication and his program's brochure.

27. These studies using standardized psychometric measurement tools, include March, Richards, and Barnes, "Multidimensional Self-Concepts: The Effect of Participation in an Outward Bound Program," *Journal of Personality and Social Psychology* 50, no. 1 (1986): 195–204. See also March and Richards, "A Test of Bipolar and Androgyny Perspectives on Masculinity and Femininity: The Effect of Participation in Vision Quest," a program for troubled urban adolescents that has been operating for some fifteen years. It uses Native American healing, empowering rituals including the sweat lodge and the sun dance as a part of living in Indian-type shelters in the woods.

28. D. Gwynn Davis, Jr., "The Effects of Appalachian Trail Hikes on Anxiety in Ministers," unpublished paper.

29. "When Psyche Meets Gaia," in *Ecopsychology*, 1.

30. This moving poem is a reminder of the unrealistic fears and public resistance that were stirred by the program to reintroduce wolves into Yellowstone National Park. The program's purpose was to restore some of the balance of that biosphere that had been lost when these beneficial predators were systematically wiped out by a misguided government bounty system.

31. Huntley Lewis, MFCC, is the therapist. See also "Animal-Assisted Therapy with Adolescents in a Psychiatric Facility," by Jacquelyn K. Banman, *The Journal of Pastoral Care*, 49, no. 3 (Fall 1995): 274–78.

32. My thanks to Jeff Clinebell for sharing these striking illustrations. On a per-

sonal level, he says that, in his early years, learning to ride horses and the accomplishment of success in horse shows gave him a sense of confidence that extended to other areas of his life.

33. "Study: Pet a Day Keeps the Doctor Away," *Santa Barbara News Press*, 2 August 1990, A19.

34. Jessica Baldwin, "Dolphin Therapy May Help Boy to Talk," *Santa Barbara News Press*, 8 January 1992, B3.

35. Gary Wilkes of the Behavior Modification and Training Services in Mesa, Arizona, is the animal trainer.

36. *The Long Shore*, 196, 200. It is noteworthy that humans have long sensed their affinity with other animals. Myths of half-human, half-animal creatures exist in many ancient cultures. One anthropologist holds that these may represent pre-Darwinian transitional forms, the "missing links." Stories of mermaids may reflect not only the loneliness of sailors long at sea, but also the fear of women's power.

37. The Delta Society's services include working with hospitals, nursing homes, and prisons to begin Pet Partner programs. It works with communities that wish to develop animal companionship and therapy programs that are safe and effective. For more information write the Delta Society, P.O. Box 1080, Renton, WA 98057.

38. Near where I live, inmates at a Lompoc, California, penitentiary cope with the depersonalization and monotony of life in a maximum-security prison by relating to wild and feral animals that live on the prison grounds. From their cell windows and the recreation yard, the prisoners regularly feed raccoons, birds, including owls and sea gulls, and feral cats. A convicted bank robber said, "It's like having a pet and being in the real world, not in prison." He enjoys it when the raccoons play, especially when one even turns on hose faucets. *Santa Barbara News Press*, 13 April 1994, 1A.

39. *The United Methodist Review*, 10 September 1993, 10.

40. *Human Scale* (New York: Coward, McCann & Geoghegan, 1980) 147.

41. Brown's essay, "Animals at Play," is illustrated with photos of playing animals, *National Geographic*, December 1994, 2–35.

42. *The Complete Writings of John Burroughs*, vol. 15, *Time and Change* (New York: Wm. H. Wise & Co., 1924), 66, 69.

43. As we flew toward the departure point in a small plane, I felt anger and sadness at the brown haze of air pollution being emitted only sixteen miles from the canyon by a large, coal-burning electrical generating plant. The good news is that the U.S. Congress subsequently required the Environmental Protection Agency to force this plant to install air scrubbers to reduce the pollutants. (Several months after our trip, we were deeply saddened by news that the lead pilot, Karen Kazan, a dynamic young woman whose love for the canyon was wonderfully contagious, had lost her life in a tragic auto accident.)

44. Quoted by Diana Butler, "Honoring All God's Creatures," *Santa Barbara News Press*, 8 October 1995, D1.

10

Ecoeducation: Teachers, Counselors, Therapists, and Parents Teach Earth-Literacy and Earth-Caring

Care for the earth. It was loaned to you by your children.
—Kenyan proverb shared by Peter Rukungah

Ask the animals, and they will teach you: the birds of the air, and they will tell you; ask the plants of the earth and they will teach you.
—Job 12:7-8

Resonance with the life forces of the natural world challenges us to nurture these same sensitivities. Extraordinary meditations as these on the ordinary rhythms and mysteries of nature are needed to reawaken our sense of intimate connection to life in all its forms. To be able to see deeply into the myriad patterns of life on earth will give us, our children, and our students a vitalizing unity with both one's own bioregion and the larger unfolding process of the universe story.
—Mary Evelyn Tucker, *Education and Ecology* [1]

Eighty-six thousand American schoolchildren were asked how they would advise newly elected President Bill Clinton on goals for his presidency. By far, their top priority goal was cleaning up the environment. This is the issue that children believe most threatens their future when their parents can no longer protect them. A fifth-grader named Dana Trudeau in the Battle Creek, Michigan, public schools wrote the president-elect: "The Earth is the only place that humans can live on. Feel the earth's pain." Adam Kudamik of Coconut Valley Elementary School in Friendsville, Pennsylvania, declared: "We fifth-graders in Mrs. Barnhart's class think that cleaning the earth is most important. . . . Come on! We can do it!" A third-grade child in Lowell, North Carolina, wrote: "I would like you to try to save the endangered animals of our world. I don't want to lose any more."[2] Children in many other countries share a passionate concern for saving a healthy environment.

What is ecoeducation? *It is any learning experience that increases earth-literacy and earth-caring, rooted in earth-bonding.* This dimension of education is the primary way of preventing earth-alienation, the fundamental cause of destructive ecological lifestyles. In our world, when the future of a viable planet hangs in the balance, it is imperative that ecological awareness become a vital part of education on all levels, in all settings, and in all countries. Lester Brown, director of the Worldwatch Institute describes the crisis well: "Ours is the first generation faced with decisions that will determine whether the Earth our children inherit is habitable."[3]

A most important key to tipping the precarious global balance toward ecosanity is equipping and inspiring more and more children, youth, and adults to become knowledgeably earth-caring. In the words of a challenging article in a recent *E* [for Environment] *Magazine*, this is "Education That Cannot Wait." Teachers, parents, counselors, health educators, and religious educators all have strategic roles in providing earth learning opportunities.

This chapter is written for these persons to provide practical resources for enabling learners of all ages to deepen their caring, informed interaction with the natural world. It will outline a model for effective environmental education aimed at enhancing positive bonding with nature. Using this model, those who teach in any setting (including families) can help learners affirm, deepen, and enjoy their connections with the natural world and learn to live in more earth-friendly ways, embodying the full ecological circle. Because therapy and education are understood in these pages as interdependent human processes functioning on one healing-growth continuum, ecotherapy and coeducation overlap at many points.[4]

This chapter will not spell out in detail the contents of ecoeducation. A variety of insightful books that do are listed in the annotated bibliography. Instead, this chapter will focus on methods and resources of ecoeducation, offering ways to enable people to express more fully in their lifestyles all three dimensions of the ecological circle.

Ecoeducational Harbingers of Hope

It is important to reemphasize the importance of keeping ecoeducation energized by reality-based hope. Many ecological developments in education today can bring the energy of hope. Teachers, in my experience, tend to be far better informed and more environmentally involved than any other professional group. More and more elementary and secondary schools and colleges, as well as adult education programs, are offering ecology courses, programs, and field trips in their curricula. Consequently, many children as well as adolescents and young adults seem to have more earth-literacy and concern than their parents. Numerous parents have laughingly "complained" with a hint of parental pride that their offspring return from school and push them to recycle, use less energy and water, and change other earth-damaging practices. The biblical words "A little child shall lead them," come to mind.

At the time political freedom was being reborn in Hungary, I took an unforgettable riverboat trip on the Danube in that country, with a Hungarian friend. The blue Danube then was very unblue. Rather it was a dirty gray-brown, discolored by toxic pollution from several East European countries through which it flows. Remembering this, the following account gives me hope: "At the edge of the Danube River in Budapest, a high school student measures sewage related bacteria, tallies her results and mails them to a classroom in Cleveland. The Cleveland kids examine her numbers, and exchange bacterial counts from their river, the infamous Cuyahoga that twice has caught fire because of the oil in the water."[5]

Around the planet, tens of thousands of teachers invite millions of students to take part in environmental education (EE) learning activities. Countless schools offer popular environmental courses, ecology clubs, and annual Earth Day celebrations. College, university, and adult education ecology courses are proliferating, drawing on interdisciplinary knowledge from the natural and social sciences, economics, and ethics. Some EE proponents build a strong case for prioritizing earth-saving education alongside the three Rs. Rather than just taking students on pleasant nature walks or occasional visits to nature centers, most EE teachers now challenge students to wrestle with the complex causes and cures of the ecojustice crisis.

Community ecoaction projects led by students are common. In one Pennsylvania school, some four hundred students generated an "Earthfest." Their whole-school celebration included food-chain pantomimes and water-cycle dances. Teaching resources are available through a variety of national organizations. Included are the Audubon Society and The National Wildlife Federation, whose motto is "Earth Day Every Day." The largest professional organization is the North American Association for Environmental Education. Their 1990 annual meeting drew members from fourteen countries. Not long ago, the U.S. Congress passed The National Environmental Education Act to provide teachers the tools they need. Several other countries are ahead of the United States in supporting ecoeducation.

Ecoeducation Opportunities for Parents, Teachers, Religious Educators, and Health Educators

Parents, health professionals, teachers, therapists, and religious professionals share the task of enabling those who trust themselves to their caring competence, to develop life-enhancing attitudes, values, commitments, and lifestyles. Ecoeducation must be included in the work that each of these groups does to achieve this crucial objective. Following is an overview of the special opportunities each of these groups has in the overall ecoeducational task.

Parents' Role

Parents, along with grandparents and aunts and uncles, literally have the future at their fingertips, as well as in their hearts. Whether they know it or not, they are major practitioners of either positive or negative ecoeducation. Even more than teachers of children and youth, parents are primary transmitters of earth-caring attitudes and behaviors to the most impressionable among us. As such they have an unequaled opportunity to prevent ecoalienation by fostering both earth literacy and earth-caring lifestyles in children and youth. They do so both by formal teaching of earth literacy and by modeling earth-caring in their attitudes and actions. The ecoeducation objectives, principles, and issues overviewed in this chapter all are relevant to these in-family teachers.

As suggested earlier, parental dreams for their children may be the most powerful motivation anyone has to become more informed and involved in ecological work.[6] Parents and other adult family members should be encouraged to understand that their attitudes and actions today help facilitate or frustrate movement toward the healthy world they desire for their children tomorrow. Parents who become aware of this often are motivated to become more involved in earth-caring causes and in more intentional ecoeducation of their offspring. They have heightened motivation to help their offspring care for the earth lovingly by doing so themselves.

As a parent and grandparent, I see the younger generation as major wellsprings of earth-saving hope. In a similar vein, the editors of *Ecopsychology* hold that "Developmental psychology, blended with the insights of Deep Ecology, may yet offer us the chance to keep our children as sane [in earth matters] as they were when they were born." My hope also stems from a perspective much like that of Paul Shepard, who holds that, in spite of the countless distortions culture piles on each newborn generation, "there is a secret person undamaged in each individual waiting to reconnect with nature."[7]

Health Educators' Role

Professional health educators who direct wellness programs in hospitals, health maintenance organizations, public health agencies, and in some companies have a strategic stake in incorporating ecoeducational content in their preventive and therapeutic teaching. Like other health professionals, they possess the technical expertise required in this field. Furthermore, they often are perceived as persons with authority in our society. This authority can open doors of opportunity to promote ecoeducation and have people listen. More than any other professional group, they have a strategic opportunity to help people reduce home and workplace environmental hazards. They also have opportunities to broaden their roles beyond issues of environmental pollution to include many other facets of ecoeducation.

Teachers' Role

Teachers in all settings have crucial opportunities to help their students to become earth-literate and earth-caring. Most of the teachers who responded to the research survey for this book showed a high degree of both ecological awareness and creative involvement in ecoeducation. Several of their statements were blossoming with insights and ecoeducational innovations. An earth-loving junior high school science teacher, my niece, Debbie Clinebell, wrote: "I enjoy animals and so always keep animals in the classroom for stress relief of troubled children as well as for teaching." She encourages students to become involved in earth-caring activities for this reason: "Without the feeling of empowerment for positive change, students can express feelings of hopelessness and despair. By working with others, they can gain a valuable sense of community. . . . Good habits in earth caretaking must be established as early as possible in a person's life."[8]

A college philosophy professor described how he explores human's relationship with nature in his courses. He also encourages all his students to include enhancing their relations with the environment as they create their "holistic personal philosophy." He also challenges business students to consider, often for the first time, how business has an environmental responsibility.[9]

A former high school teacher and principal (also a relative of mine) now lives in a Midwest farming region where he does hands-on work to help restore islands of prairie grasses whose genetic ancestors flourished for twenty-thousand years in that area and now are in danger of being lost. He also serves as a volunteer environmentalist resource person for schools and scouts, taking young people outdoors to experience and reflect on the aliveness of prairie ecosystems and streams. He encourages his students to become involved in earth-caring. He states, "The buck stops everywhere with everyone. Each person can expand the earth's potential for recovery at any given moment."[10]

Religious Educators' Role

Religious educators in all faith traditions have strategic opportunities to teach earth-caring values and earth-grounded spirituality to children, youth, and adults. Congregations can help their people learn how to generate earth-caring, spiritually energized attitudes and behaviors in themselves and the next generation. This is an unprecedented ecoeducational window of opportunity for all religious institutions. All ecoeducation, in secular as well as religious settings, should help correct the spiritual-value distortions at the roots of the ecojustice crisis. Transforming these is too large and important a task to be left only to those whose professional religious training makes such issues a special focus of their work. Religious professionals should be teachers of teachers, to enable parents and teachers touched by their ministry to understand and deal with these issues constructively. Fortunately, there is an abundance of useful resources for ecoeducation in religious settings, produced by a variety of denominational and ecumenical agencies.[11]

Religious educators have opportunities to help the next generation find the sacred in the everyday things of the earth. Medieval creation mystic Meister Eckhart caught this spirit:

> Apprehend God in all things, for God is in all things.
> Every single creature is full of God and is a book about God.
> Every creature is a word of God.
> If I spent enough time with the tiniest creature—even a caterpillar—
> I would never have to prepare a sermon.
> So full of God is every creature.[12]

Another heartening finding of the survey is the sophisticated ecological literacy of those respondents involved in training future religious leaders. For example, a Roman Catholic theological professor told of the earthy impact of growing up on a farm in Ireland. She articulated a conviction resulting both from this background and from her faith: "I regard it as potentially healing to encourage students to become involved with others to help heal the planet. This is the most basic form of community—one in which all are rooted."[13]

A brilliant seminary psychologist tells of cultivating an oriental garden as "a constant reminder of having brought the mountains to our yard in the city." Illustrating how his close relation with nature in childhood influences his practice of teaching and therapy, he wrote: "I have plants in my office and frequently use nature metaphors—for example, 'a mature plant takes three years to develop an adequate root system after being transplanted.'"[14]

A feminist teacher of pastoral counseling described her own nurturance by nature: "I find myself both enlivened and healed when I let myself be open to the natural world—lakes and forests are my most sacred natural places. Body, earth, and matter are essentially spiritual, and vice versa. The whole must be kept whole." It is noteworthy that she reports that environmental concerns of students surface most frequently in courses on understanding and healing of grief. In doing therapy with depressed, anxious, or angry people she encourages them to "open themselves up to the healing forces in nature." Explaining why she encourages both her students and counselees to become involved in earth-caring actions, she writes, "To be a healthy person in an unhealthy environment, without having access to changing it, is isolating, crazy-making, and ultimately defeating."[15]

The research survey also received encouraging responses from clergy of different denominations who are seriously involved in ecoeducation. One clergywoman has developed a pioneering model of "family cluster education" in congregations. She reports that these learning clusters of several families and single people often develop educational units studying the interrelation of peace, justice, and environmental issues in their families and communities. Their action-learning projects often include activities related to the environment. On a personal level she writes: "I love to ride my bike with the wind blowing through my hair, often singing the same (nature) hymns my mother did. I feel freedom when outside."[16]

A pastor who is deeply committed ecologically serves as a consultant to congregations on greening their programs. He reported leading numerous environmental training workshops for clergy and their congregations. In addition, when he does preparation for marriage sessions with couples, he raises lifestyle issues including earth-caring issues, seeking by this to "plant seeds which challenge their materialistic dreams."[17]

A theological seminary teacher with experience and expertise in urban ministries has developed urban ecology workshops for inner-city congregations. These focus on why Christians should be involved in earth-caring and what they can do to green their inner-city neighborhoods. He utilizes an ingenious teaching motif: a biblical passage in which the fearless prophet Jeremiah spoke as "the messenger of the Lord" to the Hebrew people enslaved in a foreign land during the Babylonian exile: "Build houses and live in them; plant gardens and eat what they produce . . . seek the welfare of the city where I have sent you into exile, and pray to the Lord on its behalf, for in its welfare you will find your welfare" (Jer. 29: 5–7, NRSV).[18]

Guidelines for Effective Ecoeducation

The following guidelines are designed to enhance the effectiveness of ecoeducation:

1. Learning in ecoeducation tends to increase when it implements the model's working principles and six transforming perspectives described in chapter 3. These basic principles and perspectives apply to both ecotherapy and ecoeducation. To the degree that students experience earth education that implements these principles, they will become more earth-literate and earth-caring. And, they will tend to understand their own relationship with the natural world in a unified, biophilic way, and to reject the anthropocentric perspectives so widespread in our society. They will be likely to adopt a more biocentric world view that provides a solid foundation for thinking ecologically about ethical guidelines for their own well-being.

2. Learning in ecoeducation increases when its knowledge base is broadly interdisciplinary, integrating accurate information from a wide range of academic disciplines.

Ecoeducation should draw on relevant knowledge from all the earth sciences and the psychosocial sciences, as well as from the humanities. The latter includes scholarly expertise in ethics, philosophy, history, psychology of religion, and religious belief systems. It is fortunate that many teachers of ecological studies seem well equipped by their training to do this challenging multidiscipline integration.

Using only rigorously accurate facts and identifying probability levels of all predictions is much more effective than what could be called "big-fear education." Scientifically validated evidence and high-probability prognoses are alarming enough. Exaggerating the dangers of the ecology crisis is educationally counterproductive, producing "Chicken-Little" or "The-Boy-Who-Cried-Wolf" responses by thoughtful people.[19]

One of the serious dangers of exaggeration and lack of rigorous attention to accuracy is that these provide ammunition for the anti-ecology backlash. In contrast to the more moderate conservatives who oppose environmental efforts because they do things like "give protection of spotted owls over lumbering jobs," the ecology-bashers include leaders of several radical right-wing religious and/or political movements who make absurdly exaggerated and attacking proclamations on their popular television and radio programs and in their books.[20] The dangerous disservice that these anti-environmental movements perform is to feed the widespread ecological denial in our society by false reassurances based on seriously understating the increasingly well documented perils all of us on the planet face.

3. *The learning and growth produced by ecoeducation are enhanced when students discover and adopt earth-caring values that motivate earth-caring actions and lifestyles.*

Support for making the teaching of earth-sustaining ethics a vital part of education has come from many sources. One of the less-expected is former Soviet leader and Nobel Peace Prize recipient Mikhail Gorbachev. In an interview he discussed his work as president of Green Cross International, an environmental network whose goal is to develop new ways of bringing nations together to protect the environment. When asked what needs to be done, Gorbachev declared:

> We must change our values. We have to educate people. We have to teach citizens and governments to be aware of the dangers to our children's future. . . . Educational systems all over the world must take up this task, international codes of law must be developed, and the practical work of environmental cleanup must be undertaken.

When pressed to explain which values have to change, Gorbachev mentioned educating people to care about the impact of the crisis on everyone, redefining national and international security to include ecological well-being; recognizing the devastating ecological impact of wars and weapons proliferation; and finding "alternatives to the consumerism dream that is attracting the world"; developing new socially oriented economic goals; and challenging the leaders of the world in religion, science, and politics to "speak out and point us in new directions, toward a new [environmental] paradigm for our civilization." He highlighted the plight of the poor with these challenging words:

> We have to care about the unique impact on the poor of the world, who live where toxic sites are most often placed. It's not always so obvious that if it hurts them, it hurts us all. The well being of all the world's poor must be an integral part of our global concerns. Without them, we cannot build the future. We must raise people out of poverty, so they will have a stake in our civilization.

Pointing to the future, Gorbachev declared: "Perhaps it will be easier to care more about each other's children if we also care more about the planet they must live on."[21]

How can teachers encourage people to substitute earth-nurturing attitudes, values, and lifestyles for those that are earth-damaging? Effective ecoeducation should seek to awaken passionate but informed commitment to protecting the health of the biosphere. Implementing this concern is dependent, for most of us in a consumer society, on revisioning and revising certain beliefs and the values that flow from these.

For many students in postsecondary education, the revisioning process must cause them to challenge their own obsessive concern with marketability and "success" defined as high levels of consumerism and consumption. People of any age who are so obsessed should be challenged to examine the impact on their own lives and bioregion of the implicit values that guide their choices of careers and everyday lifestyle choices. Young people should be encouraged to choose careers that have nondestructive impacts on the earth's health. Those with aptitudes for environmental careers may well be encouraged to consider preparing for one of these.

The highest single barrier to students' acquiring increased earth-caring values and behavior is what has been described as the "technological trance" by Thomas Berry. Bucknell University professor Mary Evelyn Tucker declares: "As educators our responsibility is to motivate students to go beyond the technological trance of a consumer society to an understanding of earth literacy."[22] The illusory belief that technology and technological "quick fixes" will solve all human problems, including ecological dilemmas, reinforces denial of the seriousness of the ecocrisis. This denial is related to the widespread belief that assumes that unlimited progress is the ultimate birthright and lifestyle goal of all humans. "Progress" in this context is defined as continuous growth in wealth and consumption, which means depletion of nonrenewable resources with no accounting for the costs of the pollution this involves. In rich countries, "progress" is valued by the vast majority of people as the ultimate economic truth. In poor countries, millions of people long for and struggle to join those in rich countries who have achieved such nonsustainable lifestyles.

The "growth fallacy" is the popular but illusory belief that unlimited economic growth is the way to make the "good life" available worldwide. This fallacy produces the "industrial bubble" that contributes in major ways to environmental deterioration. The industrial bubble eventually will burst as the earth's limited resources are depleted by spiraling populations and overconsumption by affluent persons and societies. Earth literacy must enable learners to re-vision "progress," "growth," and "development" ecologically so that these human aspirations will become sustainable, meaning that basic needs of all people today are met without robbing future generations of the same opportunity.

Challenging the collective beliefs supporting earth-damaging individual and social behavior (including the technological and growth trances) requires courage. People and groups whose profits and lifestyles are threatened often respond by attacking anyone who challenges the widely accepted beliefs and the earth-damaging practices flowing from them. Ecopioneer Rachel Carson's experience is illustrative. Until she was twenty-two, she often wandered over the farmland and woods surrounding the Pennsylvania farmhouse that was her childhood home. Earthy experiences there sensitized her to what she came to call the "intricate fabric of life." She used this phrase in her book *Silent Spring*, the powerful statement that launched the modern environmental movement. In it she documented how the continuing use of DDT and other pesticides would result in a world in which birds would be decimated because their eggshells were becoming too fragile to support the nesting parent birds. Her call for "sanity and restraint in the application of dangerous materials to the environment" was a relatively mild challenge.

Immediately she was hammered by a storm of criticism from industrial leaders whose "bottom line" (profit margin) was threatened by the truth of her assertions. They branded her a "hysterical female," "fanatic defender of the cult of nature," and other pejorative labels. In spite of all this she continued with her head bloodied but unbowed. At the time of her tragic early death from cancer, she had only one fear: that the cause for which she had labored tirelessly would be dropped after her death.

Fortunately, after her death the story took a hopeful turn. A woman named Margaret Owings raised fourteen-thousand dollars to establish a Rachel Carson Memorial Fund, to be administered by the Audubon Society. Shortly thereafter, at a meeting of that society, several people proposed that the money be used to start a fund to defend the environment. The Environmental

Defense Fund (EDF) was soon established by ten scientists and attorneys to use litigation for protecting the environment. The first lawsuit they launched resulted in the federal ban on DDT, followed through the years by numerous court victories for the earth.[23] Biographical vignettes of persons like Rachel Carson, who have made daring contributions to saving the planet, can inspire some students to become more ecologically concerned and involved.

4. *People maximize their functional learning and growth in ecoeducation when the self-motivated learning model is used (often called the adult education model)*. Many seasoned teachers, including the author, have found that this educational philosophy and methodology produces far greater long-term learning, particularly in the area of attitudes and skill development, than straight didactic approaches. Here is a summary of how to apply this model to ecoeducation:

• *Begin by recognizing that people are motivated to learn mainly by what they regard as potentially need-satisfying for themselves and those they love*. Mobilize students' inner motivation to learn by discovering their needs and interests, as well as their present knowledge and feelings about ecological issues. Ecological assessment methods (see chapter 7) such as the ecological wellness checkup or writing one's earth-story may raise interest in learning and give teachers evaluative impressions of where a student is on the path from ecoalienation to earth-bonding.

• *Awaken the inherent curiosity to explore and learn new things that is so strong in young children but often is nearly extinguished in older people by authority-centered education*. The need to explore and discover is a fundamental human need that creative education seeks to reactivate. Lew Welch communicates the ecological awareness of those with awakened curiosity: "Step out onto the Planet. Draw a circle a hundred feet round. Inside the circle are 300 things nobody understands, and, maybe nobody's ever seen. How many can you find?"[24]

• *Balance the stick and the carrot as learning motivators, with much stronger emphasis on the carrot*. In ecoeducation the stick may include motivation to avoid the real perils of the present trajectory of the ecological crisis. The carrot involves experiencing the personal satisfactions of learning to live in more earth-nurtured and earth-caring ways, and hope for more satisfying changes.[25]

• *One of the powerful positive reinforcers of ecological learning and behavior is a hope-inspiring vision of a healthier future, linked with concrete plans for taking small steps toward it*. It is vital for teachers to share with students whatever hopeful, reality-based images and possibilities for a healthier ecological future they can generate with intellectual integrity. Despair tends to extinguish long-term motivation to continue doing whatever is possible in a discouraging ecological situation. In contrast, a positive vision may well lure learners to work with others, taking small steps toward that future together.

The attraction of a hopeful future can inspire students to learn the information and skills required to move toward it.

• *Learning objectives should be developed by dialogue between teachers and students, and teaching plans should take into account learners' interests and needs.* The aim is to generate a joint learning plan or contract incorporating what the teacher regards as essential learnings but also as much as possible of what students perceive as important. This contract guides the instructor in lesson planning by suggesting relative emphases to give various ecological issues. What students need to learn to be earth-literate should be balanced by respect for their perceptions of their own learning needs. In addition to the class learning contract, students should be guided in formulating their personal learning contracts within the parameters of class learning objectives and requirements. In their own contracts, students should spell out what they want to learn and what they will do to accomplish these objectives and also satisfy the course requirement.

• *Use dialogical (rather than monological), participatory, experiential methods, in a cyclical cognition-action-reflection pattern.* Students should be encouraged to bring their present experiences and information about the environment to a class discussion. Thus they may become both learners and, to some degree, teachers of each other. Students (particularly adolescents) often are more open to learning from peers than from adult authority figures. Furthermore, their experiences of nature can enrich the learning by adding insights that are different from the teacher's because they belong to a different generation.

The following learning cycle is often used in this approach. After discovering students' ecological interests and information, teachers offer relevant ecological input via assigned reading, video tapes, mini-lectures, or written handouts. They seek to communicate knowledge in an issue-raising, open-ended challenging style, thus stimulating lots of questions and vigorous discussion. Students are then involved in direct, hands-on experiences of the ecocrisis to illuminate classroom ideas by bringing them down to earth. Following this, students learn from each other and their teacher as they report on their experiences. Then integration of cognitive input with students' experiential learnings is encouraged by inviting reflection on the meaning of their experiences in light of the initial theory. The learning cycle is then repeated focusing on other issues, including those raised by students.

A crucial missing element in didactic, lecture-oriented teaching is experiential learning. This omission diminishes the effectiveness of such teaching greatly. To encourage students to liberate their thinking from the technological and growth trances, it is crucial to involve them in firsthand experiences that bring the ecocrisis alive for them in their everyday experiences. Innovative teacher Theodore Roszak assigns his students this "lesson in applied ecology": He asks them to visit a supermarket, not to buy but to observe carefully. How much on the overstocked shelves is real food required for health and survival? How much is "nothing better than expensively packaged garbage in the making"? How much of what began as healthy food has been processed into nutritionally vacuous junk food? As they watch what people buy, even with food stamps, note how much is plastic and paper goods, alcohol and tobacco, bottles of sugar-water, and junk food.[26] If you are a parent or a teacher, such an experience could produce valuable consciousness-awakening for your children or students. Or suggest that they visit a shopping mall and look at it with eyes made keener by awareness of ecological issues. In that setting the culture's most widespread addiction—consumerism—is blatantly obvious.

Ecopsychologist Sarah Conn describes a powerful learning exercise in self-world, global connections that she uses in workshops and classrooms. She first asks students to brainstorm problems facing the world. They call out the problems at random as she writes them on the board.

> When the blackboard is full (which does not take very long), I ask the students to look at the list in silence and then draw a picture demonstrating what feelings are evoked. Then I suggest that they name their drawing and show it to one other student. This exercise leads to lively, intimate conversations about their pain for the earth. Focusing back on the board, I ask them to tell me which problems are connected to each other. They call out that Third World debt is connected to deforestation, for example, or that AIDS in Africa is connected to illiteracy, which is connected to poverty, and so on. Very soon, the board is full of connecting arrows. Finally, I ask students to pick one problem that he or she feels drawn to work on and notice the arrows attached to it. They can see, by following the arrows around the board, that working on one problem will affect every other problem. The challenge is to find one's niche, one's particular way of contributing to the world.[27]

Involving students in holistic learning means seeking to enrich their thinking, feeling, attitudes, and their relations with other people, society, nature, and the transcending aspect of their lives. The great scientist-philosopher Pierre Teilhard de Chardin once enunciated a truth that shows why ecological learning must be both holistic and experiential. He said, in effect, that knowledge alone never suffices in understanding the world. Rather, people must touch and see it and live in its presence.[28]

Ecoeducation aims at producing what Mary Elizabeth Moore, a pioneer in creative education, calls "teaching from the heart."[29] Moore and her colleague Frank Rogers co-taught an earth-centered graduate seminary course on ecology, theology, and ministry. Students were involved each week in preparing, planting, and caring for a garden. What I describe as the learning circle was used in this course, combining cognitive input by the teachers and students, assigned reading, hands-on experience, and critical reflection on the principles that students were discovering in their total experience. In addition to learning from the heart, this course shows how ecoeducation also involves learning both *from* and *with* the body, by hands-on experiences of the earth. The findings of psychologist Jean Piaget, in his studies of how children learn, confirmed that learning is grounded in the body.

Holistic ecoeducation combines left-brain intellectual wrestling with complex ecological problems, and right-brain methods of creative learning including the abundant use of pictures, imaging, and stories. An example of a moving ecostory is Marlo Morgan's report of her trek to the Australian outback to hear what a sixty-member tribe of aboriginal people desired to say to the rest of the human family. They chose to sacrifice their tribal pattern of keeping to themselves because of their fears about the future for themselves and the whole human family. As Morgan learned, to understand their lifestyle, she had to let go of her white, materialist, individualistic, Euro-American understanding of the good life. She learned from them how a mutually caring community living in intimate and sacred connectedness with the land is being destroyed by the impact of the pathologies of modern society. Their simple but poignant message was expressed by a wise old man who told Morgan that the human family is doomed because we have forgotten how to live in relation to the earth. He said: "We pray that you [the dominant cultures] will see what your way of life is doing to the water, the animals, the air, and to each other. We pray that you will find a solution to your problems without destroying this world."[30] The wisdom of these deeply concerned aboriginals, whose creation story tells how the Divine Oneness sang the world into existence, rings out for all of us to hear.

5. *Learning and growth in ecoeducation are encouraged when teachers affirm students' small steps toward increased ecological literacy and earth-caring attitudes, values, and behaviors.* This principle, confirmed by studies of the psychology of learning, is linked with another basic principle of ecoeducation: Communicating personal interest, caring, and guidance to students as they struggle to learn, helps to develop and maintain a fertile teacher-student relationship. It is important for teachers to affirm students' strengths and capacities to learn. Such a student-teacher relationship is a garden within which learning is most likely to flower.

6. *Ecological learning and growth may be facilitated when teachers occasionally share autobiographical information and use touches of humor and playfulness.* On issues about which teachers hold certain values passionately, it may be clarifying for students to know something about the background experiences that influence their teacher's views and values. It is appropriate for the teachers to share something of their own earth-story and to encourage students to do likewise. It is also important to make clear that there are various perspectives on many ecological issues and that the teacher presents opposing viewpoints fairly and respects students' rights to divergent views.

The more serious ecocrisis issues are, the more essential it is not to be solemn in teaching about them. Lightness can come in noting such things as one's own less-than-perfect environmental behavior and the absurdities of many of the earth-damaging practices in our society. Laughing at oneself and with others enlivens teacher-student relationships, enhances the well-being of both teacher and students, and improves odds that threatening truths will be heard and perhaps learned. It is much easier to look at the dark clouds of our ecological crisis if occasional rays of sunshine shine through. And, as emphasized earlier, a light touch may express an indispensable ingredient in ecoeducation: hope.

7. *It is good strategy with most people to present ecoeducation in the broader context of whole-person health rather than as an isolated issue*. Most people have at least some motivation to enhance their health and that of those they love, or at least to avoid early death. Those who are lukewarm or stone-cold toward ecological issues often become more open to considering them in the context of personal health. (Some folks who yawn when environmental issues are mentioned, actually wake up.) Reframing environmental issues in the wider health context provides a fresh, reality-based perspective on the interdependence of personal, social, and ecological well-being. This reframing should also link peacemaking and nonviolent living with earth-caring as twin essentials for a healthy future for individuals and the earth. Such an approach demonstrates the importance of a holistic, ecological view of life as the best path to a healthy personal future on a healthy planet.

8. *Learning and growth in ecoeducation are facilitated when the process is kept love-centered rather than fear-oriented or only fact-oriented*. As suggested earlier, love of life can be the most sustaining motivator for learning to love the earth. This is the central message of the great Russian writer Fyodor Dostoyevsky's moving words at the beginning of chapter 7.

He points to the way love of all life taps into the deep spiritual love of the divine Spirit, and this fosters increasing understanding of the earth that is so loved. Opening up to this love can happen in secular settings by using the language of meaning and values rather than what is usually called religious language. Grounding love of life in the love of the universe gives it a special energy for healing ecoalienation.

Some Key Issues to Emphasize in Ecoeducation

Curricula of ecoeducation should include the best current knowledge about the issues that are particularly crucial to the health of the biosphere. Among the many complex problems in this field, the following merit serious attention:

Population Growth

Ecoeducation should emphasize reversing the population explosion that is devastating the biosphere as well as contributing to countless other interrelated global problems, including poverty, violence, disease, and starvation. As I write, some one billion of the 5.5 billion people on the earth suffer from chronic hunger and malnutrition. Farmers around the planet struggle to feed 90 million more mouths each year with 24 billion fewer tons of topsoil and 15 million acres lost to desertification from overgrazing, deforestation, and destructive farming methods. Top priority in ecoeducation must be given to curtailing runaway population growth. Unless this problem is solved, all other efforts to save a livable planet eventually will be in vain.[31]

All children have an inherent right to be wanted. They also deserve the right to have their basic needs satisfied, including opportunities to develop their potential gifts. If a livable earth is to survive, the fierce controversies raging about contraception and abortion must not be allowed to delay universal access to the most effective family planning possible. If religious views make it nearly impossible for this to happen, they must be seen as an earth-damaging betrayal of humankind's most profound responsibility: the well-being of future generations.

Sustainable Economic Growth

Ecoeducation should emphasize *sustainable* economic development in both rich nations and poor. This means development that will provide all persons the basic necessities of life, in ways that do not mortgage their country's future by high pollution and overconsumption of limited natural resources. Such development must use environment-friendly technologies, transportation, agriculture, and renewable energy sources. Sustainable development should include social policies backed by laws that provide economic incentives for individual, organizational, and business practices that protect the biosphere. For example, the Japanese government now requires manufacturers to use a certain percentage of recycled materials in all their products, thus creating a market for these materials and an economic incentive for recycling, and reducing the cost of recycled goods. Such policies complete the cycle of recycling, thus preventing mountains of raw recycled materials from simply accumulating in costly storage.

Ecoeducation must emphasize that sustainable economic development is essential for the long-range health of the environment. This means that such development must be justice-based so that the growing chasm between rich and poor (individuals, groups, and nations) is bridged. A healthy earth is not possible as long as one out of every five people live in dehumanizing poverty while a tiny percentage live in extreme wealth. Furthermore, economic development, even if it is justice-based, cannot be sustainable if it aims at growth defined as continually increasing production and consumption—the goal, unfortunately, of transnational corporations and most governments. Such growth gobbles up limited natural resources while increasing the wealth of the wealthy without lessening the economic oppression of the terribly poor.

Women's Special Role

Ecoeducation must emphasize the special roles of women in earth-caring and their full liberation as essential to interrupting the vicious cycle of population growth, poverty, and pollution. Women, as life-generating mothers and potential mothers, probably have a biologically rooted empathy with life-generating Mother Earth. Their cocreating relation with nature seems to give many women special earth-caring motivation, and perhaps understanding.[32] For them to share their planet-caring insights and values with children is an invaluable educational gift to future generations. As long as child-rearing is socially defined as primarily women's work, many women will have more such opportunities with children. To accomplish its biophilia-nurturing objectives, however, ecoeducation needs to encourage shared mother-father child-rearing. This will enable fathers to become more deeply involved in children's value formation. It is very important that men also share their ecological insights and values with their children, and that couples in two-parent families model earth-caring lifestyles together.

Abundant evidence from many cultures shows that the status of women is the major key to both population control and effective environmental action. Universal access to responsible family planning, as essential as this is, by itself will not reverse our species' deadly trajectory of multiplying ourselves off a livable earth. This can only be accomplished by adding the full liberation and empowerment of women in all nations around the world.

Social activist and actress Jane Fonda spoke powerfully and prophetically to the United Nations during its preparation for the crucial 1994 Conference on Population and Development in Cairo, Egypt. She reported: "In virtually every country that has been studied, raising the level of women's education leads to declining birth rates." She observed that with as few as seven years of formal education, women marry four years later, on the average, have enhanced self-esteem and economic empowerment, and want more for their fewer children. Even a conservative Moslem country like Bangladesh, where women have little education, now has a remarkably effective and growing Women's Health Coalition, offering women's literacy classes, loans for starting income-generating projects, as well as family planning. These centers simultaneously address issues of population, poverty, and women's status. (Fonda could well have added that when more affluent women discover that they can use their gifts and gain personal satisfactions in the public sphere, they also tend to avoid centering their lives in birthing and raising many children.) She closed her UN address with a challenge that should be articulated widely in ecoeducation: "This is the defining decade. Whatever we do or don't do, the Earth will survive, life will survive. It is we who may not be around to participate. We're smarter than that—aren't we?"[33] For the health of the whole living biosphere, let us hope that our collective response as a species will be a ringing "yes!"

The late Petra Kelly, a courageous founder of the Green Party in Germany, issued this challenge to women:

> Woman must lead the efforts for peace-awareness (which includes environmental issues), because only she, I feel, can go back to her womb, her roots, her natural rhythms, her inner search for harmony and peace. . . . Our timidity must end for the earth has no emergency exit. . . . This is also a plea to all women to join those sisters who have already risen up—who helped to shape the ecology revolution. Together we can overthrow all the imposed structures of domination. . . . . The earth has been mistreated, and only by restoring a balance, only by living with the earth, by employing soft energies and soft technologies can we overcome the violence of patriarchy.[34]

Men's Special Role

Ecoeducation also must emphasize the special roles and full liberation of men in the struggle to save a viable planet. Clearly, a healthy planet will not be preserved unless men and women learn from each other and work together on this monumental task. Unfortunately, patriarchal programming causes many men to value domination, conquest, material success, rugged independence, and having many children (especially sons) as prerogatives of males. All of these contribute to damaging the biosphere. Liberation from the oppression of this "male box" involves redefining male strength at its roots and discovering the deeper male empowerment of tenderness, caring, interdependence, and nurturing in relationships with both people and the earth.[35] Richard Voss reports, from his wilderness therapy experiences with men: "I am amazed with how nurturing we (males) can be to one another" when shared experiences in wilderness enable men to let go of their patriarchal programming.[36]

We men need to fulfill our special responsibilities in the ecojustice struggle, responsibilities derived from the fact that many of our gender hold institutional power with which earth-caring changes can be made. It is also clear that pressure by husbands on wives to have too many children is a major contributor to runaway population growth, particularly in developing countries with high infant mortality rates. This male pressure is another expression of violence against women in these cultures.

Women and men have interdependent and complementary power to co-create or to co-destroy. What *liberating* ecoeducation must teach is how both genders can create a partnership of equal opportunities in which men and women bring their special gifts and strengths to earth-caring. This earth-saving partnership model needs to be claimed by young people. They have the most to give to and gain from working for a world with a long-range healthy future, simply because they probably will have longer futures.

Ecological education also needs to be transgenerational. Many seniors have both the motivation and the time to implement their concerns about the health of the world their grandchildren's generation will inherit. Adult environmental courses often enroll numerous people who are in the last fourth of their life expectancy. Furthermore, in our increasingly pluralistic world, earth-caring partnerships also need to be rainbow coalitions, drawing on the special ecological wisdom of various cultures. Because environmental problems cross all human-established boundaries, transnational and multicultural solutions are essential.

Peacemaking as Earth-caring

It is crucial in ecoeducation to teach methods of holistic peacemaking, including nonviolent conflict resolution in families and between ethnic and religious groups, as well as between nations. Long-term, there is no way to save the environment unless the human family eliminates war and the resource-squandering global preparations for war-making. Ecoeducation must emphasize the continuing threats to the biosphere of using the armaments-of-war system as a major instrument of international power plays or conflict domination. This includes the ongoing threats of future nuclear disasters that could make the Chernobyl radioactive nightmare seem like a relatively minor episode.

Nurtured by Nearby Nature

Ecoeducation should teach people how to be nurtured by nature near where they live. In our increasingly urbanized, industrialized world, finding beautiful, nurturing settings in nature is more and more difficult for many people. What is required to be nurtured by nature is earth-bonding, an inner awareness of our deep connectedness with the natural world. People who have this awareness can relate in earth-nurtured ways almost anywhere, even in the inner city. They can do so by relating to a favorite houseplant or pet, a distant cloud, or a single bush or tree outside their window. This bonding with nature often motivates earth-caring actions to clean and green the place where people live and work.

Appreciating the Gifts of the Earth

It is important in ecoeducation to help people become aware that life-sustaining nature is right under their feet, literally. Learning not to take food production for granted but rather to prize and protect the soil, is essential if the future is to provide adequate food for all. A farsighted president, Franklin D. Roosevelt, in a 1937 letter to state governors urging the passage of soil conservation laws, wrote: "The nation that destroys its soil destroys itself."[37]

Because so many methods of modern agriculture involve widespread use of chemical fertilizers and toxic pesticides, farmers often respond defensively when this is presented. But their openness may increase as they see evidence of the serious health hazards these practices pose to their families. Prizing the soil can become more experiential when city dwellers discover the welcome de-stressing that is available in caring for houseplants and gardens.

Harvest ritual feasts, food blessings, and celebrations in many cultures attest to the innate human awareness of the gifts of the earth on which all life depends. A gem of a book that can help people get in touch with the everyday miracle of food is *From the Good Earth: A Celebration of Growing Food Around the World* by Michael Ableman, a farmer-photographer-author who operates a small organic farm near Santa Barbara, California.[38] His book is full of beautiful and moving photographs taken by him of farmers and farming around the world. He points out that the language of food is universal because food is "a silent reminder of our roots and connection to each other and to the earth that provides it." Eating together is an intimate act that simultaneously can nourish our bodies, minds, spirits, and relationships. Eating food begins the miracle in which embodied sunlight in food is transformed into thoughts, feelings, creativity, and life.

Ableman writes with prophetic passion concerning modern agriculture. He observes that it

> may be the root cause of the greatest environmental damage to this planet. . . . The journey of our food from the fields to the plate was a series of assaults: on the land and the people who produce the food; on our bodies, as we consume unquantified amounts of chemical poisons in food disguised by dazzling colors, perfect shapes, or fancy packaging; and on our environment withering under the miracle of techno-chemical agriculture.[39]

It is to be hoped that increased awareness of the mystery of producing and consuming food will awaken motivation to find farming methods that are more earth-caring and sustainable. It may also reduce the unethical wasting of food in affluent countries, where it is estimated that 20 percent of nourishing food is thrown in the garbage.

Valuing the Cycles of Nature

It is important to help people learn to enjoy and value the cycles of nature, allowing these to nurture the needs of their mind-body-spirit for rhythm in their lives. In a world of unpredictable change and uncertainty, it can steady one's life to learn awareness of the dependable cycles of nature. These include day and night, produced by the earth's rotation, and the cycle of the changing seasons, caused by the earth's tipping axis and its annual trip around the sun and the cycles of the moon and constellations across the sky.

Native People's Wisdom as Resource

The ecological wisdom of native peoples can provide valuable earth-caring resources for ecoeducation. Sadly, with the widespread impact of ideological colonialism, economic oppression, alcoholism, and the cultural homogenizing of television, many Fourth World peoples seem to have lost much of their cultures. But those who have retained or recovered their earth-rooted heritages often have earth-reverencing perspectives from which we immigrants to their land can well afford to learn. Paula Gunn Allen's *The Sacred Hoop, Recovering the Feminine in American Indian Traditions*, offers a wealth of such resources.[40] Oglala Sioux lawyer Ed McGaa Eagle Man has authored *Mother Earth Spirituality: Native American Paths to Healing Ourselves and Our World*, a book particularly relevant to ecoeducation.[41]

Understanding Biocentrism and Biophilia

Ecoeducation should enable learners to adopt a life-centered, ecological self-understanding, seeing themselves as one part of the interdependent community of living beings that is the biosphere. This biocentric perspective is in contrast to the dominant anthropocentric (human-centered) understanding of humans' relation with the rest of the living world. Mary Evelyn Tucker summarizes this essential change and the way in which it moves self-perception toward seeing oneself in the context of the whole cosmos:

> A primary premise of true change and education in the future must be appreciation of our intimate relationship to the earth. This is the ultimate matrix for both survival and Sustainability. . . . It will be important to encourage students to move from an *anthropocentric* view of human domination and even stewardship over nature to an *anthropocosmic* one of reciprocity and respect between humans and all life forms on earth.[42]

Ecoeducation should enable people to understand that their own well-being depends fundamentally on the well-being of the whole, interdependent ecosystem. In discussing the urgent need for all professions to reorient their thinking radically using an ecological viability paradigm, the authors of *The Universe Story* declare:

> Education is already late in its revision, but we can expect that it will in the future be extensively altered. Education might well be defined as knowing the story of the universe, of the planet Earth, of life systems, and of consciousness, all as a single story, and recognizing the human role in the story. The primary purpose of education should be to enable individual humans to fulfill their proper role in this larger pattern of meaning.[43]

In other words, education should enable us humans to think of our species as one species among others, the ultimate well-being of any one of which depends on the well-being of them all. Inclusive cosmologies or spiritualities have a crucial role in facilitating movement away from anthropocentric to biocentric identities. The biocentric perspective is the foundation of biophilia, meaning love, respect, and reverence for all living things. Learning to love life in this holistic sense provides the motivation for accepting biocentric self-understandings.

The practical applications of this ecological reformulation of human self-understanding touch all aspects of our lives. One hopeful application is that more architects are designing buildings of all types to encourage the healthful interaction between occupants and the natural environment. Urban planners are redesigning cities with more public transportation parks, green belts, and streams flowing above ground rather than being consigned to sewers. This makes them friendly rather than alien habitats. The plants, birds, and the animals of a particular bioregion can prosper therein. All this helps make cities healthier, more beautiful habitats for our species as well as other species in our shared ecological community. Brian Swimme and Thomas Berry observe that by isolating ourselves from the biosphere, we cut ourselves off from our relationship with Planet Earth and thereby from "the entire natural order that constitutes the larger self of our own being."[44] They are pointing here to what I have called the "lost self."

Learning-by-Doing Options for Ecoeducation

To maximize earth-literacy and personal change, ecoeducation must include learning-by-doing. Al Gore enunciates a compelling principle of living and learning in the ecojustice crisis: "I have come to believe that we must take bold and unequivocal action; we must make the rescue of the environment the central organizing principle for civilization."[45] Children, youth, and adults need to be involved in environmental action as an essential part of their priority-forming or re-forming education. Here, as a sampling of countless options, are a few types of action-learning possibilities available to teachers and parents:

- *Environmental study-for-action groups aimed at enhancing earth-literacy and guiding action can be formed within families, schools, congregations, or communities.* The learning-by-doing agendas should include opportunities for learners to enjoy the mind-body-spirit pleasures of beautiful nature settings, and to face the pain of polluted places. Such study-action groups are ideal contexts in which to help students critique their lifestyles and political beliefs and to revise these to make them more earth-caring. This is an invaluable but often resisted aspect of ecological education. People tend to resist change when it involves losing some of the satisfactions of economically privileged lifestyles.

• *Another action-learning option is to do more of the relatively simple earth-caring things that require little forethought and planning, but have both practical values for the health of both one's community and oneself.* Examples include taking reusable bags for shopping; carpooling; substituting energy-efficient light bulbs and appliances; and installing water-saving plumbing. In our throwaway society, it is crucial to recycle everything possible and, when making purchases, to choose items that contain recycled materials.[46] Such simple practices are satisfying when persons know that they are making small but helpful contributions to a healthier planet.

Ecopsychology educator Elan Shapiro makes a crucial point observing that living in harmony with nature involves reciprocity—giving back to the earth for all that it provides: "Every time we learn how to join together and mend our ties with our own little place called home, we link our souls with the soils that sustain us, and nurture the network that is healing the Earth."[47]

• *One of the most useful as well as satisfying ecological actions is to plant a tree.* Tree planting is a living, growing gift to the future unfolding in the present, as the tree produces increasing quantities of life-giving oxygen for all living animals. Planting a tree (for example, a fruit tree or live Christmas tree), with a simple, family-created ceremony, can be a wonderful family ritual for children and parents. The meaning of such gifts-to-the-future is particularly poignant when older teachers or grandparents plant and help care for a fruit tree with their students or grandchildren. Such actions literally are down-to-earth ways of living out of a transgenerational perspective.

A grief recovery group member shared her experience following the suicide of her adolescent daughter, among the most crushing types of loss any parent can suffer. She told about a deeply appreciated gift from a friend in another state, who phoned on the anniversary of the daughter's death to say that she had planted a tree in loving remembrance of the girl.

I was touched when I learned about another woman who planted a tree in each place she had lived. When I heard about this, I wistfully wished that I had planted trees in the many places where our family has lived through the years. If you are considering planting a tree, this sign my pastor saw in a nursery may be relevant to your decision:

THE BEST TIME TO PLANT A TREE IS 15 YEARS AGO.
THE NEXT BEST TIME IS TODAY!

• *Another important action-learning option is to contribute your vote, as well as your time and money, to help elect political leaders committed to doing everything possible, through political processes, to protect nature's health.* Personal and private eco-peacemaking actions clearly are important but, by themselves, they will never change the laws and social practices that are threatening our global environment. A related learning-action option is to support one or more of the national environmental organizations and their local chapters. The already impressive educational and political clout of such broad-based groups needs to be increased, to offset well-financed anti-environmental organizations and politicians.

The power of political action is illustrated by the Endangered Species Act passed by the U.S. Congress in 1973. My gratitude for its impact is brought down to earth—my small piece of it—by seeing some of the twenty-five species found in Santa Barbara County that have been saved from extinction by its protection. They include the California gray whales, peregrine falcons, brown pelicans, bald eagles, Southern sea otters, Channel Island foxes, and California condors, the largest birds in North America. All this brings to mind the wise words of Aldo Leopold, pioneer ecologist and author of *Sand Country Almanac*:

> The last word in ignorance is the man who says of an animal or plant: "What good is it?" If the land mechanism as a whole is good, then every part of it is good, whether we understand it or not. If the biota, in the course of eons, has built something we like but do not understand, then who but a fool would discard seemingly useless parts? To keep every cog and wheel is the first precaution of intelligent tinkering.[48]

• *An everyday action-learning option is to put your money where your conscience is when you go shopping.* The widespread greening of business and manufacturing is encouraging. More and more companies are discovering that green can be profitable, as well as good for public relations. This movement accelerates when consumers let companies know that they are buying or not buying their products because of the company's record on environmental, peace, and justice issues, or their ecologically friendly or disastrous production and packaging.[49] A related matter is to encourage green accounting by companies, communities, and government agencies.[50]

• *A simple learning action that some families are taking is to include an earth pledge in their home and at family meetings*. Here is the "Pledge of Allegiance to the Family of the Earth" that we have posted on our refrigerator: "I pledge allegiance to the Earth, and to the flora, fauna and human life it supports, one planet, indivisible, with safe air, water and soil, economic justice, equal rights and peace for all."[51]

Ecoeducation in a Violent World

Because environmental problems transcend all boundaries and can only be solved by a transnational approach, ecoeducation must "think globally and act locally" but also think and act nationally and internationally. Although action-for-learning needs to begin in one's own neighborhood and bioregion, it should be based on wider national and global understandings. Violence against the earth near to home is inseparable from the pandemic of violence in our chaotic society and world. Attempts to understand environmental problems from the limited perspective of Western affluence, detached from their global injustice-violence context, is to misunderstand them. Robert D. Kaplan has done an astute analysis of this view. He reaches the conclusion that the natural environment and the world's poor people are primary victims of the social disintegration that is proliferating today in developing countries and in the inner cities of affluent countries. Eventually all people, including the affluent, will suffer from this.[52]

Kaplan lifts up this powerful image by Tad Homer Dixon: "Think of a stretch limo in the potholed streets of New York City, where homeless beggars live. Inside the limo are the air-conditioned post-industrial regions of North America, Europe, the emerging Pacific Rim, and a few other isolated places. . . . Outside the rest of humankind is moving in a completely different direction."[54] He points to the megacities of India, Africa, South America, China, and the chaotic inner cities of all the affluent countries, along with the savage ethnic violence erupting in many places. In his words, these are "ecological time bombs" of surging populations, ethnic conflict, and environmental disasters.

The middle-class environmental concerns of those of us inside the stretch limo are alien issues to those struggling for survival in the rest of the world.

> The intense savagery of the fighting in such diverse settings as Liberia, Bosnia, the Caucasus, and Sri Lanka—to say nothing of what obtains in American inner cities—indicates something very troubling that those of us inside the stretch limo . . . lack the stomach to contemplate. It is this: a large number of people on this planet, to whom the comfort and stability of middle-class life is utterly unknown, find war and a barracks existence a step up rather than a step down . . . in places where the Western Enlightenment has not penetrated and where there is mass poverty, people find liberation in violence.[54]

Kaplan presents a well-documented picture of tragic environmental degradation in West African countries, and elsewhere, growing from political and social anarchy, poverty, disease, and starvation. Rapid deforestation, topsoil erosion and desertification, linked with exploding populations, force tidal waves of African peasants to leave the countryside and move to the squalor of sprawling, polluted shantytowns that now circle the huge cities. Rampant criminal anarchy emerges as the consequence. He shows how runaway malaria, tuberculosis, and AIDS, scarcity of water and other limited resources and massive starvation, spiraling populations and refugee migrations, racial, tribal, and cultural conflicts and unprovoked crime, international drug cartels and the proliferation of private armies, cannot be separated from the destruction of environmental devastation. In all this chaos Kaplan declares: "It is time to understand 'the environment' for what it is: *the* national-security issue of the early twenty-first century."[55] My experiences in a variety of poor countries force me to affirm Kaplan's views.

Ecoeducation must give attention to the dark global realities Kaplan describes. But, at the same time, it must emphasize ways to interrupt this pandemic of violence by fundamental transnational political and economic changes to bridge the widening gap between the haves and the have-nots, the rich and poor, around the world. In this way ecoeducation will balance reality-based confrontations with reality-based hope.

To save a healthy planet, there can be no doubt that our species must have an explosion of innovation and creativity in many areas of life, especially education. Even with our most creative commitment to this goal, the planet's future is problematic. Perhaps you will find comfort if not courage in an image used by American novelist E. L. Doctorow as he reflected on his writing technique: "It's like driving a car at night. You never see further than your headlights, but you can make the whole trip that way."[56] This image points to the challenge—to go ahead day-by-day, as far as our limited light concerning solutions to the environmental crisis will illuminate the road. And at some places along the journey into the uncertain future, leaps of faith may be required to preserve our earth as a healthy place for future generations. As the caption of the line drawings by Kelly Frick Richards at the beginning and end of the book reminds us, *Children Ask the World of Us*.

It is important to remember that humans have always been motivated to bursts of innovative creativity by major crises. Perhaps reality-based anxiety about the fate of the earth, joined with love of that earth, can stimulate our creative juices and strengthen our commitment to learning new ways of earth-caring. As Petra Kelly declared: "If we don't do the impossible, we shall be faced with the unthinkable."[57]

Life and Death, Time, and the Environment

One of the unexpected benefits available in the ecological dimension of education and therapy is the widening of people's time horizons and enriching of their awareness of the mystery of time. This can take place, in my experience, as one's sense of continuity with the present, the past, and the future is deepened via the natural world. It is healthy for people who are obsessively focused on the past or the future to discover the incredible aliveness of nature in this here-and-now moment. Conversely, those living in a present-only perspective can enlarge their time horizons by connecting with the still-living past and, equally important, with the future that they may help shape in small but significant ways.

As I write these words, I am sitting on a small wooden platform that I built for my spouse and grandchildren in the arms of the magnificent California live oak spreading across our lot. The presence of the Chumash Native Americans seems very real to me here. From this place in the foothills, above what is now Santa Barbara, they looked out at the blue Pacific and the Channel Islands in the distance. For many centuries, as this tree grew from a tiny acorn, before the Spanish and then the Americans took their land, they lived and walked in this area. I sense the presence of Indian children laughing as they climbed in what was their tree, and frolicked together near it in the spring fields of purple lupine and golden California poppies. These sisters and brothers in the human family must have loved this tree and the beautiful views of the ocean. Somehow, they are my spiritual companions as I reflect and write. This place, and the large oak trees that grew in a grove here, were integral parts of who they were, just as this oak is becoming a part of who I am. I feel something of the cutting anger and awful sadness they must have suffered as white men robbed them of their place, despoiling their sacred heritage in nature. And I cannot forget the spoilage that continues here and throughout their tribal lands.[58]

If you have had comparable earthy experiences of connecting with the living past and the living future in the living present, you know how this can be both confronting and comforting. But it is vitally important to know that these people from the past, possessing all our human feelings, challenge our lifestyles today—we who now claim the lands that were their sacred places. They challenge us to treat the land with caring respect and protecting love for all who will live in the future on this earth.

It is also disturbing as well as challenging to visualize a place you now love after you have gone through the doors of death. This perspective awakens denial and perhaps dread in many of us, feelings rooted in our existential anxiety and our stubborn resistance to accept our own mortality. But, to the degree that we do envision the future without us, we may grow in our celebration of the living present—full of joys and wonders, in spite of its pain. It is sobering but valuable for me to picture children—my descendants or perhaps yours—playing in this oak tree or its younger offspring, to see others inspired by looking at the mountains and across the city to the sparkling sea and mountainous islands beyond. Such a perspective raises the key question again: How healthy and beautiful will this place be when we pass it on to them?

A Transgenerational Trip

Instructions. The purpose of this ecoeducational guided meditation is to experience one of the six transforming perspectives for viewing the earth and the relation of humans to it. Try it now. (If you do not have children yourself, think of nieces or nephews or other children about whom you care.) Simply read until you come to this slash mark / and then close your eyes while you do what has been described:

Using your imagination, transport yourself three or four generations into the future as a very lively ghost. / Look around and become aware of what kind of world you are in. Is it greener, cleaner, and less violent than today's world, or the opposite? / Eavesdrop now while your great-great-grandchildren discuss the world they have inherited from earlier generations. Listen as they express their thoughts and feelings about the social and natural health of their world. What are they saying about what they have heard concerning the things their ancestors did? Did those actions result in a healthier, more livable world, or the opposite? Stay in touch with how you feel as they express their feelings about all this. / Now pretend that you are no longer a ghost figure. Instead, use the time machine in your imagination to carry on a lively conversation with your great-great-grandchildren. / Listen to their questions and to your answers. Perhaps they will want to know what you yourself did to pass along their world to them. / When you are ready, tell them goodbye and return to the present time and place. Then reflect on what you experienced on this journey into the future of your family. / Jot down things you want to remember from the experience, noting particularly anything you know you need to do as a result of this trip. / If you experienced sadness, regret, or guilt (as people often do), it probably will help to decide how you will use these feelings to motivate making changes in your life or lifestyle. Becoming aware of present inadequacies can open opportunities to make changes that produce feelings of well-being concerning your responsibility to your descendants, including those not yet born. / It probably will be helpful for you to debrief your experience with someone you trust, particularly if it touched you deeply.

Exploring This Chapter's Issues Further

For fuller information about these books, see the annotated bibliography at the end of this book.

Charles Birch and John B. Cobb, Jr., *The Liberation of Life*
Herman E. Daly and John B. Cobb, Jr., *For the Common Good*
Ruth Fletcher, *Teaching Peace: Skills for Living in a Global Society*
Kathleen and James McGinnis, *Parenting for Peace and Justice*
Meadowcreek, *Education for a Sustainable World Curriculum Guide*
Mary Elizabeth Moore, *Learning from the Heart*
David W. Orr, *Ecological Literacy: Education and the Translation to a Postmodern World*
Theodore Roszak, et al. eds., *Ecopsychology* (numerous chapters, including Anita Barrows, "The Ecopsychology of Child Development," pp. 101–10, and Elan Shapiro, "Restoring Habitats, Communities, and Souls," pp. 224–39)
Brian Swimme and Thomas Berry, *The Universe Story*
Mary Evelyn Tucker, *Education and Ecology*

Notes

1. *Education and Ecology: Earth Literacy and the Technological Trance* (Chambersburg, Pa.: Anima Books, 1993), 16.

2. Carol Zimmerman, "What Should You Do Mr. President," *Parade Magazine*, 17 January 1993, p. 8

3. Quoted in a flyer announcing *State of the World 1993*, a *Worldwatch Institute Report on Progress toward a Sustainable Society* (New York: W. W. Norton, 1993), a publication that documents his statement fully.

4. Ecoeducation is one expression of what I describe as educative counseling or therapeutic education, helping processes that overlap the interface between the two disciplines. Educative counseling is a personalized form of counseling reeducation in which the goal is to blend counseling methods and educational resources to help persons handle their issues in a growth-enhancing manner. Therapeutic education is any teaching-learning experience in which unlearning of past faulty learning occurs and is replaced with reality-based knowledge, values, and attitudes.

5. This and other examples that follow are from Mike Weilbacher, "Education That Cannot Wait," *E Magazine*, March/April 1991, 29ff.

6. A valuable guidebook for parents and teachers is *Parenting for Peace and Justice, Ten Years Later* by Kathleen and James McGinnis (Maryknoll, N.Y.: Orbis Books, 1990). I wish I had read this book when our children were small. Various chapters focus on raising nonviolent, nonsexist, nonracist children in a violent, sexist, racist society; how to involve children in constructive social action by parents; and how to find spiritual resources for doing this valuable but demanding parenting. The McGinnises are founders of the Parenting for Peace and Justice Network, which has numerous chapters in North America. Other valuable resources for use by parents to help them raise earth-caring, peacemaking children are listed at the end of this chapter.

7. See Anita Barrows, "The Ecopsychology of Child Development," in *Ecopsychology: Restoring the Earth, Healing the Mind*, edited by Theodore Roszak, Mary E. Gomes, and Allen D. Kanner, 101.

8. Survey response from Debbie Clinebell, April 1993.

9. Survey response from Joe Brownrigg, November 1992.

10. Survey response from Gary Clinebell, January 1993.

11. An outstanding packet produced by the National Council of Churches of Christ in the USA is entitled "God's Earth Our Home, A Resource for Congregational Study and Action on Environmental and Economic Justice." Perhaps you are wondering, as a religious educator did at a New York City workshop recently, how poor children in public housing projects can be helped to feel connected caringly with the earth. The earth-literacy of teachers there must be expressed in innovative, hands-on learning projects. These might involve children in caring for plants and animals in the classroom, planting and caring for flowers or vegetables in gardens on nearby vacant lots, or in pots in their project homes. If parents can be involved in such projects, the children can be coached in earth-caring practices, in spite of their surroundings.

12. *Earth Prayers from Around the World*, edited by Elizabeth Roberts and Elias Amidon (San Francisco: Harper San Francisco, 1991), 251.

13. Survey response from Sister Bridget Clare McKeever, June 1992.

14. Survey response from Jim Ashbrook, October 1992.

15. Survey response from Christie Conzad Neuger, February 1993.

16. Survey response from Margaret Sawin, October 1993.

17. Survey response from Peter Moore-Kochlacs, December 1992. He has written and self-published an excellent booklet, *Caring for God's Creation, A Collection of Environmentally Focused Resource Material and Sermons* (Reseda, Calif.: Environmental Ministries, 1993).

18. Personal communication from Michael Mata.

19. It is important for students to learn healthy skepticism so as not to be misled by misinformation, including that which emanates from high places. To illustrate, Ronald Reagan declared convincingly on one occasion: "Approximately 80 percent of our air pollution stems from hydrocarbons released by vegetation, so let's not go overboard in setting and enforcing tough emission standards for man-made sources." He later "corrected" his assertion by raising 80 percent to 91 percent, thus making his statement even more inaccurate. See Mark Green and Gail MacColl, eds. *From Reagan's Reign of Error* (New York.: Pantheon Books, 1987).

20. A book by a conservative journalist is Ronald Bailey, *Eco-Scam, The False Prophets of Ecological Apocalypse* (New York.: St. Martin's Press, 1993). He uses some exaggerated prophecies of ecodoom, made by several knowledgeable, enthusiastic, and otherwise constructive environmentalists, to falsely reassure readers that things are not really as bad as some scientists say. Some of the statements Bailey debunks are in areas in which cautious scientists give only probabilities of something (global warming, for example) occurring. When such probabilities are transformed into predictions of catastrophe by strongly motivated environmentalists, they become vulnerable to being used by opponents of needed environmental protections.

21. Colin Greer, "The Well-Being of the World is at Stake," *Parade Magazine*, 23 January 1994, 4–6.

22. *Education and Ecology*, 1.

23. Andrew Revkin, "Passing the Baton: Rachel Carson and the Start of EDF," *EDF Letter*, May 1994, 7. A sad contemporary update is provided by a recent study by EDF scientists. They found that agricultural biotechnology has continued to entrench rather than reduce the use of toxic chemical pesticides in developed countries and even more so in developing countries.

24. *Earth Prayers from Around the World*, 389.

25. Studies of educational outcomes have shown that carrots (the positive reinforcement of rewards and satisfactions) generally are more effective change motivators than sticks (the negative reinforcement of punishment, threats, and dangers). Self-motivated learning is inherently rewarding because it satisfies students' needs as they understand these.

26. *The Voice of the Earth* (New York: Simon & Schuster, 1992), 23–24.

27. "When the Earth Hurts, Who Responds?" in *Ecopsychology*, 169–70.

28. Teilhard de Chardin's wisdom continues to inspire and inform many earth-caring people today.

29. See Moore, *Teaching from the Heart* (Minneapolis: Fortress Press, 1991).

30. Marlo Morgan, *Mutant Message Down Under* (New York: Harper Collins, 1994), 148. The author reported on her transforming experience in the form of a novel to protect and respect her aboriginal teachers.

31. The planet's population has tripled in the twentieth century.

32. It is my understanding that this earth relationship is an invaluable asset possessed by women in this time of ecocrisis. Their special relation with the earth must

not be misused to diminish the status of women by ignoring their intellectual and creative potentials, which are as great as men's. Their earth empathy and connectedness do not downgrade them to earth mothers, as has been done in the past by some males including Saint Augustine. Women are marvelous co-creators of new human beings in the earthy process of pregnancy and birthing, a biological process in which men have an essential but relatively minor role.

33. Jane Fonda, "High Time for Some Population Intelligence," *E Magazine*, January–February 1994, 22–24.

34. Petra Kelly, *Fighting for Hope* (Boston: South End Press, 1984), 104, 106, 107.

35. Karen Horney, pioneer pre-feminist psychoanalyst, points to the likelihood that the male creativity drive, as an expression of the need for power, is over compensatory: "Is not the tremendous strength in men of the impulse to creative work in every field precisely due to their feeling of playing a relatively small part in the creation of living things, which impels them to an overcompensation in achievement?" (New York: W. W. Norton, 1967). "The Flight from Womanhood," in *Feminine Psychology*, 1926. The exaggerated male drives to dominate others also may stem from over-compensatory needs.

36. Personal communication from Richard Voss. He points out that males are far ahead of females in rates of homicide, suicide, and other crimes of violence against people and nature. But the one-sided, disproportionate blame men receive is unfair and does not make for either change in males or cross-gender collaboration on society's ecological and social issues.

37. Robert Andrews, *The Columbia Dictionary of Quotations* (New York: Columbia University Press, 1993), 261.

38. New York: Harry N. Abrams, 1993.

39. Ibid., 12.

40. Boston: Beacon Press, 1986.

41. San Francisco: Harper San Francisco, 1990.

42. Education and Ecology, 8–9.

43. Brian Swimme and Thomas Berry, *The Universe Story: From the Primordial Flaring Forth to the Ecozoic Era* (San Francisco: Harper San Francisco, 1992), 256.

44. Ibid., 259.

45. *Earth in the Balance: Ecology and the Human Spirit* (New York: Penguin Books, 1992), 269.

46. It is estimated that some 700,000 tons of garbage are produced each day in America. This represents a tragic waste of natural resources and energy while landfills overflow. Plastic floating in the earth's oceans is estimated to average 50,000 pieces for every square mile. Our species has even cluttered outerspace with orbiting garbage and debris. Many of the planet's precious rivers are used as sewers for industrial wastes and agricultural pesticides.

47. "Restoring Habitats, Communities, and Souls," in *Ecopsychology*, 239.

48. Mike Chase, "Back from Extinction," *The Independent*, 6 July 1995, 25.

49. A handy eco-shopping guide is a little booklet entitled *Shopping for a Better World, A Quick and Easy Guide to Socially Responsible Supermarket Shopping*. It is updated and republished annually by the Council for Economic Priorities (30 Irving Place, New York, NY 10003; 800-822-6435). This council also produces a guide designed for adolescents' shopping interests and a guide evaluating the justice, peace, and ecology practices of major corporations.

50. This means including the often-ignored cost of depleted resources and of preventing and cleaning up pollution, in figuring the real cost of goods and services, especially transportation. Facing these accurate, higher costs could motivate us to reduce our overconsumption and push for earth-friendly technologies.

51. Women's Foreign Policy Council, New York, NY 10010.

52. "The Coming Anarchy," *Atlantic Monthly*, February 1994.

53. Ibid., 60.

54. Ibid., 72.

55. Ibid., 58.

56. Robert Andrews, *Columbia Dictionary of Quotations*, p. 193.

57. Quoted in *Vanity Fair*, January 1993.

58. Our beloved oak may have been a tiny seedling when the Franciscan mission fathers pressured the Chumash to move close to the mission to "save their souls" and provide free labor for building the mission and its water system in Mission Canyon. This led to the early death of countless children, women, and men. What a tragic contradiction of the spirit of Saint Francis!

Postlude: Some Parting Perspectives

Two Princeton University evolutionary biologists have for twenty years been studying the particular species of finches on the Galapagos Islands that inspired Charles Darwin's reflections on the origin of species. They have observed several generations of the birds and have actually witnessed evolving changes in their beaks and DNA as the finches have adapted to changes in that natural setting. Reflecting on the meaning of their findings, the two scientists point out that "evolution" originally meant the unfolding of a scroll. Darwin understood this process of continuing creation as the scroll being written on continually as it unfolds. They conclude with these words: "We are not complete as we stand. . . . There can be no finished form for us or for anything else alive, anything that travels from generation to generation. The Book of Life is still being written. The end of the story is not predestined. Our evolution continues."[1] The global ecojustice crisis makes clear that the human species now has unprecedented power and therefore unparalleled opportunities to influence this global process in directions that will make for either a sicker or a healthier earth home for all living creatures.

As I am finishing this book, I am teaching in Nashville. The American South is very different culturally and environmentally from the places I have lived in Illinois, New York, and California. Two people come to mind as I reflect on being in this bioregion. I think of a cherished friend and pioneer feminist educator, Nelle Morton. For many years she thought and taught in innovative ways, focusing particularly on empowering women and those suffering from the power-deprivation caused by white racism. She courageously challenged the patriarchal educational philosophy and methods that are at the roots of many forms of social alienation and violence—approaches that unfortunately still shape much education. Being in Tennessee brings an unexpected sense of quiet sadness as I remember that she is no longer on

Planet Earth, except as her creative spirit lives on in so many who knew and loved her.

Nelle's earthy roots on an east Tennessee farm profoundly influenced her life and career. Here is how she describes this influence in her book of autobiographical essays, *The Journey Is Home:* "I am an Appalachian. That means I am a mountaineer in a literal sense of the word. . . . I spent my early years in Sullivan County on a farm which had been in our family, we thought then, since the beginning of time . . . . The mountains, the winding river, the fields of corn and wheat, the people I knew as a child, will never be out of my memory." She remembers biblical images—"the mountain's a-roaring," "tree planted by the water," "sowing seed," "milk and honey"—as common in the mountaineers' everyday speech. "We lived near a cemetery. The whole countryside stopped working when a tolling bell sent out news that a neighbor had died. . . . We were never shielded from the reality or finality of death."[2]

Nelle recalls the rejoicing of her family and neighbors when a major new industry was brought to nearby Kingsport. It was heralded as the economic salvation of the mountain people who moved there to work in the cement plant, attracted by the promise of good housing and wages. In fact, most of them ended up living in an unhealthy urban slum, exploited by low, non-union wages in unhealthy workplaces. The plant's poisoning of the air and water caused sickness and death from environmental pollution to soar. Chronic bronchial disease took the life of Nelle's mother; cancer, probably also pollution related, claimed several other close relatives.

Many years later, Nelle reflected on these painful tragedies and related them to the fears of many scientists that the survival of the biosphere as a healthy human habitat may be measured in decades instead of hundreds of millions of years. She saw but two options:

> First, we can go down with the earth ship . . . or, second we can take a bold stand to effect a global survival of humankind. But that would involve a massive educational venture and a coalition of all who are victims of the patriarchal habit of ruling and the mind-set that Germaine Greer called "sexism of the mind." This mind-set spawns dualisms that split the self, alienate one people from another, and prevents a worldwide common access to knowledge and the reverent uses of the resources of the earth now available."[3]

If Nelle were here, I have no doubt that she would affirm resoundingly the urgent task of integrating ecology and justice in the planet-saving "massive educational venture" for which she called.

Nashville is known as the country music capital of the world, a place where home-grown music flourishes. Folk singer Pete Seeger comes to my mind, though country music is not his forte. Through the years, he has used his guitar and songs to inspire and energize needed social protest. In his history of folk singing, he reminds us of a great American seed planter.[4] John Chapman, known by people on the frontier in the eighteenth century as Johnny Appleseed, was born in 1776 in Massachusetts.

After attending Harvard, Chapman responded to the call of the frontier, joining the pioneers trekking by covered wagon to the wilderness of western Pennsylvania. He soon developed an ingenious scheme to supply the struggling frontier farmers with their first fruit trees. He begged apple seeds from cider mills in the East and for half a century he roamed the frontier exchanging a night's lodging for these much-needed seeds. Deeply religious, he refused to carry a gun in a region where others thought it utterly foolhardy to be unarmed. Becoming an early abolitionist, Chapman helped runaway slaves to use the underground railroad network to Canada. He was friends with the Indians, who taught him their lore of using healing herbs, knowledge that he passed on to the settlers. For the many who were illiterate, he wrote letters. It is no wonder that, as Seeger says, many thought him eccentric but thousands loved him. Certainly we can be grateful that Johnny Appleseed did not allow being thought eccentric to deter him from his seed-planting mission.

Is Johnny Appleseed's lifelong commitment to seed-planting relevant to our pained planet in this very different time? Pete Seeger uses this story to challenge folk singers to use their music for social healing. His challenge can be adapted to today's global ecojustice crisis. It is clear that the unborn future calls out for seed planters all around the planet. Countless environmental Johnny and Jane Appleseeds are needed. They are needed to plant seeds of hope and love for people and the earth, while they plant trees and gardens, and in many other ways make their lifestyles passionately earth caring. The planet also needs ecojustice Johnny and Jane Appleseeds to plant seeds of personal and social change by working to limit population, preserve parks and wilderness, pass national and international laws to protect biodiversity and the environment, work for economic justice, responsible freedom, and sustainability for the whole human family in all lands.

The wise farmer-poet Wendell Berry lifts up a powerful image from a novel by Wallace Stegner, *The Big Rock Candy Mountain*, of a foolish fir tree in a fairy tale that wanted to be cut down and dressed up with lights and ornaments and see the world as a Christmas tree. Berry appropriately applies this image to the environmentally destructive, uncaring ethics, lifestyles, and public policies that are so widespread in our society today.[5] We should all be deeply grateful that so many people know that their deep roots in the earth are the source of their life and health. They are not seduced by short-term temptations to cut their connections with the earth around them. Such people could be likened to living Christmas trees that prize their rootedness and want to continue to bring life-giving oxygen to all of us breathing beings on the earth.

The heart of what we educators, therapists, and parents do is seed planting. We share a commitment to plant seeds of healing and health, of hope and wholeness in the lives of children, youth, and adults through their life journies. An exciting, unprecedented opportunity is at the fingertips of those who practice ecotherapy and ecoeducation: to plant countless seeds of earth-grounded hope and help in the lives of those whose lives we have the privilege of touching. If we plant an abundance of seeds, some will fall on fertile, receptive soil. This seed planting is our opportunity to help tip the teetering earth toward sanity and survival, toward a new day of planetwide flowering of wholeness.

The greening of the people-nurturing arts—counseling, therapy, education, and parenting—is an idea whose time has come. Or at least it needs to come—speedily! As these healing arts literally become more down-to-earth, we have the potential to put the ounces of our collective influence on the side of steering our species toward a new day of passionate earth-caring and earth-loving.

If we join in helping this ecotransformation to occur, we can rejoice with a healthier earth. As we sing the song of the earth *with* the earth, our caring for the earth and its living creatures will be continually renewed *by* the earth. In the process of nurturing nature with loving care, we will have our own earthy self fed bountifully by the good earth. We will sing with the universe and celebrate with all the other creatures in the life-giving network of the biosphere. Many centuries ago, a wise Hebrew prophet described such rejoicing in these earthy images: "*The mountains and the hills before you shall burst into song, and the trees of the field shall clap their hands*" (Isa. 55:12). May we sing with joy the song of the earth's health and of our own. One final thought. Singing your own earth song, in the days and years ahead, probably will be difficult now and again. When this occurs, it may help to practice the wisdom of an ancient Chinese proverb: *Keep a green tree in your heart, and a singing bird may come!* Also try to bear in mind anthropologist Margaret Meade's most often quoted declaration. "Never doubt that a small group of thoughtful, committed citizens can change the world. Indeed, it is the only thing that ever has." It is true!

Notes

1. The two biologists who saw evolution happen are Peter and Rosemary Grant. Jonathan Weiner reports on their remarkable findings in *The Beak of the Finch: A Story of Evolution in Our Time* (New York: Alfred A. Knopf, 1994), 299–300.

2. Boston: Beacon Press, 1985, 184–85.

3. Ibid., 107–8.

4. David King Dinaway, *How Can I Keep Singing: Pete Seeger* (New York: McGraw-Hill, 1981).

5. Berry, "The Obligation to Care", *Sierra*, September/October 1995, 65.

For Your Further Exploration: Books and Earth-Caring Resources

This comprehensive bibliography is designed to encourage reading and research by interested individuals and by persons in academic settings. Double asterisks indicate books with special relevance to ecotherapy's theory or practice. Many valuable books do not have double asterisks. Books without annotation are either those whose contents are clear from the titles or books recommended by others.

Michael Ableman. *From the Good Earth: A Celebration of Growing Food Around the World*. New York.: Harry N. Abrams, 1993. A book by an organic farmer and photographer, illustrated by 176 of his beautiful and moving photos of food-raising in many places. Critiques the destructiveness on both humans and the soil of modern chemical-dependent methods of high-tech farming.

David Abram, *The Spell of the Sensuous, Perception and Language in a More-Than-Human World*. New York: Pantheon Books, 1996. A philosopher and ecologist shows how the most cherished human attributes—our language ability, our senses, rational intellect, and awareness of past and future—all emerge in interaction with the living, natural world and are totally dependent on that world for their coherence.

**Carol J. Adams, ed., *Ecofeminism and the Sacred*. New York: Continuum, 1993. Major thinkers in ecofeminist theory explore many facets of theology, philosophy, and spirituality as these shape earth-destroying or earth-respecting responses by the human family. The final section discusses the urgency and interrelatedness of ecological and other social issues.

Patch Adams and Maureen Mylander. *Gesundheit! Bringing Good Health to You, the Medical System, and Society through Physician Service, Complementary Therapies, Humor, and Joy*. Rochester, Vt.: Healing Arts Press, 1993. Includes a discussion of the healing power of nature.

**Paula Gunn Allen. *The Sacred Hoop: Recovering the Feminine in American Indian Traditions*. Boston: Beacon Press, 1986. A Laguna Pueblo/Sioux feminist writer and literary critic explores Native American cultures, religions, and literature, showing the central place of women in these and in attitudes toward the earth. She thanks the wind and the sky, the trees and the rocks, the sticks and the stars for being in a teaching mood when she needed to learn from them.

Douglas A. Anderson. *A New Approach to Family Pastoral Care*. Philadelphia: Fortress Press, 1980. Right-brain approaches including storytelling.

William Anderson (photography by Clive Hicks). *Green Man: The Archetype of Our Oneness with the Earth*. London and San Francisco: HarperCollins, 1990. Traces the Green Man archetype in folklore and religious art from the earliest times to the present, showing how it is the male parallel of the goddess. Holds that this image can be a resource in renewing lost oneness with a living earth.

The Animal Rights Handbook: Everyday Ways to Save Animals' Lives. Los Angeles: Living Planet Press, 1990. Facts about protecting endangered species, animal abuse, and how everyday choices about eating and dressing can protect animals.

Carl Anthony. "Why African Americans Should be Environmentalists." In Brad Erickson, ed., *A Call to Action*. San Francisco: Sierra Club Books, 1990.

Susan J. Armstrong and Richard G. Botzlerk, eds. *Environmental Ethics: Divergence and Convergence*. New York: McGraw-Hill, 1993. An insightful introduction to ecological ethics.

Richard Cartwright Austin. *Baptized into Wilderness: A Christian Perspective on John Muir*. Abingdon, Va.: Creekside Press, 1991. A pastor and environmental activist gives an engaging portrait of America's first such activist, founder of the Sierra Club, exploring his earthy spiritual roots and motivations.

**Gerald O. Barney. *Global 2000 Revisited. What Shall We Do?* Arlington, Va.: Millennium Institute, 1993. A report on the critical issues of the twenty-first century and the role of religions in their solution, prepared originally for the 1990 Parliament of World's Religions.

Gregory Bateson. *Mind and Nature, a Necessary Unity*. Toronto: Bantam Books, 1979. Shows how the mental system that governs how we think and learn is the same system that governs the evolution of all life on the planet. Holds that we must learn "to think as Nature thinks."

**Thomas Berry. *The Dream of the Earth*. San Francisco: Sierra Club Books, 1988. An ecological theologian sets forth a vision of a sustainable, healthy earth, exploring what it means to be part of an "enchanted" universe that is a living organism. (The fact that the Sierra Club published a theologian's book is of significance.)

**Thomas Berry and Thomas Clark. *Befriending the Earth: A Theology of Reconciliation Between Humans and the Earth*. Mystic, Conn.: Twenty-Third Publications, 1993. After discussing the failure of the Christian religion to contribute in significant ways to resolving the planet's ecological crisis, the authors point to paths by which individual Christians and church leaders can nourish the earth and be recipients of its bounty.

Wendell Berry. *What Are People For?* San Francisco: North Point Press, 1990. *Traveling at Home*. San Francisco: North Point Press, 1989. Prophetic essays and poems by a Kentucky farmer-author describing the importance of protecting the land and of having a "place."

**Charles Birch and John B. Cobb, Jr. *The Liberation of Life: From the Cell to the Community*. Cambridge: Cambridge University Press, 1981. A creative biologist and a down-to-earth theologian collaborate in producing an ecological model for understanding the interrelation of humankind and the natural world, as a basis for social and economic policy.

Peter Bishop. *The Greening of Psychology: The Vegetable World in Myth, Dreams, and Healing*. Dallas: Spring Publications, 1990. A Jungian exploration of the vegetative soul, the green level of the psyche. Holds that when this is neglected, the roots of our lives wither, producing addictions to plant substances (cocaine, coffee, sugar, etc.), social uprootedness, fear of rot, and worship of growth.

Frank Borman and Stephen Kellert, eds. *Ecology, Economics, and Ethics: The Broken Circle*. New Haven: Yale University Press, 1991. Bases enlightened ecoethics on a foundation of economics.

Ian Bradley. *God Is Green: Ecology for Christians*. New York: Doubleday Image Book, 1990. A minister in Scotland explores writings of early Christians, the Celtic Christian church, and mystics through the ages, showing how God is incarnate in nature and Christians should be environmentalists.

Les Brown. *Conservation and Practical Morality: Challenges to Education and Reform*. New York: St. Martin's Press, 1988. An exploration of the ethical challenges of the ecojustice crisis.

**Lester R. Brown, et al. *State of the World, 1996*. New York: W. W. Norton, 1996. The annual Worldwatch Institute valuable report on the "earth's vital signs" and progress (or lack of it) toward a sustainable society.

Noel J. Brown and Pierre Quiblier, (Eds.), *Ethics & Agenda 21, Moral Implications of a Global Consensus*. New York: United Nations Publications, 1994. Interfaith scholars illuminate the nature of the global environmental ethics needed to implement the consensus report of the Earth Summit.

Walter Bruggemann. *Living Toward a Vision: Biblical Reflections on Shalom*. New York: United Church Press, 1976. A leading scholar of the Hebrew Scriptures examines the concept of shalom and its holistic application in today's world.

Robert D. Bullard, ed., *Confronting Environmental Racism: Voices from the Grass Roots*. South End Press, 1993. A sociologist and leader in the urban environmental movement exposes the toxic cost to victims of racism and poverty.

William Cahalan. *The Earth Is Our Real Body: Cultivating Ecological Groundedness in Gestalt Therapy*. Available from the author: 603 Enright Ave., Cincinnati, OH 45205.

**Helen Caldicott. *If You Love This Planet: A Plan to Heal the Earth*. New York: W. W. Norton, 1992. A courageous physician peacemaker describes the scientific and medical consequences of human destruction of the biosphere, and sets forth a healing plan involving loving, learning, living, and legislating.

Fritjof Capra and Charlene Spretnak. *Green Politics: The Global Promise*. New York: E. P. Dutton, 1984. Highlights the hope in the holistic, ecological, and feminist Green movement in Germany and calls for an extension of this vision and movement to help transform the course of America and the world.

Sally Carrighar. *Wild Heritage*. Boston: Houghton Mifflin, 1965. From the perspective of ethnology, the science of animals' normal behavior, Carrighar shows how human behavior in such areas as sex, parenting, aggressiveness, and play is influenced by our animal heritage.

Rachel Carson. *Silent Spring*. Greenwich, Conn.: Fawcett, 1962. A classic and the most powerful influence in launching the modern environmental movement.

Forrest Carter. *The Education of Little Tree*. Albuquerque: University of New Mexico Press, 1976. A delightful fictional account by an Eastern Cherokee of a Cherokee boy growing up with his ecologically aware Native American grandparents.

Howard Clinebell. *Basic Types of Pastoral Care and Counseling*. Nashville: Abingdon Press, 1984. A textbook of holistic, growth-oriented approaches. *Contemporary Growth Therapies, Resources for Actualizing Human Wholeness*. Nashville: Abingdon Press, 1981. An overview of the major growth and spirituality resources of more than twenty contemporary psychotherapies. "Enhancing Your Well Being by Helping Heal a Wounded Planet," chapter 8 in *Well Being. A Personal Plan for Exploring and Enriching the Seven Dimensions of Life, Mind, Body, Spirit, Love, Work, Play, Earth*. San Francisco: Harper San Francisco, 1992. "Greening Pastoral Care," *Journal of Pastoral Care*, 48, no. 4 (Fall 1994), 209ff. A guest editorial describing the relevance of ecotherapy to pastoral care and counseling, and the special contributions that persons trained in these can make to planet-saving. "Healing Persons/ Healing the Earth—Pastoral Care Givers Respond to the Eco-Justice Crisis," chapter 2 in *Pastoral Care and Counseling in Pluralistic Society*, edited by Mesach Krisetya, published by the Fifth Asia Conference on Pastoral Care and Counseling, 1993. "Healing Violence Against Persons by Healing Violence Against the Earth—Pastoral Theologians and Care Givers Respond to the Global Eco-Justice Crisis." Paper presented at the theory-building conference sponsored by International Pastoral Care Network for Social Responsibility in Italy, September 1994.

Howard Clinebell, ed., *Global Peacemaking and Wholeness: Developing Justice-Based Theological, Psychological, and Spiritual Resources*. Claremont, Calif.: Institute for Religion and Wholeness, 1985. Insightful papers from an interprofessional theory-generating conference by the Institute for Religion and Wholeness.

John B. Cobb, Jr. *Is It Too Late? A Theology of Ecology*. Beverly Hills, Calif.: Bruce, 1972. Drawing on Saint Francis and Albert Sweitzer, a multi-discipline process theologian calls for a new Christianity with an ecological love of nature.

Michael J. Cohen, *How Nature Works, Regenerating Kinship With Planet Earth*. Portland: World Peace University, 1987. A psychologist and outdoor educator describes how he developed ecotherapeutic theory and methods that integrates psychology and ecology, and uses thoughtful sensory contacts with nature to catalyze wellness, spirit, and responsibility. *Reconnecting With Nature*. Friday Harbor, WA: Project Nature Connect, 1995. Describes the process for enabling people to reconnect with nature. This process has evolved over many years of teaching outdoors. It involves direct sensory contacts with nature, including nature within the person, integrated with cognitive understanding and validation of these nature feelings and sensory knowledge. =**Sustaining the Common Good, A Christian Perspective on the Global Economy*. Cleveland, OH: The Pilgrim Press, 1994. An insightful challenge of most economists' great god—growth—by a multi-disipline theologian who shows how sustainable development can save both humans and the environment from the present disastrous economic directions.

Ellen Cole, Eve Erdman, and Esther D. Rothblum, eds. *Wilderness Therapy for Women, The Power of Adventure*. New York: Harrington Park Press, 1994. Describes the theory and practice of wilderness therapy for women.

Barry Commoner. *Making Peace with the Planet*. New York: Pantheon Books, 1990.

**Joseph Cornell. *Listening to Nature: How to Deepen Your Awareness of Nature*. Nevada City, Calif.: Dawn Publications, 1987. An inspiring nature educator describes principles and methods enhancing this awareness. Includes beautiful photographs by John Hendrickson. *Sharing Nature with Children*. Nevada City, Calif.: Dawn Publications, 1979. A parents' and teachers' guidebook filled with nature activities and games. *Sharing the Joys of Nature*. Nevada City, Calif.: Dawn Publications, 1989. Imaginative nature activities for persons of all ages.

Edward Cornish, ed. *Global Solutions, Innovative Approaches to World Problems*. Bethesda, Md.: World Futures Society, 1994. Key papers from *The Futurist* journal on a variety of world problems, including the environment.

Ben Corson, et al. *Shopping for a Better World*. New York: Council on Economic Priorities, 1994. A handy guide to responsible supermarket shopping. Evaluates companies and products on the basis of several criteria, including environmental issues, treatment of women and ethnic minorities, and involvement of the company in community issues. CEP also publishes a shopping guide for teens and a guide evaluating the responsibility of their policies and practices.

Norman Cousins. *The Pathology of Power*. New York: W. W. Norton, 1987. An illuminating and impassioned exploration of the continuing threat to freedom and peace in the proliferation and misuse of governmental power and the arms race.

Charles Cummings. *Eco-Spirituality: Toward a Reverent Life*. Mahwah, N.J.: Paulist Press, 1991. A Trappist-Cistercian monk sketches an ecological spirituality and an alternative future built on universal wonder and reverence. Includes models and resources for eco-spiritual living.

Scott Cunningham. *Earth Power: Techniques of Natural Magic*. St. Paul, Minn.: Llewellyn Publications, 1992. Describes folk magic, meaning harmonizing and working with the energies of nature—trees, sun, moon, rivers, sea, and springs—to bring about needed healing and changes.

Debra Lynn Dadd. *The Nontoxic Home*. Los Angeles: Jeremy P. Tarcher, 1986. *Nontoxic, Natural and Earthwise*. Los Angeles: Jeremy P. Tarcher, 1990. How to protect yourself and your family from harmful products and life in harmony with the earth.

**Herman E. Daly and John B. Cobb, Jr. *For the Common Good: Redirecting the Economy toward Community, the Environment, and a Sustainable Future*. Boston: Beacon Press, 1989. A landmark book by a World Bank economist and a leading process theologian and ecological pioneer. After critiquing mainstream economic theories and showing how growth-oriented economic approaches produce earth devastation, the authors describe a new sustainable economics supported by public policy and social ethics—an economics supporting the well-being of both humans and the earth.

Vine Deloria, Jr., *God is Red*, New York: Grosset & Dunlap, Publishers, 1973. The former Executive Director of the National Congress of American Indians shows how only by getting reconnected with the land and its living creatures Indian cosmology, can the Christian religion respond to the religious, societal, and environmental crisis in America.

Bill Devall and George Sessions. *Deep Ecology: Living As If Nature Mattered*. Salt Lake City: Gibbs M. Smith, 1985. Explores the environmental crisis in terms of both personal and social responses, emphasizing awakening to the wisdom of nature and the oneness of humans, animals, plants, and the earth.

Calvin B. DeWitt, ed. *The Environment and the Christian: What Can We Learn from the New Testament*. Grand Rapids, Mich.: Baker, 1991.

**Irene Diamond and Gloria Feman Orenstein, eds. *Reweaving the World: The Emergence of Ecofeminism*. San Francisco: Sierra Club Books, 1990. Includes insightful papers on the well-being of the earth by thirteen theoreticians and activists in the fields of ecofeminism and feminist spirituality.

David King Dinaway. *How Can I Keep Singing: Pete Seeger*. New York: McGraw-Hill, 1981. Biography of Seeger.

Michael Dowd. *Earthspirit, A Handbook for Nurturing an Ecological Christianity*. Mystic, Conn.: Twenty-Third Publications, 1991. A prophetic integration of ecology and spirituality, offering an understanding of the Christian faith that takes into account a unified cosmology to which earth sciences point.

Alan R. Drengson. *Beyond Environmental Crisis: From Technocrat to Planetary Person.* New York: Peter Lang, 1989.

Rene Dubos. *Celebration of Life.* New York: McGraw-Hill, 1981. Challenges human beings to utilize their creativity and intelligence to live whole lives in relation to the earth.

Hans Peter Duerr. *Dreamtime: Concerning the Boundary Between Wilderness and Civilization.* Translated by Felicitas Goodman. New York: Basil Blackwell, 1985.

Alan Durning. *How Much Is Enough? The Consumer Society and the Future of the Earth.* New York: W. W. Norton, 1992.

Lee Durrell. *State of the Ark: An Atlas of Conservation in Action.* London: Gaia Books, 1986. Describes the state of the planet's ark-biosphere, the life-support systems and what can be done by humans to assure their own future.

Ed McGaa Eagle Man. *Mother Earth Spirituality: Native American Paths to Healing Ourselves and Our World.* San Francisco: Harper San Francisco, 1990. An Oglala Sioux shares a wealth of Native American philosophy, history, and rituals designed to enable healing our relations with the earth and all living things.

Earth Island Journal, International Environmental News. Quarterly publication of the Earth Island Institute, San Francisco, updating major ecological developments around the world.

The Earth Works Group. *50 Simple Things Kids Can Do to Save the Earth.* Kansas City: Andrews and McMeel, 1990. A guide for children, parents, and teachers explaining how to release "kid power" to help them keep the earth green, recycle, protect animals, and use energy wisely. *50 Simple Things You Can Do to Save the Earth.* Berkeley, Calif.: Earth Works Press, 1989. Describes ecological actions of varying levels of difficulty for adults and families.

Gretel Ehrlich. *Islands, the Universe, Home.* New York: Viking Penguin, 1991. Ten lyrical, personal explorations on the interrelationship between the human and the natural world.

Paul Ehrlich and Anne Ehrlich, *Extinction: The Causes and Consequences of the Disappearance of Species.* New York: Ballantine, 1981. Illuminates the costs of the destruction of innumerable species of animals. *The Population Explosion.* New York: Simon and Schuster, 1950. Two authorities on this issue describe its implications for humankind.

Riane Tennenhaus Eisler. *The Chalice and the Blade: Our History, Our Future.* San Francisco: Harper and Row, 1987. Traces the ancient myths and cultures in which both women and the earth were respected.

**Loren Eisley. *The Immense Journey.* Alexandria, Va.: Time-Life Books, 1981. A powerful, poetic prose book by the internationally acclaimed anthropologist and historian of science, exploring the mystery and marvels of human-nature interrelationship. *The Unexpected Universe.* New York: Harcourt, Brace, and World, 1969. Explores the wonder and mystery, the desperation and delight of the human soul's quest for meaning in our world. *The Invisible Pyramid.* New York: Charles Scribner's Sons, 1970. Describes the rise of humans as organisms devouring their environment and their first intrusion into outerspace.

Duane Elgin. *Voluntary Simplicity: Toward a Way of Life That Is Outwardly Simple, Inwardly Rich.* New York: William Morrow, 1981. A social policy analyst explores the basic changes in lifestyle that are necessary to save the planet.

John Elkington, et al. *The Green Consumer.* New York: Penguin Books, 1990. A guide to earth-friendly shopping.

William S. Ellis. "The Gift of Gardening." *National Geographic,* May 1992, pp. 52–81.

Beautifully illustrated description of the satisfactions of gardening.

Clarissa Pinkola Estes. *Women Who Run with Wolves: Myths and Stories of the Wild Woman Archetype*. New York: Ballantine Books, 1992. A Jungian therapist aims at enabling women to discover and gain empowerment from their wild strength.

**Linda Fillipi. "Place, Feminism, and Healing: An Ecology of Pastoral Counseling." *Journal of Pastoral Care*, Fall 1991, pp. 231–42. Highlights of an ecotherapeutic Ph.D. dissertation, School of Theology, Claremont, Calif.

The Findhorn Community. *The Findhorn Garden*. New York: Harper Colophon Books, 1975. Photos and text describing the intentional community in northern Scotland noted for its remarkable facilitating of growth in vegetables, fruits, flowers, and people. *Faces of Findhorn: Images of a Planetary Family*. New York: Harper and Row, 1980. Pictures of members of the Findhorn Foundation community who seek to live in harmony with each other and all living things.

Ruth Fletcher. *Teaching Peace: Skills for Living in a Global Society*. San Francisco: Harper and Row, 1983. Teacher's plans for 64 lessons in conflict management, nonviolence, cooperation, peace, and whole-earth thinking and living.

**Matthew Fox. *The Coming of the Cosmic Christ: The Healing of the Mother Earth*. San Francisco: Harper and Row, 1988. A pioneer creation theologian explores the ecospirituality of creation mysticism in a bold, inspirational book. *Original Blessing: A Primer of Creation Spirituality*. Santa Fe, N.M.: Bear and Co., 1983. A creation spirituality approach to responding to the prophetic call for erotic justice.

Warwick Fox. *Toward a Transpersonal Ecology: The Context, Influence, Meanings, and Distinctiveness of the Deep Ecology Approach to Ecophilosophy*. Boston: Shambhala Publications, 1990.

C. Dean Freudenberger. *The Gift of the Land: A Judeo-Christian Perspective on Value and Lifestyle Change in a Threatened World*. Los Angeles: Franciscans Communications, 1981. A biblically based discussion of the broken covenant in land abuse and guidelines for land-respecting values. *Global Dust Bowl: Can We Stop the Destruction Before It's Too Late*. Minneapolis: Augsburg, 1990. Shows how irresponsible land uses have resulted in the loss of half of the earth's thin topsoil on which our lives depend. Describes agroecology as a needed direction.

Albert J. Fritsch with Angela Iadavaia-Cox, *Eco-Church, An Action Manual*. San Jose, CA: Resource Publications, Inc., 1992. The co-founder of The Center of Science in the Public Interest provides a tool for enabling a congregation grow in environmental awareness and action.

M. Gerald Fromm and Bruce L. Smith, ed. *The Facilitating Environment: Clinical Applications of Winnicott's Theory*. Madison, Wisc.: International Universities Press, 1989. Discusses implications of object relations theory for issues of the human environment in therapy.

Masanobu Fukuoka. *The Natural Way of Farming*. New York: Japan Publications, 1985. Describes the theory and practice of the Green philosophy as related to ecologically sound agriculture.

Medard Gabel. *Energy, Earth, and Everyone: A Global Energy Strategy for Spaceship Earth*. New York: Doubleday, 1980. A creative plan for saving a viable planet.

William E. Gibson and Stanley R. Euston, *Gathering Hope: A Citizens' Call to a Sustainable Ethic for Guiding Public Life*. Santa Fe: The Sustainability Project, 1995. Calls for an assertive, participatory democracy in America to implement a powerful concern about developing a sustainable future in which natural and social systems thrive together.

Edward Goldsmith. *The Way: An Ecological World-view*. Boston: Shambhala, 1993.

The founder-editor of the periodical *The Ecologist* describes an ecologically sound world view, which has two fundamental principles: that the living world (the biosphere) is the basic source of all human benefits and wealth; and that the overriding goal of the behavior of ecological societies must be to preserve the critical order of the natural world so that it will be possible for these benefits to come to humanity. Mainstream economics and science contradict these principles, thereby furthering the destruction of nature.

**Al Gore. *Earth in the Balance: Ecology and the Human Spirit.* New York: Penguin Books, 1992. A challenging exploration by an American statesman and passionate environmentalist, of our ecologically dysfunctional civilization and the spiritual, ethical, economic and political roots of this problem. Proposes a global Marshall Plan as a path to an ecologically viable future.

**Larry Kent Graham. *Care of Persons, Care of Worlds: A Psychosystems Approach to Pastoral Care and Counseling.* Nashville: Abingdon, 1992. Creates a model that helps the field include the interaction between individual problems and their social context.

Nancy Sokol Green. "America's Toxic Schools," *E Magazine* 3, no. 6 (November/December 1992). A report on dangerous pollution in schools.

Susan Griffin. *Women and Nature: The Roaring Inside Her.* New York: Harper and Row, 1978. A feminist critique of the voice of Western patriarchy in contrast to the voice of women and nature, as they speak of the body, vision, and interrelatedness.

Joan Halifax. *The Fruitful Darkness: Reconnecting with the Body of the Earth.* San Francisco: Harper San Francisco, 1993.

Catharina J. M. Halkes. *New Creation: Christian Feminism and the Renewal of the Earth.* Louisville: Westminster/John Knox Press, 1989. A Dutch feminist theologian examines the false and catastrophically damaging images of domination and power, at the roots of both the rape of the earth and the oppression of women, and proposes a radical substitution of images of mutuality, connectedness, and justice.

David C. Hallman. *A Place in Creation: Ecological Visions in Science, Religion, and Economics.* Toronto: United Church Publishing House, 1992. By building bridges between religion, economics, and sciences, this book charts a hopeful path through the ecocrisis. *Ecotheology: Voices from South and North.* Maryknoll, N.Y.: Orbis Books, 1994.

Eugene C. Hargrove. *Religion and Environmental Crisis.* Athens, Ga.: The University of Georgia Press, 1986. Essays exploring the sources of sacred power in various religious traditions, power that is needed to generate environmental consciences.

Willis Harman. *Global Mind Change. The Promise of the Last Year of the Twentieth Century.* Sausalito, Calif.: Institute of Noetic Sciences, 1988. Explores the radical change in the belief structure of Western industrial society, which opens incredible potentials for human consciousness.

**D. Mark Harris. *Embracing the Earth: Choices for Environmentally Sound Living.* Chicago: Noble Press, 1990. A guide to more than 200 environmentally responsible actions, plus ecoeducational resources and a directory of environmental organizations.

George S. Hendry. *Theology and Nature.* Philadelphia: Westminster Press, 1980. Examines the relation of philosophy and religion to nature, focusing on the doctrine of creation.

**Dieter T. Hessel, ed., *Theology for Earth Community: A Field Guide.* New York:

Orbis Press, 1996. Leaders in all the disciplines of theology and ministry illuminate the implications of the eco-justice crisis.

Dieter T. Hessel, ed., *After Nature's Revolt: Eco-Justice and Theology*. Minneapolis: Fortress Press, 1992. Leading theologians and ethicists explore the intrinsic links between social justice and environmental issues.

Benjamin Hollister, et at., *Shopping for a Better World: The Quick and Easy Guide for All Your Socially Responsible Shopping*. (San Francisco: Sierra Club Books, 1994.) Revised annually by the Council for Economic Priorities based on updated research.

Jean Houston. "Awakening Your Evolutionary History" chapter 5 in *The Possible Human*. Los Angeles: Jeremy P. Tarcher, 1982. Exercises for awakening evolutionary memory.

J. Donald Hughes. *American Indian Ecology*. El Paso: Texas University Press, 1983. Describes the reverence for the earth of our Native Americans.

Fritz Hull, ed., *Earth & Spirit, The Spiritual Dimension of the Environmental Crisis*. New York: Continuum Publishing Co., 1993. Twenty-six leaders in earth-centered spirituality explore relatively untapped reservoir of wisdom, imagination, and strength, calling people into sustainable relationships with nature.

Ake Hultkantz, *Shamonic Healing and Ritual Drama, Health and Medicine in Native North American Religious Traditions*. New York: Crossroad, 1992.

Linda Mason Hunter. *The Healthy Home*. New York: Pocket Books, 1989. A guide to ecologically healthy living places.

H. Patricia Hynes. *Earth Right, Every Citizen's Guide*. Rocklin, Calif.: Prima Publishing and Communications, 1990. A guide to what anyone can do to help save their environment.

Shannon Jung. *We Are Home: A Spirituality of the Environment*. New York: Paulist Press, 1993.

Catherine Keller. *From a Broken Web: Separation, Sexism, and Self*. Boston: Beacon Press, 1986. A process theologian and ecofeminist illuminates the brokenness of society, women, and the earth.

**Stephen R. Kellert and Edward O. Wilson, eds. *The Biophilia Hypothesis*. Washington, D.C.: Island Press/ Shearwater Books, 1993. Provocative essays by scientists amplify and refine Wilson's biophilia concept from psychological, biological, cultural, and aesthetic perspectives.

Petra Kelly. *Fighting for Hope*. Boston: South End Press, 1984. A key founder of the German Green Party explores the interrelations among peacemaking, ecology, and women.

Donald Keys, *Earth at Omega: The Passage to Planetization*. Branden, 1982. An understanding of the multiple interrelated crises of the nuclear threat, overpopulation, pollution, hunger and resource depletion, and a vision of the grand opportunities confronting humankind to solve these problems by recognizing planetary citizenship.

Peter Gwillim Kreitler. *The Earth's Killer C's*. Lafayette, Calif.: Morning Sun Press, 1995. Explores major environmental threats, labeled as chopsticks, clearcutting, cows, cars, corporations, conception, consumption, chlorine, chlorofluorocarbons, chemicals, cigarettes, and Congress. Printed on 100% kenaf plant paper with soy-based ink to minimize the environmental impact.

Chung Hyun Kyung. *Struggle to Be the Sun Again*. Maryknoll, N.Y.: Orbis Books,

1993. A Korean scholar's introduction to an earth-grounded Asian women's theology.

Albert LaChance. Foreword by Thomas Berry. *Greenspirit: Twelve Steps in Ecological Spirituality*. Rockport, Mass.: Element, 1991. An individual, cultural, and planetary therapy applying the 12-step recovery program to our society's multiple addictions to comfort, ease of travel, materialism, consumerism, escapism, and overeating. Sees an ecological spirituality, expressed through the 12-steps, as a way to help ourselves and the earth.

Dolores La Chapelle. *Sacred Land, Sacred Sex, Rapture of the Deep: Concerning Deep Ecology and Celebrating Life*. Silverton, Colo.: Finn Hill Arts 1988.

Richard D. Land and Louis A Moore, eds. *The Earth Is the Lord's: Christians and the Environment*. Louisville: Broadman Press, 1992.

Scott Lewis, with the Natural Resources Defense Council. *The Rainforest Book: How You Can Save the World's Rainforests*. Los Angeles: Living Planet Press, 1990. Overviews the destruction of the rainforests that are essential to everyone's well-being, and offers practical suggestions on how to help save them.

George T. Lewith and Julian N. Kenyon. *Clinical Ecology*. Wellingborough, England: Thomas Publishing, 1985. The treatment of ill health caused by environmental factors.

J. B. Libanio. *Spiritual Discernment and Politics*. Maryknoll, N.Y.: Orbis Books, 1982. A leading liberation theologian relates a theology of justice to concrete sociopolitical options.

Robert Jay Lifton. *The Broken Connection: On Death and the Continuity of Life*. New York: Simon and Schuster, 1979. A psychiatrist explores the individual life cycle and holds that a sense of continuity, now broken by the "radical futurelessness" is necessary to full human functioning, including the sense of being fully alive.

Francis Raymond Line and Helen E. Line. *Man with a Song: Some Major and Minor Notes in the Life of Saint Francis of Assisi*. Garden City, N.Y.: Image Books, 1982. Includes a section of quotations entitled "Songs of the Earth, A Patron Saint of Ecologists."

Doris Janzen Longacre. *Living More with Less*. Scottsdale, Pa.: Herald Press, 1980. A guide to how to contribute to a healthy environment, world community, and justice by the ways we eat, travel, dress, and spend money.

Love the Earth and Be Healed. Nashville: Ecufilm, 1994. A series of six videos with study guides, for use in churches concerned about the ecojustice crisis.

James Lovelock. *Gaia, A New Look at Life on Earth*. Berkeley, Calif.: Parallax Press, 1979. Sets forth the "Gaia hypothesis" that the entire earth is a living, interdependent organism. *The Ages of Gaia*. New York: W. W. Norton, 1988.

**Joanna Rogers Macy. *Despair and Personal Power in the Nuclear Age*. Philadelphia: New Society Publishers, 1983. A "how to" book on helping people move through their denial and despair to hope and empowerment for action as peacemakers. Described by Rollo May as "the bravest book I have read since Jonathan Schell's *Fate of the Earth*." Methods can be adapted to the ecojustice crisis. *World As Lover, World as Self*. Berkeley, Calif.: Parallax Press, 1991. Draws on Macy's expertise in the field of Buddhism to show how that cosmology undergirds a nondualistic, constructive relation to the earth. Explores themes such as "The Greening of the Self."

Jerry Mander. *In the Absence of the Sacred: The Failure of Technology and the Survival of the Indian Nations*. San Francisco: Sierra Club Books, 1991.

Diane Mariechild. *Mother Wit: A Feminist Guide to Psychic Development.* A collection of exercises and psychic tools, ancient and contemporary, for facilitating feminist growth.

Freya Matthews. *The Ecological Self*. New York: Rutledge, Barnes & Noble, 1991. Explores how our personalities are earth based.

Paul McCleary and J. Philip Wogaman. *Quality of Life in a Global Society*. New York: Friendship Press, 1978. Examines the issues of hunger, environment, energy, economics, and population from a biblical perspective.

**Jay B. McDaniel. *Earth, Sky, Gods and Mortals: Developing an Ecological Spirituality*. Mystic, Conn.: Twenty-Third Publications, 1990. Describes "an ecological Christianity that is open to all horizons of human life, open to other religions, and infused with a desire to affirm our inseparability from the natural world." *With Roots and Wings: Christianity in an Age of Ecology and Dialogue*. Maryknoll, N.Y.: Orbis Books, 1995. A teacher and environmental activist trained in process theology discusses earth-saving approaches by Christians.

Seam McDonagh. *The Greening of the Church*. Maryknoll, N.Y.: Orbis Books, 1990. Discusses the growing ecological awareness and action by congregations.

**Sallie McFague. *The Body of God: An Ecological Theology*. Minneapolis: Fortress Press, 1993. Based on the thesis that "the world is our meeting place with God," a Christian theologian develops an organic, embodied theology of nature emphasizing the functional partnership of humans with God in helping life prosper on the planet. Christ is seen as the savior of the whole of creation. *Models of God: Theology for an Ecological, Nuclear Age*. Philadelphia: Fortress Press, 1987. A holistic, justice-based theology developing alternative metaphors for God—mother, lover, and friend.

James McGinnis and Kathleen McGinnis. *Christian Parenting for Peace and Justice Program Guide*. Nashville: Discipleship Resources, n.d. A guide for planning and leading a series of seven or more sessions to help parents understand and implement the vision of Shalom—God's intention of wholeness, freedom, peace, justice, and well-being.

**Kathleen McGinnis and James McGinnis. *Parenting for Peace and Justice, Ten Years Later*. Maryknoll, N.Y.: Orbis Books, 1990. Based on their own experience as parents, the cofounders of the Institute for Peace and Justice in St. Louis discuss family involvement in social action, helping children deal with a violent world, multiculturalizing family life, and how families can find spiritual resources for doing all this.

Meadowcreek. *Education for a Sustainable World Curriculum Guide.* (Available from P.O. Box 100, Fox, AR 72051; 501-363-4500.) A valuable guide for teachers interested in multidisciplinary, hands-on methods of teaching ecology to students in grades 7 through 12. Includes detailed lesson plans giving simple experiments and demonstrations in the areas of ecology, energy, and horticulture.

Carolyn Merchant. *The Death of Nature: Women, Ecology, and the Scientific Revolution*. San Francisco: Harper and Row, 1980. A feminist philosopher of science explores the history of how a sense of organic oneness with nature was lost and calls for its recovery by reorienting society's treatment of both women and nature. *Radical Ecology: The Search for a Livable World*. New York: Routledge, 1992.

Jürgen Moltmann. *God in Creation*. New York: Harper-Collins, 1991. A leading theologian explores creation theology.

Mary Elizabeth Mullino Moore, *Teaching from the Heart: Theology and Educational Method*. Minneapolis: Fortress Press, 1991. Describes case study, gestalt, incarnational, narrative, and conscientizing methods of teaching.

Peter G. Moore-Kochlacs and friends. *Caring for God's Creation: A Collection of Environmentally Focused Resource Material and Sermons*. Reseda, Calif.: Environmental Ministries, 1993. Resources for use by congregations in their ecology services.

Thomas Moore. *Care of the Soul: A Guide for Cultivating Depth and Sacredness in Everyday Life*. New York: HarperCollins, 1992. Includes a discussion of soulful ecology and the World Soul.

Nelle Morton. *The Journey Is Home*. Boston: Beacon Press, 1985. The pioneer inspirer of feminist theology traces the development of her vision, grounded in her earthy childhood on a Tennessee farm.

Robert Muller. *New Genesis, Shaping a Global Spirituality*. Garden City, N. Y.: Doubleday, 1982. A global statesman with the UN writes in the tradition of hope of Dag Hammarskjöld and Pierre Teilhard de Chardin, exploring the need to recognize the global transcendence of human values, religions, and humanity

Norman Myers. *The Gaia Atlas of Future Worlds: Challenge and Opportunity in an Age of Change*. New York: Anchor Doubleday Books, 1990. A guide to global changes and ways to participate in creating the healthiest future possible.

Jon Naar and Alex J. Naar. *This Land Is Your Land: A Guide to North America's Endangered Ecosystems*. New York: HarperCollins, 1993

Arne Naess. *Ecology, Community, and Lifestyle: Outline of an Ecosophy*. New York: Cambridge University Press, 1989. The philosopher who originated the "deep ecology" concept discusses its applications to human life in community.

James A. Nash. *Loving Nature: Ecological Integrity and Christian Responsibility*. Nashville: Abingdon Press, 1991. On Christian ecological ethics.

Roderick Nash. *The Rights of Nature: A History of the Environmental Movement*. Madison: University of Wisconsin Press, 1989. Develops the concept of a land ethic.

John G. Niehardt. *Black Elk Speaks: The Life Story of a Holy Man of the Oglala Sioux*. New York: William Morrow, 1932. Discusses the spiritual unity of humans and the rest of nature in Native American cosmologies.

Eugene P. Odum. *Ecology and Our Endangered Life-Support System*. Sunderland, Mass.: Sinauer Associates, 1989. A textbook and citizens' guide on the principles of ecology as related to the earth's life-support systems.

Carol Olwell. *Gardening from the Heart: Why Gardeners Garden* Navato, Calif.: Antelope Island Press, 1990. Persons engaged in all types of gardening reflect on gardening for beauty and happiness. Sections include "The Garden as Provider," "The Garden as Teacher," and "The Garden as Healer." This last section includes chapters on using gardening to find patience and peace, sowing community gardens, and Catherine Sneed Marcum's therapeutic gardening in prisons.

**David W. Orr. *Earth in Mind: On Education, Environment, and the Human Prospect*. Washington, D.C.: Island Press, 1994. Essays by a leading environmental educator confronting readers with the need for profound changes in our educational systems. *Ecological Literacy: Education and the Translation to a*

Postmodern World. Albany, N.Y.: SUNY Press, 1989. An insightful book on developing earth literacy in education.

P. D. Ouspensky. *Tertium Organum: A Key to the Enigmas of the World*. New York: Vintage Books, 1970. A Russian philosopher, mathematician, and mystic explores religion, science, psychology, and mystical literature in search of an understanding of humans and their place in creation. His pioneering thought suggested the concept of a living universe and a world soul.

Owen D. Owens, *Living Waters*. New Brunswick, N.J.: Rutgers University Press, 1993. A guide to rescuing a local stream from pollution.

Robert Parham. *Loving Neighbors Across Time: A Christian Guide to Protecting the Earth*. Foreword by Al Gore. Birmingham, Ala.: New Hope, 1991. Explores the Bible's call for earthkeeping and enlarges the concept of love of neighbor to include future generations.

Christopher Plant and Judith Plant, eds. *Turtle Talk: Voices for a Sustainable Future*. Philadelphia: New Society Publishers, 1990. Statements by a variety of prophetic environmental thinkers and activists including Susan Griffin, John Seed, and Starhawk, who, like turtles, model getting ahead by sticking their necks out.

**Judith Plant, ed. *Healing the Wounds: The Promise of Ecofeminism*. Philadelphia: New Society Publishers, 1989. Twenty-five leading feminist thinkers contribute stories, essays, and poems bringing together body, mind, spirit, the personal and political in generating the wholeness of the earth.

Veronica Ray. *Green Spirituality: Reflections on Belonging to a World Beyond Myself*. New York: HarperCollins, A Hazelden Book, 1992. Meditations on the spirituality that expands beyond the self to embrace our interconnectedness with the whole human family and the world.

Lewis G. Regenstein *Replenish the Earth: A History of Organized Religion's Treatment of Animals and Nature*. New York: Crossroad, 1991. Includes the Bible's message of conservation and kindness toward animals.

**Jeremy Rifkin. *Biosphere Politics—A New Consciousness for a New Age*. New York: Crown Publishers, 1991. An exploration of the nature of the dawning biospheric age and the changed consciousness that must enable it.

Jeremy Rifkin, ed. *The Green Lifestyle Handbook: 1001 Ways You Can Help Clean Up the Environment*. New York: Henry Holt, 1990.

Jeremy Rifkin and Carol Grunewald Rifkin. *Voting Green: Your Complete Environmental Guide to Making Political Choices in the 1990s*. New York: Doubleday, 1992. Gives a checklist of Green positions by which to judge candidates on sustainable economy, global environmental security, healthful food, biodiversity, clean air and water, and nontoxic workplaces.

John Robbins. *Diet for a New America*. Walpole, N.H.: Stillpoint Publishing, 1987. Exposes the high price the earth and humans pay for our species' exaggerated dependence on animal protein; proposes vegetarianism as a healthy alternative.

Elizabeth Roberts and Elias Amidon, eds. *Earth Prayers from Around the World*. San Francisco: Harper San Francisco, 1991. 365 prayers, poems, and invocations for honoring the earth, from many cultures and religions.

Steven C. Rockefeller and John C. Elder, eds. *Spirit and Nature: Why the Environment Is a Religious Issue*. Boston: Beacon Press, 1992.

Adam Rogers. *The Earth Summit, A Planetary Reckoning*. Los Angeles: Global View Press, 1993. Reports on the UN Conference on Environment and Development in June 1992, including the five official UN documents, 40-plus treaties of nongovernmental organizations, and the perspectives of youth, indigenous peoples, business, and politicians.

Holmes Rolston. *Environmental Ethics: Duties to and Values of the Natural World.* Philadelphia: Temple University Press, 1989. Ecological ethics introduction.

**Theodore Roszak. *The Voice of the Earth*. New York: Simon and Schuster, 1992. A historical analysis of the psychological factors causing the ecological crisis. Examines psychiatry and the contemporary importance of creating an ecopsychology by listening to the earth.

**Theodore Roszak, Mary E. Gomes, and Allen D. Kanner, eds. *Ecopsychology: Restoring the Earth, Healing the Mind.* San Francisco: Sierra Club Books, 1995. The key book in the field of ecopsychology, with papers by twenty-seven authors, mainly psychologists, reporting on their theories and practice.

**Rosemary Radford Ruether. *Gaia and God: An Ecofeminist Theology of Earth Healing*. San Francisco: Harper San Francisco, 1992. Examines the contribution of Christianity to the ecological crisis, and offers a reconstructed, biophilic Christian theology that can help heal the planet. Uses the name of the Greek earth goddess Gaia, as James Lovelock does, to describe the earth as a living, unified organism. *To Change the World Christology and Cultural Criticism*. New York: Crossroad, 1981. This feminist theologian critiques the relation of Christian beliefs about Christ to oppression, feminism, and abuse of the environment.

Kirkpatrick Sale. *Dwellers in the Land: The Bioregional Vision*. Philadelphia: New Society Publishers, 1991. An introduction to the bioregional vision and movement.

**Mike Samuels and Hal Zina Bennett. "How the Earth's Health and Human Health Are One," part 2 of *Well Body, Well Earth*. San Francisco: Sierra Club Books, 1983.

H. Paul Santmire. *The Travail of Nature: The Ambiguous Ecological Promise of Christian Theology*. Minneapolis: Fortress Press, 1985. Traces views of nature throughout Christian history, including "The Life and Significance of Francis of Assisi."

Stephen B. Scharper and Hilary Cunningham. *The Green Bible*. Maryknoll, N.Y.: Orbis Books, 1993 Religious ecological resources from Hebrew and Christian scriptures, plus statements by religious, political, and environmental leaders.

**John Seed, Joanna Macy, Pat Fleming, and Arne Naess. *Thinking Like a Mountain: Toward a Council of All Beings*. Santa Cruz, Calif.: New Society Publishers, 1988. An inspiring collection of deep ecology readings, meditations, poems, and guided fantasies designed to help readers reconnect with the living earth. Illustrated by line drawings of the rainforests.

**Paul Shepard. *Nature and Madness*. San Francisco: Sierra Club Books, 1982. The first book to use a psychopathological metaphor referring to what I am calling "ecoalienation." *Thinking Animals: Animals and the Development of Human Intelligence*. New York: Viking Press, 1976. A teacher of human ecology and the ecology of mind explores the view that animal images and forms are key factors in the development of human personality, speech, thought, and art, and thus a key to shaping the qualities that make humans human.

**Lesley Irene Shore, *Healing the Feminine, Reclaiming Woman's Voice*. St. Paul, MN: Llewellyn Publications, 1995. Psychologist Shore holds (similarly to ecotherapy's view), that both women and men have lost their ties with "Mother Nature, with Woman's Voice and with the feminist in our nature." This result in disowned feelings in men and symptoms such as low self-esteem, co-dependency, addictions and eating disorders in women. Describes the inner journey to reclaim this disowned part of ourselves, the path to health and wholeness. In an insightful chapter on our human alienation from nature, Shore declares: "Our well being is tied to nature. . . . Nature is life. She is the air we breathe, the food we eat, the water that quenches our thirst. She is part of us. We are the whole."

**Lesley Irene Shore. *Tending Inner Gardens: The Healing Art of Feminist Psychotherapy*. New York: Harrington Park Press, 1995. A feminist psychologist describes how she uses earth images and growth experiences integrally in her practice of psychotherapy. Shows the parallels between gardening and feminist psychotherapy.

Tom Sine. *Wild Hope: Crises Facing the Human Community on the Threshold of the 21st Century*. Dallas: Word Publishing, 1991. A wake-up call for Christians to the environmental ultimatum, and a challenge to unleash a wild hope in churches and society, a hope for our own future.

Gary Snyder. *The Practice of the Wild*. San Francisco: North Point Press, 1990.

Starhawk. *The Spiral Dance: A Rebirth of the Ancient Religion of the Great Goddess*. San Francisco: Harper and Row, 1979. A minister of the Covenant of the Goddess overviews the origins, suppression, and current reemergence of the earth-based Old Religion, and its value for both women and men.

Sara Stein. *Noah's Garden: Restoring the Ecology of Our Back Yard*. New York: Houghton Mifflin, 1993.

Taylor Stoehr, ed. *Nature Heals: The Psychological Essays of Paul Goodman*. New York: Dutton, 1977.

Bruce Stokes. *Helping Ourselves: Local Solutions to Global Problems*. New York: A Worldwatch Institute Book, 1981. Provides encouraging examples of how self-help groups around the world are empowering themselves by attacking a variety of health and quality-of-life problems.

David Sukzuki and Peter Knudtson. *Wisdom of the Elders: Honoring Sacred Native Visions of Nature*. New York: Bantam Books, 1992.

**Brian Swimme. *The Universe Is a Green Dragon*. Sante Fe, N.M.: Bear & Co., 1984. A delightful dialogical story of the exciting new understanding of continuing creation and its implications for our species.

**Brian Swimme and Thomas Berry. *The Universe Story: From the Primordial Flaring Forth to the Ecozoic Era, A Celebration of the Unfolding Cosmos*. San Francisco: Harper San Francisco, 1992. A mathematical cosmologist and an ecotheologian collaborate to present the "common creation story" derived from contemporary earth sciences and humankind's spiritual wisdom, a new cosmology to undergird the ecozoic era of our species and our planet

Paul W. Taylor. *Respect for Nature: A Theory of Environmental Ethics*. Princeton, N.J.: Princeton University Press, 1988.

Pierre Teilhard de Chardin. *The Phenomenon of Man*. New York: Harper Torchbook, 1959. A scientist-philosopher presents a cosmology that can be a foundation for ecologically creative living.

William Irwin Thompson, ed. *Gaia: A Way of Knowing, Political Implications of the New Biology*. Great Barrington, Mass.: Lindisfarne Press, 1987. Explores the cultural, political, and lifestyle implications of the new holistic biology as reflecting the Gaia model.

TreePeople, with Andy Lipkis and Katie Lipkis. *The Simple Act of Planting a Tree: Healing Your Neighborhood, Your City, and Your World*. Los Angeles: Jeremy P. Tarcher, 1990. A citizen forester's guide by the founders of TreePeople.

**Mary Evelyn Tucker. *Education and Ecology: Earth Literacy and the Technological*

Trance. Published for The American Teilhard Association for the Future of Man, Inc., by Anima Books, Chambersburg, Pa., 1993. A succinct overview of the need for and nature of education producing earth-literacy.

**Mary Evelyn Tucker and John A. Grim, eds. *Worldviews and Ecology*. Lewisburg, Pa.: Bucknell University Press, 1993. Scholars of the major world religions and several contemporary faiths discuss the ways their cosmologically based ethics relate to protecting the environment. The urgent need for a world ecological ethic is highlighted.

Steve Van Matre and Bill Weiler. *The Earth Speaks*. Warrenville, Ill.: Institute for Earth Education, 1991. A collection of brief writings by many listeners to the earth's voice, illuminated by block prints by Gwen Frostic.

Virgil J. Vogel. *American Indian Medicine*. Norman: University of Oklahoma Press, 1970. Overviews the healing methods of Native Americans.

Barbara G. Walker. *Women's Rituals, A Source Book*. San Francisco: Harper and Row, 1990. Group or individual rituals including numerous earth rituals

Alan Watts. *Nature, Man and Woman*. New York: Pantheon, 1958. A pioneer philosopher, who was open to learn from the wisdom of the East, reflects on the alienation of humans from nature as this relates to patriarchal relationships between men and women.

John Welwood, ed. *The Meeting of the Ways: Exploration in East/West Psychology*. New York: Schocken Books, 1979. Papers on resources for psychotherapy from the Taoist, Hindu, Sufi, and Buddhist traditions.

Jane Hollister Wheelwright and Lynda Wheelwright Schmidt. *The Long Shore: A Psychological Experience of the Wilderness*. San Francisco: The Sierra Club, 1991. Two Jungian analysts, a mother and daughter, explore the meaning of wilderness in modern life by telling of life on the historic Hollister ranch near Santa Barbara.

Loren Wilkinson and Mary Ruth Wilkinson. *Caring for Creation in Your Own Backyard: A Seasonal Guide*. Ann Arbor, Mich.: Servant Publications, 1992. Over 100 things Christian families can do to help the earth.

Terry Tempest Williams. "The Wild Card, A Woman's Call to Community, A Poet's Vision of Environmentalism Reinvented." In *Wilderness*, published by the Wilderness Society, Summer 1993.

**Edward O. Wilson. *Biophilia*. Cambridge, Mass.: Harvard University Press, 1984. A leading biologist-environmentalist explores the meaning and crucial importance of loving life. *The Diversity of Life*. Cambridge, Mass.: Belknap Press of Harvard University Press, 1992. Highlights the need to protect diverse species to avoid depleting the vitality of the biosphere.

World Commission on Environment and Development. *Our Common Future*. New York: Oxford University Press, 1987. An action-oriented approach to the interrelated global problems of peace, development, security and the environment.

A Few of Many Sources of Ecological Resources

American Rivers, 801 Pennsylvania Ave, SE, Suite 400, Washington, DC, 20003. Goal: Protecting the ecosystems of rivers from pollution and dams.

Center for Psychology and Social Change at Harvard University. Cambridge, MA. Developing a new psychology rooted in sustainable, mutually enhancing relationships among humans and between humans and the natural world.

The Center for Respect for Life and Environment, 2100 L Street, NW, Washington, D.C. 20037. (202) 778-6133. Publishes a quarterly *Earth Ethics*. Emphasizes concern for poor and oppressed people, and of animals, plants and the environment; and seeks to develop environmental concern in the professions and higher education.

Citizens' Clearinghouse for Hazardous Wastes. P.O. Box 926, Arlington, VA 22215; 703-276-7070.

Defenders of Wildlife. 124 19th St., NW, Washington, D.C. 20036; 202-639-9510. Goal: wildlife preservation.

Earth Island Institute, 300 Broadway, Suite 28, San Francisco, CA 94133; 415-788-3666. Goal: conservation.

Eco-Justice Working Group, National Council of Churches, 475 Riverside Dr., N.Y., N.Y. 10027. (212) 870-3342. Does education, action, research and publication concerning the U.S. and its government on peace, justice, and environmental issues.

Environmental Defense Fund, 257 Park Ave. So., New York, NY 10010; 212-505-2100. Goal: environmental legislation.

Environmental Education and Health Services, 3202 W. Anderson Lane, Austin, TX 78757; 512-288-2369. Information regarding the innovative Texas state program of environmental education and action.

Greenpeace, *USA*, 1436 U St., NW, Washington, DC 20007; 202-462-1177. Goal: preserving natural resources and wildlife.

Kerr Center for Sustainable Agriculture, Highway 271 S., P.O. Box 588, Poteau, OK 74954. Goal: developing agricultural approaches that sustain the soil and other precious natural resources, strengthen rural communities, and make our world more ecologically secure and socially responsible.

National Audubon Society, 950 Third Ave., New York, NY, 10022; 212-832-3200. Goal: conservation of wildlife and natural resources.

National Center for Environmental Health Strategies, 1100 Rural Ave., Voorhees, NJ 08043; 609-429-5358. A resource center for persons with environmentally and occupationally related illnesses.

National Parks and Conservation Assn., 1776 Massachusetts Ave., N.W., Washington, D.C. 20036. Goal: Protecting American national parks and forests.

The Nature Conservancy, 1815 N. Lynn St., Arlington, VA 22209; 703-841-5300. 398-4404. Goal: rainforest preservation around the world.

North American Coalition on Religion and Ecology, 5 Thomas Circle, NW, Washington, D.C. 20005. (202) 462-2591. An environmental education and advocacy network that is interfaith and ecumenical.

Rainforest Action Network, 450 Sansome, Suite 700, San Francisco, CA 94111.

Sierra Club, 730 Polk St., San Francisco, CA 94109; 415-776-2211. Goal: preserving and enjoying wilderness.

Superintendent of Documents, *U.S. Printing Office*, Washington, DC 20402. Offers a wide variety of pamphlets on environmental health issues.

United Nations Environmental Programme, DC 2-803, 2 United Nations Plaza, N.Y., N.Y. 10017 (212) 963-8169. The leading international publisher of environmental resources.

Index

Drawing by Kelly Frick Richards. Used by permission.

"Let us put our minds together and see what life we can make for our children."

–Lakota Sioux Sitting Bull, 1877